Microbial Biofertilizers and Micronutrient Availability

Shams Tabrez Khan • Abdul Malik
Editors

Microbial Biofertilizers and Micronutrient Availability

The Role of Zinc in Agriculture and Human Health

Springer

Editors
Shams Tabrez Khan
Department of Agricultural Microbiology
Aligarh Muslim University
Aligarh, India

Abdul Malik
Department of Agricultural Microbiology
Aligarh Muslim University
Aligarh, India

ISBN 978-3-030-76611-5 ISBN 978-3-030-76609-2 (eBook)
https://doi.org/10.1007/978-3-030-76609-2

This Springer imprint is published by the registered company Springer Nature Switzerland AG
The registered company address is: Gewerbestrasse 11, 6330 Cham, Switzerland

Dedicated to my respected and beloved father
Mr. Iltija Khan
You will be always with us

Preface

Extensive agriculture is becoming increasingly popular to meet the growing need of an ever-increasing world population, resulting in an increased burden on the soil as a renewable source and reduced product quality in terms of nutritive value. Consequently, many agricultural lands have turned barren, and much of the global population residing in developing countries are malnourished. Population in such countries, mostly dependent on cereal grain-based diet, suffers from micronutrient and vitamin deficiency. Micronutrient deficiency in plants also retards their growth and productivity, affecting their nutritive value, consequently affecting consumers at the next trophic level like humans. One such essential micronutrient is zinc, a constituent of about 3000 different proteins and various classes of enzymes. Its deficiency influences various life processes, adversely affecting the growth and reproduction of living organisms, including plants. Consumption of zinc-poor diet causes its deficiency in humans, primarily affecting children in developing countries, as these children don't have access to zinc-rich diet. Much of the world's population in developing countries depends on cereals like wheat and rice as a staple diet. An adult human's daily Zn requirement is 9–18 mg/day, which cannot be met by a cereal-based diet low in Zn content. Wheat and rice, the two most consumed cereal crops, contain only one third and one fifth of daily Zn requirement, respectively. Zn deficiency in children may lead to several health problems like recurring diarrhea, weak immune system, pneumonia, and reduced physical performance. Zn deficiency is also known to affect functioning of brain and liver, the two most important organs. However, micronutrient deficiency is not only limited to developing countries and is also becoming prevalent in countries where the population is dependent on a staple diet primarily containing grains. Micronutrient availability in food can be improved through a coordinated effort of dieticians, plant scientists, molecular biologists, microbiologists, and policymakers. Zn biofortification of food and feed crops, especially cereals, can significantly improve the Zn uptake in such countries.

Micronutrients and zinc deficiency in plants can be due to two important reasons, one being the lack of these micronutrients and Zn in soil altogether, which can also be linked to the excessive use of fertilizers, as plants concomitantly uptake micronutrients like Zn along with other nutrients added to the soil in the form of fertilizers. Another well-known reason is the presence of unavailable forms of micronutrients like insoluble forms of Zn in soil, namely smithsonite ($ZnCO_3$), sphalerite (ZnS), zincite (ZnO), franklinite ($ZnFe_2O_4$), willemite (Zn_2SiO_4), and hopeite ($Zn_3(PO_4)_2 \cdot 4H_2O$). It is estimated that one-third of the Zn problem is due to the presence of these insoluble forms of Zn in soil. Other factors also affect the bioavailability of these micronutrients to plants, including soil pH, presence of calcium carbonate, low organic carbon content, and other abiotic stresses. Various approaches are used to circumvent the bioavailability of these micronutrients to plants. These include agronomic biofortification, use of chemical fertilizers (soil and foliar application), use of biotechnology and plant breeding for improving crops, crop rotation, organic manure application, and use of microbes based biofertilizers.

Many of these approaches have inherent disadvantages, for example, the use of chemical fertilizers like Zn fertilizers is often costly and is not effective, as the soil pH and other factors again change the added soluble form of Zn to insoluble form, making it unavailable to plants. Use of chemical fertilizers is also known to cause chemical pollution in soil. One of the most effective and sustainable approaches is the use of microbial inoculants as these microbe-based biofertilizers are low cost and environment friendly. These fertilizers increase the bioavailability of various micronutrients to plants through various approaches. For instance, Zn solubilizing microorganisms' innate ability to solubilize insoluble Zn through various mechanisms increases its bioavailability to plants. This can be an effective approach to deal with at least one third of the Zn deficiency problem where the soil contains the insoluble forms of Zn. The use of biofertilizers like *Azolla*, *Azotobacter*, and other PGPRs is growing. Reports on biofertilizers for increasing the micronutrient availability are very limited. However, some commercial preparations of such biofertilizers are available in the market.

This book comprehensively addresses the issues related to micronutrient deficiency, especially zinc, and discusses the possible approaches for combating mineral deficiency. The book, however, mainly focuses on the zinc biofortification of cereal crops and contains 18 chapters on different aspects of mineral deficiency. The introductory chapter provides an overview of the mineral deficiency problem. Other chapters highlight the consequences of zinc deficiency and health risks associated with mineral deficiency, especially zinc in children and expecting mothers. Different types of food rich in minerals, especially Zinc, and the use of such food to design a diet that meets the daily Zn requirement are also discussed. Some of the chapters highlight the impact of zinc deficiency on plant health, quality of agricultural products, and the role of micronutrients in abiotic stress tolerance. The book mainly highlights the issue of Zn fortification of cereal crops and the sustainable approaches that have been used for the purpose. One of the approaches is the use of biofertilizers for the microbial solubilization of Zinc in soil, consequently improving the zinc uptake by plants. Some chapters provide a detailed account of what type of microorganisms are involved in the process. Other chapters describe the possible formulation of these microbes into a biofertilizer and the challenges involved.

Reports on the increased zinc content of the plants following the use of such fertilizers are also summarized. Problems and challenges associated with the successful application of these fertilizers and their commercialization at a wider scale are also discussed.

We thank all contributing authors who have generously given their time and expertise in preparing very informative chapters on the subject. We also thank all the reviewers for their valuable time and contribution to the peer-review process. All the technical assistance provided by Springer during the book's preparation and production is also gratefully acknowledged. We also acknowledge the generous financial support provided by the ministry of human resource development (MHRD), Government of India, under the Scheme for Promotion of Academic and Research Collaboration (SPARC P594) to prepare the book. The continuous support of the family members is also acknowledged.

Aligarh, India

Shams Tabrez Khan
Abdul Malik

Contents

Contributors

Faizan Ahmad Department of Post Harvest Engineering and Technology, Faculty of Agricultural Sciences, Aligarh Muslim University, Aligarh, Uttar Pradesh, India

Riaz Ahmad Section of Genetics, Department of Zoology, Faculty of Life Sciences, Aligarh Muslim University, Aligarh, Uttar Pradesh, India

Saghir Ahmad Department of Post Harvest Engineering and Technology, Aligarh Muslim University, Aligarh, Uttar Pradesh, India

Farzana Alim Department of Home Science, Faculty of Agricultural Sciences, Aligarh Muslim University, Aligarh, Uttar Pradesh, India

Farhana Mehraj Allai Department of Food Technology, Islamic University of Science and Technology, Awantipora, Jammu and Kashmir, India

Department of Post Harvest Engineering and Technology, Aligarh Muslim University, Aligarh, Uttar Pradesh, India

Nabat Arfi Department of Home Science, Faculty of Agricultural Sciences, Aligarh Muslim University, Aligarh, Uttar Pradesh, India

Yamshi Arif Plant Physiology Section, Department of Botany, Faculty of Life Sciences, Aligarh Muslim University, Aligarh, Uttar Pradesh, India

Z. R. Azaz Ahmad Azad Department of Post Harvest Engineering and Technology, Aligarh Muslim University, Aligarh, Uttar Pradesh, India

Anisa M. Durrani Department of Home Science, Aligarh Muslim University, Aligarh, Uttar Pradesh, India

Qazi Fariduddin Plant Physiology and Biochemistry Section, Department of Botany, Faculty of Life Sciences, Aligarh Muslim University, Aligarh, Uttar Pradesh, India

Tariq Ahmad Ganaie Department of Food Technology, Islamic University of Science and Technology, Awantipora, Jammu and Kashmir, India

Khalid Gul Department of Food Science and Engineering, School of Agriculture and Biology, Shanghai Jiao Tong University, Shanghai, China

Department of Food Process Engineering, National Institute of Technology, Odisha, India

Mahvash Haroon Department of Agricultural Microbiology, Faculty of Agricultural Sciences, Aligarh Muslim University, Aligarh, Uttar Pradesh, India

Shamsul Hayat Plant Physiology Section, Department of Botany, Faculty of Life Sciences, Aligarh Muslim University, Aligarh, Uttar Pradesh, India

Hadiya Husain Section of Genetics, Department of Zoology, Faculty of Life Sciences, Aligarh Muslim University, Aligarh, Uttar Pradesh, India

Alexandre Jousset Ecology and Biodiversity Group, Institute of Environmental Biology, Utrecht University, Utrecht, The Netherlands

Munawwar Ali Khan Department of Life and Environmental Sciences, College of Natural and Health Sciences, Zayed University, Dubai, UAE

Saba Khan Department of Home Science, Faculty of Agricultural Sciences, Aligarh Muslim University, Aligarh, Uttar Pradesh, India

Shams Tabrez Khan Department of Agricultural Microbiology, Faculty of Agricultural Sciences, Aligarh Muslim University, Aligarh, Uttar Pradesh, India

Tanveer Ahmad Khan Plant Physiology Section, Department of Botany, Faculty of Life Sciences, Aligarh Muslim University, Aligarh, Uttar Pradesh, India

Khalida Khatoon Department of Agricultural Microbiology, Aligarh Muslim University, Aligarh, Uttar Pradesh, India

Abdul Malik Department of Agricultural Microbiology, Faculty of Agricultural Sciences, Aligarh Muslim University, Aligarh, Uttar Pradesh, India

Farhana Masood Department of Post Harvest Engineering and Technology, Aligarh Muslim University, Aligarh, Uttar Pradesh, India

Romana Naaz Department of Applied Science and Humanities, Jamia Millia Islamia, New Delhi, India

Gazia Nasir Department of Post Harvest Engineering and Technology, Aligarh Muslim University, Aligarh, Uttar Pradesh, India

Heena Parveen Department of Home Science, Aligarh Muslim University, Aligarh, Uttar Pradesh, India

Fazil Qureshi Department of Petroleum Engineering, Glocal University, Saharanpur, Uttar Pradesh, India

Mohammadhossein Ravanbakhsh Ecology and Biodiversity Group, Institute of Environmental Biology, Utrecht University, Utrecht, The Netherlands

Mohd Saleem Plant Physiology Section, Department of Botany, Faculty of Life Sciences, Aligarh Muslim University, Aligarh, Uttar Pradesh, India

Shaibi Saleem Department of Agricultural Microbiology, Faculty of Agricultural Sciences, Aligarh Muslim University, Aligarh, Uttar Pradesh, India

Fareen Sami Plant Physiology Section, Department of Botany, Faculty of Life Sciences, Aligarh Muslim University, Aligarh, Uttar Pradesh, India

Husna Siddiqui Plant Physiology Section, Department of Botany, Faculty of Life Sciences, Aligarh Muslim University, Aligarh, Uttar Pradesh, India

Priyanka Singh Plant Physiology Section, Department of Botany, Faculty of Life Sciences, Aligarh Muslim University, Aligarh, Uttar Pradesh, India

Khalida Siraj Department of Home Science, Aligarh Muslim University, Aligarh, Uttar Pradesh, India

Jamjam Waqeel Department of Agricultural Microbiology, Faculty of Agricultural Sciences, Aligarh Muslim University, Aligarh, Uttar Pradesh, India

Insha Zahoor Department of Post Harvest Engineering and Technology, Aligarh Muslim University, Aligarh, Uttar Pradesh, India

Sadaf Zaidi Department of Post Harvest Engineering and Technology, Aligarh Muslim University, Aligarh, Uttar Pradesh, India

Chapter 1
The Enormity of Zinc Deficiency: An Overview

Shams Tabrez Khan and Abdul Malik

Abstract The societal cost of micronutrient deficiency (MND) or the '*hidden hunger*' calculated by different organizations including WHO is in millions of dollars per year in many countries, especially the low-income countries. MND results in reduced plant growth and production losses, increased medical care costs and loss of life, especially that of children and expecting mothers. For some nations, it can reduce the gross domestic product by as much as 11%. MNDs and malnourishment are associated with more than 6% of global morbidity and mortality burdens. Zn is an important nutrient for both plants and animals. In 2011, due to malnourishment, about 1.1 billion people were at the risk of zinc (Zn) deficiency. It has been related to adverse pregnancy outcomes, stunted growth immune system dysfunctions, neurobehavioural disorders, hair loss, cancer and millions of premature deaths. The growing evidence on health risks associated with Zn deficiency has compelled WHO and FAO to recommend Zn fortification of diet. For crops also Zn is the most important yield-limiting micronutrient after the macronutrients N, P and K causing losses in millions of dollars. Correcting Zn deficiency is a challenge due to various reasons as close to half of the agricultural soils are Zn deficient. The use of chemical Zn fertilizers to the soil is both costly and not very effective. Including Zn-dense diet is costly and cannot be implemented in the low-income region most affected by Zn deficiency. Developing Zn-rich crops through plant breeding and genetic engineering is also challenging.

Keywords Zinc deficiency · Crops · Zinc transporters · Malnutrition · Phytomicrobiome

S. T. Khan (✉) · A. Malik
Department of Agricultural Microbiology, Faculty of Agricultural Sciences, Aligarh Muslim University, Aligarh, Uttar Pradesh, India

S. T. Khan, A. Malik (eds.), *Microbial Biofertilizers and Micronutrient Availability*,
https://doi.org/10.1007/978-3-030-76609-2_1

Introduction

By 2050, the world population is expected to grow by 25% adding another 2 billion humans to world population requiring healthy food (Schroeder et al. 2013). However, the area under cultivation may shrink further with the growing population, urbanization and climate change (Satterthwaite et al. 2010; Zabel et al. 2014). The issues are even more pressing in low-income countries where farmers are increasingly resorting to extensive agricultural practices to deal with this pressure. Extensive agricultural practices and advancement of technology have although improved the availability of food, its nutritious value has decreased (Haddad et al. 2016). Malnutrition still affects every third person worldwide, and the diet is particularly becoming deficient in minerals and vitamins especially affecting children. Around 2 billion people suffer from micronutrient deficiency resulting in serious health consequences accounting for half of all the children's deaths under 5 years of age (Bailey et al. 2015). In addition to these morbidities, micronutrient deficiencies result in stunted growth in children, diminished physical and cognitive capacities and has also been linked to modern-day health problems, such as obesity, further adding to the societal cost of malnutrition (Asfaw 2007). Low-income countries mostly in Africa and Asia, where people are mainly dependent on cereal-based staple diets like wheat and rice, are suffering the most from micronutrient deficiency (Kumssa et al. 2015; Wessells and Brown 2012). The population suffers from MND even in the high-income countries due to the reduced nutritive value of the diet. The micronutrient deficiency incurs a high societal cost which includes production losses, direct medical cost and cost due to disability-adjusted life-years (DALYs). It has been estimated that Fe and Zn deficiency reduces the gross domestic product (GDP) of some developing countries by 2–11% (Darnton-Hill et al. 2005). It is also estimated that addressing micronutrient deficiency may save 35 trillion dollars for the world economy (Dimkpa and Bindraban 2016). Micronutrient deficiency due to the consumption of mineral-poor diet can be attributed to various reasons such as affordability, dependency on cereal-based crops, micronutrient deficient crops and food believes and taboos. This chapter discusses zinc deficiency and the enormity of the problem with reference to the impact on agriculture and human health.

Zinc as an Essential Element for Life

Nutrients that support life can be grouped into two categories based on their required amount, namely, macronutrients and micronutrients. Micronutrients, although required in very small amounts, are indispensable for metabolism and growth. These micronutrients cannot be synthesized by living organisms and, therefore, a balanced supply of these nutrients is required. In fact, without micronutrients survival of the living organisms will not be possible (Haddad et al. 2016). An estimated 25 micronutrients (including zinc, cobalt, copper, manganese, molybdenum and selenium) are required for maintaining the life process in plants and animals including human beings (Chellan and Sadler 2015). Zinc, having an atomic number of 30 and an atomic weight of 65.37, is a vital transition metal for the living world. Some scientists have stated that after macronutrients N, P and K, zinc is the most important micronutrient. The importance of zinc for living organisms was first reported for *Aspergillus niger* in 1869. Since then, a number of seminal studies have established the role of zinc in all forms of life. It is the second-most abundant transition metal after iron in the living organisms. The relative concentration of free zinc ions in the cytoplasm and organelles of various living cells systems varies from $\leq 10^{9}$ M to $\leq 10^{3}$ M (Fabris 1994). It is involved in all the characteristic processes of life, namely, metabolism, growth and reproduction both in plants and humans. Zinc is a constituent of about 3000 proteins contributing either to their structure (metalloproteins: e.g. protein kinases and alcohol dehydrogenases) or to their catalytic activity (metalloenzymes: e.g. carbonic anhydrase). It is present in all the six classes of enzymes including hydrolases, transferases, oxidoreductase, lyases and ligases/isomerases. These enzymes are involved in most important functions of life like translation, DNA replication and repair, inter- and intracellular signalling, maintenance of membrane integrity and photosynthesis (Andreini et al. 2006). Several proteins have been recently found to maintain the homeostasis of zinc in the cells and are involved in zinc transport (Kambe et al. 2015). Zinc is crucial for cell growth and differentiation. In humans, its role in immunity, the functioning of the brain and neurotransmission are becoming increasingly clear (Frederickson et al. 2005). Zinc participates in strengthening the defence against oxidative stress and in the synthesis, storage and release of insulin also. Zinc also plays equally important roles in plant physiology and metabolism and is essential for proper plant growth. It is involved in processes like photosynthesis, chlorophyll synthesis, plant reproduction and autophagy. The role of zinc in grain yield, seed development, protection against plant pathogens and herbivores is also reported (Cabot et al. 2019). Therefore, it is not inappropriate to call zinc as a metal of life.

Zinc and Plants

Role of Zinc in Plants

Zinc content in the plants varies between 30 and 100 mg per kg of dry matter. However, concentration varies from plant to plant, tissue to tissue, cell to cell and even from compartment to compartment within a cell. For example, in most crops, the Zn concentration in leaf required for adequate growth is in the range 15–20 mg Zn/kg of leaves (Broadley et al. 2007), whereas zinc concentration in various compartments of leaf cells varies from 74 to 3205 µg/g dry weight (Frey et al. 2000). It is present in plants both as free Zn^{2+} and as protein-bound Zn. However, it is not known what proportion of zinc is found as free Zn^{2+} and what proportion of Zn exists in the protein-bound form (Broadley et al. 2007). Some of the vital enzymes that are zinc-dependent include DNA and RNA polymerases, histone deacetylases, splicing factors and other important enzymes of mitochondria and chloroplasts (Pilon et al. 2009). It is key to various vital metabolic and physiological functions in plants such as carbohydrate metabolism including photosynthesis, synthesis of chlorophyll, sucrose and starch synthesis. It plays an important role in lipid and nucleic acid metabolism, gene expression, gene regulation, protein synthesis and maintenance of membrane integrity (Hänsch and Mendel 2009; Segal et al. 2003). In maize, it is found to be involved in the formation of functional 80S ribosomal complex and tRNA synthetases (Pilon et al. 2009). It is involved in the functioning of chloroplasts such as the repair of photosystem II by turning over photo-damaged D1 protein. The cytoplasm, lysosomes and the apoplastic spaces within plant cells also show zinc-dependent hydrolytic activities. Zinc metalloproteases are involved in destroying the signal peptides. As a cofactor of plant growth hormones like auxin, Zn regulates plant growth. It also has a special role in reproduction, as pollens are known to have a high concentration of Zn and other studies have demonstrated that Zn-binding proteins in the tapetum play an important role (Kapoor et al. 2002). Zinc is a cofactor of carbonic anhydrase enzyme involved in increasing the CO_2 content of the chloroplast, consequently increasing the activity of rubisco (Salama et al. 2006). It is also involved in signal transduction via mitogen-activated protein kinases. The role of Zn in various physiological processes has been summarized in the books by Brown et al. and Alloway (Alloway 2008; Brown et al. 1993).

Zinc Deficiency in Plants

Since zinc plays a vital role in the growth and reproduction of plants, its deficiency has varying consequences depending on the susceptibility of the plants to zinc deficiency and the extent of the deficiency. Some plants like beans, corn, grapes, maize, rice and sorghum are highly susceptible to Zn deficiency, whereas potatoes, tomatoes, sorghum and sugar beets are mildly sensitive. And crops like oat, carrot and pea may be less susceptible to Zn deficiency (Martens and Westermann 1991; Noulas et al. 2018). Even different genotypes or varieties of the same crop may vary in their ability to assimilate zinc (Mishra et al. 2019). For example, durum wheat is more sensitive to zinc deficiency than the bread wheat. Since the concentration of zinc in different plants and even within different organs of a plant varies and can be estimated using various methods like histochemical methods (dithizone and zincon), flame atomic absorption spectrometry, ultrasound-assisted sequential extraction, energy-dispersive X-ray fluorescence spectrometry (EDXRF) and laser ablation inductively coupled plasma optical emission spectrometry (LA-ICP-OES) (Cakmak et al. 2010; Leśniewska et al. 2016; Paltridge et al. 2012). The concentration estimated using different methods also vary. So far there is no general definition of critical deficiency concentration which can be used to define zinc deficiency in plants (Lindsay 1972). The critical deficiency concentration for various crops has been

Table 1.1 Critical concentration of zinc in some widely used cereals and grains

Cereals/ grains	Concentration (mg/kg)		Reference tissues	References
	Range	Average		
Spring wheat	15–70	14	Youngest tissues	Brennan and Bolland (2002)
Durum wheat	15–70	20	Youngest tissue	Brennan and Bolland (2002)
Rice	20–25	–	Leaf blade	Reuter and Robinson (1997) and Singh (1984)
Corn	20–70	–	–	Noulas et al. (2018)
Barley	–	20	Leaf blade	Genc et al. (2002)
Maize	–	22	–	Alloway (2001)
Pearl millet	–	18	–	Alloway (2008)
Oat	–	5–20	Top leaves	

determined in various studies and values for some important cereal crops are listed in Table 1.1 (Alloway 2008). Although young leaf blades have been used customarily to determine the zinc status of the plants and a concentration of 20 mg Zn/kg zinc is considered as lower critical concentration, grains can also be used to determine the zinc concentration and a value of 15 mg Zn/kg has been suggested as the critical value. Other studies suggest values of 10 mg Zn/kg and 43 mg Zn/kg as critical values for wheat and soybean, respectively (Alloway 2008). Since the currently used techniques require expensive instrumentation and expertise, easy-to-use methods/kits for on-site determination of zinc concentration in the plant tissues are highly desirable.

Since zinc is involved in carbohydrate, protein and DNA metabolism, its deficiency induces changes in carbohydrate and protein metabolism, resulting in a decreased protein content and accumulation of amino acids in zinc-deficient tissues (Kitagishi and Obata 1986). Being a structural component of the ribosome, its deficiency affects ribosome structure and consequently affects both transcription and translation. The deficiency especially affects metabolism in shoot meristems and pollen tubes where Zn concentration is 5–10 times higher than in mature leaf blades. Zinc deficiency also increases RNase activity, breaking down RNA (Pandey et al. 2002). Plasma membrane permeability increases under Zn-deficient conditions. However, symptoms of zinc deficiency are not very apparent, if the deficiency is mild to moderate. However, plants show characteristic symptoms under severe zinc deficiency that are easy to detect. Stunted growth affecting root, shoot, stem, seed and fruit formation are some of the most common zinc deficiency symptoms in plants. In leaves, it causes interveinal chlorosis resulting in the development of purple or bronze tints which is often referred to as *mottling* and *bronzing* of leaves (Fig. 1.1). The chlorosis is more noticeable on middle-aged leaves. Leaves become smaller in size and leaf lamina curls inwards called as little leaf and goblet leaves, respectively. It also results in the necrosis of root apex and stems become thin with characteristic shortened internodes. Fruit and seed formation are reduced significantly; some studies also report reduced dry matter production (Table 1.2).

Fig. 1.1 Leaf blades of wheat (**a**) and maize (**b**) showing the symptoms of zinc deficiency

Table 1.2 Some of the common symptoms of zinc deficiency in grain/cereal crops

Crop	Deficiency symptoms	References
Wheat (*Triticum aestivum* L.)	Small leaves, curved upwards, interveinal chlorosis and necrotic spots	Alloway (2008) and Cakmak et al. (1996)
Rice (*Oryza sativa* L.)	Leaf mid-rib chlorotic at base. Brown spots on older leaves. Browning of leaves. Stunted growth and reduced tillering	Wissuwa et al. (2006), Mori et al. (2016), and Alloway (2008)
Maize/corn (*Zea mays* L.)	White-to-yellow chlorosis of leaf, stunted plants with short internodes and white tips	Alloway (2008) and Mattiello et al. (2015)
Barley (*Hordeum vulgaris* L.)	Uniform chlorosis of leaves, drying up and decreased tip growth	Alloway (2008)
Oat (*Avena sativa* L.)	Bronzing and Necrosis of leaves	Alloway (2008)

However, once these symptoms appear, it becomes difficult to manage the Zn deficiency. Therefore, it is important to find out ways to timely screen the crops for such deficiency beforehand at least for crops that are sensitive to zinc deficiency. Some recent studies have reported methods for predicting Zn deficiency in plants. It has been reported that Zn deficiency causes redox imbalance in roots and shoots even before the occurrence of visible symptoms (Höller et al. 2014). Zinc deficiency enhanced NAD(P)H-dependent superoxide radical production in plasma membrane vesicles isolated from roots of bean plants (Pinton et al. 1994). One study has demonstrated that zinc deficiency induced DNA damage and is responsible for the abnormal development of leaves. And if this DNA damage is alleviated, the malformation of leaves can be controlled (Sotta et al. 2019). Early and late zinc deficiency–responsive genes in rice have also been identified using transcriptome analysis (Bandyopadhyay et al. 2017). These genes are involved in root system architecture, photosynthesis, metal transport, phytosiderophore biosynthesis and

ROS scavenging. The same study demonstrates that zinc-deficient seedlings secrete more oxalates for zinc mobilization and an increased ROS activity in zinc-deficient leaves. Increased secretion of root exudates by zinc-deficient wheat has been reported earlier also (Zhang et al. 1989). In the roots of zinc-deficient maize accumulation of lipids, tannins and oxalic acid have been reported (Alloway 2008). Identifying such biochemical markers of zinc deficiency will help in the timely management of Zn deficiency in plants. These markers coupled with machine learning using images can be used by scientific community and other stakeholders to manage Zn deficiency and to raise awareness among farmers about the significance of zinc deficiency.

Zinc and Human Health

Although the availability of food has improved, the quality of the food has declined. The status of micronutrients in food is a matter of global concern having health and economic burdens. More than 800 million people around the globe are undernourished and more than 2 billion people are known to suffer from one or more chronic micronutrient deficiencies (Kumssa et al. 2015). One such micronutrient is zinc and its use for medicinal purposes is known since 2000 BC; Egyptian papyri from 2000 BC (e.g. the Smith Papyrus 1) mentions the use of zinc in skin cream. Since then, zinc has been identified as one of the most important micronutrients. Many reviews underlining the importance of zinc for life have been published (Alloway 2001; Cakmak et al. 1996).

Role of Zinc in Metabolism and Physiology

Zinc plays a crucial role in metabolism being a constituent of all six different enzyme classes. It is estimated that the human body contains about 2.6 g of Zn. Its concentration in different tissues varies, the highest concentration is present in skeletal muscles accounting for 60% of the total body zinc (Hess et al. 2007). While blood plasma contains only 0.1% of the total body zinc (Hambidge et al. 1986; Hess et al. 2007). It can be present as protein-bound zinc as well as free zinc ion, although excess free zinc in body tissues (100–500 mg/dl) is toxic (Choi and Koh 1998). As far as protein bound to Zn is concerned, it alone constitutes 10% of the human proteome contributing to their structure or catalytic activity (Andreini et al. 2006). In addition to enzymatic catalysis, Zn plays an important role in gene regulation, intra- and intercellular signalling, stabilization of membranes and apoptosis (Cai et al. 2015; Chellan and Sadler 2015; Choi et al. 2017). The role of Zn fingers in the regulation of transcription and their role in health and disease needs a special mention here (Cassandri et al. 2017). Zn also influences hormone production, secretion and sexual development (Bjorndahl and Kvist 2011; Li 2014). It enhances the immune

Table 1.3 Multifarious role of zinc in human health

Zinc function	Examples	References
Constituent of enzymes		
Structural	Zn fingers and Zn ribbons	Lemaire et al. (2009a)
Catalytic factor		
In all six categories of enzymes		
Oxidoreductases	Alcohol dehydrogenase	Vallee and Williams (1968)
Hydrolases	Alkaline phosphatases	Bosron et al. (1975)
Lyases	Carbonic anhydrase	Christianson and Fierke (1996)
Isomerases	Phosphomannose isomerase	Bangera et al. (2019)
Transferases	Sulphur transferase	Selbach et al. (2014)
Ligases	tRNA synthetase	Sankaranarayanan et al. (2000)
Role in signalling		
Intercellular	GABA-mediated synaptic transmission	Sensi et al. (2011)
Intracellular	Zinc wave/intracellular second messenger	Yamasaki et al. (2007)
Physiological functions		
In functioning of brain and CNS	GABA synaptic transmission	Sensi et al. (2011)
Apoptosis regulation		Webb et al. (1997)
Immunity	Both innate and cellular immunity	Rink and Haase (2007) and Bonaventura et al. (2015)
Pregnancy		Chaffee and King (2012)
Antioxidant		Brieger et al. (2013)
Circadian cycle		Schmalen et al. (2014)
Endocrine system		Baltaci et al. (2019)
Anti-inflammation		Prasad (2014)

system and is involved in the proper functioning of the brain and nervous system (Frazzini et al. 2018; Fukada et al. 2019; Gammoh and Rink 2019; Portbury and Adlard 2017). Its role in the stabilization of circadian cycle proteins has also been established (Schmalen et al. 2014). Table 1.3 lists some of the important physiological processes involving zinc.

Prevalence of Zinc Deficiency and its Markers

It is estimated that 1.1 billion people or 17.3% of the total world population mainly in Africa, Asia, Andean region and Central America are at the risk of zinc (Zn) deficiency due to inadequate dietary supply (Fig. 1.2) (Brown et al. 2004a). However, the toxicity of excess zinc and the inability to store the zinc necessitates the maintenance of healthy zinc homeostasis in the body through diet or supplements. This is

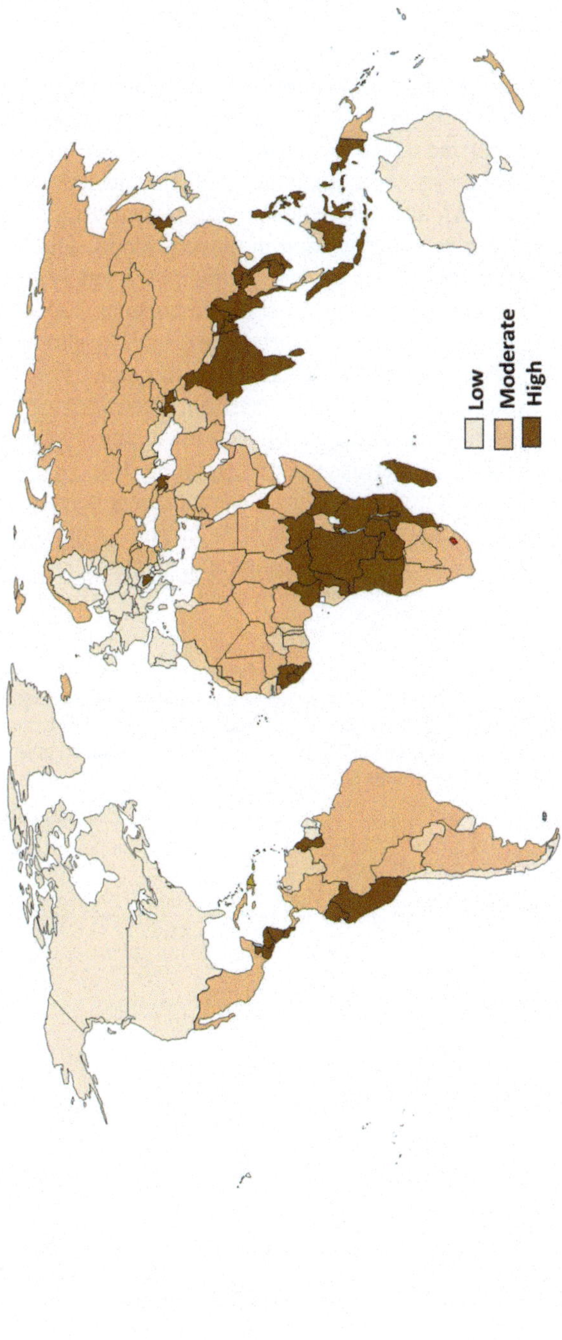

Fig. 1.2 Nation-wise Zn status based on inadequate dietary zinc and childhoodn growth stunting showing the prevalence of zinc deficiency at the global level

required for the proper functioning of cells, organs and various systems including the nervous system and immunity. Many reviews have detailed the significance of zinc in human health (Bosron et al. 1975; Vallee and Williams 1968). The zinc deficiency is a matter of concern as it has been linked to several diseases and health issues detailed below, causing financial losses and loss of lives especially affecting new born and young children. Losses due to Zn deficiency are so rampant and obvious that both WHO and FAO have recommended zinc fortification of diet (Bailey et al. 2015; Christianson and Fierke 1996). It is imperative to timely detect the zinc deficiency using standard and economic, easy to use and quick methods. Scientists are developing methods for its easy, quick and economic detection (Bangera et al. 2019; Selbach et al. 2014). Various indicators/markers of zinc deficiency/status have been used such as zinc concentration in plasma, blood cells, hair and urine (Bangera et al. 2019). But the plasma zinc concentration is a widely accepted marker of zinc deficiency. The normal human blood serum contains 800 ± 200 μg/dL of zinc which may vary with age, sex and other factors (Li 2014), although other unrelated conditions may also influence plasma zinc concentrations. For example, long-term fasting and pregnancy are known to increase and decrease zinc concentration, respectively (Sankaranarayanan et al. 2000). The lower cut-off value of the zinc ranges from 50 to 70 μg/dL for various age groups (Table 1.4).

Severe zinc deficiency is rare but mild-to-moderate zinc deficiency is widespread affecting 7.5–29.6% of the total population in different parts of the world, especially children and pregnant women (Christianson and Fierke 1996; Wessells and Brown 2012).

Table 1.4 Recommended daily intake (RDI) for Zn (Maares and Haase 2020). Range of estimated physiological requirement for absorbed zinc as recommended by World Health Organization (WHO), International Zinc Nutrition Consultative Group (IZiNCG) and Institute of Medicine (IOM). Per cent absorption and lower cut-off of morning serum zinc (Roohani et al. 2013). Per cent absorption values for different age groups (Armah 2016)

Age, sex	RDI (mg/day)	Physiological requirement (mg/day)	Per cent absorption	Lower cut-off serum Zn Conc. (mg/dl)	
				Fasting	non-fasting
6 Months–3 years	1.5–2.5	0.53–0.84	33.8–35.9	–	0.0065
3–8 years	3–4	0.83–1.2			
9–13 years	6	1.4–2.12	28.9–30.3		
10–15 years M	6	1.82		0.0074	0.007
15–18 years M		1.97–3.37			
10–15 years F	6	1.55	33.9	0.007	0.0066
15–18 years F		1.54–3.02			
Pregnancy	7–13	2.27–5.02	–	0.0056	0.0056
Lactation	11–14	2.89–4.52	–	–	–

Different researchers and institutions have defined different parameters associated with zinc status and deficiency differently

Imp Note: The age groups and values are adjusted to provide a simplified overview

Consequences of Zinc Deficiency in Humans

Undernourishment and micronutrient deficiency are associated with more than 6% of global mortality and morbidity burdens (Kumssa et al. 2015). In fact, 4.4% of childhood deaths are due to Zn deficiency, mostly in Africa, Asia and Latin America (Fischer Walker et al. 2009). Zinc deficiency is associated with anorexia, childhood mortality, cognitive dysfunction, hypogonadism, impaired immune function, the prevalence of diarrhoea, pneumonia, stunted growth and skin disorders (Kaur et al. 2014). Premature and small for gestational age infants and preschool children especially of less than 2 years of age constitute the most vulnerable group (Chaffee and King 2012). The role of zinc in various diseases like atherosclerosis, age-related degenerative neurological disorders like Alzheimer's, autoimmune diseases, diabetes, diarrhoea and several malignancies have been established as detailed in many recently published reviews (Chasapis et al. 2012; Kaur et al. 2014; Roohani et al. 2013). Some of the diseases are discussed below and are summarized in Table 1.5.

(a) *Role of zinc in pregnancy and diseases of children*: Deficiency of micronutrients like zinc may affect pregnancy outcomes and early child development (Britto et al. 2017; Flynn et al. 1981; Wang et al. 2015). Zinc affects pregnancy as it is involved in the metabolism of hormones like androgen, oestrogen and progesterone, together with the prostaglandins (Chaffee and King 2012; Favier 1992). Zinc deficiency in pregnancy can cause maternal morbidity, pre-term birth inef-

Table 1.5 Role of zinc deficiency in various human diseases

Disease	References
Diarrhoea in children	Folwaczny (1997)
Pneumonia in children and elderly	Eijkelkamp et al. (2019) and Barnett et al. (2010)
Adverse pregnancy outcomes	Kaur et al. (2014) and Chaffee and King (2012)
Low birth weight and small for gestation	Wang et al. (2015)
Stillbirth and infant mortality	Smith et al. (2017) and Jurowski et al. (2014)
Brain and nervous system	Frederickson et al. (2005) and Nuttall and Oteiza (2014)
Alzheimer and Parkinson	Adani et al. (2020)
Cognitive dysfunction in children	DiGirolamo and Ramirez-Zea (2009)
Immune system	
Hypersensitivity	Seo et al. (2017)
Skin diseases	
Acrodermatitis enteropathica	Gray et al. (2019)
Atopic dermatitis	Satria et al. (2019)
Metastasis	
Breast cancer	Jouybari et al. (2019)
Pancreatic cancer	Li et al. (2007)
Liver diseases like hepatitis C	Mohammad et al. (2012)
Diabetes	Ohta et al. (2019)

ficient labour, atonic bleeding and an increased risk to the foetus. Interestingly, it was demonstrated in a study involving 39 developed countries like Sweden and the Netherlands that zinc intervention can reduce pre-term birth by 5% (Chang et al. 2013). It is also important to note that the mortality due to pre-term birth is much higher in other parts of the world (Black et al. 2016). Infants from mother who received zinc during pregnancy show reduced risk for acute diarrhoea, dysentery and impetigo, diseases that claim the lives of about 9% of infants (Black et al. 2016; Osendarp et al. 2001). Zinc concentration in breast milk is also known to affect the weight of neonates (Doneray et al. 2017). And this deficiency in children due to breastfeeding is also referred to as acquired zinc deficiency. A detailed account of different roles of zinc in pregnancy and child health is published elsewhere (Kaur et al. 2014).

(b) *Zinc and nervous system*: The mammalian brain contains a significant amount of zinc, found as free, protein-bound and vesicular zinc. Protein-bound zinc is one of the main pools of zinc found in the forebrain. Vesicular zinc exists mainly in glutamatergic neurons accounting for 5–15% of the total brain zinc (Marger et al. 2014). There is enough evidence of the crucial role of zinc in neuronal function and signalling (Frederickson et al. 2005). Studies have demonstrated the role of Zn^{2+} transporter for the development of the nervous system, neurulation and neuronal differentiation (Chowanadisai et al. 2013). The role of zinc in memory impairment is also documented as zinc transporters like ZnT3 are known to regulate presynaptic Erk1/2 signalling and hippocampus-dependent memory (Sindreu et al. 2011). Growing evidence indicates that zinc plays a dynamic role in physiology and pathophysiology of brain function. Many studies have demonstrated a positive association between zinc intake and status with one or more measures of cognitive function. Zinc imbalances may cause brain disorders like Alzheimer's and Parkinson's disease which has been reported and reviewed by many authors (Adani et al. 2020; Frederickson et al. 2005).

(c) *Role of zinc in immunity*: Zinc also plays an important role in immunity, and imbalances in zinc concentration disturbs the functioning of the immune system especially cell-mediated immunity. Some of the recently published reviews highlight the role of zinc in immunity (Kaur et al. 2014; Prasad 2008). For example, zinc plays an important role in the development and function of T-regulatory cells involved in the prevention of auto-immune diseases (Hogstrand et al. 2009). Low zinc concentration is known to cause decreased response to vaccination and disturbs the functions of innate immune cells (Haase and Rink 2009). In a survey on higher levels of IgE, a hypersensitivity associated antibody was found to be associated with the low levels of zinc. Showing that low levels of zinc may be involved in hypersensitivity (Seo et al. 2017).

(d) *Role of zinc in skin diseases*: Acrodermatitis enteropathica (AE) is a skin disease due to zinc deficiency resulting in dry skin and skin blisters. However, this is a genetic disease resulting from a mutation in a Zn-binding membrane protein (Nistor et al. 2016). A gene, namely, 'SLC39A4', which codes for hZIP4 zinc transport protein, is found to be involved in the disease in humans (Küry et al. 2002). Zinc deficiency can also result in a hypersensitivity skin disease called atopic dermatitis (AD), while in some cases the co-occurrence of both AE and AD has also been reported (Gray et al. 2019; Satria et al. 2019). Zinc also helps in wound repair by augmenting the migration of keratinocyte (Lansdown et al. 2007). Protein MG53, a zinc-binding protein, is involved in membrane repair facilitating the repair of injury (Cai et al. 2015).

(e) *Zinc status and metastasis*: Impaired zinc homeostasis is also associated with cancer and its progression. Studies suggest impaired zinc homeostasis results in the overexpression of zinc transporter ZnT1 in at least five different cancer types (Lehvy et al. 2019). Decreased serum zinc was found in breast, prostate, liver, lung and kidney cancer. Lowered serum Zn concentrations have been shown to increase the risk of breast cancer and its onset in women (Jouybari et al. 2019). Studies describing the molecular mechanism show the involvement of zinc transporters, zinc channel ZIP7 and an anti-apoptotic kinase CK2 (Hogstrand et al. 2009; Taylor et al. 2012). In another study, it was demonstrated that zinc transporter, ZIP4 (SLC39A4), is involved in human pancreatic cancer (Li et al. 2007). Zinc also protects against prostate cancer, although the molecular mechanism is not yet well understood (Kolenko et al. 2013).

(f) *Zinc and diabetes*: Zinc plays an important role in β-cell function, glucose homeostasis, insulin action and the pathogenesis of diabetes (Ranasinghe et al. 2015). It is involved in the synthesis of insulin and determines its conformational integrity (Fukunaka and Fujitani 2018; Norouzi et al. 2017). Proinsulin is a zinc-containing hexamer that is readily converted into insulin hexamer (Emdin et al. 1980). The recently published structure of adiponectin receptors (integral membrane proteins), which controls glucose and lipid metabolism, is found to contain zinc-binding sites (Vasiliauskaité-Brooks et al. 2017). Low zinc concentration is known to affect the maturation of insulin-producing cells (Ohta et al. 2019). The expression of zinc transporter ZnT8 is known to influence the crystallization of insulin in mice, although it was not found to be involved in the homeostasis of normal glucose concentration (Lemaire et al. 2009b). However, the results of some studies to correlate zinc to diabetes remain inconclusive.

(g) *Miscellaneous health conditions and role of zinc*: Role of zinc in various other health conditions and diseases, like obesity, cardiac health, hepatitis and pneumonia, are also documented (Barnett et al. 2010; Kaur et al. 2014). Low zinc status has been regarded as a risk factor for pneumonia in the elderly (Barnett et al. 2010). Zinc intervention is known to alleviate diarrhoea and pneumonia in children, reducing child mortality (Hambidge 2006; Lassi et al. 2016). The use of zinc as a low-cost adjunct for anti-viral therapy against hepatitis C virus has been suggested (Read et al. 2018). Zinc may facilitate the anti-viral role of MTs against other viruses. Recently, the structure of hepatitis C virus replicase was found to contains a zinc-binding domain (Tellinghuisen et al. 2005). The role of zinc and zinc transporters in various health conditions is becoming clear with new findings highlighting the role of zinc as a metal of life. Table 1.5 shows a list of various diseases caused due to zinc deficiency.

Cost of Zinc Deficiency

Since zinc plays such a vital role in the growth, development and reproduction of both plants and humans as detailed above, its deficiency causes retarded growth and several diseases in plants and humans causing significantly high economic losses at a global scale (Fig. 1.3).

Losses of Crop Yield and Plant Productivity

Since zinc is vital for the growth and reproduction of the plants, its deficiency causes loss of crops or produce. Therefore, the use of zinc can help in controlling such losses and to increase productivity. For example, it was demonstrated in field experiments that the grain yield of wheat increased by 43% when zinc was used during cultivation in a zinc-deficient soil (Cakmak et al. 1996). Very few studies have estimated the actual scale of financial losses to crops and produce due to zinc deficiency. For example, it was demonstrated that under Zn deficiency, losses of up to 30% in cereal grains can occur, without any obvious visual symptoms of stress (Noulas et al. 2018). In another study by Cakmak et al., it was estimated that the use of Zn fertilizers increased yield of wheat grain in Central Anatolia, Turkey, by >600%, returning economic benefits of ca. US$ 100 million annually (Cakmak 2008). It was estimated that the use of Zn fertilizers increased the grain yield in Punjab, returning an additional US$ >800 M annually (Joy et al. 2017). A study in India reports a total increase in cereal production of about 17 metric tonnes (7.6% of the total cereal production) in India due to Zn fertilization in 2010–2011 (Prasad et al. 2013). The benefits of this increased production will be in millions of dollars. An increase in grain yields has been reported in many other studies also, but the economic benefits have not been calculated. However, more detailed studies to estimate the real scale of financial impacts are required. These localized studies provide good evidence that the use of zinc fertilizers may add agricultural benefits amounting to billions of dollars globally.

Economic Burden of Zinc Deficiency in Humans

Micronutrient deficiency, in general, and zinc deficiency, in particular, result in loss of millions of dollars annually to various countries claiming millions of lives especially affecting children. For some nations, it can reduce gross domestic product by as much as 11% (Hicks et al. 2019). In China alone, the economic burden due to the diseases caused by malnutrition is estimated to be US$ 66 billion annually (Linthicum et al. 2015). A study on Cambodia shows that malnutrition in Cambodia incurs losses of more than 400 million US dollars annually (Bagriansky et al. 2014).

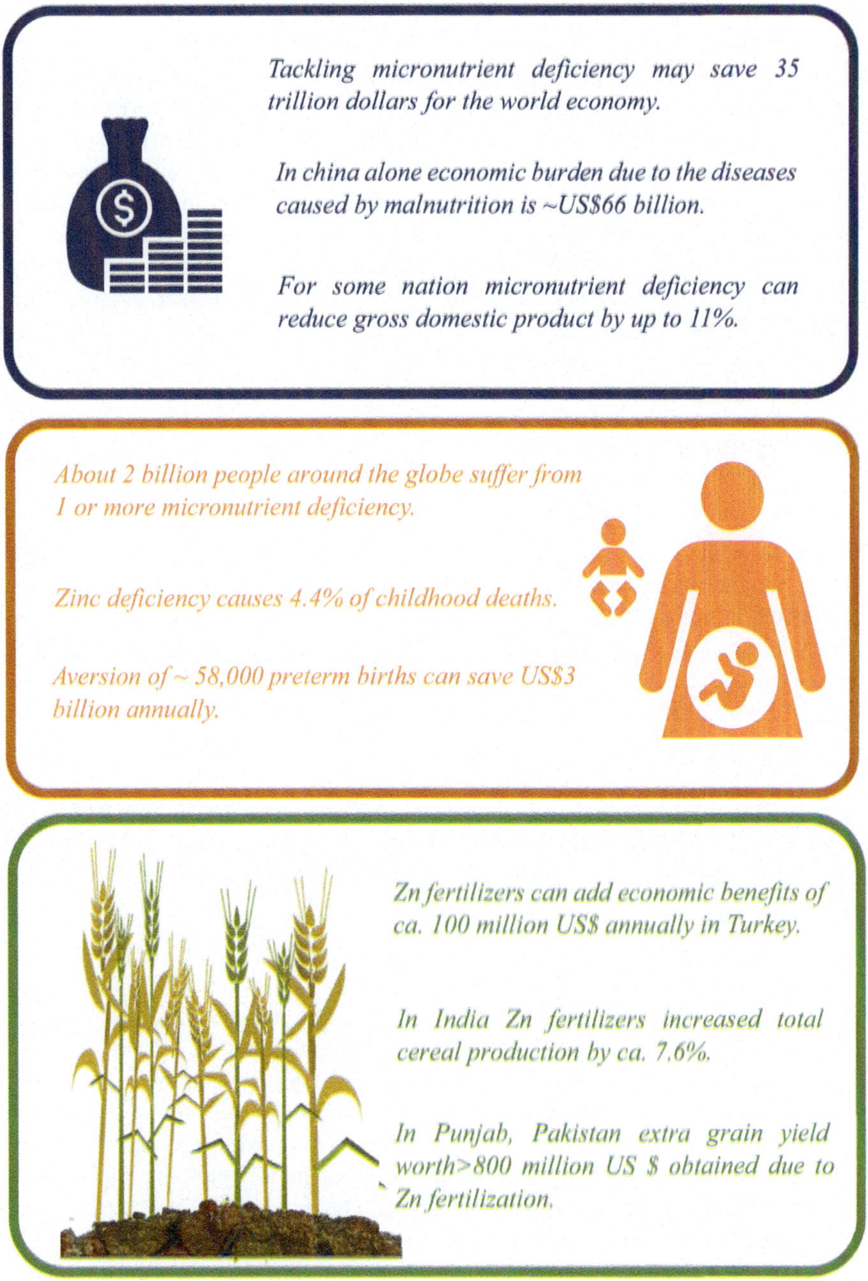

Fig. 1.3 Economic burden on the global economy due to Zinc and other micronutrient deficiency. Please refer to the text for details

These economic burdens are due to increased medical costs, cost due to losses in productivity and disability-adjusted life years (DALYs). The societal cost of micronutrient deficiency in 6- to 59-month-old children in Pakistan was estimated to be US$ 46 million in medical costs, US$ 3222 million in production losses and 3.4 million DALYs (Wieser et al. 2017). Aversion of roughly 58,000 pre-term births through zinc intervention was estimated to save about US$3 billion annually (Chang et al. 2013). The costs of micronutrient deficiency are 5 times higher in poor households than in wealthy households (Wieser et al. 2017). These studies suggest that billions of dollars can be saved annually by simply managing the micronutrient deficiency, in general, and zinc deficiency, in particular. This will require a clear understanding of how the deficiency is caused.

Reasons of Zn Deficiency

Zinc deficiency in plants and humans is caused by several factors. These include factors associated with soil, like zinc-deficient soil, and factors associated with plants, like the zinc efficiency of plants. Zinc deficiency in plants directly affects the zinc status in humans. Zinc-poor diet is one of the primary reasons for zinc deficiency in humans. Malabsorption of zinc due to disease, age and other factors also causes the deficiency of zinc. This section discusses how these factors affect zinc availability in plants and humans and how these factors are related (Fig. 1.4).

A. Reasons for Zinc Deficiency in Plants

A number of factors cause zinc deficiency in plants. These include the zinc status of soil, soil physicochemical characteristics, the inefficiency of plants to accumulate zinc and transport of zinc to various parts of plants, especially to the edible part.

(i) *Zinc deficiency in soil*: The earth's crust is made up of two types of rocks, namely, igneous and sedimentary rocks containing 48–100 and 20–200 mg/Kg of zinc, respectively (Mertens and Smolders 2013). Franklinite ($ZnFe_2O_4$), hopeite ($Zn_3(PO_4)_2{\cdot}4H_2O$), smithsonite ($ZnCO_3$), zincite (ZnO) and zinkosite ($ZnSO_4$) are some of the generally found zinc minerals in the soil. The zinc pool in the soil is present as water-soluble pool, exchangeable pool (ions bound to soil particles) and organically bound pool (chelated or complexed with organic matter). Some of the zinc may also bind to clay and other soil particles in a unexchangeable form. Zinc is generally present in the earth's crust in a concentration of 67 mg/Kg (Rudnick and Gao 2003). A detailed summary of zinc concentration in various soils is given elsewhere and in Table 1.6 (Kabata-Pendias 2010; Mertens and Smolders 2013). The normal range of total Zn in soils is 10–300 mg/kg with a mean value of 50 mg/kg (Sharma et al. 2013). Zinc which is available to plants is available either in ionic form (Zn^{2+}), which may be exchangeable or in organically complexed form. Zinc in ionic form or free form/water-soluble form can be estimated

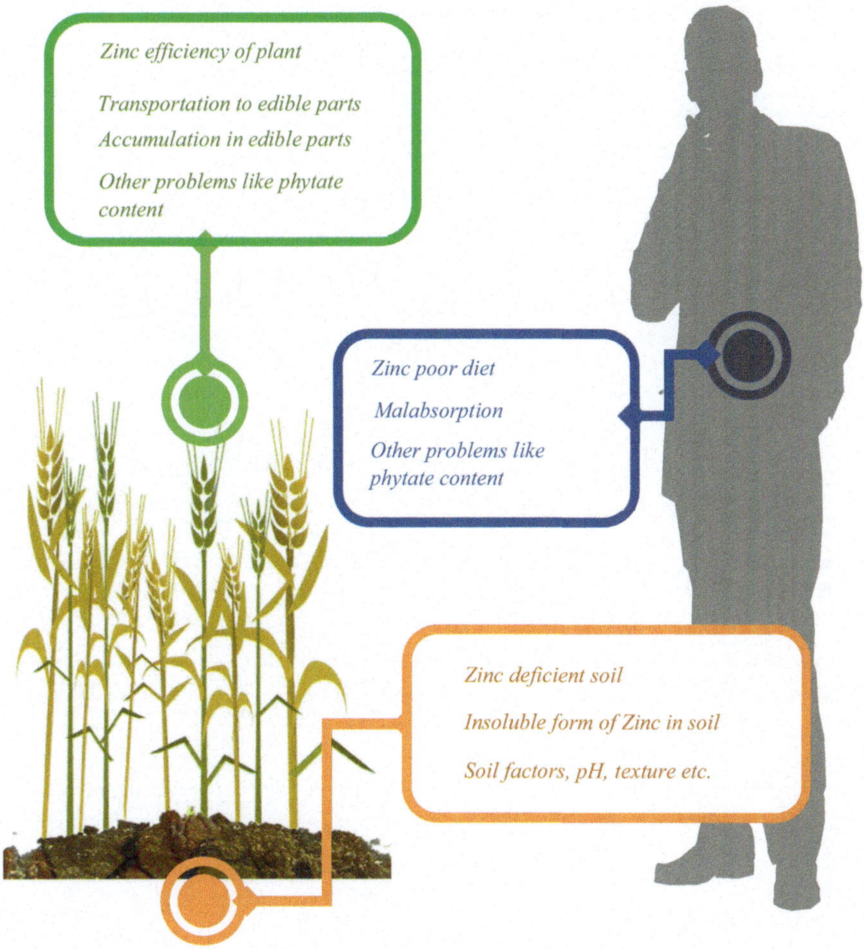

Fig. 1.4 Various problems associated with soil, plants and humans contributing to the global zinc deficiency

through different extraction methods. These include DTPA, Mehlich and HCl and, hence, different definitions for defining critical deficiency are also used. This also varies with the type of soil as shown in Table 1.7. Soils containing less than 0.5 mg of diethylenetriaminepentaacetic acid (DTPA) extractable Zn/kg of soil are classified as Zn-deficient soils (Mertens and Smolders 2013). Some studies define critical deficiency in the range 0.6–2.0 mg/kg depending on the extraction method (Singh and Singh 2005). Soils may also contain insoluble forms of zinc, which is also one of the important reasons for zinc deficiency in plants and will be discussed later. The average soil Zn concentrations and minerals are not related to zinc concentration in agricultural soils. The deficiency of zinc in the soil is a global problem and one of the most

Table 1.6 Concentration of zinc in various soils of the world (Alloway 2008; Kabata-Pendias 2010)

S. No.	Country	Zinc concentration (mg/kg soil)		Soil type
		Mean	Range	
1	USA	50	10–300	
2	Australia	34	2–180	
			4–41	Alkaline soil
			5–36	Calcareous soil
3	England and Wales	97	–	
		35	–	Sandy soil
		65	–	Coarse loamy
		90	–	Fine silty soil
		106	–	Clayey soil
4	France	17	–	Sandy soil
		40	–	Silty soil
		63.5	–	Loam
		98	–	Clayey soil
		132	–	Very clayey
6	Poland	37	–	Sandy soil
		60	–	Loess soil
		75	–	Loam
7	Germany	27.3	–	Sandy soil
		59.2	–	Loam/silt
		76.4	–	Clay
8	Sweden	65	–	
9	Japan	89	–	Agricultural soil
8	India	59	20–89	Arid/semi-arid
		52	22–74	Humid/sub-humid tropics
			69–76	Vertisols
			24–30	Oxisols
10	Brazil, Parna	73	–	
9	Philippines		63–135	Rice soils
10	Vietnam	102	40–485	
11	Indonesia		33–174	
12	Thailand	45	5–158	

Table 1.7 The critical concentration of zinc in different types of soil. The concentration varies with estimation methods as given in the table

S. No.	Zn conc. mg/Kg of soil	Method	Soil type	References
1.	0.5–1.0	DTPA	–	Cox (1987)
2.	1.1	Mehlich	–	
3.	0.1	HCl	–	
4.	0.13	DTPA	Subterranean	Brennan and Gartrell (1990)
5.	0.55	DTPA	Sand	
6.	0.55	DTPA	Clay	

important reasons for zinc deficiency as the deficiency in soil results in the deficiency of this micronutrient in various food chains. It is estimated that one-thirds of the world soil are zinc deficient. Various researchers and international organizations like FAO, WHO and IZiNCG have highlighted the issue of zinc deficiency in soil (Alloway 2008; Cakmak and Kutman 2018; Rudnick and Gao 2003). Some of the countries that have most zinc-deficient soils include countries in Asia like China, India, Indonesia and Turkey, sub-Saharan Africa and the north western region of South America. A survey of 3500 soil samples from 29 developing countries has shown that most of the zinc-deficient soils were found in Iraq (57% of samples), Turkey (35%) and Pakistan (20%) (Cakmak and Kutman 2018). No such detailed surveys have been published since then. However, some recent surveys of sub-Saharan Africa also show that zinc deficiency predominates the studied soil samples (Cox 1987). A survey of 256,000 Indian soil and 25,000 plant samples have shown that 48.5% of the soils and 44% of the plant samples are zinc deficient (Fig. 1.5) (Brennan and Gartrell 1990; Hicks et al. 2019). The frequency of low zinc is particularly higher in Maharashtra, Karnataka and Haryana (Brennan and Gartrell 1990; Hicks et al. 2019).

(ii) *Presence of zinc in insoluble/unavailable forms*: One of the reasons for zinc deficiency is the presence of zinc as insoluble complexes or as secondary minerals in the soil. These insoluble forms of zinc include Zinc carbonate ($ZnCO_3$), zinc oxide (ZnO) and zinc sulphide (ZnS). Smithsonite ($ZnCO_3$), sphalerite (ZnS), zincite (ZnO), franklinite ($ZnFe_2O_4$), willemite (Zn_2SiO_4) and hopeite ($Zn_3(PO_4)_2{\cdot}4H_2O$) are the secondary minerals containing insoluble zinc in soil. Studies claim that the insoluble form of Zn constitutes >90% of soil Zn which is not available to plants, while exchangeable, water-soluble Zn constitutes only a fraction (4×10^{-10}–4×10^{-6} M) (Gupta et al. 2016).

(iii) *Soil factors*: A number of soil factors such as pH, salinity and texture affect the solubility and availability of zinc to plants. Zn^{2+} are held on the surface of clay particles, therefore clayey and silty soils tend to contain more plant-available zinc than sandy or sandy loam soils. The median Zn concentration in sandy soils was 17 mg/kg of soil, while the concentration in clayey soil was 132 mg/kg (Mertens and Smolders 2013). Similarly, soils having alkaline pH (>6.5) reduce the Zn uptake by plants. Therefore, calcareous and sandy soils with alkaline pH tend to contain less plant-available zinc. Unfortunately, 30% of the arable lands worldwide are alkaline (Cakmak and Kutman 2018). Even if zinc fertilizer like zinc sulphate is added to such soils, these fertilizers again transform into compounds like Zn(OH) and $Zn(OH)_2$ at alkaline pH or in such soils. Under highly reduced conditions like in paddy fields, zinc sulphate may get converted into ZnS (Alloway 2008). High levels of phosphorus can also adversely affect the availability of zinc to plants due to the formation of calcium phosphate (apatite).

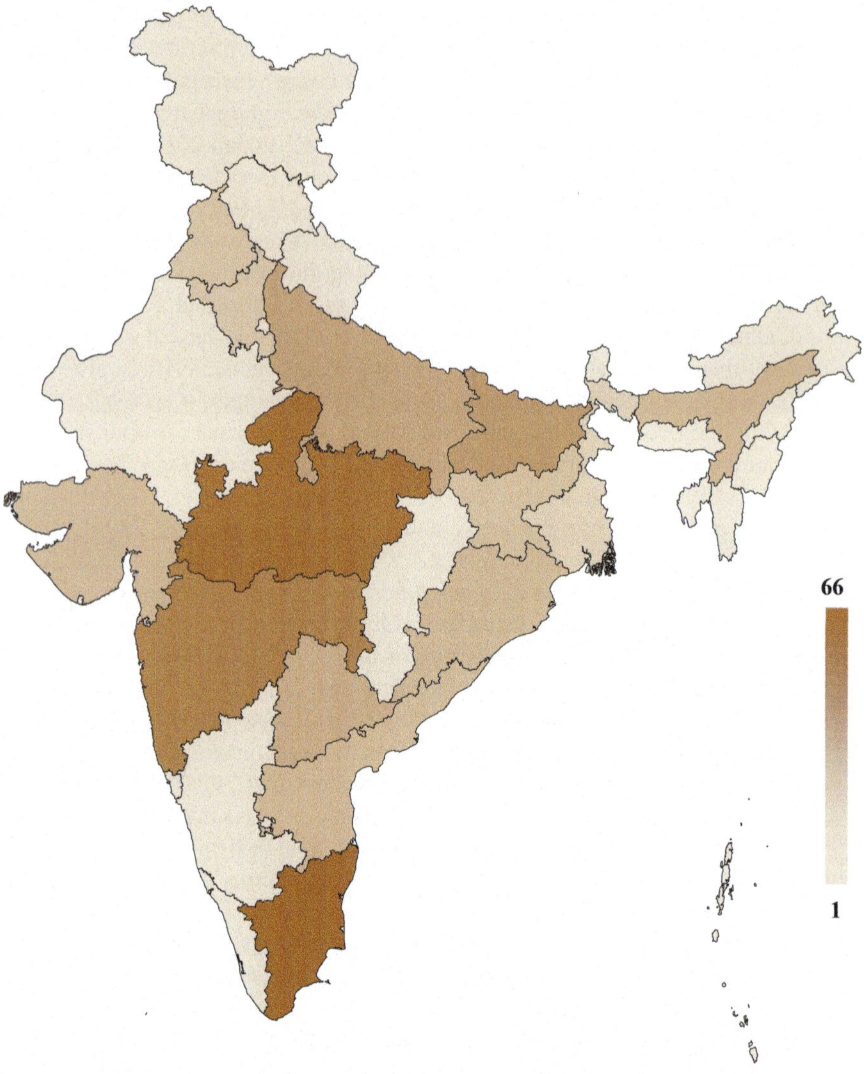

Fig. 1.5 Per cent of soil samples found to be Zn deficient in various states of India

(iv) *Zinc sensitivity and efficiency of plants*: Plants vary in their sensitivity to zinc deficiency; some are more sensitive to zinc deficiency than others. This ability of different crops or genotypes or cultivars to grow optimally and to produce good yield under zinc-deficient concentration is called as zinc efficiency (Alloway 2008; Hacisalihoglu and Kochian 2003). Zinc-efficient genotypes efficiently absorb zinc from soils to support good growth and grain yield. The mechanism behind the increased absorption of zinc may be due to the change in the root structure and physicochemical behaviour of the root. Roots that are extensive, long and fine tend to absorb more nutrients. Increased secretion of zinc-chelating siderophores by the roots can also promote the absorption (Alloway 2008). Furthermore, zinc-efficient cultivars can maintain the activity of zinc-dependent enzymes such as carbonic anhydrase and Cu/Zn superoxide dismutase. Such zinc-efficient cultivars may also have better utilization and distribution of absorbed zinc by various cell compartments and tissues. However, the zinc content of the grains may not be high. High grain zinc content is a genetic trait; therefore, cultivars that have high zinc content should be preferred. Cakmak found that wheat developed through crosses of local landraces exhibits higher zinc efficiency (Cakmak et al. 2010). On the contrary, cultivars or genotypes that are not zinc efficient will be associated with loss of productivity and decreased zinc content.

(v) *Climatic conditions*: Lower temperatures are also known to reduce the zinc uptake by plants probably due to reduced microbial activity in soil and conversion of organically bound zinc to available forms. Low temperature can affect other biogeochemical processes affecting indirectly the zinc update (Alloway 2008; Liu et al. 2019; Xue et al. 2012). For example, phosphorus-induced zinc deficiency is more severe at low temperatures.

(vi) *Interference due to other nutrients*: High soil phosphate levels also decrease zinc uptake by plants. Although high phosphorus concentrations are being related to low zinc content in many crops around the world (Barrow 1987), the underlying mechanism is not very well understood. Nitrogen is also known to affect zinc availability to crops. The availability of nitrogen increases protein formation resulting in an increased formation of zinc-protein complexes in the root. These complexes sometimes are not translocated to other parts of the plant (Alloway 2008). However, an increased zinc uptake due to nitrogen-mediated change of soil pH has been demonstrated (Liu et al. 2018). Transcriptomic studies have shown an up-regulation of N-homeostasis genes in the stems of zinc-deficient plants (Bouain et al. 2019).

(vii) *Soil microbiome/rhizobiome:* Soil microbiome not only plays a role in the nutrient uptake by plants but also in fixing the nutrient from the environment and making it available to plants. Microbes provide these nutrients to plants in a sustainable and eco-friendly way. These microorganisms are not only limited to nitrogen fixers, but several PGPR activities are also found to be associated with the soil microorganisms. With the increasing knowledge of microbiome using high-throughput sequencing and omic approaches, the multifarious role of microbes in sustaining the plant productivity and health is becoming increasingly clear (Berg et al. 2017). Some recent reports show that some plant symbionts even engineer the plant-associated microbiome or so-called phytomicrobiome through changes in root exudates, signalling and soil modification (Uroz et al. 2019). Plant-associated microbes not only control the phytomicrobiome but also are involved in root-root signalling, resulting in systematically induced root metabolite exudation or SIREM (Korenblum et al. 2020). Glycosylated azelaic as a signalling molecule carries out SIREM signalling over a much longer distance than previously thought (Korenblum et al.

2020). Microorganisms are also known to solubilize the insoluble form of zinc to make it available to plants. These microorganisms not only help in the solubilization of zinc but can also help plant growth through other plant growth-promoting activities (Kamran et al. 2017; Kumar et al. 2019). Unfortunately, chemical fertilizers even for zinc are more readily available than bio-fertilizers. Lack of awareness among farmers and the unavailability of such commercial microbe-based fertilizers have further complicated the problem. Regular and extensive use of chemical fertilizers inherently cause a number of problems jeopardizing the use of soil as a renewable resource (Wu et al. 2019).

B. Causes of Zinc Deficiency in Humans:

Zinc deficiency in humans can occur due to various reasons. Some of the important reasons include inadequate dietary intake, malabsorption, impaired utilization and increased losses during digestion.

(i) *Inadequate dietary intake*: Inadequate dietary intake of zinc is one of the main reasons for zinc deficiency. The average recommended dietary intake varies with age group, sex and with special health conditions like pregnancy. Poor dietary intake can be due to various reasons ranging from economic to food believes and taboos. Typical human zinc intake is in a range 14–30 mg/kg. In a detailed study, the zinc intake was estimated using the FAO national food balance sheet of 188 countries (Wessells and Brown 2012). It was found that an estimated 17.3% of the global population is at risk of inadequate zinc intake, while in the South-Asian region 30% of the population was at the risk. In another report, it was found that 1.1 billion people are at the risk of zinc deficiency due to inadequate dietary supply and 90% of those at risk are from Africa and Asia (Kumssa et al. 2015). Even in the United States, the elderly were at increased risk of zinc deficiency (Maret and Sandstead 2006). However, the percentage of countries at risk of Zn deficiency has improved from 35% in 1992 to 26% in 2011. The major source of zinc were animal products in the Americas, Australia, North Africa, Europe and New Zealand. While in other parts of the world, cereals served as the main source of zinc. Whole-grain cereals, lean red meat, pulses and legumes are known to provide highest concentrations of zinc 25–50 mg/kg of raw weight. However, high phytate content of cereals and vegetables or phytate-to-zinc ratio in diet was also identified as a reason of zinc deficiency as it interferes with the absorption of zinc (Gibson 2012; Raboy 2001). Therefore, animal-based food is a better source of absorbable zinc than plant-based diet (Gibson et al. 2018). However, widespread vegetarianism in countries like India need alternative ways of supplementing zinc. As far as population sub-groups are concerned, sub-groups with high growth rates and tissue synthesis, for example, infants and toddlers would require higher zinc intake (Krebs et al. 2014).

(ii) *Malabsorption of zinc*: Dietary zinc is mainly absorbed in the small intestine (duodenum and jejunum) and with the help of specific transport proteins like divalent metal-ion transporter-1 (DMT1) and zinc transporter (like ZnT-1), it enters into the circulation portal making it available to other tissues. It is estimated that about 70% of the zinc in circulation is bound to albumin, and factors that alters serum albumin concentration can effect availability of zinc to other tissues (Brown et al. 2004b). The excess zinc is excreted through gastrointestinal secretion, sloughing mucosal cells and integument. However, malabsorption of zinc also results in the losses of dietary zinc adding to the problem and requiring even higher intakes of zinc (Gibson et al. 2016). Various studies have shown that the amount of zinc absorbed from the dietary intake ranges from 25% to 93% (Abbaspour 2013). This loss can be due to various reasons, like age, diseases or due to various components of the food itself. Genetic disorder Acrodermatitis is a rare genetic disease affecting 1 person per 500,000. It is an autosomal recessive mutation of the SLC39A4 gene on chromosome 8q24.3, encoding zinc transporter Zip4. It results in the malabsorption of zinc in the diseased person (Weedon 2010). Various components of food itself may interfere with the absorption of zinc including high fat content, type of protein (animal or plant), fibres, phytates and other metals (Bel-Serrat et al. 2014; Jurowski et al. 2014; Lönnerdal 2000). A number of studies have highlighted the fact that phytate mainly found in a plant-based diet is known to interfere with the absorption of zinc (Gibson 2012). Studies have shown that higher the phytate content lower the absorption of Zinc.

Conclusion

Zinc indeed is a metal of life playing crucial roles in plant and human physiology. Its role in processes basic for life from photosynthesis in plants to functioning of the brain in humans is becoming increasingly clear. Inadequate supply of zinc in the food chain is affecting crop productivity and quality resulting in zinc-poor crop. This is having adverse implications on human health and reproductivity, claiming millions of lives, especially children. The global socio-economic cost of zinc deficiency is not calculated but is predicted to be in billions of dollars annually. This can be due to the losses of plant productivity, increased medical cost as well as due to the disabilities-adjusted life years caused by the zinc deficiency. The cost can be so huge that for some countries it can amount to 2–11% of the country's GDP. To minimize these losses, it is imperative to understand the scale of the problem and to convince stakeholders about the seriousness of the problem. The reasons for zinc deficiency are as multifarious as the scale of the zinc deficiency problem itself, starting from the prevalence of zinc-deficient soil (30% of world soils) to the problem of malabsorption of zinc in individuals. These problems can be broadly grouped into three categories, namely, (1) problems resulting in soil Zn deficiency, (2) factors contributing to Zn deficiency in plants and (3) factors contributing to Zn deficiency in humans.

Problems resulting in soil Zn deficiency include the following: (i) Unavailability of zinc (30% of the soils are Zn deficient): Various methods are used to check zinc availability in soil. And the definition of a critical lower limit varies with soil type and extraction methods. Therefore, an internationally accepted simple and easy-to-use method and definitions of zinc deficiency are required. (ii) Insoluble forms of zinc in soil: The insoluble forms of zinc like ZnO and $ZnCO_3$ are not available to plants resulting in Zn deficiency in plants. (iii) Soil properties: Various soil factors like alkalinity and soil texture influence zinc availability. (iv) Soil microbiome: An understanding of the mechanisms for microbes mediated increase in zinc bioavailability to plants is both urgent and necessary. Problems that need to be addressed to improve Zn status of plants include the following: (i) designing easy methods/kits for the timely detection of zinc deficiency in plant tissues, (ii) development of zinc-efficient genotypes or varieties, (iii) understanding the mechanism of zinc transport to the edible parts of plants and (iv) developing methods for minimizing components like phytase that interfere with zinc absorption in humans. However, issues that need attention to improve Zn status in humans are as follows: (i) Low dietary intake is one of the primary reasons for zinc deficiency in humans, which can be due to socioeconomic reasons and widespread poverty where a significant section of the world population cannot afford a zinc-rich diet. Much of the population is dependent on a cereal-based diet, which is becoming increasingly zinc deficient. Infants, children and expecting mothers constitute high-risk groups due to higher zinc demand. (ii) However, serum zinc is a widely accepted method to detect zinc deficiency. Development of still easy and economic method is a bottleneck for wider analysis and prevention. (iii) Investigations are required to understand how the absorption of Zn from the diet can be improved. The joint effort of policy-makers, scientists, agricultural scientists, farmers, dieticians, physicians and others is required to deal with the problem of zinc deficiency. This will not only save millions of lives but will also add billions of dollars to the world economy through increased crop productivity and reduced medical costs. This book discusses some of these issues.

Acknowledgement The authors acknowledge the generous financial help provided by the MHRD, Govt. of India, under the Scheme for Promotion of Academic and Research Collaboration (SPARC; P594).

References

Abbaspour, N.: Zinc and its importance for human health: an integrative review. J. Res. Med. Sci. **18**, 144–157 (2013)

Adani, G., Filippini, T., Michalke, B., Vinceti, M.: Selenium and other trace elements in the etiology of Parkinson's disease: a systematic review and meta-analysis of case-control studies. Neuroepidemiology. **54**(1), 1–23 (2020). https://doi.org/10.1159/000502357

Alloway, B.J.: Zinc-the Vital Micronutrient for Healthy, High-Value Crops. IZA, Brussels (2001)

Alloway, B.J.: Zinc in Soils and Crop Nutrition. International Zinc Association (2008)

Andreini, C., Banci, L., Bertini, I., Rosato, A.: Counting the zinc-proteins encoded in the human genome. J. Proteome Res. **5**(1), 196–201 (2006). https://doi.org/10.1021/pr050361j

Armah, S.M.: Fractional zinc absorption for men, women, and adolescents is overestimated in the current dietary reference intakes. J. Nutr. **146**(6), 1276–1280 (2016)

Asfaw, A.: Micronutrient deficiency and the prevalence of mothers' overweight/obesity in Egypt. Econ. Hum. Biol. **5**(3), 471–483 (2007). https://doi.org/10.1016/j.ehb.2007.03.004

Bagriansky, J., Champa, N., Pak, K., Whitney, S., Laillou, A.: The economic consequences of malnutrition in Cambodia, more than 400 million US dollar lost annually. Asia Pac. J. Clin. Nutr. **23**(4), 524–531 (2014). https://doi.org/10.6133/apjcn.2014.23.4.08

Bailey, R.L., West Jr., K.P., Black, R.E.: The epidemiology of global micronutrient deficiencies. Ann. Nutr. Metab. **66**(suppl 2), 22–33 (2015). https://doi.org/10.1159/000371618

Baltaci, A.K., Mogulkoc, R., Baltaci, S.: Review: the role of zinc in the endocrine system. Pak. J. Pharm. Sci. **32**, 231–239 (2019)

Bandyopadhyay, T., Mehra, P., Hairat, S., Giri, J.: Morpho-physiological and transcriptome profiling reveal novel zinc deficiency-responsive genes in rice. Funct. Integr. Genomics. **17**(5), 565–581 (2017). https://doi.org/10.1007/s10142-017-0556-x

Bangera, M., Gowda, K.G., Sagurthi, S.R., Murthy, M.R.N.: Structural and functional insights into phosphomannose isomerase: the role of zinc and catalytic residues. Acta Crystallogr. D Struct. Biol. **75**(Pt 5), 475–487 (2019). https://doi.org/10.1107/s2059798319004169

Barnett, J.B., Hamer, D.H., Meydani, S.N.: Low zinc status: a new risk factor for pneumonia in the elderly? Nutr. Rev. **68**(1), 30–37 (2010). https://doi.org/10.1111/j.1753-4887.2009.00253.x

Barrow, N.J.: The effects of phosphate on zinc sorption by a soil. J. Soil Sci. **38**(3), 453–459 (1987). https://doi.org/10.1111/j.1365-2389.1987.tb02280.x

Bel-Serrat, S., Stammers, A.L., Warthon-Medina, M., Moran, V.H., Iglesia-Altaba, I., Hermoso, M., Moreno, L.A., Lowe, N.M.: Factors that affect zinc bioavailability and losses in adult and elderly populations. Nutr. Rev. **72**(5), 334–352 (2014). https://doi.org/10.1111/nure.12105

Berg, G., Koberl, M., Rybakova, D., Muller, H., Grosch, R., Smalla, K.: Plant microbial diversity is suggested as the key to future biocontrol and health trends. FEMS Microbiol. Ecol. **93**(5), fix050 (2017). https://doi.org/10.1093/femsec/fix050

Bjorndahl, L., Kvist, U.: A model for the importance of zinc in the dynamics of human sperm chromatin stabilization after ejaculation in relation to sperm DNA vulnerability. Syst Biol Reprod Med. **57**(1–2), 86–92 (2011). https://doi.org/10.3109/19396368.2010.516306

Black, R.E., Levin, C., Walker, N., Chou, D., Liu, L., Temmerman, M.: Reproductive, maternal, newborn, and child health: key messages from Disease Control Priorities 3rd ed. Lancet. **388**(10061), 2811–2824 (2016). https://doi.org/10.1016/S0140-6736(16)00738-8

Bonaventura, P., Benedetti, G., Albarède, F., Miossec, P.: Zinc and its role in immunity and inflammation. Autoimmun. Rev. **14**(4), 277–285 (2015). https://doi.org/10.1016/j.autrev.2014.11.008

Bosron, W.F., Kennedy, F.S., Vallee, B.L.: Zinc and magnesium content of alkaline phosphatase from Escherichia coli. Biochemistry. **14**(10), 2275–2282 (1975). https://doi.org/10.1021/bi00681a036

Bouain, N., Krouk, G., Lacombe, B., Rouached, H.: Getting to the root of plant mineral nutrition: combinatorial nutrient stresses reveal emergent properties. Trends Plant Sci. **24**(6), 542–552 (2019). https://doi.org/10.1016/j.tplants.2019.03.008

Brennan, R., Bolland, M.: Relative effectiveness of soil-applied zinc for four crop species. Aust. J. Exp. Agric. **42**(7), 985–993 (2002). https://doi.org/10.1071/EA01154

Brennan, R., Gartrell, J.W.: Reaction of zinc with soil affecting its availability to subterranean clover. 1. The relationship between critical concentrations of extractable zinc and properties of Australian soils responsive to applied zinc. Aust. J. Soil Res. **28**(2), 293–302 (1990). https://doi.org/10.1071/SR9900293

Brieger, A., Rink, L., Haase, H.: Differential regulation of TLR-dependent MyD88 and TRIF signaling pathways by free zinc ions. J. Immunol. **191**(4), 1808–1817 (2013)

Britto, P.R., Lye, S.J., Proulx, K., Yousafzai, A.K., Matthews, S.G., Vaivada, T., Perez-Escamilla, R., Rao, N., Ip, P., Fernald, L.C.H., MacMillan, H., Hanson, M., Wachs, T.D., Yao, H., Yoshikawa, H., Cerezo, A., Leckman, J.F., Bhutta, Z.A.: Nurturing care: promoting early childhood development. Lancet. **389**(10064), 91–102 (2017). https://doi.org/10.1016/S0140-6736(16)31390-3

Broadley, M.R., White, P.J., Hammond, J.P., Zelko, I., Lux, A.: Zinc in plants. New Phytol. **173**(4), 677–702 (2007). https://doi.org/10.1111/j.1469-8137.2007.01996.x

Brown, P.H., Cakmak, I., Zhang, Q.: Form and function of zinc plants. In: Robson, A.D. (ed.) Zinc in Soils and Plants: Proceedings of the International Symposium on 'Zinc in Soils and Plants' held at The University of Western Australia, 27–28 September, 1993, pp. 93–106. Springer, Dordrecht (1993). https://doi.org/10.1007/978-94-011-0878-2_7

Brown, K.H., Rivera, J., Bhutta, Z., Gibson, R., King, J., Lonnerdal, B., Ruel, M., Sandtröm, B., Wasantwisut, E., Hotz, C.: Assessment of the risk of zinc deficiency in populations and options for its control. Food Nutr. Bull. **25**, S94–S203 (2004a)

Brown, K.H., Rivera, J.A., Bhutta, Z., Gibson, R.S., King, J.C., Lonnerdal, B., Ruel, M.T., Sandtrom, B., Wasantwisut, E., Hotz, C.: International Zinc Nutrition Consultative Group (IZiNCG) technical document #1. Assessment of the risk of zinc deficiency in populations and options for its control. Food Nutr. Bull. **25**(1 Suppl 2), S99–S203 (2004b)

Cabot, C., Martos, S., Llugany, M., Gallego, B., Tolrà, R., Poschenrieder, C.: A role for zinc in plant defense against pathogens and herbivores. Front. Plant Sci. **10**, 1171 (2019). https://doi.org/10.3389/fpls.2019.01171

Cai, C., Lin, P., Zhu, H., Ko, J.K., Hwang, M., Tan, T., Pan, Z., Korichneva, I., Ma, J.: Zinc binding to MG53 protein facilitates repair of injury to cell membranes. J. Biol. Chem. **290**(22), 13830–13839 (2015). https://doi.org/10.1074/jbc.M114.620690

Cakmak, I.: Zinc deficiency in wheat in Turkey. In: Alloway, B.J. (ed.) Micronutrient Deficiencies in Global Crop Production, pp. 181–200. Springer, Dordrecht (2008). https://doi.org/10.1007/978-1-4020-6860-7_7

Cakmak, I., Kutman, U.B.: Agronomic biofortification of cereals with zinc: a review. Eur. J. Soil Sci. **69**(1), 172–180 (2018). https://doi.org/10.1111/ejss.12437

Cakmak, I., Yilmaz, A., Kalayci, M., Ekiz, H., Torun, B., Braun, H.J.: Zinc deficiency as a critical problem in wheat production in Central Anatolia. Plant Soil. **180**(2), 165–172 (1996). https://doi.org/10.1007/BF00015299

Cakmak, I., Kalayci, M., Kaya, Y., Torun, A.A., Aydin, N., Wang, Y., Arisoy, Z., Erdem, H., Yazici, A., Gokmen, O., Ozturk, L., Horst, W.J.: Biofortification and localization of zinc in wheat grain. J. Agric. Food Chem. **58**(16), 9092–9102 (2010). https://doi.org/10.1021/jf101197h

Cassandri, M., Smirnov, A., Novelli, F., Pitolli, C., Agostini, M., Malewicz, M., Melino, G., Raschellà, G.: Zinc-finger proteins in health and disease. Cell Death Discovery. **3**(1), 17071 (2017). https://doi.org/10.1038/cddiscovery.2017.71

Chaffee, B.W., King, J.C.: Effect of zinc supplementation on pregnancy and infant outcomes: a systematic review. Paediatr. Perinat. Epidemiol. **26**(Suppl 1), 118–137 (2012). https://doi.org/10.1111/j.1365-3016.2012.01289.x

Chang, H.H., Larson, J., Blencowe, H., Spong, C.Y., Howson, C.P., Cairns-Smith, S., Lackritz, E.M., Lee, S.K., Mason, E., Serazin, A.C., Walani, S., Simpson, J.L., Lawn, J.E.: Preventing preterm births: analysis of trends and potential reductions with interventions in 39 countries with very high human development index. Lancet. **381**(9862), 223–234 (2013). https://doi.org/10.1016/S0140-6736(12)61856-X

Chasapis, C.T., Loutsidou, A.C., Spiliopoulou, C.A., Stefanidou, M.E.: Zinc and human health: an update. Arch. Toxicol. **86**(4), 521–534 (2012). https://doi.org/10.1007/s00204-011-0775-1

Chellan, P., Sadler, P.J.: The elements of life and medicines. Philos. Trans. A Math. Phys. Eng. Sci. **373**(2037), 20140182 (2015). https://doi.org/10.1098/rsta.2014.0182

Choi, D.W., Koh, J.Y.: Zinc and brain injury. Annu. Rev. Neurosci. **21**, 347–375 (1998). https://doi.org/10.1146/annurev.neuro.21.1.347

Choi, S.-H., Lee, K.-L., Shin, J.-H., Cho, Y.-B., Cha, S.-S., Roe, J.-H.: Zinc-dependent regulation of zinc import and export genes by Zur. Nat. Commun. **8**(1), 15812 (2017). https://doi.org/10.1038/ncomms15812

Chowanadisai, W., Graham, D.M., Keen, C.L., Rucker, R.B., Messerli, M.A.: Neurulation and neurite extension require the zinc transporter ZIP12 (slc39a12). Proc. Natl. Acad. Sci. U. S. A. **110**(24), 9903–9908 (2013). https://doi.org/10.1073/pnas.1222142110

Christianson, D.W., Fierke, C.A.: Carbonic anhydrase: evolution of the zinc binding site by nature and by design. Acc. Chem. Res. **29**(7), 331–339 (1996). https://doi.org/10.1021/ar9501232

Cox, F.R.: Micronutrient soil tests: correlation and calibration. In: Brown, J.R. (ed.) Soil Testing: Sampling, Correlation, Calibration, and Interpretation, pp. 97–117. SSSA Special Publications (1987). https://doi.org/10.2136/sssaspecpub21.c10

Darnton-Hill, I., Webb, P., Harvey, P.W., Hunt, J.M., Dalmiya, N., Chopra, M., Ball, M.J., Bloem, M.W., de Benoist, B.: Micronutrient deficiencies and gender: social and economic costs. Am. J. Clin. Nutr. **81**(5), 1198S–1205S (2005)

DiGirolamo, A.M., Ramirez-Zea, M.: Role of zinc in maternal and child mental health. Am. J. Clin. Nutr. **89**(3), 940S–945S (2009). https://doi.org/10.3945/ajcn.2008.26692C

Dimkpa, C.O., Bindraban, P.S.: Fortification of micronutrients for efficient agronomic production: a review. Agron. Sustain. Dev. **36**(1), 7 (2016). https://doi.org/10.1007/s13593-015-0346-6

Doneray, H., Olcaysu, E., Yildirim, A., Ozden, A.: The effect of the zinc concentration in breast milk on neonatal weight gain. J. Trace Elem. Med. Biol. **41**, 32–35 (2017). https://doi.org/10.1016/j.jtemb.2017.02.006

Eijkelkamp, B.A., Morey, J.R., Neville, S.L., Tan, A., Pederick, V.G., Cole, N., Singh, P.P., Ong, C.-L.Y., Gonzalez de Vega, R., Clases, D., Cunningham, B.A., Hughes, C.E., Comerford, I., Brazel, E.B., Whittall, J.J., Plumptre, C.D., McColl, S.R., Paton, J.C., McEwan, A.G., Doble, P.A., McDevitt, C.A.: Dietary zinc and the control of Streptococcus pneumoniae infection. PLoS Pathog. **15**(8), e1007957 (2019). https://doi.org/10.1371/journal.ppat.1007957

Emdin, S.O., Dodson, G.G., Cutfield, J.M., Cutfield, S.M.: Role of zinc in insulin biosynthesis. Diabetologia. **19**(3), 174–182 (1980). https://doi.org/10.1007/BF00275265

Fabris, N.: Neuroendocrine-immune aging: an integrative view on the role of zinc. Ann. N. Y. Acad. Sci. **719**, 353–368 (1994). https://doi.org/10.1111/j.1749-6632.1994.tb56842.x

Favier, A.E.: The role of zinc in reproduction. Biol. Trace Elem. Res. **32**(1), 363–382 (1992). https://doi.org/10.1007/BF02784623

Fischer Walker, C.L., Ezzati, M., Black, R.E.: Global and regional child mortality and burden of disease attributable to zinc deficiency. Eur. J. Clin. Nutr. **63**(5), 591–597 (2009). https://doi.org/10.1038/ejcn.2008.9

Flynn, A., Martier, S., Sokol, R., Miller, S., Golden, N., Del Villano, B.: Zinc status of pregnant alcoholic women: a determinant of fetal outcome. Lancet. **317**(8220), 572–575 (1981). https://doi.org/10.1016/S0140-6736(81)92029-8

Folwaczny, C.: Zinc and diarrhea in infants. J. Trace Elem. Med. Biol. **11**(2), 116–122 (1997). https://doi.org/10.1016/S0946-672X(97)80036-3

Frazzini, V., Granzotto, A., Bomba, M., Massetti, N., Castelli, V., d'Aurora, M., Punzi, M., Iorio, M., Mosca, A., Delli Pizzi, S., Gatta, V., Cimini, A., Sensi, S.L.: The pharmacological perturbation of brain zinc impairs BDNF-related signaling and the cognitive performances of young mice. Sci. Rep. **8**(1), 9768 (2018). https://doi.org/10.1038/s41598-018-28083-9

Frederickson, C.J., Koh, J.-Y., Bush, A.I.: The neurobiology of zinc in health and disease. Nat. Rev. Neurosci. **6**(6), 449–462 (2005). https://doi.org/10.1038/nrn1671

Frey, B., Keller, C., Zierold, K.: Distribution of Zn in functionally different leaf epidermal cells of the hyperaccumulator thlaspi caerulescens. Plant Cell Environ. **23**(7), 675–687 (2000). https://doi.org/10.1046/j.1365-3040.2000.00590.x

Fukada, T., Hojyo, S., Hara, T., Takagishi, T.: Revisiting the old and learning the new of zinc in immunity. Nat. Immunol. **20**(3), 248–250 (2019). https://doi.org/10.1038/s41590-019-0319-z

Fukunaka, A., Fujitani, Y.: Role of zinc homeostasis in the pathogenesis of diabetes and obesity. Int. J. Mol. Sci. **19**(2), 476 (2018). https://doi.org/10.3390/ijms19020476

Gammoh, N.Z., Rink, L.: Zinc and the immune system. In: Mahmoudi, M., Rezaei, N. (eds.) Nutrition and Immunity, pp. 127–158. Springer International Publishing, Cham (2019). https://doi.org/10.1007/978-3-030-16073-9_8

Genc, Y., McDonald, G.K., Graham, R.D.: Critical deficiency concentration of zinc in barley genotypes differing in zinc efficiency and its relation to growth responses. J. Plant Nutr. **25**(3), 545–560 (2002). https://doi.org/10.1081/PLN-120003381

Gibson, R.S.: A historical review of progress in the assessment of dietary zinc intake as an indicator of population zinc status. Adv. Nutr. **3**(6), 772–782 (2012). https://doi.org/10.3945/an.112.002287

Gibson, R.S., King, J.C., Lowe, N.: A review of dietary zinc recommendations. Food Nutr. Bull. **37**(4), 443–460 (2016). https://doi.org/10.1177/0379572116652252

Gibson, R.S., Raboy, V., King, J.C.: Implications of phytate in plant-based foods for iron and zinc bioavailability, setting dietary requirements, and formulating programs and policies. Nutr. Rev. **76**(11), 793–804 (2018). https://doi.org/10.1093/nutrit/nuy028

Gray, N.A., Dhana, A., Stein, D.J., Khumalo, N.P.: Zinc and atopic dermatitis: a systematic review and meta-analysis. J. Eur. Acad. Dermatol. Venereol. **33**(6), 1042–1050 (2019). https://doi.org/10.1111/jdv.15524

Gupta, N., Ram, H., Kumar, B.: Mechanism of zinc absorption in plants: uptake, transport, translocation and accumulation. Rev. Environ. Sci. Biotechnol. **15**(1), 89–109 (2016). https://doi.org/10.1007/s11157-016-9390-1

Haase, H., Rink, L.: The immune system and the impact of zinc during aging. Immun. Ageing. **6**(1), 9 (2009). https://doi.org/10.1186/1742-4933-6-9

Hacisalihoglu, G., Kochian, L.V.: How do some plants tolerate low levels of soil zinc? Mechanisms of zinc efficiency in crop plants. New Phytol. **159**(2), 341–350 (2003). https://doi.org/10.1046/j.1469-8137.2003.00826.x

Haddad, L., Hawkes, C., Webb, P., Thomas, S., Beddington, J., Waage, J., Flynn, D.: A new global research agenda for food. Nature. **540**(7631), 30–32 (2016). https://doi.org/10.1038/540030a

Hambidge, K.M.: Zinc and pneumonia. Am. J. Clin. Nutr. **83**(5), 991–992 (2006). https://doi.org/10.1093/ajcn/83.5.991

Hambidge, K.M., Casey, C.E., Krebs, N.F.: 1 - Zinc. In: Mertz, W. (ed.) Trace Elements in Human and Animal Nutrition, 5th edn, pp. 1–137. Academic Press, San Diego (1986). https://doi.org/10.1016/B978-0-08-092469-4.50005-4

Hänsch, R., Mendel, R.R.: Physiological functions of mineral micronutrients (Cu, Zn, Mn, Fe, Ni, Mo, B, Cl). Curr. Opin. Plant Biol. **12**(3), 259–266 (2009). https://doi.org/10.1016/j.pbi.2009.05.006

Hess, S., Peerson, J., King, J., Brown, K.: Use of serum zinc concentration as an indicator of population zinc status. Food Nutr. Bull. **28**, S403–S429 (2007). https://doi.org/10.1177/15648265070283S303

Hicks, C.C., Cohen, P.J., Graham, N.A.J., Nash, K.L., Allison, E.H., D'Lima, C., Mills, D.J., Roscher, M., Thilsted, S.H., Thorne-Lyman, A.L., MacNeil, M.A.: Harnessing global fisheries to tackle micronutrient deficiencies. Nature. **574**(7776), 95–98 (2019). https://doi.org/10.1038/s41586-019-1592-6

Hogstrand, C., Kille, P., Nicholson, R.I., Taylor, K.M.: Zinc transporters and cancer: a potential role for ZIP7 as a hub for tyrosine kinase activation. Trends Mol. Med. **15**(3), 101–111 (2009). https://doi.org/10.1016/j.molmed.2009.01.004

Höller, S., Meyer, A., Frei, M.: Zinc deficiency differentially affects redox homeostasis of rice genotypes contrasting in ascorbate level. J. Plant Physiol. **171**(18), 1748–1756 (2014). https://doi.org/10.1016/j.jplph.2014.08.012

Jouybari, L., Kiani, F., Akbari, A., Sanagoo, A., Sayehmiri, F., Aaseth, J., Chartrand, M.S., Sayehmiri, K., Chirumbolo, S., Bjørklund, G.: A meta-analysis of zinc levels in breast cancer. J. Trace Elem. Med. Biol. **56**, 90–99 (2019). https://doi.org/10.1016/j.jtemb.2019.06.017

Joy, E.J.M., Ahmad, W., Zia, M.H., Kumssa, D.B., Young, S.D., Ander, E.L., Watts, M.J., Stein, A.J., Broadley, M.R.: Valuing increased zinc (Zn) fertiliser-use in Pakistan. Plant Soil. **411**(1), 139–150 (2017). https://doi.org/10.1007/s11104-016-2961-7

Jurowski, K., Szewczyk, B., Nowak, G., Piekoszewski, W.: Biological consequences of zinc deficiency in the pathomechanisms of selected diseases. J. Biol. Inorg. Chem. **19**(7), 1069–1079 (2014). https://doi.org/10.1007/s00775-014-1139-0

Kabata-Pendias, A.: Trace Elements in Soils and Plants, 4th edn. CRC Press (2010). https://doi.org/10.1201/b10158

Kambe, T., Tsuji, T., Hashimoto, A., Itsumura, N.: The physiological, biochemical, and molecular roles of zinc transporters in zinc homeostasis and metabolism. Physiol. Rev. **95**(3), 749–784 (2015). https://doi.org/10.1152/physrev.00035.2014

Kamran, S., Shahid, I., Baig, D.N., Rizwan, M., Malik, K.A., Mehnaz, S.: Contribution of zinc solubilizing bacteria in growth promotion and zinc content of wheat. Front. Microbiol. **8**, 2593–2593 (2017). https://doi.org/10.3389/fmicb.2017.02593

Kapoor, S., Kobayashi, A., Takatsuji, H.: Silencing of the tapetum-specific zinc finger gene *TAZ1* causes premature degeneration of tapetum and pollen abortion in petunia. Plant Mol. Biol. **14**(10), 2353–2367 (2002)

Kaur, K., Gupta, R., Saraf, S.A., Saraf, S.K.: Zinc: the metal of life. Compr. Rev. Food Sci. Food Saf. **13**(4), 358–376 (2014). https://doi.org/10.1111/1541-4337.12067

Kitagishi, K., Obata, H.: Effects of zinc deficiency on the nitrogen metabolism of meristematic tissues of rice plants with reference to protein synthesis. Soil Sci. Plant Nutr. **32**(3), 397–405 (1986). https://doi.org/10.1080/00380768.1986.10557520

Kolenko, V., Teper, E., Kutikov, A., Uzzo, R.: Zinc and zinc transporters in prostate carcinogenesis. Nat. Rev. Urol. **10**(4), 219–226 (2013). https://doi.org/10.1038/nrurol.2013.43

Korenblum, E., Dong, Y., Szymanski, J., Panda, S., Jozwiak, A., Massalha, H., Meir, S., Rogachev, I., Aharoni, A.: Rhizosphere microbiome mediates systemic root metabolite exudation by root-to-root signaling. Proc. Natl. Acad. Sci. **117**(7), 3874–3883 (2020). https://doi.org/10.1073/pnas.1912130117

Krebs, N.F., Miller, L.V., Michael Hambidge, K.: Zinc deficiency in infants and children: a review of its complex and synergistic interactions. Paediatr. Int. Child Health. **34**(4), 279–288 (2014). https://doi.org/10.1179/2046905514Y.0000000151

Kumar, A., Dewangan, S., Lawate, P., Bahadur, I., Prajapati, S.: Zinc-Solubilizing Bacteria: A Boon for Sustainable Agriculture. In: Sayyed, R.Z., Arora, N.K., Reddy, M.S. (eds.) Plant Growth Promoting Rhizobacteria for Sustainable Stress Management: Volume 1: Rhizobacteria in Abiotic Stress Management, pp. 139–155. Springer, Singapore (2019). https://doi.org/10.1007/978-981-13-6536-2_8

Kumssa, D.B., Joy, E.J.M., Ander, E.L., Watts, M.J., Young, S.D., Walker, S., Broadley, M.R.: Dietary calcium and zinc deficiency risks are decreasing but remain prevalent. Sci. Rep. **5**(1), 10974 (2015). https://doi.org/10.1038/srep10974

Küry, S., Dréno, B., Bézieau, S., Giraudet, S., Kharfi, M., Kamoun, R., Moisan, J.-P.: Identification of SLC39A4, a gene involved in acrodermatitis enteropathica. Nat. Genet. **31**(3), 239–240 (2002). https://doi.org/10.1038/ng913

Lansdown, A.B.G., Mirastschijski, U., Stubbs, N., Scanlon, E., Ågren, M.S.: Zinc in wound healing. Theor. Exp. Clin. Aspects. **15**(1), 2–16 (2007). https://doi.org/10.1111/j.1524-475X.2006.00179.x

Lassi, Z.S., Moin, A., Bhutta, Z.A.: Zinc supplementation for the prevention of pneumonia in children aged 2 months to 59 months. Cochrane Database Syst. Rev. **12**, CD005978 (2016). https://doi.org/10.1002/14651858.CD005978.pub3

Lehvy, A.I., Horev, G., Golan, Y., Glaser, F., Shammai, Y., Assaraf, Y.G.: Alterations in ZnT1 expression and function lead to impaired intracellular zinc homeostasis in cancer. Cell Death Discovery. **5**(1), 144 (2019). https://doi.org/10.1038/s41420-019-0224-0

Lemaire, K., Ravier, M.A., Schraenen, A., Creemers, J.W., Van de Plas, R., Granvik, M., Van Lommel, L., Waelkens, E., Chimienti, F., Rutter, G.A., Gilon, P., et al.: Insulin crystallization depends on zinc transporter ZnT8 expression, but is not required for normal glucose homeostasis in mice. Proc. Natl. Acad. Sci. U. S. A. **106**(35), 14872–14877 (2009a). https://doi.org/10.1073/pnas.0906587106

Lemaire, K., Ravier, M.A., Schraenen, A., Creemers, J.W.M., Van de Plas, R., Granvik, M., Van Lommel, L., Waelkens, E., Chimienti, F., Rutter, G.A., Gilon, P., et al.: Insulin crystallization depends on zinc transporter ZnT8 expression, but is not required for normal glucose homeostasis in mice. Proc. Natl. Acad. Sci. **106**(35), 14872–14877 (2009b). https://doi.org/10.1073/pnas.0906587106

Leśniewska, B., Kisielewska, K., Wiater, J., Godlewska-Żyłkiewicz, B.: Fast and simple procedure for fractionation of zinc in soil using an ultrasound probe and FAAS detection. Validation of the analytical method and evaluation of the uncertainty budget. Environ. Monit. Assess. **188**(1), 29–29 (2016). https://doi.org/10.1007/s10661-015-5020-6

Li, Y.V.: Zinc and insulin in pancreatic beta-cells. Endocrine. **45**(2), 178–189 (2014). https://doi.org/10.1007/s12020-013-0032-x

Li, M., Zhang, Y., Liu, Z., Bharadwaj, U., Wang, H., Wang, X., Zhang, S., Liuzzi, J.P., Chang, S.-M., Cousins, R.J., Fisher, W.E., Brunicardi, F.C., Logsdon, C.D., Chen, C., Yao, Q.: Aberrant expression of zinc transporter ZIP4 (SLC39A4) significantly contributes to human pancreatic cancer pathogenesis and progression. Proc. Natl. Acad. Sci. **104**(47), 18636–18641 (2007). https://doi.org/10.1073/pnas.0709307104

Lindsay, W.L.: Zinc in soils and plant nutrition. In: Brady, N.C. (ed.) Advances in Agronomy, vol. 24, pp. 147–186. Academic Press (1972). https://doi.org/10.1016/S0065-2113(08)60635-5

Linthicum, M.T., Thornton Snider, J., Vaithianathan, R., Wu, Y., LaVallee, C., Lakdawalla, D.N., Benner, J.E., Philipson, T.J.: Economic burden of disease-associated malnutrition in China. Asia Pac. J. Public Health. **27**(4), 407–417 (2015). https://doi.org/10.1177/1010539514552702

Liu, H., Zhao, P., Qin, S., Nie, Z.: Chemical fractions and availability of zinc in winter wheat soil in response to nitrogen and zinc combinations. Front. Plant Sci. **9**, 1489–1489 (2018). https://doi.org/10.3389/fpls.2018.01489

Liu, D.-Y., Liu, Y.-M., Zhang, W., Chen, X.-P., Zou, C.-Q.: Zinc uptake, translocation, and remobilization in winter wheat as affected by soil application of Zn fertilizer. Front. Plant Sci. **10**, 426 (2019). https://doi.org/10.3389/fpls.2019.00426

Lönnerdal, B.: Dietary factors influencing zinc absorption. J. Nutr. **130**(5), 1378S–1383S (2000). https://doi.org/10.1093/jn/130.5.1378S

Maares, M., Haase, H.: A guide to human zinc absorption: general overview and recent advances of in vitro intestinal models. Nutrients. **12**(3), 762 (2020). https://doi.org/10.3390/nu12030762

Maret, W., Sandstead, H.H.: Zinc requirements and the risks and benefits of zinc supplementation. J. Trace Elem. Med. Biol. **20**(1), 3–18 (2006). https://doi.org/10.1016/j.jtemb.2006.01.006

Marger, L., Schubert, C.R., Bertrand, D.: Zinc: an underappreciated modulatory factor of brain function. Biochem. Pharmacol. **91**(4), 426–435 (2014). https://doi.org/10.1016/j.bcp.2014.08.002

Martens, D.C., Westermann, D.T.: Fertilizer applications for correcting micronutrient deficiencies. In: Micronutrients in Agriculture, pp. 549–592 (1991). https://doi.org/10.2136/sssabookser4.2ed.c15

Mattiello, E.M., Ruiz, H.A., Neves, J.C., Ventrella, M.C., Araujo, W.L.: Zinc deficiency affects physiological and anatomical characteristics in maize leaves. J. Plant Physiol. **183**, 138–143 (2015). https://doi.org/10.1016/j.jplph.2015.05.014

Mertens, J., Smolders, E.: Zinc. In: Alloway, B.J. (ed.) Heavy Metals in Soils: Trace Metals and Metalloids in Soils and their Bioavailability, pp. 465–493. Springer, Dordrecht (2013). https://doi.org/10.1007/978-94-007-4470-7_17

Mishra, V.K., Gupta, S., Chand, R., Yadav, P.S., Singh, S.K., Joshi, A.K., Varadwaj, P.K.: Comparative transcriptomic profiling of high- and low- grain zinc and iron containing Indian wheat genotypes. Curr. Plant Biol. **18**, 100105 (2019). https://doi.org/10.1016/j.cpb.2019.100105

Mohammad, M.K., Zhou, Z., Cave, M., Barve, A., McClain, C.J.: Zinc and liver disease. Nutr. Clin. Pract. **27**(1), 8–20 (2012). https://doi.org/10.1177/0884533611433534

Mori, A., Kirk, G.J.D., Lee, J.-S., Morete, M.J., Nanda, A.K., Johnson-Beebout, S.E., Wissuwa, M.: Rice genotype differences in tolerance of zinc-deficient soils: evidence for the importance of root-induced changes in the rhizosphere. Front. Plant Sci. **6**, 1160 (2016). https://doi.org/10.3389/fpls.2015.01160

Nistor, N., Ciontu, L., Frasinariu, O.-E., Lupu, V.V., Ignat, A., Streanga, V.: Acrodermatitis enteropathica: a case report. Medicine. **95**(20), e3553–e3553 (2016). https://doi.org/10.1097/MD.0000000000003553

Norouzi, S., Adulcikas, J., Sohal, S.S., Myers, S.: Zinc transporters and insulin resistance: therapeutic implications for type 2 diabetes and metabolic disease. J. Biomed. Sci. **24**(1), 87 (2017). https://doi.org/10.1186/s12929-017-0394-0

Noulas, C., Tziouvalekas, M., Karyotis, T.: Zinc in soils, water and food crops. J. Trace Elem. Med. Biol. **49**, 252–260 (2018). https://doi.org/10.1016/j.jtemb.2018.02.009

Nuttall, J.R., Oteiza, P.I.: Zinc and the aging brain. Genes Nutr. **9**(1), 379–379 (2014). https://doi.org/10.1007/s12263-013-0379-x

Ohta, S., Ikemoto, T., Wada, Y., Saito, Y., Yamada, S., Imura, S., Morine, Y., Shimada, M.: A change in the zinc ion concentration reflects the maturation of insulin-producing cells generated from adipose-derived mesenchymal stem cells. Sci. Rep. **9**(1), 18731 (2019). https://doi.org/10.1038/s41598-019-55172-0

Osendarp, S.J.M., van Raaij, J.M.A., Darmstadt, G.L., Baqui, A.H., Hautvast, J.G.A.J., Fuchs, G.J.: Zinc supplementation during pregnancy and effects on growth and morbidity in low birthweight infants: a randomsied placebo controlled trial. Lancet. **357**(9262), 1080–1085 (2001). https://doi.org/10.1016/S0140-6736(00)04260-4

Paltridge, N.G., Milham, P.J., Ortiz-Monasterio, J.I., Velu, G., Yasmin, Z., Palmer, L.J., Guild, G.E., Stangoulis, J.C.R.: Energy-dispersive X-ray fluorescence spectrometry as a tool for zinc, iron and selenium analysis in whole grain wheat. Plant Soil. **361**(1), 261–269 (2012). https://doi.org/10.1007/s11104-012-1423-0

Pandey, N., Pathak, G.C., Singh, A.K., Sharma, C.P.: Enzymic changes in response to zinc nutrition. J. Plant Physiol. **159**(10), 1151–1153 (2002). https://doi.org/10.1078/0176-1617-00674

Pilon, M., Cohu, C.M., Ravet, K., Abdel-Ghany, S.E., Gaymard, F.: Essential transition metal homeostasis in plants. Curr. Opin. Plant Biol. **12**(3), 347–357 (2009). https://doi.org/10.1016/j.pbi.2009.04.011

Pinton, R., Cakmak, I., Marschner, H.: Zinc deficiency enhanced NAD(P)H-dependent superoxider radical production in plasma membrane vesicles isolated from roots of bean plants. J. Exp. Bot. **45**(1), 45–50 (1994)

Portbury, S.D., Adlard, P.A.: Zinc signal in brain diseases. Int. J. Mol. Sci. **18**(12), 2506 (2017). https://doi.org/10.3390/ijms18122506

Prasad, A.S.: Zinc in human health: effect of zinc on immune cells. Mol. Med. **14**(5), 353–357 (2008). https://doi.org/10.2119/2008-00033.Prasad

Prasad, A.S.: Zinc: an antioxidant and anti-inflammatory agent: role of zinc in degenerative disorders of aging. J. Trace Elem. Med. Biol. **28**(4), 364–371 (2014). https://doi.org/10.1016/j.jtemb.2014.07.019

Prasad, R., Shivay, Y.S., Kumar, D.: Zinc fertilization of cereals for increased production and alleviation of zinc malnutrition in India. Agric. Res. **2**(2), 111–118 (2013). https://doi.org/10.1007/s40003-013-0064-8

Raboy, V.: Seeds for a better future: 'low phytate' grains help to overcome malnutrition and reduce pollution. Trends Plant Sci. **6**(10), 458–462 (2001). https://doi.org/10.1016/S1360-1385(01)02104-5

Ranasinghe, P., Pigera, S., Galappatthy, P., Katulanda, P., Constantine, G.R.: Zinc and diabetes mellitus: understanding molecular mechanisms and clinical implications. Daru. **23**(1), 44–44 (2015). https://doi.org/10.1186/s40199-015-0127-4

Read, S.A., Parnell, G., Booth, D., Douglas, M.W., George, J., Ahlenstiel, G.: The antiviral role of zinc and metallothioneins in hepatitis. J. Viral Hepat. **25**(5), 491–501 (2018). https://doi.org/10.1111/jvh.12845

Reuter, D., Robinson, J.B.: Plant Analysis: an Interpretation Manual. CSIRO publishing (1997)

Rink, L., Haase, H.: Zinc homeostasis and immunity. Trends Immunol. **28**(1), 1–4 (2007). https://doi.org/10.1016/j.it.2006.11.005

Roohani, N., Hurrell, R., Kelishadi, R., Schulin, R.: Zinc and its importance for human health: an integrative review. J. Res. Med. Sci. **18**(2), 144–157 (2013)

Rudnick, R.L., Gao, S.: 3.01 – composition of the continental crust. In: Holland, H.D., Turekian, K.K. (eds.) Treatise on Geochemistry, pp. 1–64. Pergamon, Oxford (2003). https://doi.org/10.1016/B0-08-043751-6/03016-4

Salama, Z.A., El-Fouly, M.M., Lazova, G., Popova, L.P.: Carboxylating enzymes and carbonic anhydrase functions were suppressed by zinc deficiency in maize and chickpea plants. Acta Physiol. Plant. **28**(5), 445–451 (2006). https://doi.org/10.1007/BF02706627

Sankaranarayanan, R., Dock-Bregeon, A.-C., Rees, B., Bovee, M., Caillet, J., Romby, P., Francklyn, C.S., Moras, D.: Zinc ion mediated amino acid discrimination by threonyl-tRNA synthetase. Nat. Struct. Biol. **7**(6), 461–465 (2000). https://doi.org/10.1038/nsb0600_461

Satria, B., Chen, W., Soebono, H., Radiono, S., Danarti, R.: Concurrence of acrodermatitis enteropathica and eczema herpeticum in a child with atopic dermatitis. Case Rep. Dermatol. **11**(3), 240–248 (2019). https://doi.org/10.1159/000502509

Satterthwaite, D., McGranahan, G., Tacoli, C.: Urbanization and its implications for food and farming. Philos. Trans. R. Soc. Lond. Ser. B Biol. Sci. **365**(1554), 2809–2820 (2010). https://doi.org/10.1098/rstb.2010.0136

Schmalen, I., Reischl, S., Wallach, T., Klemz, R., Grudziecki, A., Prabu, J.R., Benda, C., Kramer, A., Wolf, E.: Interaction of circadian clock proteins CRY1 and PER2 is modulated by zinc binding and disulfide bond formation. Cell. **157**(5), 1203–1215 (2014). https://doi.org/10.1016/j.cell.2014.03.057

Schroeder, J.I., Delhaize, E., Frommer, W.B., Guerinot, M.L., Harrison, M.J., Herrera-Estrella, L., Horie, T., Kochian, L.V., Munns, R., Nishizawa, N.K., Tsay, Y.F., Sanders, D.: Using membrane transporters to improve crops for sustainable food production. Nature. **497**(7447), 60–66 (2013). https://doi.org/10.1038/nature11909

Segal, D.J., Stege, J.T., Barbas, C.F.: Zinc fingers and a green thumb: manipulating gene expression in plants. Curr. Opin. Plant Biol. **6**(2), 163–168 (2003). https://doi.org/10.1016/S1369-5266(03)00007-4

Selbach, B.P., Chung, A.H., Scott, A.D., George, S.J., Cramer, S.P., Dos Santos, P.C.: Fe-S cluster biogenesis in gram-positive bacteria: SufU is a zinc-dependent sulfur transfer protein. Biochemistry. **53**(1), 152–160 (2014). https://doi.org/10.1021/bi4011978

Sensi, S.L., Paoletti, P., Koh, J.-Y., Aizenman, E., Bush, A.I., Hershfinkel, M.: The neurophysiology and pathology of brain zinc. J. Neurosci. **31**(45), 16076–16085 (2011)

Seo, H.-M., Kim, Y.H., Lee, J.H., Kim, J.S., Park, Y.M., Lee, J.Y.: Serum zinc status and its association with allergic sensitization: the Fifth Korea National Health and Nutrition Examination Survey. Sci. Rep. **7**(1), 12637 (2017). https://doi.org/10.1038/s41598-017-13068-x

Sharma, A., Patni, B., Shankhdhar, D., Shankhdhar, S.C.: Zinc - an indispensable micronutrient. Physiol. Mol. Biol. Plants. **19**(1), 11–20 (2013). https://doi.org/10.1007/s12298-012-0139-1

Sindreu, C., Palmiter, R.D., Storm, D.R.: Zinc transporter ZnT-3 regulates presynaptic Erk1/2 signaling and hippocampus-dependent memory. Proc. Natl. Acad. Sci. U. S. A. **108**(8), 3366–3370 (2011). https://doi.org/10.1073/pnas.1019166108

Singh, K.: Determination of critical limit of zinc in rice soils in India for predicting response of rice to zinc application. Field Crop Res. **9**, 143–149 (1984). https://doi.org/10.1016/0378-4290(84)90020-0

Singh, B., Singh, B.: Improving zinc efficiency of cereals under zinc deficiency. Curr. Sci. **88**, 36–44 (2005)

Smith, E.R., Shankar, A.H., Wu, L.S.F., Aboud, S., Adu-Afarwuah, S., Ali, H., Agustina, R., Arifeen, S., Ashorn, P., Bhutta, Z.A., Christian, P., Devakumar, D., Dewey, K.G., Friis, H., Gomo, E., Gupta, P., Kæstel, P., Kolsteren, P., Lanou, H., Maleta, K., Mamadoultaibou, A., Msamanga, G., Osrin, D., Persson, L.-Å., Ramakrishnan, U., Rivera, J.A., Rizvi, A., Sachdev, H.P.S., Urassa, W., West Jr., K.P., Zagre, N., Zeng, L., Zhu, Z., Fawzi, W.W., Sudfeld, C.R.: Modifiers of the effect of maternal multiple micronutrient supplementation on stillbirth, birth outcomes, and infant mortality: a meta-analysis of individual patient data from 17 randomised trials in low-income and middle-income countries. Lancet Glob. Health. **5**(11), e1090–e1100 (2017). https://doi.org/10.1016/S2214-109X(17)30371-6

Sotta, N., Sakamoto, T., Matsunaga, S., Fujiwara, T.: Abnormal leaf development of rpt5a mutant under zinc deficiency reveals important role of DNA damage alleviation for normal leaf development. Sci. Rep. **9**, 9369 (2019). https://doi.org/10.1038/s41598-019-44789-w

Taylor, K.M., Hiscox, S., Nicholson, R.I., Hogstrand, C., Kille, P.: Protein kinase CK2 triggers cytosolic zinc signaling pathways by phosphorylation of zinc channel ZIP7. Sci. Signal. **5**(210), ra11 (2012)

Tellinghuisen, T.L., Marcotrigiano, J., Rice, C.M.: Structure of the zinc-binding domain of an essential component of the hepatitis C virus replicase. Nature. **435**(7040), 374–379 (2005). https://doi.org/10.1038/nature03580

Uroz, S., Courty, P.-E., Oger, P.: Plant symbionts are engineers of the plant-associated microbiome. Trends Plant Sci. **24**(10), 905–916 (2019). https://doi.org/10.1016/j.tplants.2019.06.008

Vallee, B.L., Williams, R.J.: Metalloenzymes: the entatic nature of their active sites. Proc. Natl. Acad. Sci. U. S. A. **59**(2), 498–505 (1968). https://doi.org/10.1073/pnas.59.2.498

Vasiliauskaitė-Brooks, I., Sounier, R., Rochaix, P., Bellot, G., Fortier, M., Hoh, F., De Colibus, L., Bechara, C., Saied, E.M., Arenz, C., Leyrat, C., Granier, S.: Structural insights into adiponectin receptors suggest ceramidase activity. Nature. **544**(7648), 120–123 (2017). https://doi.org/10.1038/nature21714

Wang, H., Hu, Y.-F., Hao, J.-H., Chen, Y.-H., Su, P.-Y., Wang, Y., Yu, Z., Fu, L., Xu, Y.-Y., Zhang, C., Tao, F.-B., Xu, D.-X.: Maternal zinc deficiency during pregnancy elevates the risks of fetal growth restriction: a population-based birth cohort study. Sci. Rep. **5**(1), 11262 (2015). https://doi.org/10.1038/srep11262

Webb, S.J., Harrison, D.J., Wyllie, A.H.: Apoptosis: an overview of the process and its relevance in disease. In: Kaufmann, S.H. (ed.) Advances in Pharmacology, vol. 41, pp. 1–34. Academic Press (1997). https://doi.org/10.1016/S1054-3589(08)61052-2

Weedon, D.: 18 – metabolic and storage diseases. In: Weedon, D. (ed.) Weedon's Skin Pathology, 3rd edn, pp. 481–500, e415. Churchill Livingstone, Edinburgh (2010). https://doi.org/10.1016/B978-0-7020-3485-5.00019-X

Wessells, K.R., Brown, K.H.: Estimating the global prevalence of zinc deficiency: results based on zinc availability in national food supplies and the prevalence of stunting. PLoS One. **7**(11), e50568 (2012). https://doi.org/10.1371/journal.pone.0050568

Wieser, S., Brunner, B., Tzogiou, C., Plessow, R., Zimmermann, M.B., Farebrother, J., Soofi, S., Bhatti, Z., Ahmed, I., Bhutta, Z.A.: Societal costs of micronutrient deficiencies in 6- to 59-month-old children in Pakistan. Food Nutr. Bull. **38**(4), 485–500 (2017). https://doi.org/10.1177/0379572117720012

Wissuwa, M., Ismail, A.M., Yanagihara, S.: Effects of zinc deficiency on rice growth and genetic factors contributing to tolerance. Plant Physiol. **142**(2), 731–741 (2006). https://doi.org/10.1104/pp.106.085225

Wu, T.-Y., Gruissem, W., Bhullar, N.K.: Targeting intracellular transport combined with efficient uptake and storage significantly increases grain iron and zinc levels in rice. Plant Biotechnol. J. **17**(1), 9–20 (2019). https://doi.org/10.1111/pbi.12943

Xue, Y.-F., Yue, S.-C., Zhang, Y.-Q., Cui, Z.-L., Chen, X.-P., Yang, F.-C., Cakmak, I., McGrath, S.P., Zhang, F.-S., Zou, C.-Q.: Grain and shoot zinc accumulation in winter wheat affected by nitrogen management. Plant Soil. **361**(1), 153–163 (2012). https://doi.org/10.1007/s11104-012-1510-2

Yamasaki, S., Sakata-Sogawa, K., Hasegawa, A., Suzuki, T., Kabu, K., Sato, E., Kurosaki, T., Yamashita, S., Tokunaga, M., Nishida, K., Hirano, T.: Zinc is a novel intracellular second messenger. J. Cell Biol. **177**(4), 637–645 (2007). https://doi.org/10.1083/jcb.200702081

Zabel, F., Putzenlechner, B., Mauser, W.: Global agricultural land resources – a high resolution suitability evaluation and its perspectives until 2100 under climate change conditions. PLoS One. **9**(9), e107522 (2014). https://doi.org/10.1371/journal.pone.0107522

Zhang, F., Römheld, V., Marschner, H.: Effect of zinc deficiency in wheat on the release of zinc and iron mobilizing root exudates. Z. Pflanzenernähr. Bodenkd. **152**(2), 205–210 (1989). https://doi.org/10.1002/jpln.19891520211

Chapter 2
Zinc Malnutrition in Children and Its Consequences on Health

Nabat Arfi, Khalida Khatoon, and Farzana Alim

Abstract Zinc is a mineral that is important to all forms of life due to its fundamental role in various metabolic process such as DNA replication, gene expression and protein synthesis. The function that zinc has in human health has been acknowledged by researchers for nearly half a century since the discovery of the syndrome of zinc deficiency among adolescents in Iran. This chapter looks at both elements of zinc malnutrition in children, that is, zinc deficiency and zinc toxicity, and its consequences on health. Severe forms of zinc deficiency are considered to be quite rare, while mild zinc deficiency is more commonly found, mainly within the developing world. Controlled research trails of supplemented zinc at areas that have intrinsically low levels of zinc intake have shown that low intake of zinc has an adverse influence on child health; particularly leading to deficits in neuropsychological functioning and motor development in children. On the other hand, zinc toxicity is rare in children as compared to adults. Zinc toxicity in children is due to oversupplement of zinc intake and can lead to the obesity and other issues such as high blood pressure and cholesterol. Furthermore, this chapter looks at possible precautions and measures to handle the issues associated with zinc malnutrition along with indicating additional research work essential in the field.

Malnutrition

At a basic level, minerals are elements which are vitally important for living organisms in order to perform essential functions of life. There are four groups identified as vital nutrients and minerals form one of those groups. The others include

N. Arfi (✉) · F. Alim
Department of Home Science, Faculty of Agricultural Sciences, Aligarh Muslim University, Aligarh, UP, India

K. Khatoon
Department of Agricultural Microbiology, Faculty of Agricultural Sciences, Aligarh Muslim University, Aligarh, UP, India

S. T. Khan, A. Malik (eds.), *Microbial Biofertilizers and Micronutrient Availability*,
https://doi.org/10.1007/978-3-030-76609-2_2

essential amino acids, vitamins and essential fatty acids. However, these do not form the majority of elements found in the human body. Instead, oxygen, hydrogen, carbon and nitrogen constitute about 96% of the total body weight in human. Although this is the case, macro- and microminerals do not constitute bulk of the body but play crucial roles in both structure and function (Zoroddu et al. 2019).

Consumption of nutrient minerals through animal- and plant-based diets and drinks is critical for humans as these elements are not synthesized biochemically. The most common minerals that are present within humans are potassium, sodium, calcium, phosphorus and magnesium. The remaining available elements within the body are known as 'trace elements' or 'micronutrients'. Among these, the elements with very specific biochemical functions include zinc, chlorine, sulphur, iron, molybdenum, etc. (Berdanier et al. 2016).

The lack of balance in the availability of these minerals in human body leads to micronutrient malnutrition or deficiency. The term micronutrient malnutrition refers to the issues created by lack of minerals or vitamins within human diet. As per estimations, more than 2 billion individuals across the globe are affected by micronutrient deficiencies or 'hidden hunger'. This is mainly due to the lack of sufficient nutrients from the edibles consumed commonly. Although these micronutrients may only be needed in small quantities by the human body, they are nevertheless very necessary for maintaining proper health and avoiding other long-term illnesses.

When considering children alone, malnutrition is one of the four leading causes of child mortality in the world. The World Health Organization (WHO), in 2013, reported that malnutrition was associated with nearly 45% of childhood deaths. It is already well researched and known that deficiency caused due to lack of proper consumption of micronutrients make children prone to infections and is also a common reason for child mortality. Additionally, lack of proper nutrition consumed by a pregnant mother is a key reason contributing to poor levels of birth weight seen in babies. Babies that have a low weight are often more prone to infant mortality at the neonatal infancy period. And even if babies survive that stage, it is very likely that they will not considerably make up on the growth that was lost, and are at bigger risk of having development deficiencies. This effect does not just stick to childhood but the issues are likely to extend into adulthood as well (Sen Ray 2016).

The Triple Burden of Malnutrition

According to UNICEF, malnutrition in children can be categorized in three ways. Children who do not grow well according to their age are usually the victims of three types of nutritional imbalances. Of these imbalances, the problems associated with malnutrition is quickly spreading in societies around the world, and within some of the poorest countries in the world in particular.

Undernutrition is the most prevalent nutritional imbalance, especially in the developing countries. It is defined as dietary intakes below the minimum required levels. While there has been some decline in undernutrition over the last few decades, it still has a major effect on children across the globe. Within the first 2 years of an infant, the impact of poor nutrition is often seen in the form of stunted growth. What is even more damaging is that these infants are likely to retain many of the issues caused due to stunting even as they grow older, often leading to a situation where full recovery and achievement of physical potential is never reached. Cases of undernutrition is visible in low birth weight of infants as well. This is often due to the poor feeding practices normally amplified with poor living conditions that deny infants sufficient nutrition. In many cases such circumstances also result in death. As of 2018, it was found that nearly 149 and 50 million children were stunted and were low in weight for height, respectively.

Hidden hunger is another important nutritional imbalance. Hidden hunger, also referred to as micronutrient deficiency, can be described as insufficient consumption of vital minerals and vitamins that are generally needed by the human body. This situation deprives children of essential elements at nearly every stage of growth and can have a detrimental affect to their well-being. What makes hidden hunger more dangerous is that it is seldom recognized in children until it becomes very late. The UNICEF, as per most recently available data, estimates that just under 350 million children below the age of 5 suffer from micronutrient deficiencies.

Overweight is also an adverse consequence of nutritional imbalances. It is defined as overnourishment or excess intake of nutrients through diet resulting in obesity among children. The numbers of obese boys and girls between the ages of 5 to 19 years have soared since mid-1970s, growing by nearly 10–12 times around the world. Overweight, which was commonly believed to be an issue affecting the rich, is now being seen among the poor as well. This is an echo of the increased availability of cheap and unhealthy food items in most communities across the world. The major health impacts of it is an increased likelihood of major long-term illnesses such as diabetes and heart diseases.

Zinc Malnutrition

The ubiquitous element, zinc, is irreplaceable for several processes within the human body and also for plants and animals. Zinc is found in comparatively stronger concentration in all body tissues and secretions.

Zinc functions like a regulator for greater than 300 enzymes, as well as acting as an antioxidant. It affects the assembly of multi-protein complexes as well as the stability of biological membranes. Zinc is also useful for the synthesis of DNA and RNA. Furthermore, it takes care of the expression of hormones, insulin storage in pancreas and also plays a vital part in supporting sensory functions (Schubert et al. 2015).

These represent only a small selection of the role of zinc within the human body as there are several others bodily functions that require zinc as a vital element. The changes from optimum nutritional status of existence of zinc in the body either due to specific zinc deficiencies or dietary consumption grounded on inappropriate combinations or amounts of foods is commonly termed as zinc malnutrition. Zinc deficiency and zinc toxicity are two foremost shapes of zinc malnutrition.

Zinc Deficiency

Zinc deficiency arises from low dietary intake of zinc, which is below the minimum required levels for a healthy body.

The amount of zinc required each day depends on the age of the individual, the stage of life the individual is at as well as the condition of an individual's health (Table 2.1).

Historically, the studies on the significance of zinc can be traced back to 1869 where zinc was identified for the first time as being a vital component in the growth of *Aspergillus niger* (Raulin 1869). It was as early as 1956 when patients who were cirrhotic were seemingly identified to be carrying lower levels of zinc. By the year 1963, zinc was already established as necessary for growth in human beings. A study was publicized that discussed the condition of an Iranian man who had a stunted growth among other things, which was then helped by supplementation of zinc (Duggan et al. 2008). By 1974, the National Academy of Sciences had declared zinc as a critical element for human beings and had recommended an ideal daily allowance level. As early as the late 1970s, the Food and Drug Administration of the United States mandated the requirement of zinc as part of nutrition fluids. By the 1990s, awareness on the function of zinc deficiency in child mortality within poorer countries was already heightened (Sandstead 2013). As of 2014, more than 300 zinc-containing enzymes and more than 1000 zinc-containing transcription factors have been identified. Furthermore, numerous studies in the past have indicated a

Table 2.1 Suggested dietary intake of zinc in milligrams. (Institute of Medicine, Food and Nutrition Board 2001)

Stage of life	Suggested quantity of zinc
First 6 months of life	2 milligrams
Infants: From 7 to 12 months	3 milligrams
Children From 1 to 3 years	3 milligrams
Children: From 4 to 8 years	5 milligrams
Children: From 9 to 13 years	8 milligrams
Teenage: From 14 to 18 years (for boys)	11 milligrams
Teenage: From 14 to 18 years (for girls)	9 milligrams
Adulthood: For men	11 milligrams
Adulthood: For women	8 milligrams
Pregnant teenagers	12 milligrams
Pregnant adults	11 milligrams
Breastfeeding teenagers	13 milligrams
Breastfeeding adults	12 milligrams

strong relation between zinc deficiency and important physiological process. Some of these are described as follows:

Brain The level of zinc in the blood of individuals affected by Parkinson's disease as well as Alzheimer's disease was found to be less (Brewer et al. 2010).

Cardiovascular system The role played by zinc as part of the maintenance of arterial blood pressure was found to be notable, with women and men affected by hypertension handling the metabolism of it in a varying manner (Tubek 2007).

Liver Deficiency of zinc in liver is seen in individuals with liver cirrhosis, as well as in cases of less advanced alcoholic and non-alcoholic diseases (Bode et al. 1998).

Endocrine system A relationship is seen in studies between the deficiency of zinc among geriatric individuals and comparatively lower levels of functioning of the thymus gland and thymic hormones (Haase and Rink 2009).

Healing The deficiency of zinc was associated with the delayed healing of wounds, and was found to be particularly relevant to the recovery of gastric ulcers at early stage (Keenan and Morris 1993; Watanabe et al. 1995; Andrews and Gallagher-Allred 1999).

Age Group at High Risk of Zinc Deficiency

Specific categories of people have a heighted requirement of zinc, and are consequently at a higher risk of zinc deficiency. These include children and lactating women among others (King and Cousins 2006).

(a) Infants and children: Children, in particular the young ones, have an increased risk of issues caused by zinc deficiency due to stronger requirement of zinc in order to support growth. Infants who are entirely breast fed, and whose mothers have sufficient zinc intake will attain the needed levels of zinc for the starting 6 months of life (Brown et al. 2004). Beyond the mentioned age, additional dietary intake that contains absorbable zinc is needed to fulfil their needs. In several of the poorer countries, complementary breast feeding is deferred and other food items are consumed for nourishing. These food items can, however, have smaller levels of absorbable zinc and, therefore, falter in reaching the requirements of zinc levels. On the other hand, the use of such food items for nourishment can inhibit the absorption of breast milk–induced zinc because of the higher levels of phytate content in them (Institute of Medicine, Food & Nutrition Board 2001). The needed levels of zinc for children who already face malnutrition are amounted to be nearly 2 mg/kg and 4 mg/kg of the weight of the body. These requirements are a lot larger when compared to healthier children, most likely due to the previous zinc depletion and decreased zinc absorption as a result of changes in the intestinal tract (Müller et al. 2001).

(b) Pregnant and lactating women: The hormonal stresses of pregnancy and lactation predispose maternal and lactating women to zinc deficiencies (King 2000). The demand for zinc is extremely greater during the lactation period; however, with hormonal improvements during lactation, the need can still be satisfied. A number of studies have shown a detrimental effect of supplementary iron on zinc absorption in pregnancy (O'Brien et al. 2000) and lactation (Fung et al. 1997).
(c) Elderly people (65 years and older): It is confirmed that diets of elderly people are frequently low in zinc, even in wealthy countries. Reduced zinc consumption may be a source for low zinc absorption in older people. Age can also affect the ability of zinc to be absorbed by body (Andriollo-Sanchez et al. 2005).
(d) People with following health conditions are also at high risk of zinc deficiency:

- Patients receiving total parenteral nutrition.
- People who suffer from alcohol-induced liver disorders are very likely to have reduced liver zinc levels.
- People with inflammatory bowel disease.
- People who have severe diarrhoea (Roohani et al. 2013).

Worldwide Occurrence of Zinc Deficiency

Mild levels of zinc deficiency occur in many areas, but extreme zinc deficiency occurs in low-income countries. Zinc deficiency has been listed as one of the main contributors to infection and disease in developed nations. Considering developed and emerging nations, zinc deficiencies are ranked 11th out of the 20 leading risk factors. One billion people around the world are at risk of being malnourished because of shortage of zinc. It was estimated that zinc deficiency alone had caused 800,000 deaths with 450,000 of them occurring in children under the age of 5 (World Health Report 2002) (Fig. 2.1).

Notably, zinc consumption is linked to the development of stunting in young children. Trends and incidence of non-communicable diseases have remained stable worldwide but significant declines are seen to be demonstrated by China where the prevalence declined from 17% to 8% in 2005 (Pitchik et al. 2018). Endemic deficiency is a big problem for nearly one-thirds of the global population. Zinc deficiency is also seen in some Islamic countries, such as Iran, Egypt and Turkey, because of high phytate consumption (Vuralli et al. 2017). Countries that are deemed at high risk of zinc deficiency are those where stunting occurrence is greater than 20% and estimated iron deficiency is higher than 25%. Countries that are at low risk of zinc deficiency are those with stunting incidence less than 10% and insufficient zinc consumption under 15%. Countries which have a medium level of proneness to zinc deficiency are the ones with stunting incidence as well as sufficient dietary zinc level (Bhutta et al. 1999; Brooks et al. 2004; Fischer Walker and Black 2004; Sazawal et al. 2007).

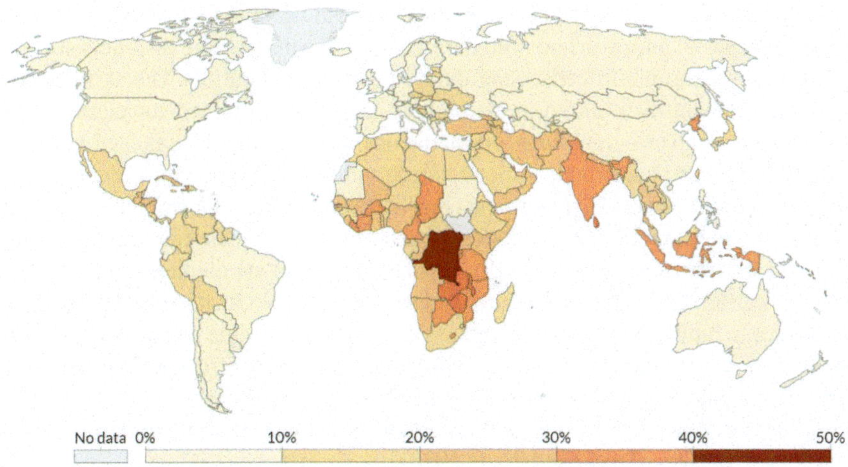

Fig. 2.1 Worldwide occurrence of zinc deficiency by country in 2005; measured as the share of total population with intakes below physiological requirements. (Source: Our World in Data, Wessells and Brown (2012))

The prevalence of zinc deficiency is seen as high in most of sub-Saharan Africa, south Asia and parts of Latin America. There is a high risk of childhood disease in countries which have low levels of zinc in the body. Unfortunately, India is one of the countries that has a great burden of malnutrition. According to one NAAS (2012) study, a large percentage of these children are significantly underweight and highly malnourished, which are avoidable by a concentrated scientific approach (Sen Ray 2016). Zinc deficiency is also widespread in children from developed countries as a result of the shortage of animal-based diets, the excessive dietary phytate content and low food intake.

Symptoms Associated with Zinc Deficiency

Medical signs of zinc deficiency are seen in cases of failure of development and reproductive ability. Significant zinc deficiency often typically contributes to delayed sexual maturity in males. There is a connection between deficiencies of zinc and numerous physical ailments. Some of these signs are non-specific and all of them are often present in other health problems, making it difficult to refer to specific nutritional deficiency (Plum et al. 2010). Zinc deficiency is considered to be caused by many symptoms. Borderline zinc deficiency can result in low immunity, neurosensory issues, memory disturbances and impairments to taste and smell and night blindness, and decreased spermatogenesis in males. Alopecia and behavioural disorders arise because of zinc deficiency in the body (Prasad 1985, 2013). Zinc deficiency is closely related to multiple sclerosis, sepsis, Alzheimer's disease,

Parkinson's disease onset and development. The clinical signs of it often range from moderate cases of anorexia and mild dermatitis to life-threatening disorders such as growth arrest, chronic infection and delayed sexual maturation. The skin lesions have predominant distribution in extremities and adjacent to orifices of the body. These rashes are vesiculobullous, erythematous and pustular. Heightened vulnerability of fungal and bacterial infection related to immune dysregulation, behavioural abnormality and cognitive impairment (Pawlitzki et al. 2018). One of the most main appearances of zinc deficiency is seen in acrodermatitis enteropathica. Both children and the subjects who reacted in a positive way to supplementation had low levels of serum zinc. Zinc deficiency is observed in some chronic diseases such as scleroderma, lupus, skin blemishes, rashes and psoriasis. Sickle cell anaemia is synonymous with zinc deficiency. Zinc helps to reduce levels of calcium in the body cells. Zinc depletion has been observed in a few cases of diabetes; one of which is juvenile type. Zinc supplements are important for proper functioning in those with adult-onset diabetes (David and Watts 1988).

People suffering from anorexia have low relative zinc and high copper levels. Apparently, protein can taste strange due to a zinc deficiency. This recent finding is seldom seen in total deficiency states. Studies have shown that individuals suffering from lack of zinc have a propensity to consume more protein. This variation may be due to the person's differing degrees of hydrochloric acid status. Zinc has also been reported to have a significant antiviral activity against several types of rhinoviruses. Irradiation has been shown to dramatically impact viral development. Zinc is also important for immunity and resistance to viral infections. In comparison, zinc deficiency can also induce minor sensory losses in vision. When a person has low tissue zinc-to-copper ratios, dyschromatopsia is common (David & Watts 1988).

Risk Factors Associated with Zinc Deficiency

Zinc deficiency may have an alimentary, iatrogenic or genetic origin, or come about when someone may have an absorption condition (pancreatic insufficiency and chronic inflammatory bowel disease), increased zinc demand (during breastfeeding, chronic disorders, high development and competitive sports) or increased zinc excretion (renal failure and liver cirrhosis).

The major cause of human zinc deficiency is a diet that is deficient in bioavailable zinc. Besides not getting enough zinc in the diet, other risk factors for zinc deficiency include ulcerative colitis, Crohn's disease, short bowel syndrome and chronic diarrhoea (iron can interfere with zinc absorption).

A case with serious pregnancy disorders may be linked with impotence, accelerated puberty, hypogonadism, dermatitis, alopecia, paronychia, learning impairment, reduced nerve conduction and nerve injury, macular degeneration, impaired iron intake, impaired taste and smell, weight loss and impaired wound healing (Hunt 2003; Hambidge and Krebs 2007) (Table 2.2).

Table 2.2 Risk factor associated with zinc deficiency. Adapted from 'Clinical aspects of zinc deficiency' (Yanagisawa 2004) and 'Zinc-Extensive blood and urine biochemistry and immunological tests' (Yanagisawa 2004)

Risk	Reasons
Insufficient intake	1. Low-zinc diets: Meals low in animal protein (vegetarians) 2. The depletion of zinc by manufacturing (desalting during production of artificial milk) 3. Extended intravenous alimentation 4. Insufficient nutrition consumption
Malabsorption	1. Congenital: Acrodermatitis enteropathica (seldom seen) 2. Acquired (a) Consumption of absorption inhibitors (b) Malabsorption syndrome: pancreatic dysfunction, liver dysfunction, inflammatory bowel disease and short bowel syndrome (c) Drugs and chelating agents: penicillamine and EDTA
Excessive loss	1. Child intractable diarrhoea, intestinal fistula and gastrointestinal illness linked to diarrhoea 2. Heightened urinary removal 3. Others: Burns and haemodialysis
Heightened demand	Pregnancy and neonates Enhanced anabolism
Other unexplained factors	Mongolism and congenital thymus defect

Impact of Zinc Deficiency on Child Development

WHO has established zinc deficiency as a significant danger to child growth, and it has specifically linked zinc deficiency to diarrhoea, lower respiratory infections and malaria. Zinc deficiency prevents the body from battling diarrhoea and pneumonia. Early-life diarrheal disorders are a major problem for many developed nations. Diarrhoea takes the lives of almost one in five infants under the age of five year (WHO and UNICEF 2009). The established inexpensive, reliable methods to manage diarrhoea exist, but in underdeveloped nations, only 39% of all children with diarrhoea obtain the necessary care. In 2012, diarrhoea took more lives than malaria, AIDS and measles together (IZA and UNICEF 2012).

Low zinc status in children can adversely affect their cognitive performance across many pathways. Malabsorption and anorexia can contribute to reduced nutritional intake, as well as the elevation of intake due to fever. Iron deficiency or anaemia can occur secondary to infections with geohelminths. It is proposed that the immune system can directly influence mood and cognition. Frequent inactivity and inactivity triggered by children may restrict their growth (Smith 1990; Neumann et al. 1991; Emond et al. 1997).

Staphylococcus infection is more likely to occur in those children who are from deprived and overcrowded households, with insufficient water supply and sanitation supply. The signs may be extended if there is a chance of serious medical treatment. Children raised in these households are more likely to grow into bad or dysfunctional ones. Concomitantly, children who are malnourished are often at a greater risk of infectious disease. This is going to impact the growth of children who are particularly vulnerable. The likelihood of infection impacting children's growth was not well understood and have not been tested at all stages of science (Mak and Saunders 2006).

Zinc deficiency is a notable issue about children's diet worldwide. Zinc is important for neuronal development as well as synapse formation. Motor defects may hamper the development of neural pathways, creating severe behaviour issues (e.g. gross and fine motor skills and social skills). Zinc supplementation among infants and children could boost both motor and mental growth in children with high threat of zinc deficiency (Gogia and Sachdev 2012).

A marginal zinc deficiency can specifically and negatively affect biological processes and structures. Typical symptoms of androgen-insensitivity syndrome are stunted development in puberty and hypogonadism in males (Tuerk and Fazel 2009).

In population surveys, the issue of zinc deficiency is known to be of elevated concern for child development when the prevalence of low-serum zinc concentrations is greater than 20% and the prevalence of excessive intakes is greater than 25%. In such cases, zinc status may be enhanced by increasing dietary zinc intake. Around 17.3% of the world's population is at risk of poor absorbable zinc consumption, with the greatest risk borne by countries in Asia, Africa and Central America (Galetti 2018).

A research performed in Egypt showed that low levels of maternal zinc intake impaired the newborn's ability to pay attention (Kirksey et al. 1994). The researchers conducting trials on 10 Bangladeshi women found that babies who were delivered to mothers who obtained zinc supplements scored substantially lower on child growth tests than other infants (Hamadani et al. 2002). Another double-blind randomized controlled study of maternal zinc supplementation in African-American mothers found little improvement in the cognitive performance of 5-year-old children, assessed on various neuropsychological measures (Tamura et al. 2003).

Depression is one of the most important issues concerned with zinc deficiency. A clinical article suggests that there was a better improvement in the mental illness of patients who were provided with zinc and antidepressants as compared to those who were provided with placebo and antidepressants. Depression is a serious condition associated with injury, high morbidity and mortality. Currently, clinical findings prove that individuals with depression have considerably lower blood levels of zinc than non-depressed people. In certain patients with major depression, a negative association between serum zinc level and symptoms was found. Cognitive function may be disturbed because of severe zinc deficiency. Moderate and extreme zinc deficiency are related to mental disorders, like restlessness, tiredness and depression (e.g. involving anhedonia). Many psychiatrists agree that low plasma zinc levels contribute to depression and anxiety. Schizophrenia have been attributed to decreased levels of zinc in the brain. Studies also suggest that zinc deficiency may be linked to depression.

Zinc deficiency also has a substantial effect on the cognitive development of infants. A zinc-deficient infant may be subject to extra external stress. For example, a child who is born with chronic zinc deficiency could have greater difficulty in dealing with maternal seperation early in life as compared to other children. Zinc deficiency may cause difficulties in cognitive processing by changes in concentration, behaviours and other facets of cognitive behavioural such as preparation or avoidance. At the same time, it should be remembered that kids with attention deficit disorder also have functional skills within the average spectrum, but there are specific deficiencies in their neuropsychological functioning. This suggests that children who are zinc deficient might have appropriate cognitive functions but they would do poorly on academic tasks (Black 1998).

There could be a link among socio-economic status of mothers, ability to learn and development of their children. Despite this, research on zinc deficiency have not considered environmental enrichment. This adjustment will vary with children according to their age. The symptoms of zinc deficiency can become worse as the body prepares for puberty (Black 1998).

Malnourished children suffer from deficient motor and cognitive growth (Simeon and Grantham-McGregor 1990). But this problem typically arises in the instances of poverty and lack of proper care. Deficits in many areas of cognitive and motor function are also a source of acute malnutrition. Regardless of the cause, early developmental disorders will dramatically impact academic achievement and the ability to create additional value to the community (Brooks-Gunn and Duncan 1997). Early developmental advantage is a good indicator of potential adaptation abilities. Bertenthal and Campos (1990) proved by observational research that crawling provides children with growing freedom. This is because kids who are mobile or capable of shifting their role will be able to concentrate on a greater range of social and physical processes without much external support.

Measures of focus, movement, motor and idea formation should be viewed with regard to the growth of the child in the relevant period. In order to explain the relationship between zinc and cognitive growth, it is necessary to identify the age when the child. Animal experiments also demonstrated that the most susceptible stages of brain development happen during adolescence. For cognitive efficiency, sub-clinical marginal zinc deficiency may not be dangerous for the general public (Bhatnagar and Taneja 2001).

Hair zinc concentration was correlated with reading ability, indicating that zinc deficiency can interfere with academic success. There are at least three randomized trials regarding the impact of zinc supplementation on cognitive performance in school-age children. A research from Canada found no variations when tested on the sub-tests of the ‘Detroit Test of Learning Abilities’. Generally, studies performed on infants and Mexican-American children revealed that zinc supplementation resulted in superior academic development as compared with the control group. From the trials seen above, it appears that the influence of zinc has stronger effects on concentration and logical processes rather than general performance tasks. Further studies are necessary in order to allow a replication and explanation of established studies on the connection among zinc status, neuropsychological functioning and academic success (Black 2003).

Zinc defects have a significant influence on the motor function of infants. The relationship between motor growth and activity is vital to recognize as those with impaired motor development may become less active because they find it too difficult to show vigorous activity. Investigators noticed that at baseline, stunted children were less involved than non-stunted children (as they are in day care centres) and that the degree of activity was related to their success on a standardized measure of growth. Nevertheless, beyond 6 months, the amount of physical and mental activity had improved for all children. The stunted children's developmental stage did not vary from non-stunted children. This can also be demonstrated by the growth in the children's capabilities. Many of the infants who were considered to be stunted were able to run after 6 months of the baseline evaluation (Gardner and Grantham-McGregor 1994).

An analysis of the connection between maternal zinc and infant activity found that the involvement of zinc in maternal blood improved the habituation cluster scores on the 'Brazelton Neonatal Behavioral Appraisal Scale' administered shortly after birth (Kirksey et al. 1994; Brazelton and Nugent 1995). Habituation is an initial sign that shows how infant attention evolves. At 6 months of age, motor activity tended to be adversely linked to poor maternal consumption of plant zinc, phytate and fibre. Toddlers who had parents with elevated plant zinc intake at the period of breastfeeding were more likely to have low score associated with motor development (Kirksey et al. 1994). In babies with a history of chronic diarrhoea, poorer motor performance was seen relative to infants from affluent families. These findings demonstrated the long-term effects of maternal diet and the understanding of psychosocial risk on infant growth (Bayley 1993). Zinc deficiency during infancy is considered to cause significant behavioural and physical defects. Researchers believe that secreted zinc is responsible for synaptic plasticity, information retrieval, learning and memory. It is true that zinc is important in the hippocampus for inducing long-term potentiation, which is a well-known model for the processes behind memory formation (Prasad 2013). Reduced dietary consumption of zinc during breastfeeding influences the resulting foetal development. In people suffering from autism spectrum disorders, substantially increased incidences of zinc deficiency have been reported. The occurrence of low levels of zinc in several instances happens along with other underlying factors that trigger complications (Prasad 2003; Hambidge and Krebs 2007; Krebs 2013).

Zinc homeostasis is associated with a wide range of neurodegenerative, psychiatric and neuropsychological diseases. Depression is evidently related to loss of zinc in the body. Zinc deficiency at the time of brain development in turn tends to result in an unlike clinical image. Mental conditions such as autism spectrum disorder and attention deficit/hyperactivity disorder may have a non-genetic factor such as a deficiency of the mineral zinc. Autistic children in Singapore have been shown to have an exceptionally high incidence of zinc deficiency, particularly in infants aged between 0 and 3 years (Sauer et al. 2016).

Zinc Supplementation

Countering zinc deficiency requires consumption of certain kinds of food on an individual level and zinc supplementation programmes on a communal level. Meat and aquatic animals have high amounts of zinc (10–50 mg/kg dry weight) and are, hence, an effective food to increase the amount of zinc in the body. In the opposite, foods such as fruits, vegetables and grains are low in zinc (<5 mg/kg) (Table 2.3).

Among the most feasible measures to enhance the zinc status in population groups include supplementation, diversification of food intake and fortification. It would be more prudent to integrate supplementation as a part of the national curriculum for health and nutrition (Akhtar et al. 2011). This technique could be more successful in combating the problematic health effects of zinc. Prioritizing and organizing efforts by governmental, corporate, foreign and customer organizations are important to make a successful strategy. Zinc supplementation is an effective way of treating a long list of diseases (Akhtar 2013).

The study by WHO and UNICEF in 2009 cited a 7-point strategy to eliminate 1.5 million infant deaths because of diarrhoeal diseases caused by zinc deficiency. The proposals included two treatment strategies and five deterrence strategies. To treat the disease, children require zinc supplement and fluid replacement. Oral rehydration treatment or ORS is the pillar of fluid replacement, and the best standard is a low-osmolarity solution. In addition to continued eating including breastfeeding, use of fluid alternatives to ORS is an important component of treating issues due to zinc deficiency. The package, detailed by the WHO and UNICEF, consists of the following:

(i) Promotion of early and sole use of breastfeeding along with additional vitamin A supplementations
(ii) Increasing awareness of washing hands with soap
(iii) Increasing the quality and quantity of water consumed as well as storage practices
(iv) Raising awareness for community-wide cleanliness
(v) Vaccinations for rotavirus and measles

United Nations International Children's Emergency Fund (UNICEF) has supported over 50 countries to integrate the use of zinc supplements into national child development plans. The actual implementation of zinc supplement services will take time if national child healthcare policy changes are needed. United Nations Foreign Children's Emergency Fund (UNICEF) urges communities to use zinc in the combined initiative in combating diarrhoea. Interventions include the provision of zinc capsules for treatment of diarrhoea for 10–14 days in kids aged 2–59 months and the provision of zinc-containing multi-micronutrient powder for 60–90 days in kids aged 6–24 months. The UNICEF is a leading provider of zinc-containing powders as well as zinc supplements to developed countries (IZA & UNICEF Report 2012). Zinc supplementation offered to low birth weight infants for a year resulted in significant decrease in mortality.

Table 2.3 Type of food and zinc-associated health benefits. (Health Benefits Food Sources and Daily Requirements, Daily Sun 2020)

Food type	Zinc-associated health benefits
Seafood	For example, lobster and crab are useful sources for zinc. Three ounces of cooked crab can supply 59% of the suggested daily levels of zinc, while small steamed lobster may contain about 43% of the daily levels
Meat and poultry	Consumption of beef, egg and chicken can deliver zinc to the body. As per USDA, one cup of chicken (roasted breast) can provide 19% of the recommended daily levels of zinc, while one egg conveys about 5% of the daily levels
Vegetables	Nutritious greens like broccoli, kale, mushrooms, garlic as well as spinach are great options for zinc intake. According to the USDA, one cup of mushrooms (raw and sliced) provides nearly 3% of the daily levels needed of zinc, while one cooked cup of kale roughly contains the same amount. Legumes and beans also contain zinc. A one-fourth cup of hummus can support the body with about 8% of the daily suggested intake of zinc, while one cup of lentils has about 11% of the daily value
Nuts and seeds	Nuts and seeds are natural sources of zinc. One ounce of either pumpkin seeds or pine nuts can provide around 20% or 16% of the daily suggested intake of zinc, respectively. One ounce of dry-roasted cashews contains 15% of the daily value. Moreover, 1 ounce of Chia seeds can offer the body nearly 12%
Whole grains	Besides fibre, minerals and vitamins, the whole grains are enriched in zinc. One cup of oats can deliver nearly 27% of the daily value, while 1 cup of brown rice (cooked) can provide 13%
Dark chocolate	Dark chocolate is a delicious source of zinc. In fact, depending upon cacao varieties, darker chocolate offers zinc in greater values. One ounce of dark chocolate can supply nearly 7–8% of the recommended daily levels of zinc
Fortified breakfast cereals	Breakfast cereals carry diverse minerals, including zinc. Generally, one cup serving can provide nearly 25% of the suggested daily levels of zinc
Milk and dairy products	Both milk and yogurt are excellent food items with zinc. A cup of milk (non-fat/low fat) contains around 9% of the suggested daily levels of zinc, while a cup of non-fat or low-fat plain yogurt can provide nearly 22% or 20% of the daily levels, respectively

Taking zinc supplements decreases the incidence and duration of diarrhoea in infants. In 2004, the United Nations and World Health Organization (WHO) published a joint statement on the clinical treatment of diarrhoea in children in developed countries, which prescribed zinc supplements to children with acute diarrhoea. It predicted that in the 1990s, more than 1 million deaths could have been avoided each year. Data from tests in countries like Bangladesh, Pakistan, India and Mali, where zinc was used for diarrhoea problems, show a noteworthy reduction in mortality, a decrease in the unnecessary use of antibiotics and non-zinc anti-diarrheal, and an exponential rise in diarrhoea care seeking. A systematic analysis of clinical studies revealed that zinc supplements decreased the risk of diarrhoea by approximately 20%, and acute lower respiratory infections by approximately 15%. Zinc supplementation had a small effect on reducing the mortality of children and its effect was highest for older children (IZA and UNICEF 2012).

Dietary diversification or modification as a successful way to preserve a balanced diet is sustainable in the long run. It is essential to increase the intake of zinc by dietary widening at the community and household scale. Such techniques typically involve (1) farm interventions, (2) cultivation and encouragement of animal products such as pigs, poultry or seafoods, and (3) manufacturing strategies in the household and commercial sphere to improve the intake of zinc from plants. Any plant-based foods can theoretically have a minor effect on the body's zinc absorption. This could benefit if supplemented by interventions to decrease the levels of dietary phytate that inhibit zinc absorption; however, this may not be sufficient to satisfy zinc needs of infants. Animal husbandry activities that contribute to the increase in red meat and liver intake may have a beneficial effect (Gibson and Anderson 2009).

A significant amount of research has shown an association between plant-based foods and zinc deficiency in several of the developing countries. The high phytic acid content in Asian diets leads to zinc-phytic acid complexes in the intestine and decreases zinc absorption in the stomach. The bioavailability of zinc improves when other inhibiting nutrients such as calcium and polyphenols are present. Many varieties of bio-fortified zinc rice and zinc wheat have now become available or are being tested in different countries (Harvest Plus 2020). The zinc that the body is able to consume from bio-fortified zinc wheat is differentiated over and above common varieties. In India, zinc supplementation decreases morbidity among mothers and their infants (Maares and Haase 2020).

Current plant breeding and genetic modification plans have brought in plans to improve the bioavailability of zinc from plant foods (Lonnerdal 2003). Bio-fortification differs from the usual fortification because it relies on the enrichment of micronutrients in areas of the crop that can be absorbed when the plants are already growing, rather than absorption of nutrients from outside supplies while the crops are being processed. This is a profitable idea because it would increase food intake in developing countries (Banuelos and Lin 2009). Bio-fortification is a potential remedy for the issue of micronutrient shortages in developed countries. Its additional advantages include higher yield due to the scarcity of micronutrients in the soil. In addition, it can increase vitality of seedlings growing from the seeds which have been enriched with zinc (Roohani et al. 2013).

While zinc supplementation is primarily to counter the issues created by zinc deficiency, there are several benefits of zinc supplementation to the human body:

1. The *European Journal of Immunology* mentioned that adding zinc to a regular diet can help the process of activating T lymphocytes (T cells) in the human body. T cells are essential to control and regulate immune responses against viruses, antigens, infected or cancerous cells. Conversely, zinc deficiency can impair the function of your immune system by increasing susceptibility to diverse pathogens.
2. Research conducted by the University of Toronto has found that zinc plays an essential role in regulating the communication network of neurons cells in the

human brain. In simple words, zinc can impact upon the process of learning and memory formation.

3. Zinc has a crucial role in preserving the integrity and structure of the skin. A Swedish study has discovered that zinc helps in the treatment of leg ulcers. Another research reveals that zinc can improve the speed of healing of wounds due to decrease in inflammation along with triggering immune cells at the place where the injury is located. For this reason, zinc is widely used in diverse skin creams for treating skin irritations and diaper rash.
4. Zinc may help in reducing the chronic diseases associated with age. A research conducted by Oregon State University has found that zinc intake may decrease the risk of inflammatory diseases. Zinc may prevent age-related macular degeneration (AMD) and prevents vision loss, according to the *Archives of Ophthalmology*.
5. Journal of Biological Chemistry boasts that zinc may assist in regulating the heartbeat which can minimize the risk of arrhythmia-related heart failure. Furthermore, zinc may treat acne, attention deficit hyperactivity disorder (ADHD), hypothyroidism and osteoporosis.
6. Zinc deficiency is a potential risk factor for the low quality of sperm and male infertility. Research has proved that zinc sulphate and folic acid supplementation can improve the sperm count of healthy adult males (United News of Bangladesh 2020).
7. Zinc supplementation is useful in the prevention of gingival disease; it helps protect against *Fusobacterium nucleatum* and *Prevotella intermedia* that are the main reasons behind gingivitis. Zinc also inhibits the proteins formed by *Porphyromonas gingivalis*. Zinc oxide nanoparticles have been shown to inhibit cell membrane and oxidative stress in gram-positive and gram-negative bacteria found in gingivitis as well as campylobacter species. A half-year randomized clinical trial comparing two dentifrices, one that contained fluoride and zinc with the other that contained fluoride only, showed that zinc-based dentifrice had a substantial clinical decrease in gingivitis and dental plaque. A research conducted by Hong and colleagues has shown that A20, an amino acid protein with zinc finger C-terminal domains, can be a possible component in the prevention of inflammatory bone-loss disorders like periodontitis related to its anti-inflammatory and anti-osteoclastogenic impact. On the other hand, studies have illustrated the link between zinc deficiency and gingivitis (Uwitonze et al. 2020).

In addition, zinc supplementation may be used as a method to treat some other illnesses and health problems as well. In paediatric practice, particularly in circumstances where dietary intake is usually assumed to be appropriate, the use of pharmacological zinc therapy is relatively rare. Two of these cases are described below.

Wilson's disease The origin of supplementation as a monotherapy for Wilson's disease (WD) is that it prevents the absorption of intestinal copper and promotes the synthesis of hepatic metallothionein that binds copper with strong potency than zinc. In 1997, the United States Food and Drug Administration approved the use of zinc. In infants, levels in the area of 50–75 mg of elemental zinc per day were used.

Two reports, including a 10-year follow-up of 22 kids, affirm the effectiveness of zinc therapy in young pre-symptomatic kids. Findings from both studies included the production of hepatic inflammation, greatly decreased hepatic transaminases and excretion of copper in urine, and (in the broader sequence) hepatic copper content. Neither of the series recorded signs of zinc toxicity or of normal growth (Marcellini et al. 2005; Mizuochi et al. 2011). A systematic review, including paediatric studies, found that zinc therapy was favoured over D-penicillamine in pre-symptomatic and neurological patients, while D-penicillamine was favoured in patients with hepatic symptoms (Wiggelinkhuizen et al. 2009). Affirmation of zinc therapy was shown to be higher than that of D-penicillamine. As for future analysis, multi-centre prospective randomized longitudinal comparative experiments are certainly needed to improve the present understanding of the effectiveness and drawbacks of zinc therapy for Wilson's disease in the paediatric population (Krebs 2013).

Common cold Use of zinc has also been thoroughly studied in advanced countries for its ability to treat 'common cold'. (URI). The use of zinc has had an unclear 'uptake' in primary care in the advanced world, but works as an indication of a pharmacological impact because most people are believed to have an acceptable zinc status at baseline. The theoretical basis for zinc efficacy against viral-induced URIs involves inhibition of viral replication or intracellular adhesion, or improved innate immune response on the mucosal surface (Geist et al. 1987; Eby 2010). Both zinc lozenges and intranasal sprays have been evaluated to shorten the time of URI signs. Results between the trials were reasonably varied, with about half of the blinded randomized trials indicating a reduced period of effects with zinc when treatment was started immediately after the symptom onset and when communication with the oropharynx was immediate (Krebs 2013). While supplementation at minimum 5 months has been correlated with a substantially reduced occurrence of URI, school absenteeism and antibiotic consumption in infants, such results have been disputed due to limited size of the sample and lack of replication. Some have recognized the ability for zinc lozenges to cause adverse effects, including poor taste and nausea. There was an acknowledgement that, despite the possibility of effectiveness, the data are inadequate to offer suggestions on dosage, composition and length (Singh and Das 2011; Science et al. 2012).

Vitamin / mineral products are commonly used in emerging regions, but there is little consensus on its effectiveness and no standard recommendation on its composition. Essentially, they contain zinc, copper, folic acid and vitamin A, and in certain instances riboflavin, magnesium and calcium. This type of supplements can also be considered for expectant mothers in the emerging countries. However, potential associations among elements of the supplement which may decrease the bioavailability of such micronutrients need to be studied. It would seem evident that the avoidance of zinc deficiency in young kids appears to be the best policy, not just on ethical grounds but as well as on economic grounds. A huge amount of work still needs to be undertaken to find an effective way to avoid zinc deficiency, but it seems that zinc supplementation or fortified foods with an acceptable zinc compound can be the solution to solving such a global nutritional crisis (Alnwick 1998). Today, there are no credible evidence to illustrate the long-term developmental importance of the different activity rates linked to zinc supplementation. It will be helpful to investigate whether the differences in cognitive function caused by zinc deficiency are lasting. Even more research is required to inspect biological pathways that explain the function of zinc in cognitive growth. Zinc deficiency is present as part of total starvation and in tandem with the other micronutrients. Additional studies are needed to research the impact of zinc supplementation on mental function in conjunction with many other nutrients, including micronutrients, and also to explore the influence of zinc deficiency independently of moderators such as social and environmental treatment (Bhatnagar and Taneja 2001).

Zinc Toxicity

Although the importance of zinc deficiency has been reported for several years, only lately has the awareness of the potential punishments for unnecessary zinc consumption being brought to light. Zinc is shown to be relatively non-toxic and less dangerous, particularly when administered orally. Even so, signs of overt toxicity (lethargy, epigastric discomfort and fatigue) may appear with exceptionally high amounts of zinc. Even with low intake, but well above the prescribed dietary allowance (100–300 mg zinc/d vs. 15 mg zinc/d RDA), signs of triggered copper deficiency with associated symptoms of anaemia and neutropenia, compromised immune function and negative impacts on the low-density lipoprotein to high-density lipoprotein (LDL/HDL)-cholesterol ratio have been recorded. Much lower amounts of zinc supplementation, similar to the prescribed daily levels, have also indicated an interference with the use of copper and iron and to negatively affect high-density lipoprotein-cholesterol concentrations (The Fosmire 1990) (Table 2.4).

The suggested daily consumption of zinc by adults is 15 mg. Effects typically are not visible unless absorption reaches about 1–2 g (Barceloux 1999). Exposures to toxicity have been found to occur through gastrointestinal, dermal, parenteral and respiratory mediums. Toxicity through inhalation varies in magnitude based on the particular substance contained, along with the time of exposure. For instance, zinc chloride-containing smoke bombs can induce airway inflammation, chest pain and sometimes also acute respiratory distress syndrome, such as clinical picture of pulmonary fibrosis as long-term sequelae (Murray et al. 2018). On the other side, respiration of zinc oxide may lead to an issue known as 'metal fume fever', which is typically secondary to exposure from work. In such instances, personnel who are exposed to gases from metal alloy manufacturing, soldering and welding may experience flu-like conditions along with cough and dyspnoea, potentially due to the direct effects of zinc on pulmonary epithelium (Greenberg and Vearrier 2015). Such exposures to zinc may also arise by ingestion of dietary supplements, pennies developed post-1981 and zinc chloride substances at concentration higher than 20%. In addition, the use of unnecessary quantities of denture cream can contribute to overdose of zinc and secondary copper deficit (Lasater 2017). Finally, zinc toxicity may

Table 2.4 Tolerable upper intake levels (ULs) for zinc. (Institute of Medicine, Food and Nutrition Board 2001)

Life period	Gender male	Gender female	Expecting mother
0–½ years	4 milligrams	4 milligrams	
7 months to 1 year	5 milligrams	5 milligrams	
1–3 years	7 milligrams	7 milligrams	
4–8 years	12 milligrams	12 milligrams	
9–13 years	23 milligrams	23 milligrams	
14–18 years	34 milligrams	34 milligrams	34 milligrams
Over 19 years	40 milligrams	40 milligrams	40 milligrams

occur upon exposure from the dermal pathway, in particular through overuse of cosmetics, sunscreen and ointment (Subramaniam et al. 2019).

The far more known problem of zinc toxicity has been its opposite impact on copper levels. Max zinc consumption of 60 mg/day (50 mg of supplementary zinc and 10 mg of dietary zinc) across approximately 10 weeks resulted in evidence of copper deficiency. Copper deficiency was also identified after excessive use of extreme levels of zinc-containing denture cream (≥2 tubes a week comprising 17–34 mg/g zinc) (Nations et al. 2008). To avoid copper deficiency, the U.S. Food and Nutrition Board developed a bearable upper intake (UL) standard of 40 mg/day for adults, comprising of dietary and supplementary zinc. A case study reported extreme nausea and vomiting inside half an hour of intake of around 4 grams of zinc gluconate (570 milligrams of elemental zinc). Consumption within the range 150–450 milligrams of zinc a day is linked to chronic effects such as low copper status, altered iron activity, decreased immune function and decreased levels of high-density lipoproteins. Reduction in copper-containing enzyme have been identified with even relatively high zinc intakes of approximately 60 milligrams per day for up to 2 months. The Fellowship National Board has identified specifically determined upper levels (ULs) for zinc intake. Intakes over the recommended UL in the longer run raise the risk of adverse health consequences. It must be remembered that ULs do not refer to people who are seeking zinc for medical care, but those people must be under the care of a physician who controls them for negative health impacts. For instance, a patient had developed a similar pattern of thrombotic microangiopathy due to taking 300 milligrams of zinc for around 2 years. In this particular case, the patient continued consuming 150 milligrams of zinc for the purpose of prostatic hypertrophy. Around one and half year later, the patient had prostatitis due to infection, and therefore raised the zinc consumption by another 150 milligrams each day. However, the prostate disorder did not change, but rather deteriorated gradually. It was not affected by antibiotics that the patient had been consuming with zinc. The results of the patient's thrombotic microangiopathy showed a high ratio of zinc to copper and the patient was advised to quit zinc and resume copper supplementation, this led to a notable change in the patient's condition. High levels of zinc are also seen to cause hypercholesterolemia, infections, anaemia and scurvy-like symptoms (David and Watts 1988).

Major Indicators of Zinc Toxicity

Manifestation of toxicity attributable to zinc due to acute absorption of zinc sulphate and concentrated zinc chloride involves dehydration, hematemesis, muscle cramping and diffuse abdominal pain, which can be associated with several potential sources. A really high degree of scepticism is, therefore, needed to diagnose zinc toxicity. Some patient groups may be at greater risk, like paediatric patients and patients with mental disorders who would potentially take high quantities of dietary supplements (David and Watts 1988).

Where zinc toxicity is alleged, a comprehensive nutritional, professional and life history is important as it could expose unknown sources of exposure to zinc. Metal smoke or 'zinc shakes' from sudden inhalation triggers flu-like signs such as cough, chills, malaise, headache, fever, body aches and nausea. The individual could have consequence periods for days of not being involved in professional work, and therefore a thorough history of employment is required. Chronic ingestion of zinc may present like a 'swayback' syndrome, contributing to a slow development of anaemia and neuropathy with increased weakness, spasticity, gait disturbances and sensory ataxia. The signs of swayback syndrome are linked to an overlaid copper deficiency. Copper supplementation and normalizing zinc consumption will reverse haematological defects within a few weeks, while neurological deficits may likely persist even after therapy (David and Watts 1988).

In general, the following symptoms can be considered as major indicators of zinc toxicity:

- Copper deficiency
- Intravascular haemolysis
- Icterus and haemoglobinuria
- Cardiac arrhythmias
- Seizures
- Pain in the body
- Feeling of burning sensations
- Cough
- Chills
- Fever
- Blood pressure being low
- A metallic taste
- Lack of urine output
- Collapse
- Feeling of breathing difficulty
- Vomiting
- Convulsions
- Diarrhoea, either bloody or watery
- Eyes or skin being yellow
- Rashes
- Sweating

Causes and Consequences of Zinc Toxicity

Several common factors contribute to zinc toxicity. One among these corresponds to the low pH level within the belly that induces the secretion of free zinc, which further produces soluble, caustic zinc salts. Such salts are ingested from the duodenum and quickly spreads to the kidneys, muscles, pancreas, liver, muscles, prostate and bones. Zinc salts has immediate corrosive impact on body tissue, interact with metabolism of other ions like copper, iron and calcium, and impede erythrocyte activity and development. The pathways through which zinc has all these harmful effects are not fully known (Khan et al. 2013).

Many studies of zinc toxicity have been associated with aspiration of zinc chloride through industrial emissions. Health professionals may prescribe zinc consumption within the range 100–300 mg in a day for different medical conditions, such as sickle cell anaemia and celiac disease. Nevertheless, continuous treatment at such significant doses induces serious copper deficiency. This degree of supplementation is believed to affect the metabolism of copper by preventing the ingestion of copper from the intestine. Zinc toxicity could also result in the deterioration of up to 50% of the iron in the liver, accompanied by further loss of copper. Both of these deficits can lead to anaemia. Other recorded effects of zinc consumption of 100–300 mg in a day entail modifications in the immune system and blood lipids. In a study, 150 mg of zinc was taken twice daily for 6 weeks, culminating in a demonstrated decrease in lymphocyte activation reactions, chemotaxis and phagocytosis of the bacteria by polymorphonuclear leukocytes (Chandra 1984). A rise in low-density lipoprotein and a reduction in high-density lipoprotein (HDL) were identified, while triglyceride and total cholesterol concentrations did not differ considerably from the baseline (Pluhator et al. 1995).

Secluded cases of acute zinc toxicity arose as a result of the ingestion of food or drinks tainted with zinc from galvanized vessels. Signs and symptoms of acute zinc toxicity involve nausea, stomach ache, vomiting and diarrhoea. Single doses of 225–450 mg of zinc typically cause vomiting. Mild gastrointestinal disturbance has also been identified at doses within the range of 50–150 milligrams per day of added zinc. Metal smoke fever has been documented as a result of swallowing zinc oxide fumes. Especially, profuse sweating, fatigue and fast breathing can develop within 8 hours of aspiration of zinc oxide and continue for 12–24 hours after termination of exposure (King and Cousins 2012; Food and Nutrition Board, Institute of Medicine 2001).

Overall, there are three main ways of zinc penetration into the body: by inhalation, through the skin and by ingestion. Each type of exposure impacts various areas of the body and requires varying levels of zinc to be absorbed.

Exposure by inhalation The far more commonly known consequence of inhalation of zinc-containing smoke is the so-called metal smoke fever (MFF), which is mainly driven by aspiration of zinc oxide. This acute syndrome is an occupational condition typically triggered by inhalation of fresh metal pollutants of a certain particle size. The formation of metal smoke is related to the exposure, but not much evidence is available on the zinc concentrations that induce this syndrome (Vogelmeier et al. 1987). A research showed that two research volunteers experienced MFF as a result of acute inhalation (10–12 mins) of 600 milligram zinc per cubic-meter as zinc oxide (Sturgis et al. 1927). Hammond and associates wrote of personnel who faced breathing issues and chest pain 2–12 hours after exposure to around 320–580 milligrams of zinc per cubic-meter of zinc oxide (Hammond 1944). After exposure to 77 milligrams of zinc per cubic-meter (15–30 mins) as zinc oxide, only small differences in forced expiration were observed. Numerous records of exposures to lower doses of zinc oxide (14 milligrams per cubic-meter for 8 hours, 8–12 milligram per cubic-meter for up to 3 hours and 0.034 milligrams per cubic-meter for 6–8 hours) did not lead to signs of metal smoke fever (Marquart et al. 1989). The acceptable intake limit under the Occupational Safety and Health Administration (OSHA) is now 5 milligram per cubic-meter for zinc oxide (smoke and dust) in the employment air during an 8-hour working day and 40-hour working week.

Dermal exposure The dermal penetration of zinc happens, but the amount of scientific studies is insufficient and the process is not yet well established. Agren and associates realized that the skin pH, the quantity of zinc added and its chemical characteristics had an effect on the penetration of zinc. Analysis has shown that a 25% zinc oxide patch, when applied on skin over 48 hours, showed no evidence of dermal inflammation. In another research that compared the dermal effects of various zinc compounds in guinea pigs and rabbits, zinc chloride was evidently the strongest irritant, followed by zinc acetate, which induced mild irritation, and zinc sulphate induced low irritation, respectively. In accordance with the Agren (1990) report, zinc oxide had no irritant impact on the skin. As described previously, zinc chloride is caustic and the inflammation does not necessarily indicate a harmful effect of zinc. In comparison to the possible negative health effect of zinc on the skin, it should be remembered that zinc is a known supplement for the cure of wounds and other dermatological disorders. Based on current evidence, it can be inferred that the exposure to zinc through dermal means does not pose a major toxicological risk (Agren et al. 1993; Lansdown 1993).

Oral exposure Owing to its function as an essential trace ingredient, the oral uptake of small quantities of zinc is vital for life. Based on a reported account of an individual who passed away after oral consumption of 28 g of zinc sulphate, it was identified that the person began to vomit and experienced tachycardia as well as hyperglycaemia after ingestion. The individual died 5 days after the incident owing to haemorrhagic pancreatitis and renal failure. Immediate symptoms of exposure to harmful levels of zinc cause blood, stomach pain and nausea. Other symptoms include anaemia, dizziness and lethargy. Specific consequences of prolonged oral exposure to zinc are gastrointestinal complications. The gastrointestinal tract is specifically impacted by the absorption of zinc until it is circulated in the bloodstream.

As a consequence, numerous gastrointestinal signs have been identified after oral zinc absorption. For instance, a 39-year-old individual showed stomach ache, vomiting and nausea 6 hours post taking 150 grams of a 10% zinc oxide lotion, but without any evidence of systemic toxicity. In addition, the person acquired a corrosive gastroduodenal fracture. Side effects lasted for 3 days and, by 4 days after admission, corrosive damage revealed no scarring (Plum et al. 2010).

Furthermore, people, particularly infants, are exposed to zinc-contained products that are usually daily household objects. Such origins of zinc comprise of batteries, zinc oxide creams, automotive parts, paint, mineral and vitamins, supplements, cooking utensils, etc. Among the most common forms of zinc that induces toxicity post-consumption is the USA's Lincoln Penny. Any pennies produced during 1983, and all pennies manufactured since then, are 97.5% zinc in weight (~2.440 mg of elemental zinc per coin).

The effects of zinc toxicity differ and are, therefore, highly dependent on the extent and severity of the exposure to zinc. Zinc toxicity pathophysiology refers to the medium of exposure as well as the dosage administered. The toxicity to zinc-containing gases causes inflammation and possibly direct cell injury to the respiratory tract, leading to a condition known as metal smoke fever. While pathophysiology is not well known, cell injury triggered by zinc nanoparticles is believed to induce an inflammatory immune response that causes flu-like effects and local tissue harm. Chronic ingestion of zinc and the resulting copper deficiency will contribute to sideroblastic anaemia, granulocytopenia and myelodysplastic syndrome. In addition, an ascending, sensorimotor polyneuropathy condition was seen with high levels of copper and zinc deficiency. The exact reason for this condition is unclear to date (Bennett et al. 1997; Gregg et al. 2002).

The intake of extra zinc in food also raises the likelihood of adverse consequences for patients with diabetes. High-zinc supplementation in such persons contributes to high-serum zinc levels and inhibits cell insulin receptors, leading to reduced glucose tolerance. Another possible adverse impact of zinc supplementation in such individuals is that larger quantities of zinc in normal adults increase low-density lipoprotein and decrease high-density lipoprotein cholesterol. This adds to a state in patients culminating in a rise in the incidence of hypertension, coronary heart disease and diabetes mellitus, respectively. In addition, zinc toxicity may have significant consequences for pregnant mothers. It can cause foetal copper deficiency and adversely affect pregnancy. Lengthy copper deficiency in expectant mothers and animals leads to early foetal mortality, serious structural defects involving skeletal, respiratory and cardiovascular deficiencies and chronic neurological, immunological and biological anomalies. Respiratory distress syndrome can be the primary cause of morbidity in premature babies. A research in Nepal also stated that a large quantity of zinc provided during gestation inhibits the positive impact of iron and decreases the level of haemoglobin resulting in neonate malformation (Singh and Taneja 2009).

Zinc Toxicity and Child Development

Exposure to zinc toxicity is uncommon but is most likely to happen in grown humans than in kids. It is typically linked to workplace hazards and has been documented to happen in metal workers who are exposed to zinc-containing fumes. Some cases of zinc toxicity were identified in individuals who eat acidic food or liquids preserved in galvanized zinc vessels. Consuming excess supplementary zinc can lead to nausea, diarrhoea and vomiting. Chronic consumption of surplus zinc supplements may contribute to a deficit of copper, as zinc inhibits the absorption of copper. Inappropriate levels of copper in humans is usually the result of certain genetic diseases. Significant differences in copper metabolism appear in two inherited disorders, which are Wilson disease and Menkes disease. They include mutations in proteins that carry copper, which in special channels allow copper ions to pass across cell membranes. Wilson's disease initially causes effects in teens and young adults. Copper accumulates in the brain, liver and kidneys, causing liver injury and damage to nervous system. Menkes disease is a disorder in which impaired transmission of copper from the digestive tract results in reduced amounts of copper in the blood, while copper accumulates in the kidneys, pancreas and skeletal muscles. Menkes disease is typically initially seen in childhood, with children facing Menkes disease having stereotypical kinky hair, seizures, developmental failures and gradual brain degeneration (Chang 2002; Spanierman 2004).

Too much zinc in the diet promotes the synthesis of nutrients and cell differentiation, working by genes, leading to the development of humans. A study found that increased zinc intake during the infant development process raises the blood growth hormone level that raises the growth rate, raises the amount of fat cells in the body, facilitates fat absorption and increases insulin activity. The growth hormone is removed when it comes to maturity, and the ingested nutrients are guided to fat cells within the influence of heightened insulin production where they are stored as fat (Singh and Taneja 2009).

Other investigations also concentrate on the obese point of view of zinc toxicity. Zinc-charged food products, when eaten, help children develop quicker and often add fat to the tissues that make them look stable but face significant health concerns when they reach adulthood. Constant consumption of excess nutrients, such as zinc, across a period of time leads to obesity in adulthood. Obesity represents a daunting threat to the increasing population as it is etiologically related to insulin resistance, insulin-dependent diabetes mellitus, hypertension and coronary artery disease.

Not only this, the additional free zinc ions in food impede the uptake of copper as a result of the comparable physiochemical properties of these two components and the copper-zinc antagonistic reaction at the intestinal stage resulting in copper deficiency. The relationship between copper and zinc is of functional interest since it can happen with relatively low levels of zinc supplementation, and zinc-induced copper deficiency is relatively easy to develop in adult humans (Fosmire 1990).

As a consequence of zinc toxicity, copper deficiency causes additional determent to the antioxidant system enzymes, namely catalase, superoxide dismutase and glutathione peroxidase, and raises oxidative stress (Patterson et al. 1985). Copper deficiency is also reported to raise blood cholesterol (hypercholesterolemia), cause hypertension and lower density lipoprotein fraction increase in blood resulting in a heart attack. Excessive zinc ions, considering the other side, either render insulin inactive or decrease insulin binding to its receptor owing to copper deficiency and intensify the current risk factor for non-insulin-dependent diabetes mellitus (NIDDM). Thus, the consumption of zinc-fortified food for longer periods of time can make rising children more susceptible to these diseases. Possible prevention steps must also be taken to monitor the use of significant zinc in diets and zinc-fortified foods (Arsenault and Brown 2003).

To validate this point, a study stated that a rise in zinc-rich diet to twice the level of RDA raises the body fat portion of healthy children and eventually leads to obesity (Prentice 1993). A survey, conducted randomly, of nutritionally appropriate young healthy individuals (25–35 years) done in India found a significant positive linear association between body mass index (BMI) and tissue concentration of zinc.

Overweight and obese people have a higher amount of zinc in tissues than average or when compared to lean body weight persons. The children of NIDDM and ischaemic heart disease patients (destined to inherit NIDDM and ischaemic diseases) had two to four times the concentration of zinc tissue and less than half the concentration of copper than their equivalents of non-diabetic parents. This zinc and copper disproportion has persisted for some time, which after reaching the threshold level manifests itself as illness. The disturbance caused by additional quantity of zinc in daily diet was linked with obesity and obesity-related diseases among Indians (Taneja and Singh 2009).

Overcoming Zinc Toxicity

One of the commonly described treatment of zinc toxicity is chelation therapy. The therapy complexes the zinc ion and allows the discharge of unnecessary zinc from the body, making it instantly non-toxic and reducing the late effects. Specific drugs and medicines that retain metals in the blood are used. During the therapy, the medication is injected in the individual using an intravenous (IV) tube in the arm and, when the medication is connected to the zinc ions, the body expels unwanted zinc via the urinary tract. This therapy is also the favoured method to reduce the toxic effects of some other metals as well.

However, many chelators have disadvantages with numerous negative impacts, non-specific binding and administration inconveniences. Given the ever-increasing exposure to zinc in the industrialized world, the chelation therapy acts as an essential tool to fight against metal storage disorders. However, the lack of major clinical trials still offers disagreements about its clinical therapeutic benefits. Nevertheless, despite all the disadvantages, it is critical to comprehend the need for more precise and advanced chelation molecules not just to fix unresponsive poisonings, such as cadmium toxicity, but also to achieve full clinical recovery in cases of other metal disorders. In addition, more recent therapeutic measures that can deliver improved outputs must be analysed (Flora and Pachauri 2010).

Future Direction

Zinc malnutrition, especially zinc deficiency, is now considered as a major public health problem particularly when it comes to child development. The issue is even more widespread in developing countries, mainly in South Asia and Sub-Saharan Africa. However, no concrete evidence is available to suggest that particular healthcare policies seem to be working in tackling the issue and reaching a long-term sustainable solution in these regions. Likewise, the lack of quality data on the magnitude of zinc deficiencies over time in these regions leads to weakly informed decision-making regarding strategies. In many of the studies, the sample sizes used are smaller than required to represent the respective populations. Therefore, the survey process used to identify the status of zinc availability needs to be modified or re-engineered to cover a much larger sample size that would be more representative of the population.

Additionally, efforts in the past to improve the status of zinc among vulnerable populations have shown that there is a trend to receive traditional medical treatment for immediate relief more than using the zinc supplementation programmes available for long-term benefits. This indicates that more promotional effort is needed to spread the awareness regarding zinc deficiency and the importance of zinc supplementation programmes, especially among the poorer communities. Research has shown that there are convincing scientific evidences that propose several ways to make programmes targeting supplementation effective. Such measures include the combination of supplementation along with oral rehydration salt (ORS) meant to deal with diarrhoea, training and education activities for all stakeholders in the programme, regular availability of zinc products at a consistent standard, etc. Ultimately, a strong level of committed involvement is needed from the key players in the ecosystem, that is, governments, healthcare industry, donors, etc. This should be complemented with powerful social advocacy, political will, budget and infrastructure development to successfully implement zinc supplementation programmes (Akhtar 2013).

Furthermore, more research is needed in the field of associating cognitive and motor development with zinc deficiency. Studies of child development at the infancy stage can also incorporate effects of actions of dietary intake through the stages of an expectant mother's timeline, with particular focus on the procurement of zinc within the body. Deficiencies due to the lack of micronutrients during pregnancy, especially when coupled with psychosocial factors, can weaken early infant growth. Further to the standard indices used for child development and academic prowess, researchers must also examine the way in which neuropsychological functioning may influence the overall performance from a developmental point of view. In early teenage, it would be a useful addition to existing literature to investigate some of the additional neuropsychological methods that may show a strong association with deficiencies due to the lack of zinc intake and can have an influence on cognitive performance in high-pressure settings such as school and shift to early adulthood. To add to it, it would also be interesting to dig deeper into understanding the effectiveness of care-providing environments in terms of providing better protections towards zinc malnutrition.

Finally, zinc malnutrition usually comes along with other micronutrient deficiencies and, therefore, the research linking the casual relationship between cognitive development and zinc deficiency alone is still not concrete. Hence, further research will only go on to strengthen it. Also, on the other side, further research is still required to understand the deeper effects of zinc supplementation in combination with the other nutrients (Black 1998).

References

Agren, M.S.: Percutaneous absorption of zinc from zinc oxide applied topically to intact skin in man. Dermatologica. **180**(1), 36–39 (1990) https://pubmed.ncbi.nlm.nih.gov/2307275/

Agren, M.S., Franzén, L., Chvapil, M.: Effects on wound healing of zinc oxide in a hydrocolloid dressing. J. Am. Acad. Dermatol. **29**(2 Pt 1), 221–227 (1993). https://doi.org/10.1016/0190-9622(93)70172-p

Akhtar, S.: Zinc status in South Asian populations--an update. J. Health Popul. Nutr. **31**(2), 139–149 (2013). https://doi.org/10.3329/jhpn.v31i2.16378

Akhtar, S., Anjum, F.M., Rehman, Z.U., Riaz, M., Arshad, M., Basit, A., Ismail, T.: Effect of zinc and iron fortification of the feed on liver and thyroid function in rats. Biol. Trace Elem. Res. **144**(1–3), 894–903 (2011). https://doi.org/10.1007/s12011-011-9050-9

Alnwick, D.J.: Combating micronutrient deficiencies: problems and perspectives. Proc. Nutr. Soc. **57**(1), 137–147 (1998). https://doi.org/10.1079/pns19980020

Andrews, M., Gallagher-Allred, C.: The role of zinc in wound healing. Adv. Wound Care: the journal for prevention and healing. **12**(3), 137–138 (1999) Retrieved from https://pubmed.ncbi.nlm.nih.gov/10655793/

Andriollo-Sanchez, M., Hininger-Favier, I., Meunier, N., Toti, E., Zaccaria, M., Brandolini-Bunlon, M., Polito, A., O'Connor, J.M., Ferry, M., Coudray, C., Roussel, A.M.: Zinc intake and status in middle-aged and older European subjects: the ZENITH study. Eur. J. Clin. Nutr. **59**(Suppl 2), S37–S41 (2005). https://doi.org/10.1038/sj.ejcn.1602296

Arsenault, J.E., Brown, K.H.: Zinc intake of US preschool children exceeds new dietary reference intakes. Am. J. Clin. Nutr. **78**(5), 1011–1017 (2003). https://doi.org/10.1093/ajcn/78.5.1011

Banuelos, G.S., Lin, Z.-Q.: Use and Development of Biofortified Agricultural Products. CRC Press (2009) Retrieved from https://books.google.de/books?hl=en&lr=&id=n89nrVZF0GIC&oi=fnd&pg=PP1&ots=cSsuaKnWl1&sig=FVYjvmEsbNRNfLVXz9i-fN2z6g0&redir_esc=y#v=onepage&q&f=false

Barceloux, D.G.: Zinc. J. Toxicol. Clin. Toxicol. **37**(2), 279–292 (1999). https://doi.org/10.1081/clt-100102426

Bayley, N.: Bayley Scales of Infant Development, 3rd edn. The Psychological Corporation, San Antonio, TX (1993)

Bennett, D.R., Baird, C.J., Chan, K.M., Crookes, P.F., Bremner, C.G., Gottlieb, M.M., Naritoku, W.Y.: Zinc toxicity following massive coin ingestion. Am J Forensic Med Pathol. **18**(2), 148–153 (1997). https://doi.org/10.1097/00000433-199706000-00008

Berdanier, C.D., Dwyer, J.T., Heber, D.: Handbook of Nutrition and Food, 3rd edn.). (eISBN 978-1-4665-0572-8), (2016). https://doi.org/10.1201/b15294

Bertenthal, B.I., Campos, J.J.: A systems approach to the organizing effects of self-produced locomotion during infancy. Adv. Infancy Res. **6**, 1–60 (1990)

Bhatnagar, S., Taneja, S.: Zinc and cognitive development. Br. J. Nutr. **85**(Suppl 2), S139–S145 (2001). https://doi.org/10.1079/bjn2000306

Bhatnagar, S., Bhandari, N., Mouli, U.C., Bhan, M.K., IAP National Task Force: Consensus Statement of IAP National Task Force: status report on management of acute diarrhea. Indian Pediatr. **41**(4), 335–348 (2004) Retrieved from https://pubmed.ncbi.nlm.nih.gov/15123862/

Bhutta, Z.A., Black, R.E., Brown, K.H., et al.: Zinc Investigators' Collaborative Group: prevention of diarrhea and pneumonia by zinc supplementation in children in developing countries: pooled analysis of randomized controlled trials. J. Pediatr. **135**(6), 689–697 (1999). https://doi.org/10.1016/s0022-3476(99)70086-7

Black, M.M.: Zinc deficiency and child development. Am. J. Clin. Nutr. **68**(2 Suppl), 464S–469S (1998). https://doi.org/10.1093/ajcn/68.2.464S

Black, M.M.: The evidence linking zinc deficiency with children's cognitive and motor functioning. J. Nutr. **133**(5 Suppl 1),1473S–6S (2003). Retrieved from https://doi.org/10.1093/jn/133.5.1473S

Bode, C., Schäfer, C., Bode, J.C.: The role of gut-derived bacterial toxins (endotoxin) for the development of alcoholic liver disease in man. In: Blum, H.E., Bode, C., Bode, J.C., Sartor, R.B. (eds.) Gut and the Liver (1998) Retrieved from https://scholar.google.com/scholar_lookup?title=The%20role%20of%20gut-derived%20bacterial%20toxins%20%28endotoxin%29%20for%20the%20development%20of%20alcoholic%20liver%20disease%20in%20man&author=C.%20Bode&author=C.%20Sch%C3%A4fer&author=J.%20Bode&pages=281–298&publication_year=1998

Brazelton, T.B., Nugent, J.K.: Neonatal Behavioral Assessment Scale, 3rd edn. Mac Keith Press, London (1995)

Brewer, G.J., Kanzer, S.H., Zimmerman, E.A., Molho, E.S., Celmins, D.F., Heckman, S.M., Dick, R.: Subclinical zinc deficiency in Alzheimer's disease and Parkinson's disease. Am. J. Alzheimers Dis. Other Dement. **25**(7), 572–575 (2010). https://doi.org/10.1177/1533317510382283

Brooks, W.A., Yunus, M., Santosham, M., Wahed, M.A., Nahar, K., Yeasmin, S., Black, R.E.: Zinc for severe pneumonia in very young children: double-blind placebo-controlled trial. Lancet (London, England). **363**(9422), 1683–1688 (2004). https://doi.org/10.1016/S0140-6736(04)16252-1

Brooks-Gunn, J., Duncan, G.J.: The effects of poverty on children. Futur. Child. **7**(2), 55–71 (1997) Retrieved from https://pubmed.ncbi.nlm.nih.gov/9299837/

Brown, K.H., Rivera, J.A., Bhutta, Z., et al.: Assessment of the risk of zinc deficiency in populations and options for its control. Food Nutr. Bull. **25**(1 Suppl 2), S99–S203 (2004) Retrieved from https://pubmed.ncbi.nlm.nih.gov/18046856/

Chandra, R.K.: Excessive intake of zinc impairs immune responses. JAMA. **252**(11), 1443–1446 (1984) https://pubmed.ncbi.nlm.nih.gov/6471270/

Chang, C.H.: Menkes disease. (2002). Retrieved from https://emedicine.medscape.com/article/1180460-overview

David, L., Watts, D.C.: The relationship of zinc. J. Orthomol. Med. **3**(2) (1988) Retrived from http://webcache.googleusercontent.com/search?q=cache:xdxdMRmo5G4J:teikorea.com/03_edu/The%2520Nutritional%2520Relationships%2520of%2520Zinc.pdf+&cd=2&hl=en&ct=clnk&gl=de

Duggan, C., Watkins, J.B., Walker, W.A.: Nutrition in Pediatrics: Basic Science, Clinical Application, 4th edn. BC Decker, Hamilton (2008)

Eby 3rd, G.A.: Zinc lozenges as cure for the common cold--a review and hypothesis. Med. Hypotheses. **74**(3), 482–492 (2010). https://doi.org/10.1016/j.mehy.2009.10.017

Emond, A.M., Howat, P., Evans, J.A., Hunt, L.: The effects of housing on the health of preterm infants. Paediatr. Perinat. Epidemiol. **11**(2), 228–239 (1997). https://doi.org/10.1046/j.1365-3016.1997.d01-15.x

Fighting Child Malnutrition (2012) Policy Paper 55. NAAS, New Delhi, India

Fischer Walker, C., Black, R.E.: Zinc and the risk for infectious disease. Annu. Rev. Nutr. **24**, 255–275 (2004). https://doi.org/10.1146/annurev.nutr.23.011702.073054

Flora, S.J., Pachauri, V.: Chelation in metal intoxication. Int. J. Environ. Res. Public Health. **7**(7), 2745–2788 (2010). https://doi.org/10.3390/ijerph7072745

Fosmire, G.J.: Zinc toxicity. Am. J. Clin. Nutr. **51**(2), 225–227 (1990). https://doi.org/10.1093/ajcn/51.2.225

Fung, E.B., Ritchie, L.D., Woodhouse, L.R., Roehl, R., King, J.C.: Zinc absorption in women during pregnancy and lactation: a longitudinal study. Am. J. Clin. Nutr. **66**(1), 80–88 (1997). https://doi.org/10.1093/ajcn/66.1.80

Galetti, V.: Zinc deficiency and stunting. In: Handbook of Famine, Starvation, and Nutrient Deprivation, pp. 1–23 (2018). 10.1007%2F978-3-319-40007-5_93-1

Gardner, J.M., Grantham-McGregor, S.M.: Physical activity, undernutrition and child development. Proc. Nutr. Soc. **53**(1), 241–248 (1994). https://doi.org/10.1079/pns19940025

Geist, F.C., Bateman, J.A., Hayden, F.G.: In vitro activity of zinc salts against human rhinoviruses. Antimicrob. Agents Chemother. **31**(4), 622–624 (1987). https://doi.org/10.1128/aac.31.4.622

Gibson, R.S., Anderson, V.P.: A review of interventions based on dietary diversification or modification strategies with the potential to enhance intakes of total and absorbable zinc. Food Nutr. Bull. 30(1 Suppl), S108–43 (2009). https://doi.org/10.1177/15648265090301S107

Gogia, S., Sachdev, H.S.: Zinc supplementation for mental and motor development in children. Cochrane Database Syst. Rev. **12** (2012). https://doi.org/10.1002/14651858.CD007991.pub2

Greenberg, M.I., Vearrier, D.: Metal fume fever and polymer fume fever. Clin. Toxicol. (Phila.). **53**(4), 195–203 (2015). https://doi.org/10.3109/15563650.2015.1013548

Gregg, X.T., Reddy, V., Prchal, J.T.: Copper deficiency masquerading as myelodysplastic syndrome. Blood. **100**(4), 1493–1495 (2002). https://doi.org/10.1182/blood-2002-01-0256

Haase, H., Rink, L.: The immune system and the impact of zinc during aging. Immun. Ageing. **6**, 9 (2009). https://doi.org/10.1186/1742-4933-6-9

Hamadani, J.D., Fuchs, G.J., Osendarp, S.J., Huda, S.N., Grantham-McGregor, S.M.: Zinc supplementation during pregnancy and effects on mental development and behaviour of infants: a follow-up study. Lancet (London, England). **360**(9329), 290–294 (2002). https://doi.org/10.1016/S0140-6736(02)09551-X

Hambidge, K.M., Krebs, N.F.: Zinc deficiency: a special challenge. J. Nutr. **137**(4), 1101–1105 (2007). https://doi.org/10.1093/jn/137.4.1101

Hammond, J.W.: Metal fume fever in crushed stone industry. J. Ind. Hyg. Toxicol. **26**(4), 117–119 (1944) Retrieved from https://www.cabdirect.org/cabdirect/abstract/19442701642

Harvest Plus.: Better crops better nutrition [CGIAR Research Program on Agriculture for Nutrition and Health & A4NH]. Better crops, Better nutrition. (2020). Retrieved from https://www.harvestplus.org/what-we-do/nutrition

Hunt, J.R.: Bioavailability of iron, zinc, and other trace minerals from vegetarian diets. Am. J. Clin. Nutr. **78**(3 Suppl), 633S–639S (2003). https://doi.org/10.1093/ajcn/78.3.633S

Institute of Medicine (US) Panel on Micronutrients: Dietary Reference Intakes for Vitamin A, Vitamin K, Arsenic, Boron, Chromium, Copper, Iodine, Iron, Manganese, Molybdenum, Nickel, Silicon, Vanadium, and Zinc. National Academies Press (US) (2001) Retrieved from https://pubmed.ncbi.nlm.nih.gov/25057538/

Keenan, J.M., Morris, D.H.: How to make sure your older patients are getting enough zinc. Geriatrics. **48**(10), 57–65 (1993) Retrieved from https://pubmed.ncbi.nlm.nih.gov/8406052/

Khan, K.N.M., Hard, G.C., Alden, C.L.: Kidney. In: Haschek & Rousseaux (ed.) Handbook of Toxicologic Pathology, 3rd edn, pp. 1667–1773 (2013). https://doi.org/10.1016/B978 012415759-0.00047-9

King, J.C.: Determinants of maternal zinc status during pregnancy. Am. J. Clin. Nutr. **71**(5), 1334S–1343S (2000). https://doi.org/10.1093/ajcn/71.5.1334s

King, J.C., Cousins, R.J.: Zinc. In: Shils, M., Shike, M., Ross, C., Caballero, B., Cousins, R. (eds.) Modern Nutrition in Health and Disease, pp. 271–285. Lippincott Williams & Wilkins, Philadelphia (2006)

King, J.C., Cousins, R.J.: Zinc. In: Ross, A.C., Caballero, B.H., Cousins, R.J., Tucker, K.L., Ziegler, T.R. (eds.) Modern Nutrition in Health and Disease, 11th edn. Wolters Kluwer Health Adis (ESP) (2012)

Kirksey, A., Wachs, T.D., Yunis, F., Srinath, U., Rahmanifar, A., McCabe, G.P., Galal, O.M., Harrison, G.G., Jerome, N.W.: Relation of maternal zinc nutriture to pregnancy outcome and infant development in an Egyptian village. Am. J. Clin. Nutr. **60**(5), 782–792 (1994). https://doi.org/10.1093/ajcn/60.5.782

Krebs, N.F.: Update on zinc deficiency and excess in clinical pediatric practice. Ann. Nutr. Metab. **62**(Suppl 1), 19–29 (2013). https://doi.org/10.1159/000348261

Lansdown, A.B.: Influence of zinc oxide in the closure of open skin wounds. Int. J. Cosmet. Sci. **15**(2), 83–85 (1993). https://doi.org/10.1111/j.1467-2494.1993.tb00072.x

Lasater, M.: Zinc toxicity because of denture adhesive. J. Neurosci. Nurs.: journal of the American Association of Neuroscience Nurses. **49**(1), 23–24 (2017). https://doi.org/10.1097/JNN.0000000000000203

Lonnerdal, B.: Genetically modified plants for improved trace element nutrition. J. Nutr. **133**, 1490S–1493S (2003). https://doi.org/10.1093/jn/133.5.1490S

Maares, M., Haase, H.: A guide to human zinc absorption: general overview and recent advances of in vitro intestinal models. Nutrients. **12**(3), 762 (2020). https://doi.org/10.3390/nu12030762

Mak, T.W., Saunders, M.E.: The immune response: Basic and clinical principles. (2006). Amsterdam: Elsevier/Academic

Marcellini, M., Di Ciommo, V., Callea, F., Devito, R., Comparcola, D., Sartorelli, M.R., Carelli, G., Nobili, V.: Treatment of Wilson's disease with zinc from the time of diagnosis in pediatric patients: a single-hospital, 10-year follow-up study. J. Lab. Clin. Med. **145**(3), 139–143 (2005). https://doi.org/10.1016/j.lab.2005.01.007

Marquart, H., Smid, T., Heederik, D., Visschers, M.: Lung function of welders of zinc-coated mild steel: cross-sectional analysis and changes over five consecutive work shifts. Am. J. Ind. Med. **16**(3), 289–296 (1989). https://doi.org/10.1002/ajim.4700160307

Mizuochi, T., Kimura, A., Shimizu, N., Nishiura, H., Matsushita, M., Yoshino, M.: Zinc monotherapy from time of diagnosis for young pediatric patients with presymptomatic Wilson disease. J. Pediatr. Gastroenterol. Nutr. **53**(4), 365–367 (2011). https://doi.org/10.1097/MPG.0b013e31821d5abe

Müller, O., Becher, H., van Zweeden, A.B., Ye, Y., Diallo, D.A., Konate, A.T., Gbangou, A., Kouyate, B., Garenne, M.: Effect of zinc supplementation on malaria and other causes of morbidity in west African children: randomised double blind placebo controlled trial. BMJ (Clinical research ed.). **322**(7302), 1567 (2001). https://doi.org/10.1136/bmj.322.7302.1567

Murray, B.P., Ralston, S.A., Dunkley, C.A., Carpenter, J.E., Geller, R.J., Kazzi, Z.: Pneumonitis and respiratory failure secondary to civilian exposure to a smoke bomb in a partially enclosed space. J. Spec. Oper. Med: a peer reviewed journal for SOF medical professionals. **18**(4), 24–26 (2018) Retrieved from https://europepmc.org/article/med/30566719

Nations, S.P., Boyer, P.J., Love, L.A., Burritt, M.F., et al.: Denture cream: an unusual source of excess zinc, leading to hypocupremia and neurologic disease. Neurology. **71**(9), 639–643 (2008). https://doi.org/10.1212/01.wnl.0000312375.79881.94

Neumann, C., McDonald, M.A., Sigman, M., Bwibo, N., Marquardt, M.: Relationships between morbidity and development in mildly to moderately malnourished Kenyan toddlers. Pediatrics. **88**(5), 934–942 (1991) https://pubmed.ncbi.nlm.nih.gov/1945634/

O'Brien, K.O., Zavaleta, N., Caulfield, L.E., Wen, J., Abrams, S.A.: Prenatal iron supplements impair zinc absorption in pregnant Peruvian women. J. Nutr. **130**(9), 2251–2255 (2000). https://doi.org/10.1093/jn/130.9.2251

Patterson, W.P., Winkelmann, M., Perry, M.C.: Zinc-induced copper deficiency: megamineral sideroblastic anemia. Ann. Intern. Med. **103**(3), 385–386 (1985). https://doi.org/10.7326/0003-4819-103-3-385

Pawlitzki, M., Uebelhör, J., Sweeney-Reed, C.M., Stephanik, H., Hoffmann, J., Lux, A., Reinhold, D.: Lower serum zinc levels in patients with multiple sclerosis compared to healthy controls. Nutrients. **10**(8), 967 (2018). https://doi.org/10.3390/nu10080967

Pitchik, H.O., Fawzi, W.W., McCoy, D.C., Darling, A.M., Abioye, A.I., Tesha, F., Smith, E.R., Mugusi, F., Sudfeld, C.R.: Prenatal nutrition, stimulation, and exposure to punishment are associated with early child motor, cognitive, language, and socioemotional development in Dar es Salaam, Tanzania. Child Care Health Dev. **44**(6), 841–849 (2018). https://doi.org/10.1111/cch.12605

Pluhator, M.M., Thomson, A.B.R., Fedorak, R.N.: Clinical aspects of trace elements: zinc in human nutrition – zinc metabolism. Can. J. Gastroenterol. Hepatol. **9** (1995). https://doi.org/10.1155/1995/745794

Plum, L.M., Rink, L., Haase, H.: The essential toxin: impact of zinc on human health. Int. J. Environ. Res. Public Health. **7**(4), 1342–1365 (2010). https://doi.org/10.3390/ijerph7041342

Prasad, A.S. Clinical manifestations of zinc deficiency. Annu. Rev. Nutr. **5**, 341–363 (1985)

Prasad, R.: Protein-energy malnutrition and fertilizer use in India. Fertilizer News. **48**(8), 13–26 (2003). Retrieved from https://www.pphouse.org/upload_article/6_IJBSM_March_2014_Shivay_et_al.pdf

Prasad, A.S.: Impact of the discovery of human zinc deficiency on health. J. Am. Coll. Nutr. **28**(3), 257–265 (2013). https://doi.org/10.1080/07315724.2009.10719780

Prentice, A.: Does mild zinc deficiency contribute to poor growth performance? Nutr. Rev. **51**(9), 268–270 (1993). https://doi.org/10.1111/j.1753-4887.1993.tb03118.x

Raulin, J.: "Chemical studies on vegetation". *Annales des Sciences Naturelles*,11, 293–299. In: Hughes, D.A., Darlington, L.G., Bendich, A. (eds.) Diet and Human Immune Function. Nutrition and Health (1869). https://doi.org/10.1007/978-1-59259-652-2_13

Roohani, N., Hurrell, R., Kelishadi, R., Schulin, R.: Zinc and its importance for human health: an integrative review. Journal of research in medical sciences : the official journal of Isfahan University of Medical Sciences, 18(2), 144–157 (2013). https://doi.org/10.1146/annurev.nu.05.070185.002013

Sandstead, H.H.: Human zinc deficiency: discovery to initial translation. Adv. Nutr. **4**(1), 76–81 (2013). https://doi.org/10.3945/an.112.003186

Sauer, A.K., Hagmeyer, S., Grabrucker, A.M.: Zinc deficiency, nutritional deficiency. Open access peer reviewed chapter. (2016). https://doi.org/10.5772/63203

Sazawal, S., Black, R.E., Ramsan, M., Chwaya, H.M., Dutta, A., Dhingra, U., Stoltzfus, R.J., Othman, M.K., Kabole, F.M.: Effect of zinc supplementation on mortality in children aged 1–48 months: a community-based randomised placebo-controlled trial. Lancet (London, England). **369**(9565), 927–934 (2007). https://doi.org/10.1016/S0140-6736(07)60452-8

Schubert, C., Guttek, K., Reinhold, A., Grüngreiff, K., Reinhold, D.: The influence of the trace element zinc on the immune system. J. Lab. Med. **39**(s1), 000010151520150060 (2015). https://doi.org/10.1515/labmed-2015-0060

Science, M., Johnstone, J., Roth, D.E., Guyatt, G., Loeb, M.: Zinc for the treatment of the common cold: a systematic review and meta-analysis of randomized controlled trials. CMAJ: Can. Med. Assoc. J = journal de l'Association medicale canadienne. **184**(10), E551–E561 (2012). https://doi.org/10.1503/cmaj.111990

Sen Ray, K.: Zinc under-nutrition in India. Vitam. Miner. **5**(2) (2016). https://doi.org/10.4172/2376-1318.1000e148

Simeon, D.T., Grantham-McGregor, S.M.: Nutritional deficiencies and children's behaviour and mental development. Nutr. Res. Rev. **3**(1), 1–24 (1990). https://doi.org/10.1079/NRR19900004

Singh, M., Das, R.R.: Zinc for the common cold. Cochrane Database Syst. Rev. (2), CD001364 (2011). https://doi.org/10.1002/14651858.CD001364.pub3

Singh, K.B., Taneja, S.K.: Hazard effects of excess of zinc in diet. Sci. Vis. **9**(4), 159–165 (2009) Retrieved from https://www.sciencevision.org/storage/journal-articles/February2019/PmX4dl8G18LhESDfMzhp.pdf

Smith, A.P.: Respiratory virus infections and performance. Philos. Trans. R. Soc. Lond. Ser. B Biol. Sci. **327**(1241), 519–528 (1990). https://doi.org/10.1098/rstb.1990.0095

Spanierman, C.: Toxicity, iron. (2004). Retrieved from http://www.healthofchildren.com/M/Mineral-Toxicity.html#ixzz6S9rRbxa2

Sturgis, C.C., Drinker, P., Thompson, R.M.: Metal fume fever: I. Clinical observations on the effect of the experimental inhalation of zinc oxide by two apparently normal persons. J. Ind. Hyg. **9**, 88–97 (1927) Retrieved from https://www.cabdirect.org/cabdirect/abstract/19272701395

Subramaniam, V.D., Prasad, S.V., Banerjee, A., Gopinath, M., Murugesan, R., Marotta, F., Sun, X.F., Pathak, S.: Health hazards of nanoparticles: understanding the toxicity mechanism of nanosized ZnO in cosmetic products. Drug Chem. Toxicol. **42**(1), 84–93 (2019). https://doi.org/10.1080/01480545.2018.1491987

Tamura, T., Goldenberg, R.L., Ramey, S.L., Nelson, K.G., Chapman, V.R.: Effect of zinc supplementation of pregnant women on the mental and psychomotor development of their children at 5 y of age. Am. J. Clin. Nutr. **77**(6), 1512–1516 (2003). https://doi.org/10.1093/ajcn/77.6.1512

Taneja, S.K., Singh, K.B.: Beneficial effects of modified egg* on oxidative stress in F1- generation of metabolic syndrome-X induced Wistar rat. Indian J. Exp. Biol. **47**(2), 104–112 (2009) https://pubmed.ncbi.nlm.nih.gov/19374165/

The National Institute of Agricultural Sciences.: Fighting Child Malnutrition. (2012). Retrieved from http://www.esocialsciences.org/Download/Download.aspx?fname=A20141016155138_35.pdf&fcategory=Articles&aid=6180

Tubek, S.: Role of zinc in regulation of arterial blood pressure and in the etiopathogenesis of arterial hypertension. Biol. Trace Elem. Res. **117**, 39–51 (2007). https://doi.org/10.1007/BF02698082

Tuerk, M.J., Fazel, N.: Zinc deficiency. Curr. Opin. Gastroenterol. **25**(2), 136–143 (2009). https://doi.org/10.1097/MOG.0b013e328321b395

United News of Bangladesh.: Zinc: Health Benefits, Food sources, and Daily Requirements (2020). Retrieved from https://unb.com.bd/category/Lifestyle/zinc-health-benefits-food-sources-and-daily-requirements/54349

Uwitonze, A.M., Ojeh, N., Murererehe, J., Atfi, A., Razzaque, M.S.: Zinc adequacy is essential for the maintenance of optimal oral health. Nutrients. **12**(4), 949 (2020). https://doi.org/10.3390/nu12040949

Vogelmeier, C., König, G., Bencze, K., Fruhmann, G.: Pulmonary involvement in zinc fume fever. Chest. **92**(5), 946–948 (1987). https://doi.org/10.1378/chest.92.5.946

Vuralli, D., Tumer, L., Hasanoglu, A.: Zinc deficiency in the pediatric age group is common but underevaluated. World J. Pediatr. **13**(4), 360–366 (2017). https://doi.org/10.1007/s12519-017-0007-8

Watanabe, T., Arakawa, T., Fukuda, T., Higuchi, K., Kobayashi, K.: Zinc deficiency delays gastric ulcer healing in rats. Dig. Dis. Sci. **40**(6), 1340–1344 (1995). https://doi.org/10.1007/BF02065548

Wessells, K.R., Brown, K.H.: Estimating the global prevalence of zinc deficiency: results based on zinc availability in national food supplies and the prevalence of stunting. PLoS One. **7**(11), e50568 (2012). https://doi.org/10.1371/journal.pone.0050568

Wiggelinkhuizen, M., Tilanus, M.E., Bollen, C.W., Houwen, R.H.: Systematic review: clinical efficacy of chelator agents and zinc in the initial treatment of Wilson disease. Aliment. Pharmacol. Ther. **29**(9), 947–958 (2009). https://doi.org/10.1111/j.1365-2036.2009.03959.x

World Health Report.: Reducing risks, promoting healthy life. (2002). Retrieved from https://apps.who.int/iris/bitstream/handle/10665/42510/WHR_2002.pdf

World Health Organization and United Nations International Children's Emergency Fund.: Diarrhoea: Why children are still dying and what can be done. (2009). Retrieved from https://www.unicef.org/media/files/Final_Diarrhoea_Report_October_2009_final.pdf

World Health Report.: Zinc save kids: an International Zinc Association (IZA) initiative in support of UNICEF (2012). Retrieved from https://www.zinc.org/wpcontent/uploads/sites/24/2015/01/zn.broch-02-12-rev2_final_030212.pdf. https://www.who.int/whr/2002/annex/en/

Yanagisawa, H.: Zinc deficiency and Clinical practice. Japan Med. Assoc. **47**(8), 359–364 (2004). Retrieved from https://www.med.or.jp/english/pdf/2004_08/359_364.pdf

Zoroddu, M.A., Aaseth, J., Crisponi, G., Medici, S., Peana, M., Nurchi, V.M.: The essential metals for humans: a brief overview. J. Inorg. Biochem. **195**, 120–129 (2019). https://doi.org/10.1016/j.jinorgbio.2019.03.013

Chapter 3
Zinc Deficiency and Its Consequences During Pregnancy

Anisa M. Durrani and Heena Parveen

Abstract Zinc is an important microelement that promotes immunity, proper development and growth of the nervous system and resistance to infection. It is a constituent of more than 200 enzymes involved in digestion, metabolism, synthesis of RNA and DNA, reproduction and is also essential for healthy pregnancy outcomes. An estimated 17.3% of global population is at risk for zinc deficiency due to inadequate dietary supply and bioavailability. Inadequate intake of zinc during pregnancy has been related to adverse pregnancy outcomes including abortion, foetal neural tube defects, membrane rupture, preterm delivery, prolonged labour and stillbirth and risk of maternal and infant mortality. Other risk factors of zinc deficiency are diarrheal disease, malabsorption syndromes, parasitosis, hot and humid climate, rapid multiplicative cell growth (pregnancy, infancy, and adolescence), lactation and genetic disease. Zinc deficiency is primarily related to the poor zinc status of the diet. Animal protein-based diet is a rich source of zinc, whereas vegetable and cereal-based diet is not only poor in zinc content but also interferes with its absorption due to high content of phytates. Zinc supplementation during pregnancy and lactation plays a role in improving both maternal and foetal health. Zinc supplementation reduces the frequency of preterm birth by 14%, decreases respiratory infections and childhood diarrhoea, lowers all-cause mortality and increases growth and weight gain among infants and young children. While there are no separate programmes for zinc supplementation in India, zinc tablets are provided along with oral rehydration solution (ORS) for treatment of diarrhoea as per WHO guidelines. More research is required to assess the benefits of the large-scale zinc intervention programmes during pregnancy. This chapter reviews the importance of zinc in pregnancy and its effect on foetal and infant growth.

Keywords Deficiency of zinc · Gestation · Associated risk factors · Supplements of zinc

A. M. Durrani (✉) · H. Parveen
Department of Home Science, Aligarh Muslim University, Aligarh, UP, India

S. T. Khan, A. Malik (eds.), *Microbial Biofertilizers and Micronutrient Availability*,
https://doi.org/10.1007/978-3-030-76609-2_3

Introduction

Zinc is a required mineral and very important for various genetic functions such as cellular dissection, synthesis of protein and metabolism of nucleic acid (King 2000). Because of its fundamental role in gene expression, replication and cell development, it is also known as an essential trace mineral to all forms of life (Hambidge 2000). Zinc also preserves the intracellular organelles, stabilises the protein structure and plays a role in immunological responses and the transfusion (Burits and Ashwood 1994). Facts are submitted that in the developed and developing countries, deficiency of zinc is one of the vital problems. A report published in 1991 highlighted that in some countries such as China, Iran, Egypt, Turkey, Canada and Yugoslavia, zinc deficiency was prevalent because of the low intake of red meat and high intake of fibre (W.B Sanders and Sandstead 1991). In 1961 in Iran and in 1963 in Egypt, the deficiency of zinc in human was stated for the first time (Prasad 1961, 1963). Growth stunting, cognitive dysfunction, skin disorders, impaired immune function, hypogonadism, and anorexia were reported due to clinical deficiency of zinc (Prasad 1991). Although mild-to-moderate zinc deficiency is prevalent throughout the world, severe zinc deficiency is rare (Sandstead 1991). Despite this, randomised controlled traits of zinc supplementation have begun to demonstrate how low dietary intake of zinc adversely affects child health in low zinc intake areas (Brown et al. 1998; Zinc Investigators' Collaborative Group 1999, 2000). The goal of this chapter is to describe the deficiency of zinc and its consequences during pregnancy. Severe deficiency of zinc during gestation has a difficult effect on pregnancy outcome. Researches on humans and trial beasts indicate that zinc deficiency among mothers may be reason for infertility, prolonged labour, teratogenesis, retardation of intrauterine growth and death of embryo.

Consequences of Zinc Deficiency During Pregnancy

The Severe deficiency of zinc is rare in individuals, but mild to moderate inadequacy is very common, particularly in populations with low intake of zincrich animal-sources, high intake of phytate rich sources, prevents the absorption of zinc in the intestine (Chaffee and King 2012). Zinc deficiency in humans primarily results from poor absorption, decreased dietary intake, increased zinc loss or increased zinc requirements of the body. However, insufficient dietary intake is the most common cause. In the United States, the RDA (Recommended Dietary Allowance) for zinc is 8 mg/day for women and 11 mg/day for men (Bales and Ritchie 2009). Nearly, 95% population of the South East Asia suffered from zinc deficiency as a major public health problem due to low dietary zinc intake (Brown et al. 2001). As the expectant mothers has more risk of zinc insufficiency which may affect the health of both mother and the child. High zinc requirement during pregnancy is due to the high growth rate of the foetus as zinc plays an important role in DNA translation, transcription and protein synthesis required for rapid growth (Black and Sazawal 2001; Walsh et al. 1994). Low dietary intake therefore exposes both mother and the foetus to zinc deficiency (Raman & Shatrugna 2001). Following pregnancy, also inadequate dietary intake of zinc might be a high risk for both the women and the foetus (Yasodhara 1996; Raman & Shatrugna 2001). It has been advised that lack of zinc in mothers may lead to deprived birth outcomes. Plasma with low zinc concentrations reduce transportation of placental zinc that affects the zinc supply to the foetus (King 2000). The level of circulation and number of hormones associated with the onset of labour also alters by zinc deficiency, and zinc deficiency may result in intrauterine infections, which is a major cause of pre-term birth (Chaffee and King 2012). Zinc insufficiency during pregnancy can adversely affect both the foetus and mother and thus complicate natal outcomes. Main complications related to zinc deficiency are increased morbidity in mothers, eclampsia, defects of neural tube in foetus, membrane rupture, low birth weight, reduced neurobehavioural growth, cerebral weakening, intrauterine development obstruction, decreased glucose tolerance, abortions, congenital defects, preterm labour, postnatal haemorrhage, prolonged or no progressed labour and delivery, inept uterine contraction, anomalies in gene replication, late immune system development which leads to increased foetal morbidity and infant death rate (Brown et al. 2004; Karimi et al. 2012).

Pregnancy is related with increased metabolic demands and physiological changes that increased plasma volume of blood, reduced nutrient circulation concentrations proteins binding and low bio-absorption of micronutrients. All these situations are observable among expecting women (Tamura et al. 2000; Brown et al. 2004). Amended nutritional requirements during gestation influence expectant women to develop zinc deficiency. As zinc requirement is generally high for adequate foetal development, expecting women are more vulnerable to zinc deficiency, mostly in second and third trimesters (Karimi et al. 2012).

Zinc deficiency has also serious health outcomes including the immune system disorder and a high prevalence of childhood infections such as pneumonia and diarrhoea, impaired maternal health and deprived pregnancy consequence (Shah and Sachdev 2001; Caulfield and Black 2014). The expectant mothers in developing countries are vulnerable to various micronutrient insufficiencies with zinc. Various studies have also shown that animal food sources, haemoglobin level, gestational age (trimester), short birth interval and serum albumin level were recognised as risk factors of zinc scarcity during pregnancy. The difference in demographic and socioeconomic factors and nutritional status earlier than and during pregnancy advanced age, lactation, are also reported. Alcoholism, high consumption of iron, and folic acid are known to raise the possibility of zinc insufficiency (Caulfield et al. 1998; Brown et al. 2001). Many studies have revealed that the growth of foetus in zinc shortage ends up with disorders of central nervous systems (Shah and Sachdev 2001).

The mother and foetus would put at risk because of dietary factors that prevent zinc absorption, or any gastrointestinal infection that delays the capacity of zinc absorption. Through the plasma, zinc is transmitted to the foetal-placenta. Therefore, any factor that changes the transport function could also put the gestation at risk.

Circumstances that affect the absorption and transportation of zinc through plasma cause zinc deficiency and are discussed below briefly.

(i) *Dietary Phytate*

The zinc status of the diet depends on both the bioavailability of the zinc and its content in the diet. Phytate is one of the greatest inhibitory factors that affect zinc bioavailability. The zinc to phytate ratio is often used as an indicator of zinc bioavailability in the diet (WHO 1996). There is some disagreement about the exact zinc to phytate ratio which can be defined as the cut-off value, some have used 15 as the cut-off value and others have used 20 (Fitzgerald et al. 1993). WHO recommends another procedure for valuing zinc bioavailability which is created by zinc consumption and its obtainability feature. The accessibility factor, i.e. the percentage of zinc available in diet, is expected to range from 10 to 50% depending on the diet's phytate-zinc molar ratio. Obtainability of estimated zinc is to be 10% if the phytate-zinc ratio is more than 30, 15% if the ratio is 15–30, 30% if the ratio is 5–15, and more than 50% if the ratio is higher than 15 as in unprocessed, ungerminated, unfermented cereal grain, particularly when the zinc stimulated with inorganic calcium salts and when consumption of protein from animal sources is minor. In developing countries, several poor women consumed such diets. A high-phytate diet consumption through gestation can restrict the zinc absorption in pregnant women and can lead to a secondary insufficiency of zinc. The association of zinc nutriture among mothers and pregnancy outcome in a group of females living in a village of Egypt were studied. The zinc intake of these women was not particularly low. It was related to a better neonatal habituation behaviour, a measure of early information processing (Kirksey et al. 1994).

A current research in Malawi revealed that the zinc status of pregnant women is reduced by malarial infections and high intake of phytate along with frequent pregnancies (Gibson and Huddle 1998). The average ratio of phytate-zinc was 17 in that population. Zinc concentrations of plasma in women at 24 weeks of pregnancy with a below average of phytate-zinc ratio did not vary from an above average ratio in women. The average value of plasma zinc was 7.9 mmol/L. It is projected that 37% of women had values below the cut-off value particularly in gestational stages. The women which had molar ratios of phytate-zinc more than the medium were knowingly elder, had additional gestations and had considerably lesser concentrations of hair zinc. Mineral analysis of hair is an important nutritional balancing assessment method. The insufficiency of zinc is defined as zinc level of hair below 16 mg/DL. Thus, hair zinc is a more stable indicator than the plasma zinc which reveals chronic zinc status over the growth period of hair. An acute index of zinc status is plasma zinc; during pregnancy zinc status is hard to infer because in this period there is a deterioration in plasma zinc concentration that rises in the first trimester and remains all over the gestation. These data show that the absorption of zinc limits and maternal zinc status declines by chronic intake of diets with high phytate to zinc molar ratios. It is remarkable that, the plasma zinc concentrations and dietary phytate zinc ratios was no associated among those women. Practically, preserve to plasma zinc concentrations in the occurrence of low zinc absorption followed by additional compensatory changes in zinc metabolism. Furthermore, it may be more difficult to detect the relation in plasma zinc concentration and phytate intake because of the various factors besides zinc status that effect the zinc of plasma, containing fasting, diurnal variation, disease state, sample refrigeration conditions and interval of time before separating the plasma from cells. The availability of zinc may decrease by high intake of calcium in the existence of high phytate. In a women group of Guatemala, the intake of zinc was equivalent to that of healthy North American populations consuming omnivorous diets. However, the diet of Guatemalan's women is largely plant-based and has high amounts of calcium, phytate and fibre. Though the acute ratio of phytate, calcium to zinc which decreases zinc bioavailability has not been defined. It has been recommended that values >22/MJ (mega Joule), to account for differences in the quantity of food consumed, may result in marginal zinc deficiency. The ratios of periurban Guatemalan women during the third trimester of pregnancy averaged 40/MJ (Fitzgerald et al. 1993). More researches on zinc status of pregnant women, pregnancy outcome and physiologic adjustments in zinc metabolism in populations that consume diets with low zinc obtainability are essential to describe onsets for supporting good pregnancy outcomes.

(ii) *Iron Supplementation*

Basically, around the world, supplemental iron is given to all pregnant women who are receiving prenatal care. In the United States, maximum pregnant women take prenatal supplements of vitamin-mineral. Though the American College of Obstetrics and Gynaecology and Institute of Medicine mention 30 mg Fe/d for all expectant women, commercial prenatal supplements of vitamin-mineral have an average of 54 mg Fe; in supplements the concentration of iron ranges from 36 to 65 mg (PDR Electronic Library 1997). Furthermore, in developing countries, the World Health Organization mentioned 60–120 mg zinc/day for women, also recommended the folic acid supplementation, but zinc supplementation zinc is not. Hence, in developing countries, women usually consume 100 mg supplemental iron without any additional zinc during their pregnancies. As stated above, the dietary zinc consumptions of these women remain naturally low and the zinc is poorly accessible (Viteri 1997). In gastrointestinal absorption, interaction occurs among zinc and iron. The toxic influence of iron supplementation on zinc absorption and a reduction in the plasma zinc concentrations were revealed in researches of investigational animals and humans (Dawson et al. 1989; Solomons and Ruz 1997). The arrangement of intraluminal and intracellular effects probably involved the mechanisms for this iron-zinc interaction (Solomons et al. 1983). In the absorption process, iron and zinc may participate jointly, competing for pathways through the mucosal cell into the blood stream, transferring one another on the molecule essential for their acceptance from the lumen into the intestine, interacting with one another and a third substance to form an insoluble complex and spoiling the absorption of both minerals. Researches assess the outcome of iron fortification in weaning foods but fail to show an effect on absorption of zinc (Solomons and Ruz 1997). Perhaps, the variance in the effects of iron-fortified food compared with iron supplementation are because of the variances in the dose and form of iron given. The supplements provided a greater amount of iron than that is present in one serving of fortified cereal or formula and is in a highly soluble form (ferrous sulphate), facilitating the iron to readily fix with ligands in the intestine (Davidsson et al. 1995).

(iii) *Smoking*

The opposing effect of smoking on foetal growth are relative to the frequency of smoking. When all factors affecting birth weight were organised for, infants born to mothers who smoked constantly weighed less than did infants who were born to non-smoking mothers. Further reducing foetal growth, experience of cigarette smoke in uterus rises the risk of perinatal mortality, preterm delivery and spontaneous abortion. Smoke of cigarette has >2000 diverse compounds. The mechanism is not known through which exposure to cigarette smoke decreases uterine growth, but two commonly cited mechanisms are intrauterine hypoxia and reduced uteroplacental blood flow (Lonnerdal et al. 1990).

The foetal growth may also reduce by a secondary zinc deficiency. Many researches have shown that zinc is trapped in the smoker's placentas, leading to a less zinc transfer to the foetus and reduced growth (Kuhnert et al. 1992). Moreover, the zinc status of infants born to non-smoker mothers is higher than that of infants born to smoker mothers. The smoker mother's infants had considerably lower plasma zinc concentrations, cord vein erythrocyte zinc and alkaline phosphatase. Probably, tobacco smoke produces cadmium that may induce synthesis of higher concentrations of metallothionein in maternal tissues, such as liver, and reduce the zinc availability in the foetus. The data show that placental cadmium and zinc was not associated with birth weight (Kuhnert et al. 1987a, b). The function of alpha-2-macroglobulin is to bind to a variety of compounds, like lectins, proteolytic enzymes, lymphokines and ions such as nickel and zinc. In a study, 289 women were included who had serum samples drawn at 18 and 30 weeks of gestation, an association was found between high alpha-2-macroglobulin concentration as early as 18 weeks of pregnancy and significant rise in foetal growth obstruction. In smokers, influence was greater and 1.0 g/L alpha-2-macroglobulin increase in serum was associated with a 227 g decrease in birth weight. Albumin and alpha-2-macroglobulin are two major zinc binding proteins present in the plasma, and alpha 2-macroglobulin has a higher binding constant for zinc, constructing zinc bound to that protein, making it less obtainable than that of albumin-bound zinc (Goldenberg et al. 1991). Concentrations of plasma zinc did not vary amongst the smokers and non-smokers. Although smoke of tobacco can change the zinc metabolism and decrease its availability to the foetus, appropriate dietary intake of zinc can overcome these effects. However, most researches revealed that the quantity of nutrients and the quality of the diet intakes are low in smokers than in non-smokers.

(iv) *Alcohol Abuse*

It has been theorised that deficiency of zinc is a teratogen by alcohol. Alcohol is an effective teratogen; however, the exact mechanism by which alcohol intake harmfully affects the growth and morphogenesis of foetus is not known. Urinary loss of zinc increases with alcohol use, and zinc deficiency is also known as a potent teratogen. Many studies on rats have shown that the deficiency of zinc with intake of alcohol had additional severe effects on growth of foetal than did either low zinc intakes or alcohol only. The weight of foetus was lower in the group who fed diet deficient in zinc by either 15% or 20% of energy from alcohol. Moreover, the alcohol blend with a Zn-deficient diet improved the amount of reabsorption and exterior deformities more than the diet which is deficient in zinc or ethanol. The results suggest that alcohol reduces the placental transport and may also reduce the intestinal absorption of zinc (Antonson and Vanderhoof 1983). There are rare researches on the effect of alcohol on maternal zinc status; one study compared the zinc status among pregnant women from which 25 women was alcoholic and 25 women was non-alcoholic; the umbilical cord blood zinc concentrations were significantly lesser in the alcoholic women than in the non-alcoholic women. The birth defect cases were correlated with plasma zinc concentrations of mothers but not associated with cord blood concentrations (Flynn et al. 1981). It has been suggested that zinc supplementation is a way to prevent the adverse effects of alcohol. But, a research on rats failed to show that zinc supplementation increased to zinc placental transport in presence of a diet with high ethanol (Ghishan and Greene 1983).

(v) *Acute Maternal Stress*

Metallothionein, an intracellular protein, is known to bind with copper and zinc. An acute level response brings a secondary Zn deficiency in pregnant animals by stimulating the synthesis of metallothionein due to a number of various drugs and environmental challenges. Various organic compounds like endogenous hormones (adrenocorticosteroids) and cytokines can also encourage the synthesis of metallothionein. While the divalent cations are classic indicators of its. As a result of its synthesis, zinc is isolated in the liver and plasma zinc concentrations drop. Decrease in plasma zinc concentrations reduces the availability of zinc to the foetus and could signify a developmental risk. During mid-gestation, many developmental toxins given to experimental animals elevated maternal liver metallothionein and concentrations of zinc and reduced maternal plasma zinc concentrations and embryonic zinc uptake (Keen et al. 1993). For example, maternal liver zinc intake increased by 8% and the plasma and embryonic content reduced by 38% and 57%, respectively, through administration of tumour necrosis factor to a pregnant rat group on eighth day of gestation. These researches recommend, any state that rises the release of cytokines for a chronic period of time can decrease the accessibility of zinc to the foetus and causes developmental defects. Strenuous exercise throughout gestation leads to an increase in ejection of cytokine and associated with a reduction in foetal birth weight. Probably, the reduced birth weight of foetus is because of zinc hepatic isolating and a reduction in plasma zinc concentrations.

Prevalence of Zinc Deficiency in Pregnancy

Projected worldwide, above 82% of women during gestation have insufficient intake of zinc (Caulfield et al. 1998). Globally, in Ethiopia the prevalence of zinc deficiency among pregnant women was 55.3% (Mekonnen et al. 2018), in India 64–72% (Pathak et al. 2008), in Sidama 53% (Gebremedhin et al. 2011), in Gondar 57.4% (Kumera et al. 2015), in China 51.5% (Ma et al. 2004), in Nigeria 97.2% (Oguizu AD 2015), in Kenya 66.9% (Mitheko 2013), in Turkey 72.3% (Açkurt 1995) in Pakistan 27.8% (Lutfullah 2014), in Sudan 45% (Shommo et al. 2013), in Malawi 46% (Gibson and Huddle 1998), in Iran 29% and 49% (Roshanravan et al. 2015), in Vietnam 29% (Nguyen et al. 2013) and in Bangladesh 14.7% (Shamin et al. 2013). Several researches involved different population by culture and geographical area which may contribute to the differences in the incidence of zinc deficiency during pregnancy (Severi et al. 2013). Worldwide, the frequency of zinc insufficiency is unknown amongst expecting and lactating women (King and Turnlund 1989). Furthermore, more studies are required to expose that potential details could be similar in zinc levels of soil and feeding style.

Dietary Zinc Requirements

In children and infants during rapid growth periods and in women during pregnancy and lactation, the requirements of zinc become advanced. In the third trimester of gestation, the physiological necessity of zinc is around twice as high as that in non-pregnant women. It is considered that up to 80% of zinc from mother's milk was absorbed by breastfed infants.

Diets, which has a high obtainability of zinc (50%), are characterised by high protein content (in fish and meats) and low phytate content (in cereal fibre). The daily dietary zinc requirement is shown in Table 3.1. In general, daily requirement ranges between 3.5 and 5 mg per day based on World Health Organization/Food and Agricultural Organization evaluation for obtainability of zinc in three types of diet.

Dietary Sources of Zinc

The protein content and the zinc content of food are interrelated. The animal products such as red meat, oysters, liver, shellfish and poultry are the richest dietary sources of zinc. Beans, lentils, chickpeas and peas are plant protein sources which are relatively zinc rich. The other rich sources are roasted peanuts, dried pumpkin seeds, dried sunflower seeds, dried sesame and unsalted cashew nuts. The zinc is contained in the bran and germ portions of whole cereal grains, which are usually gone in milling process. The zinc content of some of the food sources are shown in Table 3.2. Even in developed countries and urban population of developing

Table 3.1 Dietary zinc requirement

Recommended nutrient intakes (RNIs) for dietary zinc (mg/day) to meet the requirements of normative storage for different population groups from diets different in zinc bioavailability:

Population group	Assumed body weight (kg)	High bioavailability	Moderate bioavailability	Low bioavailability
Infants and children				
0–6 months	6	1.1[a]	2.8[b]	6.6[b]
7–12 months	9	0.8[a], 2.5[b]	4.1	8.4
1–3 years	12	2.4	4.1	8.3
4–6 years	17	12.9	4.8	9.6
7–9 years	25	3.3	5.6	11.2
Adolescents				
Females, 10–18 years	47	4.3	7.2	14.4
Males, 10–18 years	49	5.1	8.6	17.1
Pregnant women				
First trimester	–	3.4	5.5	11.0
Second trimester	–	4.2	7.0	14.0
Third trimester	–	6.0	10.0	20.0
Lactating women				
0–3 months	–	5.8	9.5	
3–6 months	–	5.3	8.8	
6–12 months	–	4.3	7.2	

Source: WHO/FAO/IAEA (International Atomic Energy Agency), 1996

[a]Infant fed exclusively human milk

[b]Formula-fed infant

Table 3.2 Some common foods and their zinc content

Category of zinc (mg/1000 kcl)	Foodstuffs
Very-poor (0–2)	Alcoholic drinks, butter, cheese, cream, chocolates, fats, sweets, oils, soft drinks and preserves
Poor (1–5)	Chips, sausages, fruits, pastries, fish, refined cereal products, biscuits, puddings, cakes and tube plants
Rich (4–12)	Egg, poultry, low fat cheese milk, nuts whole grains and yogurt
Very rich- (12–882)	Beef, lamb, leafy and root vegetables, heart, kidney and liver.

Source: H. E. Zinc deficiency, DNA damage and cancer risk. J Nutr Biochem. 2004;15(10):572–578

countries (non-vegetarian population), the consumption of zinc is commonly inadequate because of the high consumption of refined breads, soft drinks, sweets, potato chips and cakes, which are poor sources of zinc.

The body generally absorbs nearly 15–30% of dietary zinc, and bioavailability of zinc in foods of animal origin is more than vegetable. Human and animal researches revealed that the plant foods that contain phytate can prevent the absorption of zinc. Furthermore, the plant foods that contain fibre, tannin oxalate and lignin can also decrease the zinc absorption. The high amount of these factors is present in legumes, leafy vegetables and whole grains. Cooking processes of plant foods such as soaking and germination of plant foods can reduce the phytic acid content. The zinc stored in body tissues does not function as zinc reserves, so the dietary intake can only ensure the daily requirement of zinc.

Zinc Supplementation During Pregnancy

Supplementation of zinc in gestation may play a role in enlightening the health of both mother and child and can decrease the risk of preterm births. While there are no separate programmes for zinc supplementation in India, zinc tablets are provided alongside ORS (oral rehydration solution) for diarrhoea treatment as per WHO (World Health Organization) protocol. Studies have shown that the use of zinc supplements by expectant mother increases the birth weight of newborn and decreased the child mortality rate (Caulfield and Black 2014).

WHO (World Health Organization), IZiNCG (International Zinc Nutrition Consultative Group), IAEA (International Atomic Energy Agency), WHO and UNICEF (United Nations Children's Fund) approved that under 5 years of age children the growth stunting as an indicator of zinc deficiency. It is projected that about one-third population of the world lives in countries which have a high risk of zinc deficiency (de Benoist et al. 2007). In gestation and lactation, the requirements of zinc have been assessed from the zinc content of tissues increased in gestation and the zinc content of milk produced during lactation (King and Turnlund 1989). In addition to the zinc accumulated by the foetus, amniotic fluid, zinc is deposited in the placenta, and mammary and uterine tissue. Approximately, 40% of the total zinc is accumulated in the maternal tissue and 60% in the concepts. During gestation, the additional daily need of zinc increases to fulfil the demands for foetal growth, increasing from ~0.1 mg/day additional zinc in the first trimester of pregnancy to ~0.7 mg/day additional zinc in the fourth trimester (Brown et al. 2009). Studies on the effects of supplemental zinc in gestation have exposed unpredictable outcomes, because of the challenges in founding baseline zinc status among populations (de Benoist et al., 2007). Through between 1977 and 2008, the recent meta-analyses report of randomised controlled trials of supplementation of zinc in gestation conducted, largely among low socioeconomic status women, on a variety of maternal and neonatal outcomes (Chaffee and King 2012; Mori et al. 2012). The analyses determined that supplementation of zinc was linked with a 14% decrease in preterm birth. However, supplementation of zinc had no obvious effect on other infant consequences including mean age of gestational, neonatal mortality or any factor of foetal growth such as possibility of mean weight or low birth weight, head circumference or length at the time of birth or on major maternal results such as hypertension (Mori et al. 2012). The effect of supplemental zinc on preterm birth may be because of the decreased frequency or harshness of maternal infections, which is a recognised risk factor for preterm birth.

The mother's nutritional status in gestation is an important supporter for both maternal and perinatal morbidity and mortality, refining the quality, or nutrient density of the mother's diet is identified to increase gestational outcomes (Koblinsky 1995). The limited influence perceived of supplementation of zinc on gestational outcomes, it is careful to emphasis on research to recognising the ways of improving the overall nutritional status of women in low-income areas (Shrimpton et al. 2005). In developing kingdoms, UNICEF is currently promoting supplements of micronutrient for pregnancy which include folic acid, iron and zinc for all pregnant women, given that they are expected to have low of micronutrients from diet only (UNICEF/UNU/WHO 1999). However, the public health benefit by supplementation of zinc during gestation presently seems inadequate (Chaffee and King 2012; Bhutta et al. 2013).

Conclusion

The prevalence of zinc deficiency during gestation is linked with anaemia, low consumption of animal sources, short birth interval, low dietary variety score and hypoalbuminemia. Hence, effort must be prepared to motivate women to take a balanced diet during gestational period containing food sources by animal mainly during third trimester, observer albumin levels of serum and haemoglobin and for the prevention of zinc deficiency, support to the family planning implementations. However, more prospective population-based studies and prospective follow-up studies should be accompanied to assess the degree of severity and risk factors of zinc deficiency amongst pregnant women.

It has been reported that maternal zinc status and birth weight are significantly associated. However, the low status of zinc in mothers is because of the existence of a factor that modifies the use of zinc rather than to the consumption of a low-zinc diet. Maternal zinc status can decrease by conditions like alcohol-smoking, abuse, dependence on high phytate cereal-based diet, consumption of iron supplementation in high amounts, and a severe trauma or infections. Poor plasma zinc concentrations and decrease in the zinc volume accessible to the foetus may occur due to these conditions. In these conditions, supplements could be given to women.

The existing data are limited on lactation performance and zinc supplementation, and the growth of infant and zinc status fail to show any reliable use for the child or mother.

However, the available studies are only from relatively healthy women in developed countries. The effect of zinc supplementation on mothers during pregnancy or in lactation is essentials to be studied in malnourished women.

References

Açkurt, F., Wetherict, H., Löker, M.: Biochemical assessment of nutritional status in pre-and post-natal Turkish women and outcome of pregnancy. Eur. J. Clin. Nutr. **49**(8), 613–622 (1995)

Antonson, D.L., Vanderhoof, J.A.: Effect of chronic ethanol ingestion on zinc absorption in rat small intestine. Dig. Dis. Sci. **28**, 604–608 (1983)

Bales, C.W., Ritchie, C.S. (eds.): Handbook of Clinical Nutrition and Aging. Humana Press, New York, NY, USA (2009)

Bhutta, Z.A., et al.: Evidence-based interventions for improvement of maternal and child nutrition: what can be done and at what cost? Lancet. **S0140-6736**(13), 60996–60994 (2013)

Black, R., Sazawal, S.: Zinc and childhood infectious disease morbidity and mortality. Br. J. Nutr. Brit J Nutr. 85(S2), S125–S129 (2001). https://doi.org/10.1079/BJN2000304

Brown K.H.: Effect of infections on plasma zinc concentration and implications for zinc status assessment in low-income. Am. J. Clin. Nutr. **68**(2), 425S–429S (1998), https://doi.org/10.1093/ajcn/68.2.425S

Brown, K.H., Wuehler, S.E., Peerson, J.M.: The importance of zinc in human nutrition and estimation of the global prevalence of zinc deficiency. Food Nutr. Bull. **22**, 113–125 (2001)

Brown, K.H., Rivera, J.A., Bhutta, Z., Gibson, R.S., King, J.C., Lönnerdal, B., Ruel, M.T., Sandtröm, B., Wasantwisut, E., Hotz, C., de Romaña, D.L.: International Zinc Nutrition Consultative Group (IZiNCG) technical document# 1. Assessment of the risk of zinc deficiency in populations and options for its control. Food Nutr. Bull. **25**(1 Suppl 2), S99 (2004)

Brown, K.H., Engle-Stone, R., Krebs, N.F., Peerson, J.M.: Dietary intervention strategies to enhance zinc nutrition: promotion and support of breastfeeding for infants and young children. Food Nutr. Bull. **30**, S142–S169 (2009)

Burits, C.A., Ashwood, E.D.: Tietz Text Book of Clinical Chemistry, 2nd edn, (1994)

Caulfield, L.E., Zavaleta, N., Shankar, A.H., Merialdi, M.: Potential contribution of maternal zinc supplementation during pregnancy to maternal and child survival. Am. J. Clin. Nutr. **68**, 499S–508S (1998)

Caulfield, L.E., Black, R.E.: Zinc Deficiency. In Comparative Quantification of Health Risks. WHO, World Health Organization. Geneva (2014)

Chaffee, B.W., King, J.C.: Effect of zinc supplementation on pregnancy and infant outcomes: a systematic review. Paediatr Perinat Epidemiol. Suppl 1(01):118–37 (2012). https://doi.org/10.1111/j.1365-3016.2012.01289.x. PMID: 22742606; PMCID: PMC3787719.

Davidsson, L., Almgren, A., Sandström, B., Hurrell, R.F.: Zinc absorption in adult humans: the effect of iron fortification. Br. J. Nutr. **74**, 417–425 (1995)

Dawson, E.B., Albers, J., McGanity, W.J.: Serum zinc changes due to iron supplementation in teen-age pregnancy. Am. J. Clin. Nutr. **50**, 848–842 (1989)

de Benoist, B., Darnton-Hill, I., Davidsson, L., Fontaine, O., Hotz, C.: Conclusions of the joint WHO/UNICEF/IAEA/ IZiNCG interagency meeting on zinc status indicators. Food Nutr. Bull. **28**, S480–SS79 (2007)

Fitzgerald, S.L., Gibson, R.S., Quan de Serrano, J., et al.: Trace element intakes and dietary phytate/Zn and Ca 3 phytate/Zn millimolar ratios of periurban Guatemalan women during the third trimester of pregnancy. Am. J. Clin. Nutr. **57**, 195–201 (1993)

Flynn, A., Martier, S.S., Sokol, R.J., Miller, S.I., Golden, N.L., Del Villano, B.C.: Zinc status of pregnant alcoholic women: a determinant of fetal outcome. Lancet. **1**, 572–575 (1981)

Gebremedhin, S., Enquselassie, F., Umeta, M.E., Umeta, M.: Prevalence of prenatal zinc deficiency and its association with socio-demographic, dietary and health care related factors in Rural Sidama, Southern Ethiopia: a crosssectional study. BMC Public Health. **11**(1), 898 (2011)

Ghishan, F.K., Patwardhan, R., Greene, H.L.: Fetal alcohol syndrome: inhibition of placental zinc transport as a potential mechanism for fetal growth retardation in the rat. J. Lab. Clin. Med. **100**, 45–42 (1983)

Gibson, R.S., Huddle, J.-M.: Suboptimal zinc status in pregnant Malawian women: its association with low intakes of poorly available zinc, frequent reproductive cycling, and malaria. Am. J. Clin. Nutr. **67**, 702–709 (1998)

Goldenberg, R.L., Tamura, T., Cliver, S.P., Cutter, G.R., Hoffman, H.J., Davis, R.O.: Maternal serum alpha2-macroglobulin and fetal growth retardation. Obstet. Gynecol. **78**, 594–598 (1991)

Hambidge M.: Human Zinc Deficiency, Nutr. J., 130(5), 1344S–1349S. https://doi.org/10.1093/jn/130.5.1344S

Karimi, A., Bagheri, S., Nematy, M., Saeidi, M.: Zinc deficiency in pregnancy and fetal-neonatal outcomes and impact of the supplements on pregnancy outcomes. Iran. J. Neonatol. **3**(2), 77–83 (2012)

Keen, C.L., Taubeneck, M.W., Daston, G.P., Rogers, J.M., Gershwin, M.E.: Primary and secondary zinc deficiency as factors underlying abnormal CNS development. Ann. N. Y. Acad. Sci. **678**, 3–47 (1993)

King, J.C.: Determinants of maternal zinc status during pregnancy. Am. J. Clin. Nutr. **71**, 1334S–1343S (2000)

King, J.C., Turnlund, J.R.: Human zinc requirements. In: Mills, C.F. (ed.) Zinc in Human Biology, pp. 335–350. SpringerVerlag, London (1989)

Kirksey, A., Wachs, T.D., Yunis, F., Srinath, U., Rahmanifar, A., McCabe, G.P., Galal, O.M., Harrison, G.G., Jerome, N.W.: Relation of maternal zinc nutriture to pregnancy outcome and infant development in an Egyptian village. Am J Clin Nutr. 60(5):782–92 (1994). https://doi.org/10.1093/ajcn/60.5.782. PMID: 7942587

Koblinsky, M.A.: Beyond maternal mortality - magnitude, interrelationship, and consequences of women's health, pregnancy related complications and nutritional status on pregnancy outcomes. Int. J. Gynecol. Obstet. **48**, S21–S32 (1995)

Kuhnert, B.R., Kuhnert, P.M., Debanne, S., Williams, T.G.: The relationship between cadmium, zinc, and birth weight in pregnant women who smoke. Am. J. Obstet. Gynecol. **157**, 1247–1251 (1987a)

Kuhnert, P.M., Kuhnert, B.R., Erhard, P., Brashear, W.T., Groh-Wargo, S.L., Webster, S.: The effect of smoking on placental and fetal zinc status. Am. J. Obstet. Gynecol. **157**, 1241–1246 (1987b)

Kuhnert, B.R., Kuhnert, P.M., Groh-Wargo, S.L., Webster, S., Erhard, P., Lazebnik, N.: Smoking alters the relationship between maternal zinc intake and biochemical indices of fetal zinc status. Am. J. Clin. Nutr. **55**, 981–984 (1992)

Kumera, G., Awoke, T., Melese, T., Eshetie, S., Mekuria, G., Feleke, M., Ewunetu, T., Gedle, D.: Prevalence of zinc deficiency and its association with dietary, serum albumin and intestinal parasitic infection among pregnant women attending antenatal care at the University of Gondar Hospital, Gondar, Northwest Ethiopia. BMC Nutr. **1**(1), 31 (2015)

Lonnerdal, B., Keen, C.L., Hendrickx, A.G., Golub, M.S., Gershwin, M.E.: Influence of dietary zinc and iron on zinc retention in pregnant rhesus monkeys and their infants. Obstet. Gynecol. **75**, 369–374 (1990)

Lutfullah, G.: Prevalence of zinc deficiency among rural women during childbearing age in Peshawar. Pakistan. Pak. J. Pharm. Sci. **27**(1), 173–177 (2014)

Ma, A.G., Chen, X.C., Xu, R.X., Zheng, M.C., Wang, Y., Li, J.S.: Comparison of serum levels of iron, zinc and copper in anemic and non-anemic pregnant women in China. Asia Pac. J. Clin. Nutr. **13**(4), 348 (2004)

Mekonnen, A.: Prevalence and associated factors of zinc deficiency among pregnant women attending antenatal care at Gambella Hospital, South West of Ethiopia: Institution based cross-sectional study. (2018)

Mitheko, A.N.: Dietary, socio-economic and demographic factors influencing serum zinc levels of pregnant women at Naivasha level 4 hospital Nakuru County, Kenya (2013)

Mori, R., et al.: Zinc supplementation for improving pregnancy and infant outcome. Cochrane Database Syst. Rev. (7), CD000230 (2012)

Nguyen, V.Q., Gotot, A., Nguyen, T.V.T., Vo, K.T., Ta, T.N.T., Nguyen, T.N.T., et al.: Prevalence and correlates of zinc deficiency in pregnant Vietnamese women in Ho Chi Minh City. Asia Pac. J. Clin. Nutr. **22**(4), 614–619 (2013)

Oguizu, A.D.: Assessment of iron, selenium and zinc status of pregnant women in Obio-Akpor LGA Rivers State. Pak. J. Nutr. **14**(1), 1–5 (2015)

Pathak, P., Kapil, U., Dwivedi, S.N., Singh, R.: Serum zinc levels amongst pregnant women in a rural block of Haryana state, India. Asia Pac. J. Clin. Nutr. **17**(2), 276–279 (2008)

PDR Electronic Library: Physician's Desk Reference. Medical Economics, Montvale (1997)

Prasad, A.S.: Syndrome of iron deficiency, anemia, hepatic, hypogonadism, dwarfism and geophagia. Am. J. Med. **31**, 532–546 (1961)

Prasad, A.S.: Zinc metabolism inpatients, syndrome of iron deficiency anemia. Am. J. Lab. Ed. **61**, 537? (1963)

Prasad, A.S.: Discovery of human zinc deficiency and studies in an experimental human model. Am J Clin Nutr. Feb;53(2):403–12 (1991). https://doi.org/10.1093/ajcn/53.2.403. PMID: 1989405

Raman, L., Shatrugna, V.: Nutrition during pregnancy and lactation. In: Bamji, M.S., Rao, N.P., Reddy, V. (eds.) Textbook of Human Nutrition, pp. 220–232. Oxford and IBH Publishing Co. Pvt. Ltd, New Delhi (2001)

Roshanravan, N., Alizadeh, M., Hedayati, M., Asghari-Jafarabadi, M., Alamdari, N.M., Anari, F., Tarighat-Esfanjani, A.: Effect of zinc supplementation on insulin resistance, energy and macronutrients intakes in pregnant women with impaired glucose tolerance. Iran. J. Public Health. **44**(2), 211 (2015)

Severi, C., Hambidge, M., Krebs, N., Alonso, R., Atalah, E.: Zinc in plasma and breast milk in adolescents and adults in pregnancy and pospartum: a cohort study in Uruguay. Nutr. Hosp. **28**(1), 223 (2013)

Shah, D., Sachdev, H.P.: Effect of gestational zinc deficiency on pregnancy outcomes: summary of observation studies and zinc supplementation trials. Br. J. Nutr. **85**(2), 101–108 (2001)

Shamin, A.A., Merrill, R.D., Ali, A.H., Rashid, M., Schulze, K.: Plasma zinc, vitamin B 12 and α-tocopherol are positively and plasma γ-tocopherol is negatively associated with Hb concentration in early pregnancy in north-west Bangladesh. Public Health Nutr. **16**(8), 1354–1361 (2013)

Shommo, S., Zumrawi, F., Saeed, A.: Micronutrients dietary intakes of pregnant Sudanese women. Int. J. Sci. Res. **4**(4), 2319–7064 (2013)

Shrimpton, R., et al.: Zinc deficiency: what are the most appropriate interventions? Br. Med. J. **330**, 347–349 (2005)

Solomons, N.W., Jacob, R.A., Pineda, O., Viteri, F.: Studies on the bioavailability of zinc in man. II. Absorption of zinc from organic and inorganic sources. J. Lab. Clin. Med. **94**, 335–343 (1979)

Tamura, T., Goldenberg, R.L., Johnston, K.E., Dubard, M.: Maternal plasma zinc concentrations and pregnancy outcome. Am. J. Clin. Nutr. **71**(1), 109–113 (2000)

UNICEF/UNU/WHO.: Composition of a multi-Micronutrient supplement to be used in pilot programmes among pregnant women in developing countries Report of a Workshop. New York: UNICEF (1999)

Viteri, F.E.: Iron supplementation for the control of iron deficiency in populations at risk. Nutr. Rev. **55**, 195–209 (1997)

W.B Sanders, Sandstead, H.H.: Zinc deficiency, a public health problem? Am. J. Dischild. **145**(8), 853–859 (1991)

Walsh, C.T., Sandstead, H.H., Prasad, A.S., Newberne, P.M., Fraker, P.J.: Zinc: health effects and research priorities for the 1990s. Environ. Health Perspect. **102**, 425s–459s (1994)

World Health Organization: Trace Elements in Human Nutrition and Health. World Health Organization, Geneva (1996)

Yasodhara, P., Ramaraju, L.A., Raman, L.: Trace minerals in pregnancy 1. Copper Zinc Nutr. Res. **11**, 15–21 (1996)

Zinc Investigators' Collaborative Group (1999) Prevention of diarrhea and pneumonia by zinc supplementation in children in developing countries: pooled analysis of randomized controlled trials. Journal of Pediatrics 135, 689–697

Zinc Investigators' Collaborative Group (2000) Therapeutic effects of zinc in acute and persistent diarrhea in children in developing countries: pooled analysis of randomized controlled trials. American Journal of Clinical Nutrition 72, 1516–1522

Chapter 4
Malnutrition: Impact of Zinc on Child Development

Farhana Mehraj Allai, Khalid Gul, Insha Zahoor, Tariq Ahmad Ganaie, Gazia Nasir, and Z. R. Azaz Ahmad Azad

Abstract Malnutrition is a condition that results from dependency on diet which is deficient in one or more essential nutrients causing health problems. Zinc is known to be an essential micronutrient in humans that contributes to the structure and proper functioning of important organs including brain. It has been reported that malnutrition can interfere with the functioning of brain by reducing growth of neuronal and brain development by altering the function of neurotransmitter, delays in cognitive functioning and enhancing neurotoxicity. Child malnutrition can be reduced by interfering foetal development and early childhood. Children of age group under 24 months require special care as this stage is the most important phase for growth and development of children. During growth, deficiency of zinc leads to growth failure. Zinc deficiency clinically affects most of the organs in humans like gastrointestinal, skeletal, nervous, immune and reproductive system. There are four main intervention steps for improving deficiency of zinc that include dietary supplementation, modification and fortification of essential micronutrient. This chapter provides a review on physiological and biochemical functions of zinc, bioavailabil-

F. M. Allai
Department of Food Technology, Islamic University of Science and Technology, Awantipora, Jammu and Kashmir, India

Department of Post Harvest Engineering and Technology, Aligarh Muslim University, Aligarh, Uttar Pradesh, India

K. Gul
Department of Food Process Engineering, National Institute of Technology, Rourkela, Odisha, India

Department of Food Science Engineering, School of Agriculture and Biology, Shanghai Jiao Tong University, Shanghai, China

I. Zahoor · G. Nasir · Z. R. Azaz Ahmad Azad (✉)
Department of Post Harvest Engineering and Technology, Aligarh Muslim University, Aligarh, Uttar Pradesh, India

T. A. Ganaie
Department of Food Technology, Islamic University of Science and Technology, Awantipora, Jammu and Kashmir, India

S. T. Khan, A. Malik (eds.), *Microbial Biofertilizers and Micronutrient Availability*,
https://doi.org/10.1007/978-3-030-76609-2_4

ity of zinc, causes and consequences of zinc deficiency and how to tackle deficiency of minerals in children with special focus on zinc.

Keywords Malnutrition · Zinc deficiency · Neurotransmitter · Cognitive functioning · Children

Introduction

Zinc is the second most essential micronutrient after iron involved in multiple essential roles in all living organisms. It has a fundamental role in cellular growth and metabolism as a component of various enzymes and in the synthesis of DNA and RNA (Kambe et al. 2015). Dietary uptake of zinc depends on the quantity of zinc in the diet and the dietary phytate (Engle-Stone et al. 2014). In developing countries, deficiency of zinc in children is prevalent due to less intake of animal food and high consumption of plant-based diet and cereals (Brown 2007). As a result, the absorption of zinc decreases and thus causes zinc deficiency (Gibson 2006). Roohani et al. (2012a) reported that consumption of higher quantity of phytate-rich diet decreases the bioavailability of zinc which is also indicative of Zn deficiency risk in children. It is well established that phytate interferes with the absorption of zinc. It has been also reported that supplementation of zinc is extremely important for newborn, children, pregnant ladies for normal foetal growth, in prevention and treatment of pneumonia and diarrhoea and during lactation for production of milk (Brown and Pollit 1996; Fung et al. 1997). Zinc actively participates in cell growth and division, transmission of neuro signals, immune response, modification in membrane protein and enzymatic stabilization (King 2011; Terrin et al. 2015; Wardlaw and Hampl 2007). In brain, zinc is the most prevalent element where it associates with the proteins and thus contributing to the function and structure of the brain. Another important concern is the change in diet: frequent intake of soy milk instead of breast feeding can cause severe diarrhoea which further leads to loss of zinc which in turn may cause half of deaths in children due to diarrhoea (Godana and Mengistie 2013; Siregar et al. 2018). There exists a co-relationship between diarrhoea and breast milk (Ugboko et al. 2020). Mother's feed has both long- as well as short-term effect on the health of children as it improves the immune and gut system (Bartick et al. 2017; Ogbo et al. 2018). Children above 6 months of age can be given combination of vitamin A, oral zinc therapy and ORS, which strengthen the immune system, enhance the mucosal regeneration and increase the levels of brush border enzymes (Liu et al. 2018). Thus, supplementation of zinc may be fundamental intervention to enhance the children's growth and weight.

According to WHO, malnutrition is defined as the imbalance between the supply of the essential nutrients, both micro as well as macronutrients, energy and the demand of the body to provide proper growth, development and maintenance. Micronutrients are required in very small quantity by organisms throughout their entire life in order to perform a variety of different physiological functions. These micronutrients include iron, zinc, iodine and calcium and are also called "magic wands" as described by the World Health Organization (WHO), which enable the human body to synthesize different hormones, enzymes and other essential components for proper growth and development. Among all of them, zinc is very important for cellular metabolism and for providing protection for the body to prevent against the oxidative damage (Mittal and Dipti 2016). It has also proved that its deficiency leads to the retardation of growth of both linear and ponderal and has adverse effect on the reproductive system, central nervous and immune system (Bhatnagar and Natchu 2004). Brown (2003) investigated that zinc significantly influences the young children's growth especially in under-developing countries where the resources are not adequate enough to meet their dietary requirement, or intake of zinc might be less or due to frequent enteric infections that causes excessive loss of zinc. The initial period of life for children is very crucial for their growth and development (Black et al. 2013). Diverse diet rich in essential nutrients are important for children to support the neurobehavioural development, to meet the nutrient needs and for optimal growth (Nguyen et al. 2013).

Relation Between Zinc and Growth

Nutritional deficiency is a critical issue across the globe that leads to long-term deficits in academic performance, behavioural changes, cognitive and motor senses, and immune function; for example, deficiencies of micronutrient like iodine and iron have been directly associated with motor and cognitive development (Bleichrodt and Born 1994). Similarly, deficiency of zinc commonly occurs in children that may be linked with the decrease in attention, activity and motor delay. During the first year of life, the investigative-orientation system for infants is active as they start exploring novel and new aspects of their surroundings (Ruff and Rothbart 1996). Exploration at early stage depends mostly on the developmental skills. The environmental risk has significant effect on neurodevelopment of children that include insufficient cognitive stimulation, poverty, and malnutrition. This exploration, investigation, learning and observing new and novel things around the environment often enhance the infant's growth and development in a much better way (Maureen 1998).

Growth rate has been reduced in children due to the deficiency of zinc that may increase the morbidity or other severe diseases, decreases the scholastic skills and negatively affects the physical activity (Hautvast et al. 2000). Hautvast et al. (2000) investigate that the retardation of growth starts in the beginning of life when the breast feeding declines or stops, weaning foods are given and vulnerability to infection starts. Once the environmental condition becomes unfavourable and the height of these children starts decreasing, then most of them will remain in their stage of growth period which they have acquired till then. Thus, supplementation of one single micronutrient will have very little beneficial effect on overall growth and development of the child. Hence it is advisable to have different varieties of nutrients, so that there is improvement in the linear growth, but that growth is still less than the expected. Zinc and iron are the two major minerals that have been found to be very much critical during the early period of life, that is during infancy and childhood, where the rapid growth may occur. Several studies reported that these two micronutrients are in association with the increased height of children and birth weight, but it was very difficult to find out the exact role of these minerals and other micronutrients on the growth of child (Salgueiro et al. 2002).

Children at their early age are at high risk of zinc deficiency as they require higher amount of zinc for their growth. Infants up to 6 months of age are especially dependent on the breast feed that contains high level of zinc (Brown et al. 2004). After this age, weaning food contains absorbable zinc that satisfies their daily requirements. In many countries, mostly low-income group of people used to feed their babies with cereal foods. Thus, the complementary feeding is delayed as these foods contain low level of absorbable zinc and cannot meet their requirement. Additionally, introduction of such food at an early stage may also interfere with the zinc absorption from breast feed because of their enhanced phytate level (Russell et al. 2001). Zinc requirement for healthy children is estimated to be 0.17 mg/kg for up to 1–3 years. These requirements are much lesser than those for malnourished children (2 mg/kg -4 mg/kg body weight) (Muller et al. 2001).

Biochemical and Physiological Function

Zinc is found everywhere within the cell in comparison to iron, which is present in defined cellular components and has defined physiological roles. There are three major functional classes of zinc namely structural, regulatory, and catalytic groups (Roohani et al. 2013). Zinc is nontoxic in nature, but if it is found in large quantity in diet and is readily available, then it can interfere with the absorption of other minerals like copper, calcium, cadmium and iron and can cause anaemia (Bhagavan 2002).

Metabolism

In biological processes, zinc plays a vital role for cell growth, metabolism and differentiation. This micronutrient deficiency restricts the childhood growth and reduces the resistance to infections which together significantly leads to morbidity and mortality in young generation (Brown 2003). Although severe deficiency of zinc is very rare in human beings, only little to moderate deficiency can occur and this mild to moderate deficiency is also difficult to check as symptoms and signs are not much visible like susceptibility to infections and retardation of growth can also be shared due to other mineral deficiencies.

Zinc Transporters

It means for zinc to perform its diverse bioactive roles, a number of specific systems to transport zinc across the biological membrane are required. Thus, transportation of zinc protein is necessary for the physiology of zinc (Kambe et al. 2014). Zinc transporters, especially Zrt-Irt-ZIP proteins and Zn transporter (ZnT), are widely used to perform several cellular and physiological functions like endocrine, growth, immune, reproductive, neuronal and skeletal (Fukada and Kambe 2011). In human cells, there are approximately 15 ZIP and 10ZnT transporters (Cousins et al. 2006). The change in zinc level regulates the expression and cellular distribution of ZnT transporters (Devergnas et al. 2004). ZnT decreases the availability of intracellular zinc, while ZIP transporters enhance the availability of intracellular zinc by promoting the uptake of extracellular zinc and possibly releases the vesicular zinc into the cytoplasm (Sekler et al. 2007).

Zinc is redox neutral unlike copper and iron, and in biological reactions, zinc is highly reactive as a Lewis acid. Due to all these features, zinc plays a vital role as a signalling component. Zinc is required to perform the 300 different functions of zinc metalloenzymes. Among these 300 enzymes, zinc metalloenzymes are essential to perform the different physiological functions. These metalloenzymes are oxidoreductases, lyases, hydrolases, transferases, ligases and isomerases and are involved in the regulatory, structural, catalytic and non-catalytic functions. The structural and functional properties of metalloproteases have been reported to have several functions like processing of hormone, blood clotting, immune and inflammatory activity (Auld 2013).

Absorption

Dietary zinc is absorbed as a saturable process and is determined by two factors; first the amount of zinc ingested and second the intake of dietary phytate (Hambidge et al. 2010). In the human body, the primary site for absorption of zinc is to be in the small bowel, either the proximal jejunum or distal duodenum, by carrier-mediated mechanism (Krebs et al. 1998a, b). Under normal physiological conditions, the absorption of zinc includes the presence of dietary promoters like protein derived from animal source, breast milk and the amount of zinc absorbed in the intestinal lumen, or the inhibitors such as phytates (inositol hexaphosphate), fibre, tannin, lignin, oxalate and other minerals like copper, iron, cadmium and calcium can reduce the absorption of zinc if present in abundant quantities. About 70% of zinc belongs to the blood protein albumin. If any such circumstances arise that alter the concentration of albumin, then it might have a secondary effect on the zinc level present inside the human body (Saper 2009). If the concentration of albumin is low, then in that case the body does not properly digest and absorbs the albumin protein like in case of Celiac disease, Crohn's disease or other form of diseases where the protein is lost in large quantity through diarrhoea. Albumin is synthesized by liver, so any form of disease or inflammation in liver can have negative effect on zinc levels. Moreover, the fraction of absorbed zinc is not easy to determine as the zinc is also secreted into the human gut.

Generally, infancy, pregnancy and during lactation the requirement of zinc is quite high and is associated with the enhanced efficiency of fractional absorption. Breast milk infant is the best example of high fractional absorption as the average absorption level is approximately 0.55 which is equal to the rate of zinc absorption administered alone with water (Krebs 2000). Zinc administered in solid form is absorbed very less in contrast to aqueous solutions that are absorbed efficiently approximately 60–70%, and this absorption varies depending on the composition of diet and the amount of zinc content present in the food substance (FAO/WHO n.d.). In addition, zinc absorption may be influenced by the zinc status. This element is absorbed with increased efficiency in case of zinc-deprived humans, as zinc-rich diet shows decreased absorption efficiency (Roohani et al. 2013). During fasting, for gluconeogenesis, mobilization of amino acids from muscles to liver takes place where excess of zinc is passed through urine. Zinc loss also takes place through skin, hair, nail, semen, menstrual blood loss and through sweat. During the first several weeks for postpartum, zinc loss may occur through lactation process in which almost 2–3 mg of zinc is lost per day; then after 2–3 months of postpartum, it reduces to 1 mg/day and then progressively decline over the period of time (Shrimpton and Shankar 2008).

During digestion, zinc is released from food as free ions. These free ions are first attached to endogenously secreted legends and are then transported in to the duodenum and jejunum (Tubek 2007a, b). The portal system directly delivers the absorbed zinc to the liver, and then released it to the systemic circulation for delivery to other tissue. In circulation almost 70% of zinc is linked to the protein albumin. If the concentration of serum albumin is altered by any condition, then it might have secondary effect on serum zinc levels. However, whole body zinc contains only 0.1% of serum zinc that is needed for circulating zinc to meet the tissue requirement (Brown et al. 2004). The presence of unabsorbed fat or dietary phytate may hinder the efficiency of reabsorption process and thus can cause oozing of zinc from the human body, for example in infants malabsorption takes place and many studies are still under investigation to determine the possible adverse effects caused by inhibitors on reabsorption of zinc.

Homeostasis

In animals and humans, maintaining a constant rate of cellular zinc is important for livelihood and is known as zinc homeostasis. Homeostasis is maintained by adjusting the absorption of total zinc content and excretion of endogenous zinc (Hambidge and Krebs 2001). Zinc homeostasis is needed to maintain the antioxidant capacity, immune system and healing of wounds. It is achieved by several processes like influx and efflux of zinc into the cell or out of cells respectively and storage and compartmentalization within the cells. These processes are done by using several different protein families such as zinc binding and zinc transporter proteins (Zhong and Zhou 2019). Zinc is also needed for many replication and transcription processes. In critically ill children, the concentration of plasma zinc is very low that leads to the inflammation of different organs which ultimately leads to organ failure and death (Bass and Steinhorn 2011).

Bioavailability

Bioavailability of zinc means consumption of zinc that can be absorbed into the blood stream and is used to perform several different physiological functions of human body. Lonnerdal (2000) reported that intake of low zinc diet enhances the absorption of zinc in all age groups like young men was fed by a diet having 15 mg of zinc/day, and the zinc content was determined in a starvation state after 6 days and found that the 81% were absorbed from high zinc diet and 92% from low zinc diet. In a healthy person, zinc is measured by three important factors: total zinc content present in the diet, presence of soluble zinc and zinc status in an individual. The daily recommended allowance is different at every different stage of life, that is, it varies according to the age as given in Table 4.1. Human body absorbs approximately 20–40% of consumed zinc that depends on the bioavailability in the food items, for example, bioavailability of zinc from animal source is higher than that from grains and plant foods. Red meat, oysters and poultry contain high quantity of zinc per serving in comparison to any other food.

Zinc is considered to be essential for DNA stabilization, and it resides in the nucleus for longer period than any other compartment in cell. Thus, probably the level of intracellular zinc enhances and more iron will be displaced and low OH driven will cause DNA damage. The dislocation of zinc from transcription factor of zinc finger and the release of radicals of oxygen onto the DNA can be the cause of mechanism for carcinogenicity.

Causes and Consequences of Zinc Deficiency

Deficiency of zinc is more common in developing countries because consumers mostly take phytate-rich diet, calcium and other trace elements such as copper, cadmium and mercury. There are several factors that influence the absorption of zinc content (Table 4.2) like inositol hexa- and penta-phosphate that limits the bioavailability of zinc binding in gastrointestinal tract as they form insoluble complexes with cations, and thus phytate-bound minerals will be passed through stool (Hambidge et al. 2011). People who consume only vegetarian food are more prone to zinc deficiency. Food that contains too much of iron can interfere with the

Table 4.1 Recommended daily allowance (RDA) of zinc for different age groups

S. no	Age	RDA of zinc/day	References
	Infant		
1.	0–6 months	2 mg	Roohani et al. (2013)
2.	7–12 months	3 mg	Abrams et al. (1997)
	Children	3 mg	Institute of Medicine (US) Panel on Micronutrients (2001)
3.	1–3 years		
4.	4–8 years	5 mg	Gibson et al. (1989) and Walravens et al. (1983)
5.	9–13 years	8 mg	Davidsson et al. (1996)
6.	*Adolescent* 14–18 years	9 mg for girls 11 mg for boys	Kuczmarski et al. (2000)
7.	*Adult* 19–30 years 31–50 years 51–70 years >70 years	11 mg for men	Wada and King (1986)
8.	*Adult* 19–30 years 31–50 years 51–70 years >70 years	8 mg for women	Gibson et al. (2000)
9.	*For pregnancy*		
	14–18 years	12 mg	Caulfield et al. (1999)
	19–30 years	11 mg	Hambidge et al. (1983)
	31–50 years	11 mg	Ortega et al. (1997)
10.	*Lactation*		
	14–18 years	13 mg	Krebs et al. (1985)
	19–30 years	12 mg	Krebs et al. (1995)
	31–50 years	12 mg	Moser-Veillon and Reynolds's (1990)

Table 4.2 Major causes of deficiency of zinc

S. no.	Cause	References
1.	*Less consumption* (i) Diet deficit in zinc Vegetarian food: animal protein is very poor (ii) During processing, loss of zinc takes place In artificial milk, desalting is done (iii) Shortage of micronutrient consumption (iv) Poor resources (v) Less income	Yanagisawa (2004)
2.	*Malabsorption* (i) Genetic disorder: Aacrodermatitis enteropathica (ii) Consumption of inhibitors Decreases the absorption process like phytic acid and fibres	Yanagisawa and Wada (2002)
3.	*Loss due to* Diarrhoea, increased urination in case of diabetes mellitus, haemolytic anaemia, renal problem, diuretic, etc.	Yanagisawa et al. (1999)

absorption of zinc. Staple food such as rice, corn, cereals and legumes mostly contain phosphorus storage compounds as they are negatively charged and has high-density phosphate groups which play an essential role in storage of minerals (King and Cousins 2006). In infant formula milk, soy-based milk has low zinc bioavailability in comparison to human milk and milk formula. Additionally, when the phytate content is added to formula milk, the concentration of absorption of zinc is reduced to the same level to milk formula as the inhibitors reduce the absorption of zinc in soy milk. Absorption of zinc was significantly improved when the phytate was eliminated from soy protein by the process of precipitation (Lonnerdal 2000). There are various techniques that can reduce the phytic acid content in several foods like germination, fermentation and soaking. In cereals and legumes, phytate content is reduced by milling and germination process. In case of cereal products such as bread, phytate content is decreased by the process of fermentation. In whole grain cereals or legumes, enzymatic hydrolysis of phytic acid is promoted by the fermentation process that enhances the endogenous or exogenous activity of phytase enzyme (Hurrell 2004). Extrusion and thermal processing may cause less phytate loss as compared to enzymatic and non-enzymatic methods (Schlemmer et al. 2009).

Deficiency of zinc is associated with retardation of growth, alopecia, susceptibility to infections, hyposmia, delayed wound healing, dermatitis and hypogeusia. Zinc deficiency is diagnosed by determining the most reliable level, that is, i.e serum zinc level. It has been reported that level of serum copper can also be measured and then calculate the ratio of level of serum copper to zinc that can be used as a reference point for determining the deficiency of zinc (Tomita 2002). In developing countries, the zinc deficiency is measured to be around 4% of the global mortality and morbidity of young generation (Penny 2013). Zinc is present in all body tissues, and around 85% of zinc is found in bones and muscles with 11% in liver and skin and the remaining portion of this element in other tissues (Calesnick and Dinan 1988). Burn patients are mostly dressed with zinc oxide. The absorption and excretion of zinc reduces with age. Therefore, the overall homeostasis of zinc is not with deregulated age (Semrad 1999).

Role of Zinc in Human Body

Zinc has indispensable role in human body regarding to both human health and diseases. It has several health implications, pharmacological effects and biological functions that makes it an attractive element regarding its human health and nutrition as shown in Table 4.3.

Table 4.3 Effect of zinc on different parts of body

Organ	Effect of zinc	References
Endocrine system Testes, ovaries, adrenal, thyroid, pituitary and parathyroid gland	Metabolism of oestrogen, androgen and progesterone. Synthesis, storage, secretion of insulin Regulation of thymic hormones, sex hormones in both males and females Disruption of the oestrous cycle, recurrent abortion, delayed pregnancy	Jameson (1993) and Favier (1992)
Immune system	Humoral immunity, haematopoiesis and secretion of cytokine Role in basic cellular function like activation of immune cell RNA transcription, DNA replication, proliferation and differentiation Enhanced resistance to infections	Fraker and King (2004) and Foster and Samman (2012)
Brain	Serves as neuromodulator and signalling ion During neonatal, highest level of zinc is found in cerebellum, which is associated with rapid growth and development Cytosolic zinc plays a vital function in cognitive and learning performance	Brion et al. (2021)
Liver	Supplementation of zinc has proved to reduce the level of blood ammonia and protects against encephalopathy Improves weight loss, gastrointestinal disturbance and anaemia in case of chronic Hepatitis C virus	Takuma et al. (2010) andGrungreiff and Reinhold (2010)

Zinc Deficiency

Deficiency of zinc affects almost 50% of the population around globe (Golub et al. 1995). It involves the most significant pathological state like metabolism of metal that causes abnormalities in the human body (Takeda 2011). These abnormalities and deficiencies (Table 4.4) are caused in those consumers who take powerful chelator diet, cereal-based phytate concentration diet and less protein-containing diet (Brown et al. 2001; Tapiero and Tew 2003). Zinc deficiency has been well reported in premature and low weight newborns if supplementation of zinc was not given. Deficiency of zinc includes growth retardation, both linear and ponderal, dermatitis and low concentration of serum/plasma zinc (Barbarot et al. 2010). Zinc is found in abundance in the organelles of nucleus, cytosol and vesicles. However, a small deficiency is a disaster to the health of human being (King and Keen 2003).

Table 4.4 Effect of deficiency of zinc

S. no.	Disease	Effect of zinc deficiency	References
1.	Acrodermatitis enteropathica	Infections due to reduced count of lymphocyte caused by atrophy of the thymus function and size, and ultimately death can take place It is hereditary deficiency disease manifested by hypertrophic skin, baldness, ulcerated and chronic diarrhoea It is a rare autosomal inherited recessive disorder caused due to reduced intestinal absorption of zinc	Barceloux (1999)
2.	Inflammatory bowel syndrome	Affects several signalling and cellular pathways involved in oxidative stress as well as alters the gut microbiota	Gilca-Blanariu et al. (2018)
3	Celiac disease	Decreased growth rate, anorexia Reduced concentration of serum zinc Insoluble zinc complexes were lost with phosphate and fat	Altuntaş et al. (2000) and Fathi et al. (2013)
4	Hypogonadal dwarf syndrome	Retardation in growth, hypogonadism associated with dysosmia, dysgeusia, anaemia and poor wound healing	Jurowski et al. (2014)
5	Diarrhoeal disease	Reduced dietary intake initially and then followed by increased losses via stool	Semrad (1999)
6	Depression	Psychiatric disorder It is characterized by lower serum zinc levels	Maes et al. (1994)
7	Cardiovascular	Arterial hypertension Atherosclerosis, sudden deaths due to heart attack or heart failure In children if serum zinc level is extremely lower than congestive heart failure may take place. This disease is linked with deficiency of zinc that causes reduction in dietary intake, decreased absorption and impaired motility, enhanced loss of zinc and excessive loss through urine	Tomat et al. (2011), Tubek (2007a, b), Little et al. (2010), and Cohen and Golik (2006)
8	Renal disease	Renal function is associated with zinc Severe zinc deficiency induces reduced function of renal blood flow and glomerular filtration rate Hypozincaemia increased excretion of urinary zinc	Bao et al. (2010)
9	Alzheimer's disease	Oxidant damage can take place in the brain, and the tangles in neurofibrillary and amyloid plaques can produce toxic oxidant radicals with With age, zinc serum concentration reduces, and patients having this disease are more prone to decline the zinc level more rapidly Extracellular amyloid plaques are known for binding zinc that further reduces the availability of zinc for neurons	Curtain et al. (2001), Sayre et al. (2000), Grant (1997), Brewer et al. (2010), and Smart et al. (2004)

(continued)

Table 4.4 (continued)

S. no.	Disease	Effect of zinc deficiency	References
10	Wilson's disease	It is recessive inherited autosomal inborn disease caused due to abnormal handling of copper by liver Neuropsychiatric symptoms	Corona et al. (2010)
11	Cystic fibrosis	Intestinal mucosal abnormalities Hypozincaemia has been reported in young infants, especially in newborns Flattering of growth is observed in older children	Krebs et al. (1998a, b)

Interventional Strategies for Zinc Deficiency

Supplementation of zinc has shown numerous health-promoting benefits (Maret and Sandstead 2006). The consumption of zinc where the diets contain inadequate quantity of this element should adopt major strategies for improvement of zinc-based diet like dietary modification, fortification, supplementation or bio-fortification. Dietary diversification is used to improve the consumption of wide range of micronutrients. This technique has various potentials to enhance the consumption of zinc bioavailability. These interventional strategies include promotion of animal source diet, agricultural practices and processing practices to increase absorption of zinc from plant-based foods (Gibson and Anderson 2009). During processing, the bioavailability of zinc is enhanced by reducing the levels of other compounds that inhibit the absorption of zinc like phytate. However, this is not enough to meet the zinc needs of babies and young generation. Diet rich in red meat and liver intake has positive effect on the infants and young children. Cheese and milk are also an essential source of zinc (Roohani et al. 2012b). Special care must be taken in order to avoid the adverse effects of the contaminants present in the germinated cereals like aflatoxin and loss of water-soluble essential nutrients by soaking cereal.

Food fortification is increasingly recognized and practiced as an effective approach to improve a population's micronutrient status. Fortification of food is a cost-economical and sustainable technology to overcome the malnutrition of micronutrients in comparison to supplementation. As supplementation process includes the additional/special programmes on daily or weekly basis to deliver the lectures for improving the deficiency of iron such as anaemia and other micronutrient deficiencies. Supplemental zinc is also given to the children suffering from diarrhoea as an adjunct therapy (Fontaine 2001). Fortification programmes are also needed to enhance the consumption of zinc in high-risk groups like newborn babies and young children who only consume particular type of food. Complementary food like infant formula milk is fortified with several micronutrients including zinc. Commercially available infant formula milk contains 1 mg/L of zinc concentration. Fortification is done for selected foods that are consumed widely in stable and predictable amounts. $ZnSO_4$ and zinc oxide are most commonly used as it is cost-effective and is mostly used by food industry. Staple food, such as wheat and corn, is also fortified with $ZnSO_4$ that provides more absorbable zinc as it has greater solubility, but it is a bit expensive.

Lonnerdal (2003) reported that bioavailability of zinc is improved by enhancing the concentration of zinc and decreases the inhibitor content or enhances the absorption of zinc containing compounds like amino acids. Bio-fortification is different from fortification process as it focuses on the internal enrichment of essential nutrients in still growing plant parts, rather than adding micronutrients to the food during processing (Banuelos and Lin 2009). This bio-fortification technique is an improvement strategy for poor rural people for providing essential nutrients to them. It also includes the application of zinc fertilizers that help in the development of crop rapidly and acquires large quantity of zinc from the soil and thus accumulates it in the edible part of the plant (White and Broadley 2011).

Conclusion

Zinc deficiency remains as an existential threat to the people around the globe. In developing and developed countries, malnutrition during the critical stage of life, that is, during the growth period, has become a vital health issue, especially among the young children, infants and pregnant ladies having an imbalanced diet. The consequences of zinc deficiency are the impairment of cognitive functions and retardation of children's growth and development. This threat is more prone to the children of age group below 5 and low- and middle-income countries of the world. Zinc transporters play a critical role in understanding the physiological, molecular and biochemical role of zinc. There are some potential strategies that overcome the deficiency of zinc like supplementation, fortification and dietary modification. These techniques have very well explained the beneficial effects of zinc on children health all around the globe.

Future Recommendations

Deficiency of zinc mostly occurs in association with other micronutrient deficiencies. Further research is needed regarding their interaction with each other and with other micronutrients and with the absorption of zinc from mixed diet. There is a need of an hour to examine and plan effective strategies in order to improve the zinc nutrition of malnutrition groups. Fortifying food with zinc rather than in combination with other micronutrients may improve serum zinc status. Mandatory fortification of staple food with zinc is likely to reach all people. It will also ensure access and equity of zinc adequacy in children and women especially those belonging to low socio-economic background.

References

Abrams, S.A., Wen, J., Stuff, J.E.: Absorption of calcium, zinc, and iron from breast milk by five- to seven-month-old infants. Pediatr. Res. **41**(3), 384–390 (1997)

Altuntaş, B., Filik, B., Ensari, A., Zorlu, P., Teziç, T.: Can zinc deficiency be used as a marker for the diagnosis of celiac disease in Turkish children with short stature? Pediatr. Int. **42**, 682–684 (2000)

Brion, L.P., Heyne, R., Lair, C.S.: Role of zinc in neonatal growth and brain growth: review and scoping review. Pediatr. Res. **89**(7), 1627–1640 (2021)

Jurowski, K., Szewczyk, B., Nowak, G., Piekoszewski, W.: Biological consequences of zinc deficiency in the pathomechanisms of selected diseases. J. Biol. Inorganic. Chem. **19**(7), 1069–1079 (2014)

Auld, D.S.: Metalloproteases. In: Biomedical Sciences Encyclopedia of Biological Chemistry, pp. 86–89 (2013)

Banuelos, G., Lin, Z.Q.: Phytoremediation of selenium-contaminated soil and water produces biofortified products and new agricultural byproducts. In: Development and Uses of Biofortified Agricultural Products, pp. 57–70. CRC Press, Boca Raton (2009)

Bao, B., Prasad, A., Beck, F., Fitzgerald, J., Snell, D., Bao, G., Singh, T., Cardozo, L.: Zinc decreases C-reactive protein, lipid peroxidation, and inflammatory cytokines in elderly subjects: a potential implication of zinc as an atheroprotective agent. Am. J. Clin. Nutr. **91**(6), 1634–1641 (2010)

Barbarot, S., Chantier, E., Kuster, A., Hello, M., Roze, J.C., Blouin, E., Stalder, J.F.: Symptomatic acquired zinc deficiency in at-risk premature infants: high dose preventive supplementation is necessary. Pediatr. Dermatol. **27**(4), 380–383 (2010)

Barceloux, D.: Zinc. Clin. Toxicol. **37**(2), 279–292 (1999)

Bartick, M.C., Jegier, B.J., Green, B.D., Schwarz, E.B., Reinhold, A.G., Stuebe, A.M.: Disparities in breastfeeding: impact on maternal and child health outcomes and costs. J. Pediatr. **181**, 49–55 (2017)

Bass, L.M., Steinhorn, D.M.: Chapter 85 – Gastrointestinal structure and function. In: Pediatric Critical Care, 4th edn, pp. 1207–1220 (2011)

Bhagavan, N.V.: Chapter 37 – mineral metabolism. In: Medical Biochemistry, 4th edn, pp. 873–900 (2002)

Bhatnagar, S., Natchu, U.C.: Zinc in child health and disease. Indian J. Pediatr. **71**(11), 991–995 (2004)

Black, R.E., Victora, C.G., Walker, S.P., Bhutta, Z.A., Christian, P., de Onis, M., Ezzati, M., Grantham-McGregor, S., Katz, J., Martorell, R., Uauy, R.: Maternal and child undernutrition and overweight in low-income and middle-income countries. Lancet. **382**(9890), 427–451 (2013)

Bleichrodt, N., Born, M.: A meta-analysis of research on iodine and its relationship to cognitive development. In: Stansbury, J.B. (ed.) The Damaged Brain of Iodine Deficiency, pp. 195–200. Cognizant Communication Corporation, New York (1994)

Brewer, G., Kanzer, S., Zimmerman, E., Molho, E., Celmins, D., Heckman, S., Dick, R.: Subclinical zinc deficiency in Alzheimer's disease and Parkinson's disease. Am. J. Alzheimers Dis. Other Dement. **7**, 572–575 (2010)

Brown, K.H.: Commentary: zinc and child growth. Int. J. Epidemiol. **32**(6), 1103–1104 (2003)

Brown, K.H.: Information gaps for scaling-up programs to improve zinc nutrition. Clin. Nutr. **72**, 1516–1522 (2007)

Brown, L., Pollit, E.: Malnutrition, poverty and intellectual development. Sci. Am. **274**(2), 38–43 (1996)

Brown, K., Wuehler, S., Peerson, J.: Handbook on toxicology of metals. Food Nutr. Bull. **22**, 113–125 (2001)

Brown, K.H., Rivera, J.A., Bhutta, Z., Gibson, R.S., King, J.C., Lönnerdal, B., Ruel, M.T., Sandtröm, B., Wasantwisut, E., Hotz, C.: International Zinc Nutrition Consultative Group

(IZiNCG) technical document #1. Assessment of the risk of zinc deficiency in populations and options for its control. Food Nutr. Bull. **25**, S99–S203 (2004)

Calesnick, B., Dinan, A.: Zinc deficiency and zinc toxicity. Am. Fam. Physician. **37**, 267–270 (1988)

Caulfield, L.E., Zavaleta, N., Figueroa, A.: Adding zinc to prenatal iron and folate supplements improves maternal and neonatal zinc status in a Peruvian population. Am. J. Clin. Nutr. **69**, 1257–1263 (1999)

Cohen, N., Golik, A.: Zinc balance and medications commonly used in the management of heart failure. Heart Fail. Rev. **11**, 19–24 (2006)

Corona, C., Masciopinto, F., Silvestri, E., Viscovo, A., Lattanzio, R., Sorda, R., Ciavardelli, D., Goglia, F., Piantelli, M., Canzoniero, L., Sensi, S.: Dietary zinc supplementation of 3xTg-AD mice increases BDNF levels and prevents cognitive deficits as well as mitochondrial dysfunction. Cell Death Dis. **1**(10), e91 (2010)

Cousins, R.J., Liuzzi, J.P., Lichten, L.A.: Mammalian zinc transport, trafficking, and signals. J. Biol. Chem. **281**(34), 24085–24089 (2006)

Curtain, C.C., Ali, F., Volitakis, I., Cherny, R.A., Norton, R.S., Beyreuther, K., Barrow, C.J., Masters, C.L., Bush, A.I., Barnham, K.J.: Alzheimer's disease amyloid-β binds copper and zinc to generate an allosterically ordered membrane-penetrating structure containing superoxide dismutase-like subunits. J. Biol. Chem. **276**(23), 20466–20473 (2001)

Davidsson, L., Mackenzie, J., Kastenmayer, P., Aggett, P.J., Hurrell, R.F.: Zinc and calcium apparent absorption from an infant cereal: a stable isotope study in healthy infants. Br. J. Nutr. **75**(2), 291–300 (1996)

Devergnas, S., Chimienti, F., Naud, N., Pennequin, A., Coquerel, Y., Chantegrel, J., Favier, A., Seve, M.: Differential regulation of zinc efflux transporters ZnT-1, ZnT-5 and ZnT-7 gene expression by zinc levels: a real-time RT-PCR study. Biochem. Pharmacol. **68**(4), 699–709 (2004)

Engle-Stone, R., Ndjebayi, A.O., Nankap, M., Killilea, D.W., Brown, K.H.: Stunting prevalence, plasma zinc concentrations, and dietary zinc intakes in a nationally representative sample suggest a high risk of zinc deficiency among women and young children in Cameroon. J. Nutr. **144**(3), 382–391 (2014)

FAO/WHO: Expert consultation on human vitamin and mineral requirements, vitamin and mineral requirements in human nutrition: report of joint FAO/WHO expert consolation 341 (n.d.)

Fathi, F., Ektefa, F., Tafazzoli, M., Rostami, K., Nejad, M.R., Fathi, M., Rezaei-Tavirani, M., Oskouie, A.A., Zali, M.R.: A concentration of serum zinc in celiac patients compare to healthy subject in Tehran. Gastroenterol. Hepatol. Bed Bench. **6**(2), 92–95 (2013)

Favier, A.E.: The role of zinc in reproduction. Hormonal mechanisms. Biol. Trace Elem. Res. **32**, 363–382 (1992)

Lonnerdal, B.: Dietary factors influencing zinc absorption. J. Nutr. **130**(5), 1378S–1383S (2000)

Liu, E., Pimpin, L., Shulkin, M., Kranz, S., Duggan, C.P., Mozaffarian, D., Fawzi, W.W.: Effect of zinc supplementation on growth outcomes in children under 5 years of age. Nutrients. **10**(3), 377 (2018)

Ogbo, F.A., Nguyen, H., Naz, S., Agho, K.E., Page, A.: The association between infant and young child feeding practices and diarrhoea in Tanzanian children. Trop. Med. Health. **46**(1), 2 (2018)

Fontaine, O.: Effect of zinc supplementation on clinical course of acute diarrhoea. J. Health Popul. Nutr. **19**(4), 339–346 (2001)

Foster, M., Samman, S.: Zinc and regulation of inflammatory cytokines: implications for cardiometabolic disease. Nutrients. **4**(7), 676–694 (2012)

Fraker, P.J., King, L.E.: Reprogramming of the immune system during zinc deficiency. Annu. Rev. Nutr. **24**, 277–298 (2004)

Fukada, T., Kambe, T.: Molecular and genetic features of zinc transporters in physiology and pathogenesis. Metallomics. **3**(7), 662–674 (2011)

Fung, E.B., Ritchie, L.D., Woodhouse, L.R., Roehi, R., King, J.C.: Zinc absorption in women during pregnancy and lactation: a longitudinal study. Am. J. Clin. Nutr. **66**(1), 80–88 (1997)

Gilca-Blanariu, G.E., Diaconescu, S., Ciocoiu, M., Stefanescu, G.: New insights into the role of trace elements in IBD. Biomed. Res. Int. **24**, 1–9 (2018)

Gibson, R.S., Anderson, V.P.: A review of interventions based on dietary diversification or modification strategies with the potential to enhance intakes of total and absorbable zinc. Food Nutr. Bull. **30**(1), S108–S143 (2009)

Gibson, R.S., Heath, A.L., Prosser, N., Parnell, W., Donovan, U.M., Green, T., McLaughlin, K.E., O'Connor, D.L., Bettger, W., Skeaff, C.M.: Are young women with low iron stores at risk of zinc as well as iron deficiency? In: Roussel, A.M., Anderson, R.A., Favrier, A. (eds.) Trace Elements in Man and Animals, vol. 10, pp. 323–328. Kluwer Academic, New York (2000)

Gibson, R.S., Vanderkooy, P.D., MacDonald, A.C., Goldman, A., Ryan, B.A., Berry, M.: A growth-limiting, mild zinc-deficiency syndrome in some southern Ontario boys with low height percentiles. Am. J. Clin. Nutr. **49**(6), 1266–1273 (1989)

Gibson, R.S.: Zinc: the missing link in combating micronutrient malnutrition in developing countries. Proc. Nutr. Soc. **65**(1), 51–60 (2006)

Godana and Mengistie: Determinants of acute diarrhoea among children under five years of age in Derashe District, Southern Ethiopia. Rural Remote Health. **13**(3), 2329 (2013)

Golub, M., Keen, C., Gershwin, M., Hehdrickx, A.: Developmental zinc deficiency and behavior. J. Nutr. **125**(8), 2263S–2271S (1995)

Grant, W.: Dietary links to Alzheimer's disease. Alzheimers Dis. Rev. **2**, 42–55 (1997)

Grungreiff, K., Reinhold, D.: Zinc: a complementary factor in the treatment of chronic hepatitis C? Mol. Med. Rep. **3**(3), 371–375 (2010)

Hambidge, K.M., Krebs, N.F., Jacobs, M.A., Favier, A., Guyette, L., Ikle, D.N.: Zinc nutritional status during pregnancy: a longitudinal study. Am. J. Clin. Nutr. **37**, 429–442 (1983)

Hambidge, K.M., Miller, L.V., Westcott, J.E., Sheng, X., Krebs, N.F.: Zinc bioavailability and homeostasis. Am. J. Clin. Nutr. **91**(5), 1478S–1483S (2010)

Hambidge, K.M., Miller, L.V., Krebs, N.F.: Physiological requirements for zinc. Int. J. Vitamin Nutr. Res. **81**(1), 72–78 (2011)

Hambidge, M., Krebs, N.F.: Interrelationships of key variables of human zinc homeostasis: relevance to dietary zinc requirements. Annu. Rev. Nutr. **21**, 429–452 (2001)

Ugboko, H.U., Nwinyi, O.C., Oranusi, S.U., Oyewale, J.O.: Childhood diarrhoeal diseases in developing countries. Heliyon. **6**(4), e03690 (2020)

Yanagisawa, H.: Zinc deficiency and clinical practice. J. Jpn. Med. Assoc. **47**(8), 613–616 (2004)

Yanagisawa, H., Wada, O.: Clinical aspects of zinc deficiency. J. Jpn. Med. Assoc. **127**(2), 261–268 (2002)

Yanagisawa, H., Kurihara, N., Wada, O.: Zinc—Extensive blood and urine biochemistry and immunological tests (2). Jpn. J. Clin. Med. **57**(extra issue), 282–286 (1999)

Hurrell, R.F.: Phytic acid degradation as a means of improving iron absorption. Int. J. Vitam. Nutr. Res. **74**(6), 445–452 (2004)

Institute of Medicine: Dietary Reference Intakes for Vitamin a, Vitamin K, Arsenic, Boron, Chromium, Copper, Iodine, Iron, Manganese, Molybdenum, Nickel, Silicon, Vanadium, and Zinc. The National Academies Press, Washington, DC (2001). https://doi.org/10.17226/10026

Jameson, S.: Zinc status in pregnancy: the effect of zinc therapy on perinatal mortality, prematurity, and placental ablation. Ann. N. Y. Acad. Sci. **678**, 178–192 (1993)

Hautvast, J.L.A., Tolboom, J.J.M., Kafwembe, E.M., Musonda, R.M., Mwanakasale, V., van Staveren, W.A., van't Hof, M.A., Sauerwein, R.W., Willems, J.L., Monnens, L.A.H.: Severe linear growth retardation in rural Zambian children: the influence of biological variables. Am. J. Clin. Nutr. **71**(2), 550–559 (2000)

Kambe, T., Hashimoto, A., Fujimoto, S.: Current understanding of ZIP and ZnT zinc transporters in human health and diseases. Cell. Mol. Life Sci. **71**(17), 3281–3295 (2014)

King, J.C., Cousins, R.J.: Zinc. In: Shils, M.E., Shike, M., Ross, A.C., Caballero, B., Cousins, R.J. (eds.) Modern Nutrition in Health and Disease, 10th edn, pp. 271–285. lippincott williams and wilkins, Baltimore (2006)

King, J.C., Keen, C.L.: Zinc. In: Shils, M.E., Olson, J.A., Shike, M., Ross, C.A. (eds.) Modern Nutrition in Health and Disease, 9th edn, pp. 223–239. Lippinkott Williams & Wilkins, New york (2003)

King, J.C.: Zinc: an essential but elusive nutrient. Am. J. Clin. Nutr. **94**(2), 679S–684S (2011)

Krebs, N.F., Hambidge, K.M., Jacobs, M.A., Rasbach, J.O.: The effects of a dietary zinc supplement during lactation on longitudinal changes in maternal zinc status and milk zinc concentrations. Am. J. Clin. Nutr. **41**(3), 560–570 (1985)

Krebs, N.F., Reidinger, C.J., Hartley, S., Robertson, A.D., Hambidge, K.M.: Zinc supplementation during lactation: effects on maternal status and milk zinc concentrations. Am. J. Clin. Nutr. **61**(5), 1030–1036 (1995)

Krebs, N.F., Sontag, M., Accurso, F.J., Hambidge, K.M.: Low plasma zinc concentrations in young infants with cystic fibrosis. J. Pediatr. **133**(6), 761–764 (1998a)

Krebs, N.F., Westcott, J.E., Huffer, J.W., Miller, L.V.: Absorption of exogenous zinc and secretion of endogenous zinc in the human small intestine. FASEB J. **12**, A345 (1998b)

Kuczmarski, R.J., Ogden, C.L., Grummer-Strawn, L.M., Flegal, K.M., Guo, S.S., Wei, R., Mei, Z., Curtin, L.R., Roche, A.F., Johnson, C.L.: CDC Growth Charts: United States. Advance data from vital health statistics. No. 314:1–27. National Center for Health Statistics, Hyattesville (2000)

Little, P.J., Bhattacharya, R., Moreyra, A.E., Korichneva, I.L.: Zinc and cardiovascular disease. Nutrition. **26**(11–12), 1050–1057 (2010)

Lonnerdal, B.: Genetically modified plants for improved trace element nutrition. J. Nutr. **133**(5 suppl 1), 1490S–1493S (2003)

Maes, M., D'Haese, P., Scharpe, S., D'Hondt, P., Cosyns, P., De Broe, M.: Hypozincemia in depression. J. Affect. Disord. **31**(2), 135–140 (1994)

Maret, W., Sandstead, H.H.: Zinc requirements and the risks and benefits of zinc supplementation. J. Trace Elem. Med. Biol. **20**(1), 3–18 (2006)

Salgueiro, M.J., Zubillaga, M.B., Lysionek, A.E., Caro, R.A., Weill, R., Boccio, J.R.: The role of zinc in the growth and development of children. Nutrition. **18**(6), 510–519 (2002)

Maureen, M.: Black zinc deficiency and child development. Am. J. Clin. Nutr. **68**(2 Suppl), 464S–469S (1998)

Mittal, P., Dipti, B.: Role of zinc in malnutrition. Ann. Nutr. Disord. Ther. **3**(1), 1029 (2016)

Moser-Veillon, P.B., Reynolds, R.D.: A longitudinal study of pyridoxine and zinc supplementation of lactating women. Am. J. Clin. Nutr. **52**(1), 135–141 (1990)

Muller, O., Becher, H., van Zweeden, A.B., Ye, Y., Diallo, D.A., Konate, A.T., ... Garenne, M.: Effect of zinc supplementation on malaria and other causes of morbidity in west African children: randomised double blind placebo controlled trial. Bmj. **322**(7302), 1567 (2001)

Krebs, N.F.: Overview of zinc absorption and excretion in the human gastrointestinal tract. J. Nutr. **130**(5), 1374S–1377S (2000)

Roohani, N., Hurrell, R., Kelishadi, R., Schulin, R.: Zinc and its importance for human health: an integrative review. J. Res. Med. Sci. **18**(2), 144–157 (2013)

Nguyen, P.H., Avula, R., Ruel, M.T., Saha, K.K., Ali, D., Tran, L.M., Frongillo, E.A., Menon, P., Rawat, R.: Maternal and child dietary diversity are associated in Bangladesh, Vietnam, and Ethiopia. J. Nutr. **143**(7), 1176–1183 (2013)

Ortega, R.M., Andres, P., Martinez, R.M., Lopez-Sobaler, A.M., Quintas, M.E.: Zinc levels in maternal milk: the influence of nutritional status with respect to zinc during the third trimester of pregnancy. Eur. J. Clin. Nutr. **51**(4), 253–258 (1997)

Penny, M.E.: Zinc supplementation in public health. Ann. Nutr. Metab. **62**(Suppl 1), 31–42 (2013)

Shrimpton, R., Shankar, A.H.: Chapter 15 zinc deficiency. In: Nutrition and Health in Developing Countries, 2nd edn, pp. 455–478. Human Press (2008)

Roohani, N., Hurrell, R., Wegmueller, R., Schulin, R.: Zinc and phytic acid in major foods consumed by a rural and a suburban population in central Iran. J. Food Compos. Anal. **28**(1), 8–15 (2012a)

Roohani, N.: Human zinc nutrition in arid regions with zinc deficiency in soils and crops: a case study in central Iran. Doctoral dissertation, ETH Zurich (2012b)

Ruff, H.A., Rothbart, M.K.: Attention in Early Development. Oxford University Press, New York (1996)

Russell, R., Beard, J.L., Cousins, R.J., Dunn, J.T., Ferland, G., Hambidge, K.M., ... Yates, A.A.: Dietary reference intakes for vitamin A, vitamin K, arsenic, boron, chromium, copper, iodine, iron, manganese, molybdenum, nickel, silicon, vanadium, and zinc. A Report of the Panel on Micronutrients, Subcommittees on Upper Reference Levels of Nutrients and of Interpretation and Uses of Dietary Reference Intakes, and the Standing Committee on the Scientific Evaluation of Dietary Reference Intakes Food and Nutrition Board Institute of Medicine (2001)

Saper, R.: Zinc: An Essential Micronutrient. Am. Fam. Physician. **79**(9), 768–772 (2009)

Sayre, L., Perry, G., Harris, P., Liu, Y., Schubert, K., Smith, M.: In situ oxidative catalysis by neurofibrillary tangles and senile plaques in Alzheimer's disease: a central role for bound transition metals. J. Neurochem. **74**(1), 270–279 (2000)

Schlemmer, U., Frølich, W., Prieto, R.M., Grases, F.: Phytate in foods and significance for humans: food sources, intake, processing, bioavailability, protective role and analysis. Mol. Nutr. Food Res. **53**(2), S330–S375 (2009)

Sekler, I., Sensi, S.L., Hershfinkel, M., Silverman, W.F.: Mechanism and regulation of cellular zinc transport. Mol. Med. **13**(7–8), 337–343 (2007)

Semrad, C.: Zinc and intestinal function. Curr. Gastroenterol. Rep. **1**, 398–403 (1999)

Siregar, A.Y., Pitriyan, P., Walters, D.: The annual cost of not breastfeeding in Indonesia: the economic burden of treating diarrhea and respiratory disease among children (<24 mo) due to not breastfeeding according to recommendation. Int. Breastfeed J. **13**(1), 10 (2018)

Smart, T.G., Hosie, A.M., Miller, P.S.: Zn^{2+} ions: modulators of excitatory and inhibitory synaptic activity. Neuroscientist. **10**(5), 432–442 (2004)

Kambe, T., Tsuji, T., Hashimoto, A., Itsumura, N.: The physiological, biochemical, and molecular roles of zinc transporters in zinc homeostasis and metabolism. Physiol. Rev. **95**(3), 749–784 (2015)

Takeda, A.: Zinc signaling in the hippocampus and its relation to pathogenesis of depression. Mol. Neurobiol. **44**, 166–174 (2011)

Takuma, Y., Nouso, K., Makino, Y., Hayashi, M., Takahashi, H.: Clinical trial: oral zinc in hepatic encephalopathy. Aliment. Pharmacol. Ther. **32**(9), 1080–1090 (2010)

Tapiero, H., Tew, K.D.: Trace elements in human physiology and pathology: zinc and metallothioneins. Biomed. Pharmacother. **57**(9), 399–411 (2003)

Terrin, G., Berni Canani, R., Di Chiara, M., Pietravalle, A., Aleandri, V., Conte, F., De Curtis, M.: Zinc in early life: a key element in the fetus and preterm neonate. Nutrients. **7**(12), 10427–10446 (2015)

Tomat, A.L., Costa Mde, L., Arranz, C.T.: Zinc restriction during different periods of life: influence in renal and cardiovascular diseases. Nutrition. **27**(4), 392–398 (2011)

Tomita, H.: Taste Disorder and Diet, pp. 3–140. Kodansha Ltd, Tokyo (2002) (in Japanese)

Tubek, S.: Role of zinc in regulation of arterial blood pressure and in the etiopathogenesis of arterial hypertension. Biol. Trace Elem. Res. **117**(1–3), 39–51 (2007a)

Tubek, S.: Selected zinc metabolism parameters in premenopausal and postmenopausal women with moderate and severe primary arterial hypertension. Biol. Trace Elem. Res. **116**(3), 249–256 (2007b)

Wada, L., King, J.C.: Effect of low zinc intakes on basal metabolic rate, thyroid hormones and protein utilization in adult men. J. Nutr. **116**(6), 1045–1053 (1986)

Walravens, P.A., Krebs, N.F., Hambidge, K.M.: Linear growth of low income preschool children receiving a zinc supplement. Am. J. Clin. Nutr. **38**(2), 195–201 (1983)

Wardlaw, G.M., Hampl, J.S.: Perspectives in Nutrition, 5th edn. McGraw-Hill, Boston (2007)

White, P.J., Broadley, M.R.: Physiological limits to zinc biofortification of edible crops. Front. Plant Sci. **2**, 80 (2011)

Zhong, W., Zhou, Z.: Chapter 8 – sealing the leaky gut represents a beneficial mechanism of zinc intervention for alcoholic liver disease. In: Dietary Interventions in Gastrointestinal Diseases, Foods, Nutrients, and Dietary Supplements, pp. 91–106. Academic Press (2019)

Chapter 5
Role of Zinc in Liver Pathology

Hadiya Husain and Riaz Ahmad

Abstract Liver diseases have a direct impact on zinc metabolism as it is involved in functions related to zinc homeostasis. Role of Zn in vital cellular processes like DNA synthesis, cell division and RNA transcription is irreplaceable. Therefore, Zn is crucial to the normal functioning of various organs, especially liver. It has been demonstrated that Zn intervention abrogates alcoholic liver diseases through processes like abatement of apoptotic hepatocyte death, reduction in proinflammatory cytokine and reduction of endotoxaemia. Nevertheless, Zn deficiency results in a number of health consequences including loss of appetite, changes in sense of smell and taste, reduction in drug expulsion ability and impairment of immune and cerebral functions. Mechanisms resulting in these health consequences and metabolic derangements due to Zn deficiency especially those involving liver health are being worked out. Some studies documenting the significance of Zn supplementation and its hepatoprotective role against alcohol-induced liver injury are available. This chapter therefore defines various metabolic anomalies caused due to Zn deficiency and also the beneficial ramifications of Zn supplementation during a variety of liver diseases.

Keywords Zn homeostasis · Zn absorption · Liver pathology · Zn and liver diseases · Hepatic encephalopathy

Zinc is one of the essential micronutrients having various properties like anti-inflammatory, anti-apoptotic and anti-oxidant activity (Leoni et al. 2014). It is one of those trace elements which is actively involved in basic physiological processes at cellular level, wherein it interacts and regulates the activity of several proteins involved in various metabolic reactions (Leoni et al. 2014). The symptoms of zinc deficiency in animals or subjects suffering from the ailment acrodermatitis enteropathica, which is the hereditary disease of malabsorption of zinc, contribute to the

H. Husain · R. Ahmad (✉)
Section of Genetics, Department of Zoology, Faculty of Life Sciences, Aligarh Muslim University, Aligarh, Uttar Pradesh, India

S. T. Khan, A. Malik (eds.), *Microbial Biofertilizers and Micronutrient Availability*,
https://doi.org/10.1007/978-3-030-76609-2_5

information regarding the various functions of zinc in humans (McClain et al. 1986). Gradually it has become evident that symptoms of the body related to zinc deficiency clinically and biochemically appear as an after effect of some kind of adversity like malabsorption and diarrhoea on the organism (Gaetke et al. 1997; McClain et al. 1993). Much work has been done in various areas related with zinc in the context of transportation, cellular processes in the physiology, absorption, excretion, etc. These realms hold the potential of becoming the major prospective sectors that concentrate on zinc and related health.

Zinc is a trace element known since a long time for its multiple functions which remain crucial in many processes of the body and also for the healing properties it possesses. It has also been recognized in Ayurveda and is utilized in several forms including zinc metal form (Yasada), calcified form (Sodhana and Marana), zinc oxide (Pushpanjana), zinc carbonate (Kharpara) or alloy form (Pittala) as therapeutic agents for the treatment of diseases. All these configurations have been documented in various places since fourteenth century, for their remedial attributes related to liver diseases, gastrointestinal disorders, bacterial and microbial infections, diabetes, etc. (Kaur et al. 2014). In the subjects suffering from geriatric diseases, the oral administration of Zn has been utilized for its therapeutic properties (Haase et al. 2006). In the environment, zinc comes in the second place after iron for its abundance as a trace element which is also essential for all organisms (Kambe et al. 2015). However, its essentiality was not recognized in the form of a micronutrient for the health of humans until 1961 which after deficiency causes skin deformities, serious anaemia, growth reduction, hypogonadism, mental illness, etc. (Prasad et al. 1961). Various manifestations of zinc deficit have been documented after this ground breaking discovery which includes liver-related maladies, alopecia, severe diarrhoea, brain deformities, inflammation, immunity-related disorders and also several psychological deteriorations ranging from stress, irritability, emotional uncertainty, etc. (Devirgiliis et al. 2007; Hambidge 2000; Maret and Sandstead 2006; Prasad 1985; Sandstead 2013; Takeda and Tamano 2009; Vallee and Falchuk 1993). Thus, it is evident that a range of physiological symptoms relate with zinc-related deficit. It can be said that liver is a crucial organ on which the metabolism of zinc depends, and zinc also affects liver leading to various ailments as it decreases in the body. Zinc has various functions that it performs in the cell signalling and physiology of several proteins by modifying their molecular roles and is associated with vigorous growth and repair of cells (Jackson 1989). The fact that zinc is an essential micronutrient for human health was found by Prasad in 1960s. Therefore, the work onward has resulted in the establishment of various zinc-related information regarding diseases, aetiologies and manifestations. Major portion of the total concentration of zinc present in the body is mainly in the form of intracellular ions. It is present in all types of tissues of the body, in which muscles and bones have 85% and liver and skin possess 11%. The zinc, approximately 2–3% which is left, is present in all other tissues of the body (Grüngreiff 2002). The habit of consuming alcohol by people varies tremendously depending upon the choice and craving. Some of those who drink a lot but out of them only a small percentage lead to dangerous configurations of the liver ailments. This hints towards the possibility of reforming agents that lead to the ailment or abrogation of the malady. These agents can either be related to genes or to diet. Thus in simple terms, diet may also be a crucial reforming agent which has attributes that give the power to the subject and the clinician to abate the disease in question. Zinc is an essential micronutrient which is a significant dietary agent (McClain et al. 2017). However, the disproportion of the amount of ROS generation and antioxidant activity instigates disparate changes in context to the pathology of liver (Ahmad and Ahmad 2012; Ahmad et al. 2009; Husain et al. 2018; Latief and Ahmad 2018).

There have been several studies regarding zinc, and in recent times, there has been literally a boom in the research related with zinc in the body diseases, aetiologies, molecular mechanisms, roles of zinc at the molecular level, etc. (Prasad 1995). This has led to the information about the significance of zinc in the physiological actions, and also this information has been promoted by the finding of a huge amount of factors associated with absorption and excretion of zinc which are involved in the moderation of molecular mechanisms (Liuzzi and Cousins 2004). The deficit of zinc has been reported in various ailments (Stamoulis et al. 2007). Subjects suffering from cirrhosis together with alcoholic liver disease exhibited low levels of zinc in the body (Vallee and Wacker 1956). Over the past few years, this decrease in the serum level of zinc has been documented in various pathologies leading to disparate liver ailments (Gür et al. 1998). Some of these ailments are related with decrease in the serum zinc levels despite of the variation in pathologies, for example the viral hepatitis (Kalkan et al. 2002). The association between viral disease and zinc homeostasis may be discerned with the inference that the viral diseases lead to serious oxidative damage in the hepatic cells that causes apoptosis and further leading to a reduction in the protein formation. As a matter of fact, low amount of zinc in the serum corresponds with the low albumin amount in subjects undergoing hepatitis (Faa et al. 2008) (Fig. 5.1).

Role of Zinc in the Physiology

(i) *Zinc Homeostasis and Its Task*:

Various proteins which have peculiarly special roles in the physiology have zinc as their crucial constitutional components. One of those proteins are matrix metalloproteinases needing zinc for catalysis (Tapiero and Tew 2003) and which have the function of tearing down cartilages in rheumatoid arthritis (Faa et al. 2008) and fibrosis abatement in liver ailments (Arthur 1997). Moreover, 300 reactions of enzymes need zinc in their constitutional and functional parts (McCall et al. 2000). The absorption, transportation, stowage and discarding of zinc are possible with the help of several proteins that cause zinc homeostasis by a synchronized modulation (Hambidge and Krebs 2001). The main function of these proteins is to proffer the zinc metal ions to the proteins that have been freshly manufactured. These proteins perform crucial part in the constitutional conservation of chromatin and membranes of the cell, homeostasis and defence against oxidative deterioration (Powell 2000). The function of zinc is immensely pertinent in all the areas of cell physiology and

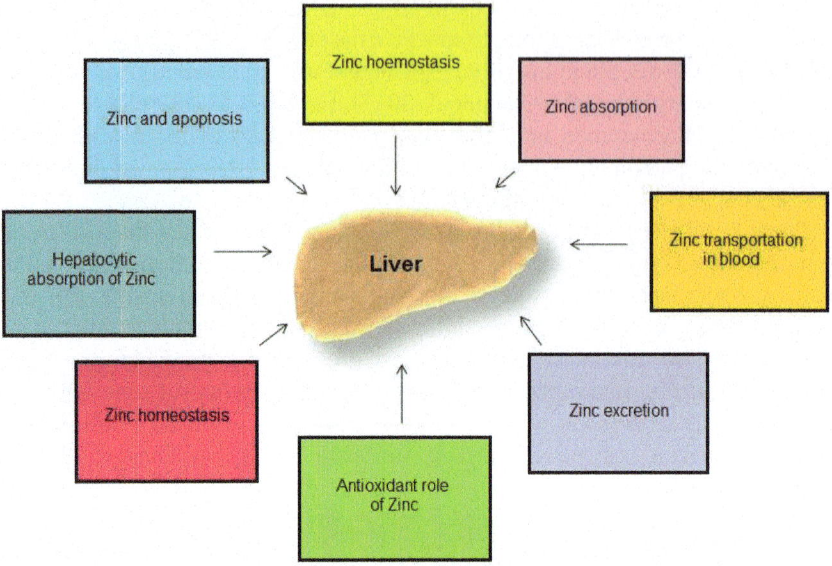

Fig. 5.1 Role of zinc in the physiology related with liver

modulation of various processes related to metabolism. As a matter of fact, minute fluctuations in the zinc levels in the body have the potential to influence the health in several physiological and molecular aspects mainly in aged patients (Chandra 2004).

(ii) *Zinc Absorption*:

The absorption of zinc in terms of quantity is adhered to various aspects which may importantly influence the absorption. The amount of zinc present in the food, existence of dietary influencers, physiology of the body all contribute to the need of zinc absorption (Krebs 2000). Zinc intake also depends upon the bulk of proteins present in a meal as the content of protein is directly proportional to the bioavailability and absorption of Zn in the intestine; a few types of proteins, for example soy protein and milk protein, have been documented to suppress the process of intake of zinc (Lönnerdal et al. 1984).

(iii) *Transportation of Zinc in Blood*:

In plasma, albumin plays the role of major transporting agent of zinc (Chester and Will 1981). The amino acids, for example cysteine and histidine, in proteins also create complexes in combination with zinc (Gachot et al. 1991). The amount of zinc in plasma is approximately 15 mol/L, out of which maximum percentage is in attachment with albumin, rest of it is bound with an alpha-2-macroglobulin and 1% to amino acids (Tapiero and Tew 2003).

(iv) *Absorption of Zinc by Hepatocytes*:

The transportation of zinc in the hepatocytes is not a simple process and includes various types of carrier proteins. In cultured liver cells, there have been two major pathways for the absorption of zinc which consist of a low affinity straight mechanism and a high affinity saturable mechanism (Taylor and Simons 1994). The major stowage of zinc in the enterocytes, hepatocytes and other cells is by way of its attachment to metalloproteins like gene-modulating molecules, metallozymes, zinc transporters and storage proteins (Vallee and Falchuk 1993). In hepatocytes, zinc is attached to metallothioneins and stored away.

(v) *Excretion of Zinc*:

Gastrointestinal secretions are the major path for discarding zinc from the body which is approximately 2.5–5.5 mg/day (Tapiero and Tew 2003). It takes place through the epithelial cells of the intestinal mucosa undergoing apoptosis or by the secretions of pancreas, intestine, salivary glands, biliary secretion, etc. Liver establishes zinc homeostasis by absorption of the micronutrient from plasma, stowage, metabolization of zinc and attaching them to several proteins, and reallocating zinc into disparate configurations into bile or blood (Ballatori 1991).

(vi) *Antioxidant Function of Zinc*:

In aerobic cell metabolism, there is a production of reactive oxygen species that emanates from the reactions of electron transport occurring in endoplasmic reticulum and mitochondria (Stehbens 2003). Zinc has a major attribute relating to the defence against reactive oxygen species in the process of shielding the biological parts; it also retains the amount of metallothioneins, and it is a mandatory constituent of superoxide dismutase (Stehbens 2003). It also provides defence for thiol groups by interceding with the reaction between thiol groups and other chemicals and stopping the creation of disulphide bonds (Gibbs et al. 1985). It also performs the task of intervening with the peroxidation of lipids in mitochondria and membranes of microsomes; it provides firmness to cell membrane (Tapiero and Tew 2003).

(vii) *Zinc and Apoptosis*:

It has been reported that fluctuations in zinc homeostasis directly affects and increases the vulnerability of cells towards apoptosis. This apparently is modulated by a distinct set of intracellular zinc (Truong-Tran et al. 2001). Apoptosis can be initiated by the instigation of caspase-3 and caspase-8 which is as a matter of fact caused by low amount of zinc (Duffy et al. 2001). In cell culture, some experiments related to the addition of zinc has exhibited to avert apoptosis instigated by tumour necrosis factor alpha (TNF alpha) (Seve et al. 2002). Zinc also has the function of maintaining the balance between anti-apoptotic protein Bcl-2 and proapoptotic protein Bax. This is attained by elevating the Bcl-2/Bax ratio which further causes an elevation in the cell defence against apoptosis (Fukamachi et al. 1998) (Fig. 5.2).

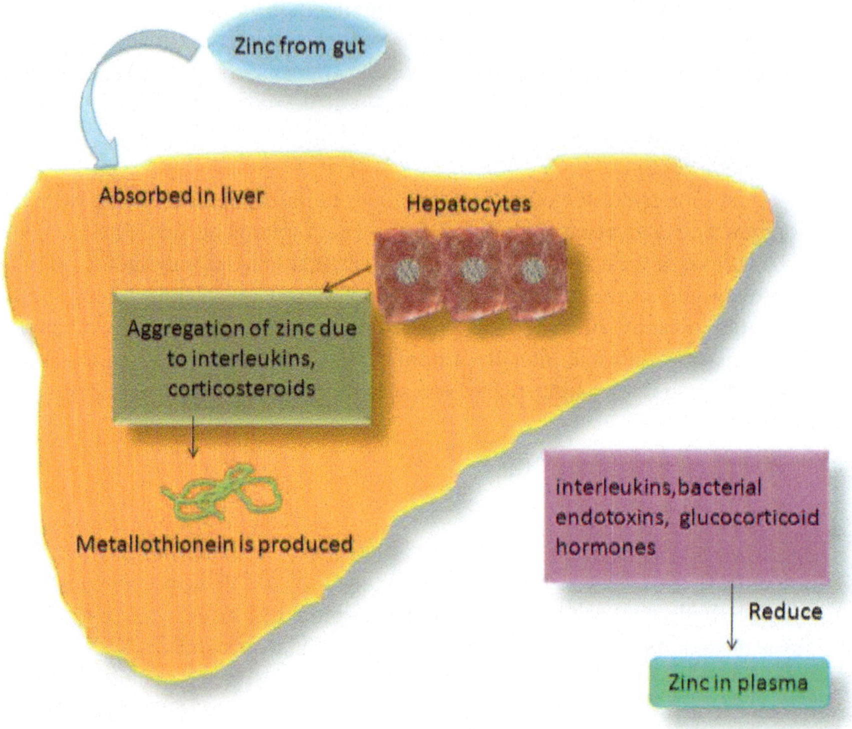

Fig. 5.2 Metabolism of zinc in liver

Metabolism of Zinc in the Liver

Out of the total concentration of zinc present in the body, only a small portion of it takes part in the actual metabolism of the body, and liver is the crucial organ playing a significant role in this process (Miller et al. 1994). Liver actively absorbs zinc after it is taken up from the gut (Cousins 1985). The presence of zinc in plasma is reduced by interleukins, bacterial endotoxins, glucocorticoid hormones, etc., while zinc is reallocated in the liver (Etzel et al. 1979; Klasing 1984). In cultured hepatocytes, corticosteroids and interleukins cause elevation in aggregation of zinc remarkably, and the manufacturing of metallothionein (MT) occurs which attaches to zinc (Failla and Cousins 1978; Schroeder and Cousins 1990). Zinc modulates metallothionein gene and subsequently metallothionein carries the task of an indicator of dietary zinc level in the human body (Davis and Cousins 2000). Increase in the levels of dietary zinc further increases amount of metallothionein present in the intestine (Cousins and Lee-Ambrose 1992). The synthesis of metallothionein apparently effects the aggregation of intestinal zinc (Richards and Cousins 1975). It has been documented that liver metallothionein instigation by endotoxin is modulated by interleukin-6 (Lee et al. 1999).

There is a sufficient amount of expeditiousness in the transportation of zinc between plasma and liver.

Deficiency of Zinc and Liver Diseases

Since in the past few years, several researches related to zinc in vivo and in vitro have significantly enhanced the apprehension of pathways in which zinc deficit may aggravate alcohol-induced liver injury and the process by which zinc may act as a therapeutic agent against this disease. These studies have provided proofs exhibiting a reduction in the zinc levels of tissues in mice supplemented with alcohol. This demonstrated that administration of alcohol directly affected zinc transportation which has a crucial function in the maintenance of zinc homeostasis (Sun et al. 2014). It has also been reported that alcohol administration led to a derangement in the intestinal-barrier role that was revamped by the administration of zinc (Zhong et al. 2010a, b). Deficiency of zinc is not to be taken so lightly as it has gone to become a potentially global health concern (Hambidge 2000; Prasad 2003; Sandstead 1991). Around 4% of the children mortality is caused by the zinc deficit and problems arising from it in the developing countries (Penny 2013). On the contrary, older age group is seen to be affected by the zinc deficit in the developed countries. Zinc administration in humans has the capability to reduce diarrhoea morbidity in children and occurrence of infections and also to enhance immunity (Haase et al. 2006; Penny 2013; Prasad et al. 2007; Tuerk and Fazel 2009). In terms of process, the fatty acid metabolic reaction in liver may be influenced by zinc deficit. It has been documented that a lipid-free intake of food directly affected the activity of delta9 desaturase (Kudo et al. 1990). The opposite function of zinc and cAMP on glycolysis, along with the significant reduction in free zinc amount which is instigated by glucagon, points out towards the potential role of zinc as a major mediator in the metabolism of carbohydrates (Brand and Kleineke 1996). Zinc is also related with obesity as leptin present in the serum is found to be indirectly proportional to the amount of zinc present in adipose tissues in mice administered with fatty diet (Tallman and Taylor 2003).

Excess of Zinc and Its Consequences Related with Liver

It has been reported that the elevation in glycosylated haemoglobin levels in blood indicate that surplus of zinc aids obesity and other ailments in adolescents and also patients suffering from diabetes become more vulnerable (Singh and Taneja 2009). Zinc overabundance also relates with the incidence of anaemia (Fiske et al. 1994). Zinc excess is harmful as it has been found that myeloneuropathy in some victims is caused by overabundance of zinc. Hyperzincaemia is a malady related with metabolism, and copper deficit follows the suit (Kumar and Ahlskog 2004).

Undersupply of Zinc and Liver Cirrhosis

Liver diseases, when they become fatal, take the form of liver cirrhosis. Various other maladies follow the suit together with the disease, for example hypoalbuminaemia. In this pathology, there is a parallel elevation of a specific macroglobulin which has a tendency to form strong bonds with zinc (Schechter et al. 1976) effectuating the discard of zinc through the urinary passage (Kiilerich and Christiansen 1986). Over and above, those subjects suffering from liver cirrhosis in many cases also had a deteriorated absorption system from the small intestine (Karayalcin et al. 1988; Riggio et al. 1992). This gives rise to dire zinc deficiency in the body of a liver cirrhotic patient.

Relation Between Zn Deficiency and Hepatic Encephalopathy

Generally, subjects suffering from liver cirrhosis also have this ailment of hepatic encephalopathy which consists of symptoms related with psychiatry. The actual aetiology of the disease has not yet been delineated; however, ammonia is recognized as one of the crucial factors. Out of all the subjects suffering from both hepatic encephalopathy and liver cirrhosis, most of the patients have high ammonia concentration (Grüngreiff 2002). An elevation in the deficit of zinc is a common factor found in hepatic encephalopathy patients suffering from liver cirrhosis (Yoshida et al. 2001). This deficit of zinc led to the disruption of the metabolic process of nitrogen by causing a decrease in the activity of glutamyl synthetase in muscle and ornithine transcarbamylase in liver (Grüngreiff et al. 1989; McClain et al. 1986). Furthermore, in subjects suffering from liver cirrhosis, the amount of ammonia in blood was inversely proportional to zinc content in the serum (Miatto et al. 1985).

Zinc and Liver Fibrosis

Zinc has a direct effect on the structural and physiological reorganization of liver fibrosis and cirrhosis which very much contributes to its part in the transmogrification of several liver-related pathologies (Dashti et al. 1997). The zinc concentrations in serum and liver are substantially decreased in the patients suffering from liver cirrhosis in comparison to normal humans (Goode et al. 1990).

Therapeutic Interventions of Zinc in Human Body

Zinc has attributes related to the attenuation of inflammation in subjects suffering from viral hepatitis C. It has been reported that the supplementation of a complex of zinc known as polaprezinc has demonstrated restorative influence over inflammation in liver (Sugino et al. 2008). This complex was also administered in a mouse model of NASH which resulted in abrogation of fibrosis but not steatosis in liver (Katayama et al. 2014). The symptoms of hepatic encephalopathy and high amount of ammonia were both evidently reduced by the administration of zinc for a considerable length of time in subjects suffering from cirrhosis (Himoto et al. 2007; Murakami et al. 2007; Takagi et al. 2001). Zinc also has been shown to have a positive effect in the abatement of hepatocellular carcinoma as subjects having viral hepatitis C were able to avert hepatocellular carcinoma from occurring by the supplementation of polaprezinc (Marchesini et al. 1996).

Conclusions

Metabolism of zinc is influenced very rapidly by one main organ, that is, liver, as it consists of a dynamic reservoir of zinc. Due to the possession of various attributes related with therapeutics including anti-apoptotic, anti-oxidant, etc., it can be said that zinc is crucial for a smooth functioning of liver. However, the role of zinc has not been fully explored, and it has been recommended for abrogation of a few diseases which include Wilson's disease, hepatic encephalopathy with cirrhosis in combination. It is evident that more information and research in context to the function of zinc in the aetiologies of various ailments of liver will suffice that zinc administration can abate more number of these maladies. Hence it is a much needed step to gauge the role of zinc further in order to restore the deficit of zinc in brain-related disorders as well. This chapter has marshalled the role of zinc in liver ailments and vice versa the part played by liver in zinc metabolism and recommends more work to be done on these lines for a better understanding and therapeutic vistas.

References

Ahmad, A., Ahmad, R.: Understanding the mechanism of hepatic fibrosis and potential therapeutic approaches. Saudi J. Gastroenterol. **18**(3), 155 (2012)

Ahmad, R., Ahmed, S., Khan, N.U., Hasnain, A.: *Operculina turpethum* attenuates N-nitrosodimethylamine induced toxic liver injury and clastogenicity in rats. Chem. Biol. Interact. **181**(2), 145–153 (2009)

Arthur, M.J.: Matrix degradation in liver: a role in injury and repair. Hepatology. **26**(4), 1069–1071 (1997)

Ballatori, N.: Mechanisms of metal transport across liver cell plasma membranes. Drug Metab. Rev. **23**(1–2), 83–132 (1991)

Brand, I.A., Kleineke, J.: Intracellular zinc movement and its effect on the carbohydrate metabolism of isolated rat hepatocytes. J. Biol. Chem. **271**(4), 1941–1949 (1996)

Chandra, R.K.: Impact of nutritional status and nutrient supplements on immune responses and incidence of infection in older individuals. Ageing Res. Rev. **3**(1), 91–104 (2004)

Chester, J.K., Will, M.: Zinc transport proteins in plasma. Br. J. Nutr. **46**, 111 (1981)

Cousins, R.J.: Absorption, transport, and hepatic metabolism of copper and zinc: special reference to metallothionein and ceruloplasmin. Physiol. Rev. **65**(2), 238–309 (1985)

Cousins, R.J., Lee-Ambrose, L.M.: Nuclear zinc uptake and interactions and metallothionein gene expression are influenced by dietary zinc in rats. J. Nutr. **122**(1), 56–64 (1992)

Dashti, H.M., Mathew, T.C., Jadaon, M.M., Ashkanani, E.: Zinc and liver cirrhosis: biochemical and histopathologic assessment. Nutrition. **13**(3), vi–212 (1997)

Davis, S.R., Cousins, R.J.: Metallothionein expression in animals: a physiological perspective on function. J. Nutr. **130**(5), 1085–1088 (2000)

Devirgiliis, C., Zalewski, P.D., Perozzi, G., Murgia, C.: Zinc fluxes and zinc transporter genes in chronic diseases. Mutat Res. **622**(1–2), 84–93 (2007)

Duffy, J.Y., Miller, C.M., Rutschilling, G.L., Ridder, G.M., Clegg, M.S., Keen, C.L., Daston, G.P.: A decrease in intracellular zinc level precedes the detection of early indicators of apoptosis in HL-60 cells. Apoptosis. **6**(3), 161–172 (2001)

Etzel, K.R., Shapiro, S.G., Cousins, R.J.: Regulation of liver metallothionein and plasma zinc by the glucocorticoid dexamethasone. Biochem. Biophys. Res. Commun. **89**(4), 1120–1126 (1979)

Faa, G., Nurchi, V.M., Ravarino, A., Fanni, D., Nemolato, S., Gerosa, C., Van Eyken, P., Geboes, K.: Zinc in gastrointestinal and liver disease. Coord. Chem. Rev. **252**(10–11), 1257–1269 (2008)

Failla, M.L., Cousins, R.J.: Zinc accumulation and metabolism in primary cultures of adult rat liver cells. Regulation by glucocorticoids. Biochim. Biophys. Acta. **543**(3), 293–304 (1978)

Fiske, D.N., McCoy III, H.E., Kitchens, C.S.: Zinc-induced sideroblastic anemia: report of a case, review of the literature, and description of the hematologic syndrome. Am. J. Hematol. **46**(2), 147–150 (1994)

Fukamachi, Y., Karasaki, Y., Sugiura, T., Itoh, H., Abe, T., Yamamura, K., Higashi, K.: Zinc suppresses apoptosis of U937 cells induced by hydrogen peroxide through an increase of the Bcl-2/Bax ratio. Biochem. Biophys. Res. Commun. **246**(2), 364–369 (1998)

Gachot, B., Tauc, M., Morat, L., Poujeol, P.: Zinc uptake by proximal cells isolated from rabbit kidney: effects of cysteine and histidine. Pflugers Arch. **419**(6), 583–587 (1991)

Gaetke, L., McClain, C.J., Talwalkar, R., Shedlofsky, S.: Effects of endotoxin on zinc metabolism in human volunteers. Am. J. Phys. **272**, E952–E956 (1997). [PubMed: 9227437]

Gibbs, P.N., Gore, M., Jordan, P.M.: Investigation of the effect of metal ions on the reactivity of thiol groups in human 5-aminolaevulinate dehydratase. Biochem. J. **225**(3), 573–580 (1985)

Goode, H.F., Kelleher, J., Walker, B.E.: Relation between zinc status and hepatic functional reserve in patients with liver disease. Gut. **31**(6), 694–697 (1990)

Grüngreiff, K.: Zinc in liver disease. J. Trace Elem. Exp. Med. **15**(1), 67–78 (2002)

Grüngreiff, K., Presser, H.J., Franke, D., Lössner, B., Abicht, K., Kleine, F.D.: Correlations between zinc, amino acids and ammonia in liver cirrhosis. Z. Gastroenterol. **27**(12), 731–735 (1989)

Gür, G., Bayraktar, Y., Ozer, D., Ozdogan, M., Kayhan, B.: Determination of hepatic zinc content in chronic liver disease due to hepatitis B virus. Hepato-Gastroenterology. **45**(20), 472–476 (1998)

Haase, H., Mocchegiani, E., Rink, L.: Correlation between zinc status and immune function in the elderly. Biogerontology. **7**, 421–428 (2006)

Hambidge, M.: Human zinc deficiency. J. Nutr. **130**(5), 1344S–1349S (2000)

Hambidge, M., Krebs, N.F.: Interrelationships of key variables of human zinc homeostasis: relevance to dietary zinc requirements. Annu. Rev. Nutr. **21**(1), 429–452 (2001)

Himoto, T., Hosomi, N., Nakai, S., Deguchi, A., Kinekawa, F., Matsuki, M., Yachida, M., Masaki, T., Kurokochi, K., Watanabe, S., Senda, S.: Efficacy of zinc administration in patients with hepatitis C virus-related chronic liver disease. Scand. J. Gastroenterol. **42**(9), 1078–1087 (2007)

Husain, H., Ahmad, R., Khan, A., Asiri, A.M.: Proteomic-genomic adjustments and their confluence for elucidation of pathways and networks during liver fibrosis. Int. J. Biol. Macromol. **111**, 379–392 (2018)

Jackson, M.J.: Physiology of zinc: general aspects. In: Zinc in Human Biology, pp. 1–14. Springer, London (1989)

Kalkan, A., Bulut, V., Avci, S., Celik, I., Bingol, N.K.: Trace elements in viral hepatitis. J. Trace Elem. Med. Biol. **16**(4), 227–230 (2002)

Kambe, T., Tsuji, T., Hashimoto, A., Itsumura, N.: The physiological, biochemical, and molecular roles of zinc transporters in zinc homeostasis and metabolism. Physiol. Rev. **95**(3), 749–784 (2015)

Karayalcin, S., Arcasoy, A., Uzunalimoglu, O.: Zinc plasma levels after oral zinc tolerance test in nonalcoholic cirrhosis. Dig. Dis. Sci. **33**(9), 1096–1102 (1988)

Katayama, K., Saito, M., Kawaguchi, T., Endo, R., Sawara, K., Nishiguchi, S., Kato, A., Kohgo, H., Suzuki, K., Sakaida, I., Ueno, Y.: Effect of zinc on liver cirrhosis with hyperammonemia: a preliminary randomized, placebo-controlled double-blind trial. Nutrition. **30**(11–12), 1409–1414 (2014)

Kaur, K., Gupta, R., Saraf, S.A., Saraf, S.K.: Zinc: the metal of life. Compr. Rev. Food Sci. Food Saf. **13**(4), 358–376 (2014) 19(suppl 1):132–133

Kiilerich, S., Christiansen, C.: Distribution of serum zinc between albumin and α2-macroglobulin in patients with different zinc metabolic disorders. Clin. Chim. Acta. **154**(1), 1–6 (1986)

Klasing, K.C.: Effect of inflammatory agents and interleukin 1 on iron and zinc metabolism. Am. J. Phys. Regul. Integr. Comp. Phys. **247**(5), R901–R904 (1984)

Krebs, N.F.: Overview of zinc absorption and excretion in the human gastrointestinal tract. J. Nutr. **130**(5), 1374S–1377S (2000)

Kudo, N., Nakagawa, Y., Waku, K.: Effects of zinc deficiency on the fatty acid composition and metabolism in rats fed a fat-free diet. Biol. Trace Elem. Res. **24**(1), 49–60 (1990)

Kumar, N., Ahlskog, J.E.: Myelopolyneuropathy due to copper deficiency or zinc excess? Arch. Neurol. **61**(4), 604–605 (2004)

Latief, U., Ahmad, R.: Herbal remedies for liver fibrosis: a review on the mode of action of fifty herbs. J. Tradit. Complement. Med. **8**(3), 352–360 (2018)

Lee, D.K., Carrasco, J., Hidalgo, J., Andrews, G.K.: Identification of a signal transducer and activator of transcription (STAT) binding site in the mouse metallothionein-I promoter involved in interleukin-6-induced gene expression. Biochem. J. **337**(1), 59–65 (1999)

Leoni, G., Rosato, A., Perozzi, G.: Zinc proteome interaction network as a model to identify nutrient-affected pathways in human pathologies. Genes Nutr. **9**, 436 (2014)

Liuzzi, J.P., Cousins, R.J.: Mammalian zinc transporters. Annu. Rev. Nutr. **24**, 151–172 (2004)

Lönnerdal, B., Cederblad, Å., Davidsson, L., Sandström, B.: The effect of individual components of soy formula and cows' milk formula on zinc bioavailability. Am. J. Clin. Nutr. **40**(5), 1064–1070 (1984)

Marchesini, G., Fabbri, A., Bianchi, G., Brizi, M., Zoli, M.: Zinc supplementation and amino acid-nitrogen metabolism in patients with advanced cirrhosis. Hepatology. **23**(5), 1084–1092 (1996)

Maret, W., Sandstead, H.H.: Zinc requirements and the risks and benefits of zinc supplementation. J. Trace Elem. Med. Biol. **20**(1), 3–18 (2006)

McCall, K.A., Huang, C.C., Fierke, C.A.: Zinc and health: current status and future directions. J. Nutr. **22**, 1437–1446 (2000)

McClain, C.J., Antonow, D.R., Cohen, D.A., Shedlofsky, S.I.: Zinc metabolism in alcoholic liver disease. Alcohol. Clin. Exp. Res. **10**(6), 582–589 (1986). [PubMed: 3544908]

McClain, C.J., ML, M.C., Boosalis, M.G., Hennig, B.: Zinc and the stress response. Scand. J. Work Environ. Health. **19**, 132 (1993)

McClain, C., Vatsalya, V., Cave, M.: Role of zinc in the development/progression of alcoholic liver disease. Curr. Treat. Options Gastroenterol. **15**(2), 285–295 (2017)

Miatto, O., Casaril, M., Gabrielli, G.B., Nicoli, N., Bellisola, G., Corrocher, R.: Diagnostic and prognostic value of serum copper and plasma fibrinogen in hepatic carcinoma. Cancer. **55**(4), 774–778 (1985)

Miller, L.V., Hambidge, K.M., Naake, V.L., Hong, Z., Westcott, J.L., Fennessey, P.V.: Size of the zinc pools that exchange rapidly with plasma zinc in humans: alternative techniques for measuring and relation to dietary zinc intake. J. Nutr. **124**(2), 268–276 (1994)

Murakami, Y., Koyabu, T., Kawashima, A., Kakibuchi, N., Kawakami, T., Takaguchi, K., Kita, K., Okita, M.: Zinc supplementation prevents the increase of transaminase in chronic hepatitis C patients during combination therapy with pegylated interferon α-2b and ribavirin. J. Nutr. Sci. Vitaminol. **53**(3), 213–218 (2007)

Penny, M.E.: Zinc supplementation in public health. Ann. Nutr. Metab. **62**(Suppl. 1), 31–42 (2013)

Powell, S.R.: The antioxidant properties of zinc. J. Nutr. **130**(5), 1447S–1454S (2000)

Prasad, A.S.: Clinical manifestations of zinc deficiency. Annu. Rev. Nutr. **5**(1), 341–363 (1985)

Prasad, A.S.: Zinc: an overview. Nutrition. **11**(1 Suppl), 93–99 (1995)

Prasad, A.S.: Zinc deficiency: has been known of for 40 years but ignored by global health organisations. BMJ. **330**(7487), 347–349 (2003)

Prasad, A.S., Halsted, J.A., Nadimi, M.: Syndrome of iron deficiency anemia, hepatosplenomegaly, hypogonadism, dwarfism and geophagia. Am. J. Med. **31**, 532–546 (1961)

Prasad, A.S., Beck, F.W., Bao, B., Fitzgerald, J.T., Snell, D.C., Steinberg, J.D., Cardozo, L.J.: Zinc supplementation decreases incidence of infections in the elderly: effect of zinc on generation of cytokines and oxidative stress. Am. J. Clin. Nutr. **85**(3), 837–844 (2007)

Richards, M.P., Cousins, R.J.: Mammalian zinc homeostasis: requirement for RNA and metallothionein synthesis. Biochem. Biophys. Res. Commun. **64**(4), 1215–1223 (1975)

Riggio, O., Merli, M., Capocaccia, L., Caschera, M., Zullo, A., Pinto, G., Gaudio, E., Franchitto, A., Spagnoli, R., D'aquilino, E., Seri, S.: Zinc supplementation reduces blood ammonia and increases liver ornithine transcarbamylase activity in experimental cirrhosis. Hepatology. **16**(3), 785–789 (1992)

Sandstead, H.H.: Zinc deficiency: a public health problem? Am. J. Dis. Child. **145**(8), 853–859 (1991)

Sandstead, H.H.: Human zinc deficiency: discovery to initial translation. Adv. Nutr. **4**(1), 76–81 (2013)

Schechter, P.J., Giroux, E.L., Schlienger, J.L., Hoenig, V., Sjoerdsma, A.: Distribution of serum zinc between albumin and α2-macroglobulin in patients with decompensated hepatic cirrhosis. Eur. J. Clin. Investig. **6**(1), 147–150 (1976)

Schroeder, J.J., Cousins, R.J.: Interleukin 6 regulates metallothionein gene expression and zinc metabolism in hepatocyte monolayer cultures. Proc. Natl. Acad. Sci. **87**(8), 3137–3141 (1990)

Seve, M., Chimienti, F., Favier, A.: Role of intracellular zinc in programmed cell death. Pathol. Biol. **50**(3), 212–221 (2002)

Singh, K.B., Taneja, S.K.: Hazard effects of excess of zinc in diet. Sci. Vis. **9**, 159–165 (2009)

Stamoulis, I., Kouraklis, G., Theocharis, S.: Zinc and the liver: an active interaction. Dig. Dis. Sci. **52**(7), 1595–1612 (2007)

Stehbens, W.E.: Oxidative stress, toxic hepatitis, and antioxidants with particular emphasis on zinc. Exp. Mol. Pathol. **75**(3), 265–276 (2003)

Sugino, H., Kumagai, N., Watanabe, S., Toda, K., Takeuchi, O., Tsunematsu, S., Morinaga, S., Tsuchimoto, K.: Polaprezinc attenuates liver fibrosis in a mouse model of non-alcoholic steatohepatitis. J. Gastroenterol. Hepatol. **23**(12), 1909–1916 (2008)

Sun, Q., Li, Q., Zhong, W., Zhang, J., Sun, X., Tan, X., Yin, X., Sun, X., Zhang, X., Zhou, Z.: Dysregulation of hepatic zinc transporters in a mouse model of alcoholic liver disease. Am. J. Physiol. Gastrointest. Liver Physiol. **307**(3), G313–G322 (2014)

Takagi, H., Nagamine, T., Abe, T., Takayama, H., Sato, K., Otsuka, T., Kakizaki, S., Hashimoto, Y., Matsumoto, T., Kojima, A., Takezawa, J.: Zinc supplementation enhances the response to interferon therapy in patients with chronic hepatitis C. J. Viral Hepat. **8**(5), 367–371 (2001)

Takeda, A., Tamano, H.: Insight into zinc signaling from dietary zinc deficiency. Brain Res. Rev. **62**(1), 33–44 (2009)

Tallman, D.L., Taylor, C.G.: Effects of dietary fat and zinc on adiposity, serum leptin and adipose fatty acid composition in C57BL/6J mice. J. Nutr. Biochem. **14**(1), 17–23 (2003)

Tapiero, H., Tew, K.D.: Trace elements in human physiology and pathology: zinc and metallothioneins. Biomed. Pharmacother. **57**, 399 (2003)

Taylor, J.A., Simons, T.J.: The mechanism of zinc uptake by cultured rat liver cells. J. Physiol. **474**(1), 55–64 (1994)

Truong-Tran, A.Q., Carter, J., Ruffin, R.E., Zalewski, P.D.: The role of zinc in caspase activation and apoptotic cell death. In: Zinc Biochemistry, Physiology, and Homeostasis, pp. 129–144. Springer, Dordrecht (2001)

Tuerk, M.J., Fazel, N.: Zinc deficiency. Curr. Opin. Gastroenterol. **25**(2), 136–143 (2009)

Vallee, B.L., Falchuk, K.H.: The biochemical basis of zinc physiology. Physiol. Rev. **73**(1), 79–118 (1993)

Vallee, B.L., Wacker, W.E.: Zinc metabolism in hepatic dysfunction. I. Serum zinc concentrations in Laennec's cirrhosis and their validation by sequential analysis. N. Engl. J. Med. **255**, 403 (1956)

Yoshida, Y., Higashi, T., Nouso, K., Nakatsukasa, H., Nakamura, S.I., Watanabe, A., Tsuji, T.: Effects of zinc deficiency/zinc supplementation on ammonia metabolism in patients with decompensated liver cirrhosis. Acta Med. Okayama. **55**(6), 349–356 (2001)

Zhong, W., McClain, C.J., Cave, M., Kang, Y.J., Zhou, Z.: The role of zinc deficiency in alcohol-induced intestinal barrier dysfunction. Am. J. Physiol. Gastrointest. Liver Physiol. **298**(5), G625–G633 (2010a)

Zhong, W., Zhao, Y., McClain, C.J., Kang, Y.J., Zhou, Z.: Inactivation of hepatocyte nuclear factor-4α mediates alcohol-induced downregulation of intestinal tight junction proteins. Am. J. Physiol. Gastrointest. Liver Physiol. **299**(3), G643–G651 (2010b)

Chapter 6
Role of Micronutrients in Providing Abiotic Stress Tolerance

Husna Siddiqui, Priyanka Singh, Yamshi Arif, Fareen Sami, Romana Naaz, and Shamsul Hayat

Abstract Minerals and organic materials present in soil are responsible for regulatory activity of cell organelles. Micronutrients such as zinc, boron, copper, iron, manganese, molybdenum and chlorine are readily absorbed and found in lower concentrations in plant tissues. Micronutrients reduce ion toxicity, preserve water balance, improve mineral uptake and assimilation, modify various gas exchange traits and reduce oxidative stress in plants. Plants contain sufficient micronutrients that possess greater tolerance capacity towards abiotic stress. The response of micronutrient application to various abiotic stresses depends on the crop, growth stage and concentration of the nutrient solution. This chapter explains the significance of micronutrients in crop production and their ability against the abiotic stress in plants.

Keywords Abiotic stress · Antioxidants · Mineral deficiency · Photosynthesis · Plant nutrition

Introduction

Plant nutrition is the study of the availability of nutrients and their role in plant growth and development, and plants are the primary producer for nutrition which influences the growth of plants as well as other living beings (Maathuis 2009). Optimum level of nutrient supply from soil is essential for the production of healthy crop, and any sort of imbalance in this level would alter the growth and development of crop. In extreme situations, unavailability of nutrients may even result in the nutrient deficiency or toxicity. The condition arises due to poor supply of nutrients

H. Siddiqui · P. Singh · Y. Arif · F. Sami · S. Hayat (✉)
Plant Physiology Section, Department of Botany, Faculty of Life Sciences, Aligarh Muslim University, Aligarh, India

R. Naaz
Department of Applied Science and Humanities, Jamia Millia Islamia, New Delhi, India

S. T. Khan, A. Malik (eds.), *Microbial Biofertilizers and Micronutrient Availability*,
https://doi.org/10.1007/978-3-030-76609-2_6

from soil or insufficient application of fertilizers. When a particular nutrient occurs above the sufficient or tolerable range than what is required by plant, it may lead to toxicity which diminishes the quality and quantity of crop productivity (McCauley et al. 2011). Although a number of nutrients are required by the plants, some are classified as essential nutrients.

Role of Micronutrients in Plants and Consequences of Their Deficiency

Micronutrients play an important role in plant growth and physiology (Fig. 6.1) and are vital to process ranging from photosynthesis to production and regulation of plant growth hormones. The role of some important micronutrients and the consequences of their deficiency are discussed in the following text.

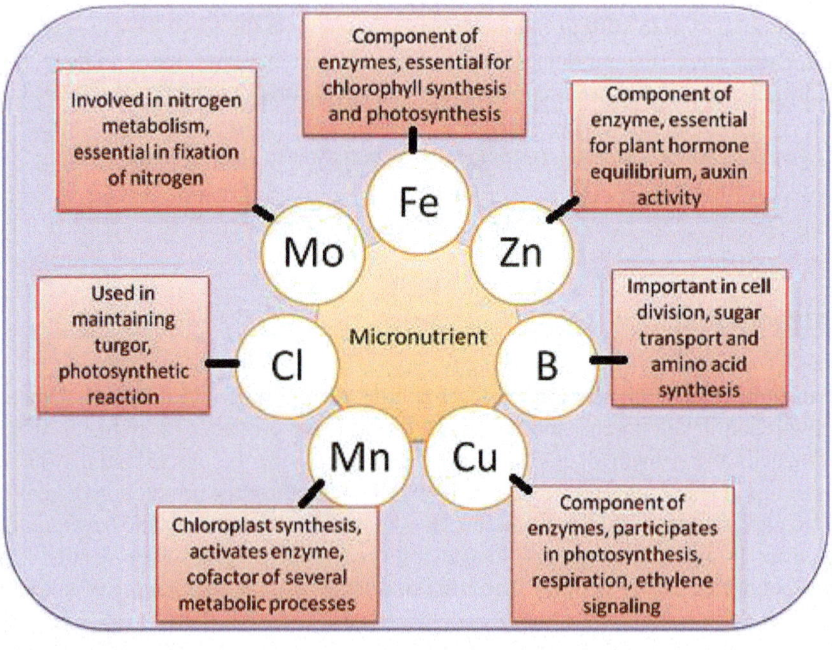

Fig. 6.1 Roles of different micronutrients in plants [iron (Fe), zinc (Zn), boron (B), copper (Cu), manganese (Mn), chlorine (Cl) and molybdenum (Mo)]

Boron

Boron (B) is mainly taken up by the plants in the form of boric acid. Boron is transported in plants via phloem by forming complex with polyols. Boron is known to regulate wide range of physiological processes in plants, mainly respiration, protein synthesis, sugars translocation, carbohydrate and RNA metabolism and phytohormone like indole acetic acid (Hansch and Mendel 2009). It is essential for the growth and development of pollen tube (Rattan 2015). Boron is also required for growing tips of plant and bud development (Rattan 2015). Deficiency of boron has a number of consequences on plant health and physiology. Its deficiency can be first detected in younger leaves and growing tips at the apex due to its immobility in plants. Leaves exhibiting boron deficiency turns bluish green in colour. Since boron deficiency damages the growing plant organs, that is, root and shoot tips, it leads to shortening of internodes and plant acquires a rosette-like appearance, showing an overall stunted growth of the plant.

Copper

The requirement of copper (Cu) in plants is generally fulfilled from the soil which is available in the form of Cu^{2+}. As copper is mostly immobile, it is transported from older to younger leaves and is required for various processes like photosynthesis and mitochondrial respiration. It is an important constituent of enzymes like cytochrome C oxidase, polyphenol and ascorbic oxidase, superoxide dismutase (Cu-Zn-SOD) etc. (Taiz and Zeiger 2002). It plays an important role in oxidoreductase metabolic reactions. Symptoms of copper deficiency include necrosis, developing as spots at the tips of younger leaves, later extending along the margins towards the base of the leaf leading to reduced growth of internodes. The severe deficiency of Cu also tends to premature loss of leaves (Taiz and Zeiger 2002).

Iron

Iron (Fe) is an essential micronutrient and is available to plant from the soil in the form of Fe^{2+}. It is involved in various processes such as photosynthesis, respiration, osmoprotection, ROS scavenging and nitrogen assimilation. It is also an important regulatory factor during phytohormone biosynthesis such as jasmonic acid, gibberellic acid and ethylene (Hansch and Mendel 2009). Besides, its presence in cytochrome also plays an important role during electron transfer in redox reactions (Taiz and Zeiger 2002). Due to the immobile nature of Fe, the deficiency symptoms occur first and prolonged deficiency results in the necrosis of leaf and ultimately the whole plant turns necrotic.

Manganese

Manganese (Mn) in the form of Mn^{2+} is absorbed by the plants and its biochemical roles are the same as that of magnesium (Rattan 2015). Mn plays an important role to split the water for the release of oxygen during photosynthesis. There are several enzymes like phenylalanine ammonia lyase, isocitrate dehydrogenase, malic enzyme and PEP carboxykinase whose activities are influenced by the presence of Mn (Hansch and Mendel 2009).

As Mn plays an important role in photosynthesis, its deficiency has adverse effect on photosynthetic apparatus and photosynthesis. Mn is an important part of oxygen-evolving complex (OEC) and their inadequate supply hinders the photosynthetic electron transport (Ono et al. 1992; Umena et al. 2011). There is strong evidence that deficiency of Mn leads to photochemical disturbances and the visible symptoms include interveinal chlorosis in leaves especially affecting the younger leaves. This may be due to the low phloem mobility of Mn. Prolonged Mn deficiency causes leaf necrosis and even oxidative stress which appears as brown spots between the veins (Hannam and Ohki 1988). Mn also plays an important role in the lignin polymerization, hence its deficiency results in a lower rate of lignin accumulation, especially in exposing plants to an increased risk of fungal infections (Rengel et al. 1993).

Molybdenum

Molybdenum (Mo) is categorized as a trace element which is available in the soil and is required by most living organisms including plants and animals. This transition element can be present in several oxidation states ranging from zero (0) to VI and VI is considered as the most commonly found in agriculture soils. Molybdate (MoO_4^{2-}) is considered as the bioavailable form of Mo. The entrance of Mo into the cell is correlated with subsequent incorporation into metal cofactors by a complex biosynthetic system (Dos Santos et al. 2004). Being a redox-active metal, Mo takes part in many enzymatic reactions where it acts as cofactors and successfully catalysed the redox reactions (Hoke et al. 2004). Mo is present predominantly as organic 'pterin' complex, known as molybdenum cofactor (MoCo). This MoCo binds with the Mo-requiring enzymes (molybdoenzymes) present in different organisms including plants (Williams and Frausto da Silva 2002). Bioavailability of Mo depends on several soil factors including pH, organic matter content, water availability and amount of adsorbing oxides (e.g. iron oxides). Alkaline soil pH is known to enhance the bioavailability of Mo, making it soluble and accessible for the plants.

Average amount of Mo available in the lithosphere is 23 mg/kg, which is available in various forms such as molybdenite (MoS_2), ferrimolybdenite [Fe_2 (MoO_4)] and wulfenite ($PbMoO_4$) in agriculture land (Reddy et al. 1997). Lime application in the agricultural soil adjusts the pH and enhances the soluble molybdate. Soluble MoO_4 can also form complexes with several ions, including K, Na, Mg and Ca, and organic matter, mainly fulvic and humic acids (Jenne 1977). These complexes reduce the quantity of metal oxides bound MoO_4 and enhance the quantity of MoO_4 which is easily available in the solution (Reddy et al. 1997).

The requirement of Mo for the plant growth in hydroponic culture of tomato was demonstrated by Arnon and Stout (1939b). It was also reported that characteristic symptoms such as involute shape of leaf lamellae, a phenotype well referred to as 'whiptail' and mottling lesions of leaves were noted in the plants grown in Mo-deficient solutions (Arnon and Stout 1939b). Furthermore, Mo supply to these symptomatic plants transformed them back to normal healthy leaves. Enzymes having Mo, such as nitrate reductase (NR), xanthine dehydrogenase (XDH), sulphite oxidase (SO) and aldehyde oxidase (AO), are vital for plant physiology as they participate in reactions catalysing global cycling of C, N and S. Dissociation of Mo from protein part of the enzyme makes it unstable (Basu and Burgmayer 2011). Mo is mainly important for the nitrogen metabolisms, as enzymes involved in the process require Mo as cofactor, and hence Mo deficiency results in inadequate N fixation. Moreover, Mo deficiency also influences plant growth and other metabolic activities. Mo reduced the amount of pollen, indirectly decreasing the plant viability (Skrzypczak et al. 2020).

Chlorine

Chlorine (Cl) is an essential micronutrient that plays diverse roles in plants and is acquired from soil as Cl^- ion. Average amount of Cl^- found in plants is 2–20 g/kg of dry matter (DM). Influx of Cl^- across the membrane is carried out via two methods: (i) $2H^+/Cl^-$ symporter and (ii) antiporter using ATP-activated hydroxyl ion. ATP activation depends on the pH gradient generated by Cl^- (White and Broadley 2001). This Cl^- is further transported to different cells and tissues through transporters like chloride channel (CLC) proteins, electroneutral cation-chloride co-transporters (CCCs) and through chloride influx-efflux process. Cl regulates roles, influences process like transpiration, affecting nutrient uptake, consequently affecting the overall growth of the plant (Li et al. 2017). It synchronizes the enzymatic activities in the cytoplasmic region, acts as an important cofactor during photosynthesis, and takes part in pH regulation, turgor maintenance (using anion counteraction) and membrane potential (Teakle and Tyerman 2010). Cl^- regulates the stomatal movement; influx-efflux of Cl^- in the guard cell regulates the opening and closing of stomata. Photosynthesis is a process of prime importance not only for plants but also for other living organisms, as it is the basis for the primary producers. Furthermore, it also results in the release of oxygen which is used by other living organisms. Optimal concentrations of Cl^- and Mn are crucial for photosynthesis and imbalances in its concentration will influence the process. Cl^- also acts as an osmoregulator and hence is important for maintaining osmotic balance in the plant cells. Optimum Cl^- concentration in the phloem sap is responsible for the loading/unloading of the sugars (Fromm and Eschrich 1988). Cl^- also affects the uptake of several micronutrients including N, K, P, Ca, Si, Mn, Zn, S, Cu, Fe and Mg (Zhong and Ma 1993). The deficiency of Cl^- may cause chlorosis, necrosis, wilting of leaves and also unusual bronze discolouration (Li et al. 2017).

Zinc

Zinc (Zn) is listed among one of the essential micronutrients and is taken up by plants mainly as Zn^+. Its concentration in plants varies from 25 to 150 mg/kg of dry matter weight. It is a phloem mobile micronutrient. Zn has multifarious roles in plants mainly in protein synthesis and plant metabolism. Zn is essential constituent of proteins and enzymes and it is indispensable catalytic component of more than 300 enzymes. Amount of zinc required by the plants is way greater than any other micronutrient. Three major enzymes having Zn ions are the superoxide dismutase (SOD), alcoholic dehydrogenase (AD) and carbonic anhydrase (CA) (Rattan 2015). Zn-Cu-SOD is an antioxidant enzyme which protects the membrane proteins and lipids from oxidation stress by detoxifying the superoxide radicals. Alcoholic dehydrogenase catalyses the conversion of acetaldehyde into ethanol under anaerobic conditions. CA is a cytoplasmic enzyme which facilitates the transfer of CO_2/bicarbonates CO_2 fixation in photosynthesis. Enzymes having Zn regulate the RNA processing, translation, DNA transcription and are required during the seed development (Hansch and Mendel 2009). Carbohydrate and nitrogen metabolism also rely on the Zn supply in plants and it also enhances the synthesis of cytochrome C. Phosphorus transport and translocation is also affected by Zn (Rattan 2015). Zn deficiency is defined as Zn concentration below 15 mg/kg of dry matter. Low Zn concentration causes shortened internodes and interveinal chlorosis. Shoots of zinc-deficient plants bear small stiff leaves which are malformed, arranged as bushy rosette (Mengel and Kirkby 2006). Plant facing Zn deficiency show stunted growth, late maturity and reduced crop yield. Zn deficiency causes different diseases in plants, among the well-noted diseases are khaira disease of rice, little leaf of cotton, mottle leaf or frenching of citrus, white bud of maize, crown choking in arecanut and rosette disease of apple (Tangle Agro 2019). Zn deficiency causes button shedding, shortening of crown, reduced leaf with chlorosis and even the malformed flower (TNAU 2015). In areca palm, Zn deficiency causes crown choking and changes the leaf size and leaf colour.

Exogenous Fertilization of Micronutrients and Its Effect on Plants

The essentiality of micronutrients to the plants has already been studied and confirmed by various researchers. The first ever known experiment regarding the exogenous supply of micronutrients was conducted by Arnon and Stout (1939a), which proved the role of these micronutrients in the plants. Further experiments were also done to confirm the results or to further investigate the details of the role played by essential elements. The plants generally receive nutrient from the soil, and root is the major organ by which it has been taken and transported through the xylem to the whole plant body. Additionally, plant leaves also have the capability to absorb the nutrients and transport them to other plant tissues. The ability of plants to acquire exogenously applied nutrients both from the leaves and roots provides better option for nutrient management and helps in improving the mineral density in the crops. For example, it was observed that soil application increased the grain yields, while foliar application improved the grain quality. Soil application of minerals is more helpful in improving plant growth, but is required in large quantities compared to foliar application. While foliar application required in comparatively less amount is known to enhance mineral level in the plants adequately. Droplets in the sprayer enter into the leaves via cuticle, stomata, ectodesmata and aqueous pores (Hussain et al. 2020).

All the elements directly or indirectly affect plant growth and physiology, requiring supplementation of these nutrients in adequate amount. Nutrients are applied in the form of fertilizers. These nutrients can be supplied in various forms such as in nanoforms (NPs) or as other conjugates of elements. Nanoformulations of fertilizers including that of micronutrients have certain advantage over their conventional form. For example, nanoparticles (NPs) are easily taken up by the plants, therefore are more bioavailable (Rizwan et al. 2019). However, higher doses of NPs may be harmful for the plant and are known to disturb nutrient uptake and metabolic reactions at higher concentration (Du et al. 2019). Exogenous application of various micronutrients for improving plant growth and development (Table 6.1) is discussed below briefly.

Zinc

Different studies have investigated the effect of Zn fertilization on plant in various forms, such as $ZnSO_4$ and Zn organic acid chelates (Zn-DTPA, Zn-EDTA, Zn-arginine and Zn-glycine), Zn NPs and Zn nanochelates and were found to improve plant health and growth. Weisany et al. (2012) found that Zn application improves shoot and root length, its dry and fresh weight and even the activities of the antioxidant enzymes when required in soybean plants. The application of $ZnSO_4$ to the wheat plant is found to increase the grain Zn content and a decrease in its

Table 6.1 Beneficial effects of exogenously applied micronutrients on plants

Micronutrient	Application form	Beneficial effects	Plant	Reference
Zn	$ZnSO_4$	Improved growth and antioxidant machinery	*Glycine max*	Weisany et al. (2012)
Zn	$ZnSO_4$	High Zn content in grain, low phytic acid content	*Triticum aestivum*	Wang et al. (2019)
B	Solubor ($Na_2B_8O_{13}$ $4H_2O$, US Borax, Boron, CA)	Increased seed quantity and quality	*Medicago sativa*	Dordas (2006)
B	Not available	Improved plant health	*Solanum melongena*	Saha et al. (2020)
Cu	$CuSO_4$	Promoted photosynthetic efficiency	*Oryza sativa*	Mo et al. (2020)
Fe	$FeSO_4$	High growth and more nodule formation	*Vigna radiata*	Saini and Singh (2017)
Mn	Mn-EDTA	More growth and high oil content	*Linum usitatissimum*	Emam (2020)
Mo	Molybdate	High nitrogen supply in pods	*Phaseolus vulgaris*	Vieira et al. (1998)
Mo	Not available	High yield	*Cicer arietinum*	Montenegro et al. (2010)

phytic acid content (Wang et al. 2019). Zhang et al. (2010) reported that the application of $ZnSO_4$ increased the grain Zn content and it was observed that it can be easily transported to the seed endosperm. Another study investigated the effect of foliar spray of three micronutrients: Zn as $ZnSO_4$, Mn as $MnSO_4$ and B as H_3BO_3 on the olive grove (Chatzistathis et al. 2017). A significant increase in the concentration of Ca, Mg, P, N and Fe in olive grove leaves was reported. However, a reduction in the chlorophyll fluorescence activities was also observed. Joy et al. (2017) used Zn fertilizer for wheat crop and found that the Zn fertilization increased gross yield and grain zinc content and hence should be used to improve dietary Zn content.

Boron

It is proved from various experimental evidences that the foliar application of boron (B) reduced the deficiency symptoms of B and other beneficial artefacts. The application of B in *Medicago sativa* is found to improve both the quantity and quality of seeds (Dordas 2006). The growth of brinjal in the soil possesses low levels of Zn and B and can be improved when supplemented with farm yard manure; Zn and B at different combinations successfully promoted the fruit yield and improved other agro-morphological parameters. Furthermore, application of only B proved beneficial for brinjal plant health (Saha et al. 2020). However, different crops respond differently to boron fertilization. Different methods of fertilization are also known to influence crop productivity and growth differently. For example, B oxides improved plant health and yield of cauliflower, while organic forms of B were better fertilizers for okra. Residual B application was found to promote the yield of cowpea crop (Bhupenchandra et al. 2020).

Copper

Copper (Cu) is mainly applied as $CuSO_4$ and Cu NPs to the plants, but the information is very limited. The effect of foliar application of Zn, Mn and Cu on the wheat was studied by Stepien and Wojtkowiak (2016) and reported that the application enhanced the growth as well as yield characteristics and combined application proved more beneficial. The research also indicated that the foliar application of $CuSO_4$ on the ultrasonic-treated rice promoted the photosynthetic efficiency along with other yield parameters (Mo et al. 2020).

Iron

Iron (Fe) was applied on the plant in different forms, that is, $FeSO_4$, $C_6H_5FeO_7$, Fe_2O_3 NPs and Fe chelates (e.g. Fe-EDTA, Fe-HEDTA-Fe, Fe-DTPA, Fe-Na EDTA, Fe-amino acids, Fe-nicotianamine). The foliar spray of Fe significantly improves the growth of different crops tested either solely or in combinations of other nutrients and also applied as fertilizers. Enhanced grain Fe content following $FeSO_4$ fertilization is reported in many studies (Manasa and Devaranavadagi 2015; Arabhanvi and Hulihalli 2018). Combination of Fe with nicotinamine, a metal ligand, gave even better results in enhancing the Fe content in the grains (He et al. 2013; Yuan et al. 2013). The research also indicated that the foliar spray of $FeSO_4$ on *Vigna radiata* was found to promote plant height, branch per plant, number of grains per pods and number of nodules in the roots (Saini and Singh 2017). In another report, it was noted that spray of Fe was found to increase the number of pods per plant in *Vigna radiata* (Soni and Kushwaha 2020).

Manganese

The effect of Mn (an important micronutrient) was also checked in many studies using a spray of $MnSO_4$ or chelated Mn (e.g. Mn-EDTA). The leaf of linseed was sprayed with Mn-EDTA in combination with Zn and Fe and noted that it significantly improved plant height, straw yield, number of fruiting branches per plant, fruiting zone length, number of capsule/seeds per plant and oil yield (Emam 2020).

Earlier, Soleymani et al. (2012) also conducted the experiment with the same nutrients on the forage sorghum and found positive impact on the yield, plant weight and plant height.

Molybdenum

Various compounds containing Mo such as ammonium molybdate, sodium molybdate and Mo NPs have been tested for plants to evaluate the effect Mo on plant growth and productivity. Nanoformulations of various nutrients including that of Mo are being recently evaluated for their efficacy. Mo is well known for its role in nitrogen fixation; it is a Mo cofactor in the N_2-fixing enzymes (nitrate reductase and nitrogenase). Hence, Mo status of the plants is directly related with the amount of N_2 fixation. The plants of *Phaseolus vulgaris* were sprayed with ammonium molybdate and it was predicted that Mo can be used instead of nitrogen fertilizers (Vieira et al. 1998). The experimental evidence also suggests that Mo spray increased the size of berry in grapevines and also influenced the seed formation and petiolar nutrient composition (Williams et al. 2005). A pot experiment was conducted by Montenegro et al. (2010) on the chickpea plant, sprayed with Mo along with Zn and B. It was reported in this experiment that combined dose with appropriate concentration of each supplement was most fruitful in increasing the yield. A field experiment using same plant and nutrients showed that Mo was capable of counteracting the effects of B application (Valenciano et al. 2011). In one of the recent study, it was noted that foliar application of Mo NPs increased the growth of spinach as well as the height, weight, chlorophyll activities, NO_3 concentration and NR activities of the plant (Abbasifar et al. 2020).

Effect of Micronutrients During Abiotic Stress Tolerance

Micronutrients (Fe, Zn, Cu, Mn, B, Mo, Cl) play a crucial role in enhancing plant growth, morpho-physiological and biochemical traits under normal as well as in stressed conditions. Application of different micronutrients individually or in combination is highly beneficial in ameliorating different environmental stresses. Additionally, mineral nutrient enhances the crop yield and quality and provides crop tolerance which promotes sustainable agriculture during various abiotic stresses (Babaeian et al. 2011). Micronutrient promotes antioxidant enzyme, proline, total sugars and phenols production which scavenges ROS and maintains cell turgor. Thus, it enhances physio-biochemical and yield attributes, and provides plant tolerance against environmental stress (Askary et al. 2017; Fahad et al. 2019).

Zinc

Zn is an important micronutrient for plant which participates as metal component for classes of enzymes. Further it enhances various plants physiological and cellular metabolic activities under normal and stressed conditions. It plays a vital role during photosynthesis, protein synthesis, auxin synthesis and cell division under normal and stressed environment (Said-Al Ahl and Mahmoud 2010). Zn application mitigated the salt stress in maize by positively regulating physiological and biochemical traits and also enhanced the production of glycine betaine, proline, sugars and free amino acids which serve as ROS scavengers (Iqbal et al. 2018). Exogenous application of Zn alleviated the Cd stress toxicity by promoting antioxidant enzymatic machinery (SOD and POX) and also by increasing the plants morpho-physiological parameters (Wu and Zhang 2002). Exogenous application of Zn on *Ocimum basilicum* L. under salt stress enhanced the photosynthesis, plant height, biomass, oil content and yield parameter and helped in diminishing the toxic effect of salt (Said-Al Ahl and Mahmoud 2010). The application of ZnO NPs mitigated salt stress in *Moringa* by significantly increasing the chlorophyll, carotenoid, proline, protein and antioxidant enzyme production and also increased the plant height, leaf area and number and shoot and root fresh and dry mass (Soliman et al. 2015). Water stress decreased the grain yield and biochemical activity in sunflower. However, Zn application decreased the stress by adjusting osmotic potential and turgor, which further increased the morphological and yield traits (Babaeian et al. 2011). Zn has same physical and chemical property as Cd which further enhances Cd detoxification from plant cell by the application of Zn (Rizwan et al. 2016). In rice, Zn supplementation mitigated the Cd stress and further enhanced the growth traits, panicle number, spikelet number in each panicle and grain quality and quantity (Fahad et al. 2015; Fahad et al. 2019). In Cd-stressed plant, Zn application also helped in Cd mobilization from roots to shoots and to other vegetative parts of the plant. This helps in Cd compartmentalization to several plant tissues. Zn application under Cd stress induced antioxidant enzyme activity and reduced MDA content (Basnet et al. 2014; Rizwan et al. 2016). The application of ZnO NPs on wheat induced the growth and yield and also promoted the chlorophyll synthesis and gaseous exchange which helps in mitigating the Cd stress (Rizwan et al. 2019). In *Zea mays* L., Zn application enhanced the growth and production of plant and made plant tolerant to Cd by up-regulating several metabolic activities (Sajedi et al. 2011). Application of Zn made wheat Cd tolerant by enhancing the nutrient homeostasis, growth traits, morpho-physiological and biochemical traits as well as the grain yield and weight (Rizwan et al. 2017). The supplementation of Zn also mitigated the drought stress by promoting growth, seed yield and kernel weight (Monjezi et al. 2013). Further, it was also reported that Zn application on *Glycine max* L. reduced the toxic effect of salt by increasing the calcium and potassium contents in plant cell and also increased the growth and yield traits (Weisany et al. 2014). In lavender, application of Zn showed detrimental effect on salinity as it increased the fresh biomass, total phenolic content, total sugars, antioxidant capacity and essential oil yield (Chrysargyris et al. 2018).

Boron

Boron has a significant role in root and shoot elongation, sugar translocation, nucleic acid synthesis, pollen tube growth and development (Azeem and Ahmad 2011). Furthermore, it is also a part of enzyme indole-3-acetic acid oxidase and participates in cell wall integration and lignifications (Azeem and Ahmad 2011). Boron plays a significant role in enhancing water use efficiency in plants under various stresses (Hrishikesh et al. 2012). Foliar application of B (boric acid) on tomato mitigates the deleterious effect of salinity by enhancing root elongation and fruit yield (Azeem and Ahmad 2011). Combined application of B with selenium mitigates the salt stress in stevia by improving growth parameters, leaf area and glycoside production (Shahverdi et al. 2018). Application of B came out to be highly beneficial in diminishing the drought conditions in different crops as it regulates the stomatal conductance, transpiration, membrane integrity, photosynthetic machinery and chlorophyll synthesis and also reduced the ROS content by activating the antioxidant enzyme. It is also reported to maintain the cell osmoticum and turgor for proper growth and development (Waraich et al. 2011). Foliar application of B on tea plant under drought conditions enhanced the leaf area, leaf turgor, leaf water content, stomatal conductance and photosynthesis and further induced the antioxidant enzymes (CAT, APX and SOD) which helps to increase the morpho-physiological and biochemical traits, plant growth and yield and ultimately makes the tea tolerant to drought (Hrishikesh et al. 2012). Exogenous application of B ameliorate the salinity in *Trigonella foenum-graecum* L. by reducing the Na ions accumulation in plant cell and through improving the nutrient uptake and equilibrium which further enhanced the production by improving the yield attributes (Ibrahim and Faryal 2014).

Copper

Cu is the mineral nutrient as well as transition metal which participates in several physio-biochemical activities of plants (photosynthesis, respiration, antioxidant machinery and ethylene signal transduction). Cu also acts as an important micronutrient which increases the carbohydrate distribution and protein metabolism which are responsible for the enhanced plant growth and developmental processes (Sağlam et al. 2016). Cu serves as redox-active transition metal, so its application on *Zea mays* L. during water scarcity condition enhanced photosynthesis, carbohydrate metabolism and antioxidant enzyme which helps to increase the growth and yield attributes even under stress conditions (Sajedi et al. 2011). The application of Cu on wheat under soil salinity increased the growth, morpho-physiological traits and biochemical parameters as well as grain quality and quantity (El-Fouly et al. 2011). In *Catharanthus roseus* L., the application of Cu is highly beneficial even under the Cd stress and enhanced the root elongation, growth and biomass (Chen et al. 2018). Corn was made salinity tolerant by the application of Cu in lower doses where it regulates the photosynthesis, nutrition uptake and improved antioxidant system by accumulating the proline, glycine betaine, sugars and amino acids, which further helped to increase the plant growth and yield traits (Iqbal et al. 2018). The research also indicates that the application of Cu positively regulates the morpho-physiological and biochemical traits under the salt stress in tomato. It also improved production of phenols, vitamin C and glutathione as well as antioxidant machinery (Pérez-Labrada et al. 2019). The application of Cu NPs increases the drought tolerance in maize by regulating the leaf water content, leaf area, anthocyanin, chlorophyll and carotenoid contents which helps to increase the total seed number and grain yield (Van Ha et al. 2020). The drought stress in several crops was partially and completely overcome by enhancing the photosynthesis, leaf water content, cell turgor and membrane stability. It also promotes protein synthesis and antioxidant synthesis for proper growth and development of plant (Waraich et al. 2011). In another study, it was reported that drought stress can be overcome by the application of Cu by enhancing the antioxidant enzyme (CAT, APX and GSH) which quenches ROS from cell and also increased the soluble sugars and proline content (Zhang et al. 2019). Cu treatment also diminished the soil salinity in rosemary plant by increasing the total phenolic content, sugars, relative water content and membrane permeability, whereas it reduced the lipid peroxidation and ROS content (Hejazi et al. 2012).

Iron

Fe is highly essential micronutrient and transition metal involved in oxygen transport and metabolism, cellular proliferation, electron transfer, nucleic acid synthesis, thylakoid synthesis and also serves as a cofactor of several enzymes. Furthermore, Fe plays a crucial role during the photosynthesis process and helps in electron transfer process (Said-Al Ahl and Mahmoud 2010; Sah et al. 2017). Apart from this, Fe plays a crucial role during various abiotic stresses by increasing the plant growth and development, and also by regulating several metabolic activities. The application of Fe increased plant height, number of branches and other morphological traits in basil even under salinity (Said-Al Ahl and Mahmoud 2010). The application of Fe on chamomile plant during salinity stress modulated plant height, biomass, antioxidant machinery, proline and carbohydrate content. Furthermore, it also decreased Na^+ concentration in root and shoots and increased K^+ concentration; thus, Fe also maintains ion homeostasis and makes plant salt tolerant (Heidari and Sarani 2012). The study indicated that supplementation of appropriate dose of Fe NPs on *Mentha* helped in mitigating toxic effect of NaCl by increasing height, weight, nutrient content and yield parameters. It promoted antioxidant production which scavenges

ROS from plant cell and maintained proper osmotic potential (Askary et al. 2017). In *Moringa*, salt stress is alleviated by Fe application in the form of nanofertilizer (Fe_3O_4), it induced the growth and yield traits and also enhanced the chlorophyll synthesis, protein synthesis as well as the production of enzymatic and non-enzymatic antioxidants (Soliman et al. 2015). In *Lycopersicon esculentum* L., foliar application of Fe combated soil salinity by enhancing vegetative growth, biochemical traits and yield parameter (Azeem and Ahmad 2011). Fe application mitigated the Cd stress on wheat by enhancing antioxidant enzymatic and non-enzymatic machinery and also increased the growth traits, biomass and yield attributes (Hussain et al. 2019). The application of Fe also ameliorated the toxic effect of metal on spinach by enhancing the photosynthesis, stomatal conductance, carbon dioxide assimilation, plant growth and other developmental processes (Ouzounidou et al. 1998). The water stress in *Sesamum indicum* L. was overcome by the application of Fe by increasing the plant growth, biomass and yield traits (Heidari et al. 2011). In sunflower also the water stress was reduced by regulating photosynthetic activity, carbohydrate and proline contents with Fe and reflected in increased seed and biological yields (Babaeian et al. 2011). The toxicity of Cd was also reduced by the spray of Fe and growth of the plant was increased (Rizwan et al. 2016). The treatment of Fe helped in reducing the toxic effect of Cd in rice by increasing the antioxidant enzyme production, which serves as ROS scavenger, and decreasing MDA content. Furthermore, some reports also suggested that Fe application modulated photosynthetic electron transport chain to stressed one (Sebastian and Prasad 2016; Rizwan et al. 2016). Drought conditions are mitigated by Fe application on corn, which further enhances photosynthesis, antioxidant enzymatic system, growth, biomass and yield traits (Sajedi et al. 2011). The application of Fe made wheat tolerant to water stress by inducing the yield and productivity (Monjezi et al. 2013). NPs of Fe in oxide form mitigated deleterious effect of environmental stress (Saif et al. 2016). The NPs of Fe also mitigate the stress, and in one of the study, Fe NPs mitigated the deleterious effect of Cd by inducing antioxidant enzymatic machinery which reduced oxidative stress and further enhanced the growth traits and yield parameter in *Triticum aestivum* L. (Rizwan et al. 2019). The drought stress was also overcome in corn by Fe NPs by improving relative water content, kernel number and weight, grain quality and quantity (Rafsanjani et al. 2019).

Manganese

Mn is an essential micronutrient and transition metal, which acts as an activator and cofactor of about 30 enzymes such as Mn SOD, oxalate oxidase (OxOx), catalase, malic enzyme and phosphoenol, playing a vital role in oxido-reduction, decarboxylation and hydrolytic reactions. Four atoms of Mn are required in OEC located at PSII. Thus, this makes Mn an important metal which is required for oxygen evolution (DalCorso et al. 2014; Schmidt and Husted 2019). During water stress in sunflower, Mn application induced highest positive effect on yield parameters and further enhanced the proline and carbohydrate content (Babaeian et al. 2011). An application of Mn combated the drought stress in corn by increasing the proline and antioxidant enzyme production for maintaining the osmotic balance and scavenging ROS, respectively, because of Mn involvement in Mn SOD (Sajedi et al. 2011). The foliar application of Mn mitigated the salinity stress in rice by enhancing growth and development; it further increased grain yield, straw yield, harvest index and yield traits (Zayed et al. 2011). The salt stress is also mitigated in wheat by the foliar application of Mn (El-Fouly et al. 2011). The toxicity generated by Cd is mitigated by Mn and biomass production, root and shoot growth were increased. Besides this, biochemical and physiological traits (photosynthesis, respiration and restoration of photosystem and chloroplast structure) was also enhanced. Furthermore, it also increased the grain yield and quality (Sarwar et al. 2010). The application of Mn ameliorates the drought in *Camellia sinensis* L. by improving the photosynthesis, CAT, SOD, APX and glutathione synthesis. Mn also enhanced the yield attributes (Hrishikesh et al. 2012). The supplementation of Mn with Cd leads to improved antioxidant enzyme system and glyoxalase system and reduced the ROS and methylglyoxal content in *Oryza sativa* L. (Rahman et al. 2016). In *Trigonella*, foliar application of Mn diminished the salt toxicity by improving the ion homeostasis and compartmentalization which leads to improved performance of the crop (Ibrahim and Faryal 2014).

Molybdenum

Mo is the important essential nutrient for proper growth and metabolism of plant with and without stress. Mo serves as the cofactor for several enzymes [nitrate reductase (NR), xanthine dehydrogenase (XDH) and aldehyde oxidase (AO)], which are important part of the antioxidant system. NR plays a key role in the nitrogen metabolism and has the antioxidative role in scavenging ROS during stress. XDH plays a significant role in the synthesis of purine catabolism and ureide synthesis and both are scavengers of ROS and participate in stress mitigation. AO plays a vital role during environmental stress and participates in the synthesis of IAA and ABA, which are important for mitigating the toxic effects of stress. Thus, Mo maintains the proper plant growth and development during environmental stress (Babenko et al. 2015). In crested wheatgrass, the application of Mo prevents the oxidative damage under salinity stress and also improves the plant health by improving the growth and development (Babenko et al. 2015). The application of Mo mitigated the salt stress by inducing antioxidant enzyme (CAT, SOD and POX) and non-enzymatic antioxidants such as ascorbic acid, glutathione and carotenoid in *Brassica campestris* L. Mo also reduced the Na ion inside the cell and helped to increase the potassium content (Zhang et al. 2012). Mo application in Chinese cabbage ameliorated the salinity by improving the ion homeostasis and photosynthetic machinery (Zhang et al. 2014). The application of Mo makes wheat drought tolerant by enhancing the growth and yield attributes as well as quality and quantity of grain (Kareem et al.

2017). Mo mitigated the drought stress in *Triticum* by increasing the kernel number per spike, kernel weight, length of peduncle, grain yield, biological yield, protein yield, protein content and harvest index (Ghafarian et al. 2013). The salt stress was also overcome by the application of Mo which assessed in the form of increased seedling germination, growth and development (Batyrshina et al. 2018). The supplementation of Mo also improved the cold tolerance in cauliflower seeds (Rihan et al. 2014) and in wheat (Al-Issawi et al. 2016). The application of Mo reduced the salt stress in *Phaseolus vulgaris* L., *Cicer arietinum* L. and *Lens culinaris* M. by enhancing the seedling growth, development, nitrogen metabolism and yield attributes (Salha and Chaabane 2016).

Chlorine

Cl is a mobile anion in plants and have significant role in stomatal regulation. Chloride enhances the water use efficiency (WUE) by reducing the wilting and restricting the highly branched root systems in cereal crops (Ahanger et al. 2016). Cl plays a significant role in plant growth and development; furthermore, it also improves the osmotic potential and stomatal conductance under normal and stressed conditions. Cl plays a key role in oxygen evolution in photosynthesis during abiotic stress. Cl application improves the growth, yield and quality of several crops (Chen et al. 2010). Gaseous chlorine dioxide (ClO_2) fumigation decreased the ROS production and membrane damage in longan fruit (Chomkitichai et al. 2014). Cl regulates the WUE in leaves and controls the stomatal movement which makes plant drought tolerant by maintaining osmoticum (Ahanger et al. 2016).

Conclusions and Future Perspective

To increase productivity of crop and sustainable agriculture system, the level of micronutrients plays a very crucial role. Micronutrients are required in trace amount and are highly beneficial in sustaining proper growth and development of plant. They participate in several metabolic activities and enzymatic reactions. Furthermore, they serve as a cofactor and participate in the redox reactions and also up-regulates the photosynthesis, stomatal conductance, nitrogen fixation, nucleic acid synthesis, protein synthesis and carbohydrate metabolism. The growth of the plant as well as metabolic activities gets hampered if they are not present in soil or they are not properly homogenized in the cell. The deficiencies and toxicity of micronutrients show the impaired growth, reduced photosynthesis and cause large reduction in yield. Deficiency symptoms of mineral nutrient are seen on older leaves generally, while toxicity symptoms are seen on younger leaves. Therefore, low doses of micronutrient are highly essential. This chapter highlighted the role of micronutrient in increasing the photosynthesis, gaseous exchange, stomatal

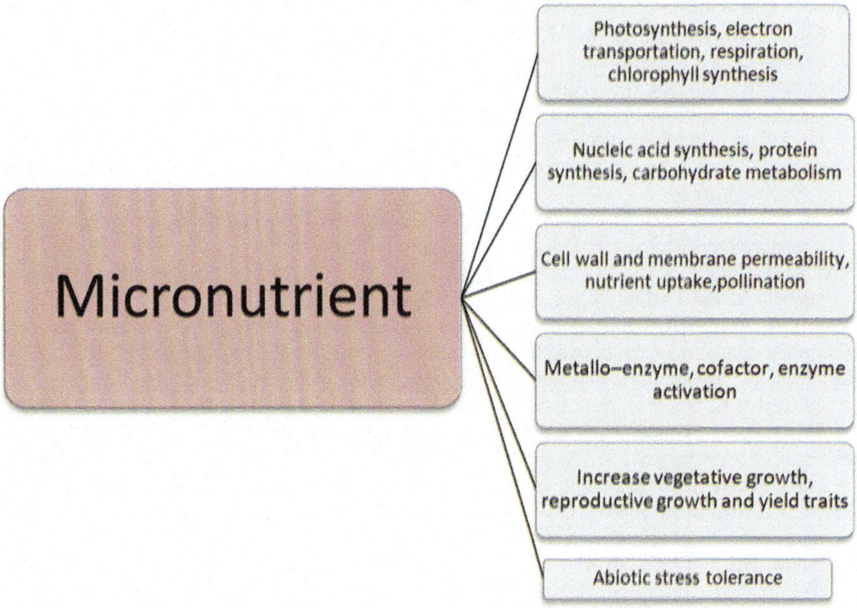

Fig. 6.2 Illustration of role of micronutrients

conductance and other vital cellular activities required in proper growth and development (Fig. 6.2). Furthermore, the role of micronutrients in mitigating the abiotic stress is also discussed. Micronutrients in proper doses alone or in combination play a significant role in diminishing the toxic effect of salinity, drought, heavy metal and cold; they increased level of antioxidants, total sugar, phenolics and proteins which help in scavenging the ROS and maintaining the cell turgor. Therefore, this helps in enhancing the yield traits and improving the quality and quantity of crop.

There is still not well-established research done on few micronutrients deficiency and toxicity. Furthermore, unexplored areas such as mineral nutrient homeostasis and compartmentalization still need to be explored. Physiological and molecular studies are needed for elucidating the role of micronutrients. Therefore, more cellular, biochemical, molecular and physiological studies should be explored in mitigation of environmental stresses by the application of mineral nutrient.

References

Abbasifar, A., Valizadeh Kaji, B., Iravani, M.A.: Effect of green synthesized molybdenum nanoparticles on nitrate accumulation and nitrate reductase activity in spinach. J. Plant Nutr. **43**(1), 13–27 (2020)

Ahanger, M.A., Morad Talab, N., Fathi Abd-Allah, E., Ahmad, P., Hajiboland, R.: Plant growth under drought stress: significance of mineral nutrients. In: Water Stress and Crop Plants: A Sustainable Approach, vol. 2. John Wiley & Sons, Ltd., Chichester (2016)

Al-Issawi, M., Rihan, H.Z., Al-Shmgani, H., Fuller, M.P.: Molybdenum application enhances antioxidant enzyme activity and COR15a protein expression under cold stress in wheat. J. Plant Interact. **11**(1), 5–10 (2016)

Arabhanvi, F., Hulihalli, U.K.: Agronomic fortification with zinc and iron to enhancing micronutrient concentration in sweet corn grain to ameliorate the deficiency symptoms in human beings. Int. J. Curr. Microbiol. Appl. Sci. **7**, 333–340 (2018)

Arnon, D.I., Stout, P.R.: Molybdenum as an essential element for higher plants. Plant Physiol. **14**, 599–602 (1939a)

Arnon, D.I., Stout, P.R.: The essentiality of certain elements in minute quantity for plants with special reference to copper. Plant Physiol. **14**, 371–375 (1939b)

Askary, M., Talebi, S.M., Amini, F., Bangan, A.D.B.: Effects of iron nanoparticles on *Mentha piperita* L. under salinity stress. Biologija. **63**(1), 65–75 (2017)

Azeem, M., Ahmad, R.: Foliar application of some essential minerals on tomato (*Lycopersicon esculentum*) plant grown under two different salinity regimes. Pak. J. Bot. **43**(3), 1513–1520 (2011)

Babaeian, M., Tavassoli, A., Ghanbari, A., Esmaeilian, Y., Fahimifard, M.: Effects of foliar micronutrient application on osmotic adjustments, grain yield and yield components in sunflower (Alstar cultivar) under water stress at three stages. Afr. J. Agric. Res. **6**(5), 1204–1208 (2011)

Babenko, O.N., Brychkova, G., Sagi, M., Alikulov, Z.A.: Molybdenum application enhances adaptation of crested wheatgrass to salinity stress. Acta Physiol. Plant. **37**(2), 14 (2015)

Basnet, P., Amarasiriwardena, D., Wu, F., Fu, Z., Zhang, T.: Elemental bioimaging of tissue level trace metal distributions in rice seeds (*Oryza sativa* L.) from a mining area in China. Environ. Pollut. **195**, 148–156 (2014)

Basu, P., Burgmayer, S.J.N.: Pterin chemistry and its relationship to the molybdenum cofactor. Coord. Chem. Rev. **255**(9–10), 1016–1038 (2011)

Batyrshina, Z., Yergaliyev, T.M., Nurbekova, Z., Moldakimova, N.A., Masalimov, Z.K., Sagi, M., Omarov, R.T.: Differential influence of molybdenum and tungsten on the growth of barley seedlings and the activity of aldehyde oxidase under salinity. J. Plant Physiol. **228**, 189–196 (2018)

Bhupenchandra, I., Basumatary, A., Dutta, S., Singh, L.K., Datta, N.: Impact of boron fertilization on boron fractions at different crop growth stages in cauliflower, cowpea, okra sequence in an inceptisols of North East India. J. Plant Nutr. **43**(8), 1175–1188 (2020)

Chatzistathis, T., Delviniotis, A., Panagakos, A., Giannakoula, A., Tranaka, V., Molassiotis, A.: Foliar manganese, zinc and boron application effects on mineral nutrition of an experimental olive grove (cv. "Chondrolia Chalkidikis"). J. Plant Nutr. **40**(12), 1728–1742 (2017)

Chen, W., He, Z.L., Yang, X.E., Mishra, S., Stoffella, P.J.: Chlorine nutrition of higher plants: progress and perspectives. J. Plant Nutr. **33**(7), 943–952 (2010)

Chen, Q., Lu, X., Guo, X., Pan, Y., Yu, B., Tang, Z., Guo, Q.: Differential responses to Cd stress induced by exogenous application of Cu, Zn or Ca in the medicinal plant *Catharanthus roseus*. Ecotoxicol. Environ. Saf. **157**, 266–275 (2018)

Chomkitichai, W., Chumyam, A., Rachtanapun, P., Uthaibutra, J., Saengnil, K.: Reduction of reactive oxygen species production and membrane damage during storage of 'Daw' longan fruit by chlorine dioxide. Sci. Hortic. **170**, 143–149 (2014)

Chrysargyris, A., Michailidi, E., Tzortzakis, N.: Physiological and biochemical responses of *Lavandula angustifolia* to salinity under mineral foliar application. Front. Plant Sci. **9**, 489 (2018)

DalCorso, G., Manara, A., Piasentin, S., Furini, A.: Nutrient metal elements in plants. Metallomics. **6**(10), 1770–1788 (2014)

Dordas, C.: Foliar boron application improves seed set, seed yield, and seed quality of alfalfa. J. Agron. **98**(4), 907–913 (2006)

Dos Santos, P.C., Dean, D.R., Hu, Y., Ribbe, M.W.: Formation and insertion of the nitrogenase iron-molybdenum cofactor. Chem. Rev. **104**, 1159–1173 (2004)

Du, J., Zhang, Y., Guo, R., Meng, F., Gao, Y., Ma, C., Zhang, H.: Harmful effect of nanoparticles on the functions of freshwater ecosystems: insight into nanoZnO-polluted stream. Chemosphere. **214**, 830–838 (2019)

El-Fouly, M.M., Mobarak, Z.M., Salama, Z.A.: Micronutrients (Fe, Mn, Zn) foliar spray for increasing salinity tolerance in wheat *Triticum aestivum* L. Afr. J. Plant Sci. **5**(5), 314–322 (2011)

Emam, S.: Estimation of straw, seed and oil yields for flax plants (*Linum usitatissimum* L.) cultivars of foliar application of Mn, Fe and Zn under dry environment. Egypt. J. Agron. **42**(1), 35–46 (2020)

Fahad, S., Hussain, S., Khan, F., Wu, C., Saud, S., Hassan, S., et al.: Effects of tire rubber ash and zinc sulfate on crop productivity and cadmium accumulation in five rice cultivars under field conditions. Environ. Sci. Pol. **22**(16), 12424–12434 (2015)

Fahad, S., Rehman, A., Shahzad, B., Tanveer, M., Saud, S., Kamran, M., et al.: Rice responses and tolerance to metal/metalloid toxicity. In: Advances in Rice Research for Abiotic Stress Tolerance, pp. 299–312. Woodhead Publishing, Duxford (2019)

Fromm, J., Eschrich, W.: Transport processes in stimulated and non-stimulated leaves of *Mimosa pudica*. III. Displacements of ions during seismonastic leaf movements. Trees. **2**, 65–72 (1988)

Ghafarian, A.H., Zarghami, R., Zand, B., Bayat, V.: Wheat performance as affected by foliar application of molybdenum (Mo) under drought stress condition. Int. J. Plant Prod. **4**(11), 3050–3056 (2013)

Hannam, R.J., Ohki, K.: Detection of manganese deficiency and toxicity in plants. In: Manganese in Soils and Plants, pp. 243–259. Springer, Dordrecht (1988)

Hansch, R., Mendel, R.R.: Physiological functions of mineral macronutrients (Cu, Zn, Mn, Fe, Ni, Mo, B, Cl). Curr. Opin. Plant Biol. **12**, 259–266 (2009)

He, W., Shohag, M.J.I., Wei, Y., et al.: Iron concentration, bioavailability, and nutritional quality of polished rice affected by different forms of foliar iron fertilizer. Food Chem. **141**, 4122–4126 (2013)

Heidari, M., Sarani, S.: Growth, biochemical components and ion content of chamomile (*Matricaria chamomilla* L.) under salinity stress and iron deficiency. J. Saudi Soc. Agric. Sci. **11**(1), 37–42 (2012)

Heidari, M., Galavi, M., Hassani, M.: Effect of sulfur and iron fertilizers on yield, yield components and nutrient uptake in sesame (*Sesamum indicum* L.) under water stress. Afr. J. Biotechnol. **10**(44), 8816–8822 (2011)

Hejazi, M.M., Shariatmadari, H., Khoshgoftarmanesh, A.H., et al.: Copper effects on growth, lipid peroxidation, and total phenolic content of rosemary leaves under salinity stress. J. Agric. Sci. Technol. **14**(1), 205–212 (2012)

Hoke, K.R., Cobb, N., Armstrong, F.A., Hille, R.: Electrochemical studies of arsenite oxidase: an unusual example of a highly cooperative two-electron molybdenum center. Biochem. **43**(6), 1667–1674 (2004)

Hrishikesh, U., Biman, K.D., Lingaraj, S., Sanjib, K.P.: Comparative effect of Ca, K, Mn and B on post-drought stress recovery in tea [*Camellia sinensis* (L.) O Kuntze]. Am. J. Plant Sci. **3**(4), 18 (2012)

Hussain, A., Ali, S., Rizwan, M., ur Rehman, M.Z., Qayyum, M.F., Wang, H., Rinklebe, J.: Responses of wheat (*Triticum aestivum*) plants grown in a Cd contaminated soil to the application of iron oxide nanoparticles. Ecotoxicol. Environ. Saf. **173**, 156–164 (2019)

Hussain, S., Umar, A., Amir, M., Aon, M.: Biofortification of cereals through foliar application of minerals. In: Vitamins and Minerals Biofortification of Edible Plants, pp. 191–221. John Wiley & Sons Ltd., Hoboken (2020)

Ibrahim, S., Faryal, S.: Augmentation of *Trigonella foenum-graecum* L. (methi) growth under salinity stress and allelochemical stress through Mn^+ B^+ Zn mixture foliar spray. J. Pharmacogn. Phytochem. **3**(2), 39–44 (2014)

Iqbal, M.N., Rasheed, R., Ashraf, M.Y., Ashraf, M.A., Hussain, I.: Exogenously applied zinc and copper mitigate salinity effect in maize (*Zea mays* L.) by improving key physiological and biochemical attributes. Environ. Sci. Pol. **25**(24), 23883–23896 (2018)

Jenne, E.A.: Trace element adsorption by sediments and soils: sites and processes. In: Chappell, W.R., Peterson, K.K. (eds.) Molybdenum in the Environment. Marcel Dekker, Inc., New York (1977)

Joy, E.J., Ahmad, W., Zia, M.H., Kumssa, D.B., Young, S.D., Ander, E.L., Watts, M.J., Stein, A.J., Broadley, M.R.: Valuing increased zinc (Zn) fertiliser use in Pakistan. Plant Soil. **411**(1–2), 139–150 (2017)

Kareem, F., Rihan, H., Fuller, M.: The effect of exogenous applications of salicylic acid and molybdenum on the tolerance of drought in wheat. Agric. Res. Technol. **9**(4), 1–10 (2017). https://doi.org/10.19080/ARTOAJ.2017.09.555768

Li, B., Tester, M., Gilliham, M.: Chloride on the move. Trends Plant Sci. **22**, 236–248 (2017)

Maathuis, F.J.M.: Physiological functions of mineral macronutrients. Curr. Opin. Plant Biol. **12**, 250–258 (2009)

Manasa, L.P., Devaranavadagi, S.B.: Effect of foliar application of micronutrients on growth, yield and nutrient uptake of maize. Karnataka J. Agric. Sci. **28**, 474–476 (2015)

McCauley, A., Jones, C., Jacobsen, J.: Plant nutrient functions and deficiency and toxicity symptoms. In: Nutrient Management Module 9. Montana State University Extension, Bozeman (2011)

Mengel, K., Kirkby, E.A.: Principles of Plant Nutrition, First Indian Reprint, 5th edn. Rashtriya Printers, New Delhi (2006)

Mo, Z., Liu, Q., Xie, W., Ashraf, U., Abrar, M., Pan, S., Duan, M., Tian, H., Wang, S., Tang, X.: Ultrasonic seed treatment and Cu application modulate photosynthesis, grain quality, and Cu concentrations in aromatic rice. Photosynthetica. **58**(3), 682–691 (2020). https://doi.org/10.32615/ps.2020.009

Monjezi, F., Vazin, F., Hassanzadehdelouei, M.: Effects of iron and zinc spray on yield and yield components of wheat (*Triticum aestivum* L.) in drought stress. Cercetari Agronomice Moldova. **46**(1), 23–32 (2013)

Montenegro, J.B.V., Fidalgo, J.A.B., Gabella, V.M.: Response of chickpea (*Cicer arietinum* L.) yield to zinc, boron and molybdenum application under pot conditions. Span. J. Agric. Res. **3**, 797–807 (2010)

Ono, T.A., Noguchi, T., Inoue, Y., Kusunoki, M., Matsushita, T., Oyanagi, H.: X-ray detection of the period-four cycling of the manganese cluster in photosynthetic water oxidizing enzyme. Science. **258**(5086), 1335–1337 (1992)

Ouzounidou, G., Ilias, I., Tranopoulou, H., Karataglis, S.: Amelioration of copper toxicity by iron on spinach physiology. J. Plant Nutr. **21**(10), 2089–2101 (1998)

Pérez-Labrada, F., López-Vargas, E.R., Ortega-Ortiz, H., Cadenas-Pliego, G., Benavides-Mendoza, A., Juárez-Maldonado, A.: Responses of tomato plants under saline stress to foliar application of copper nanoparticles. Plan. Theory. **8**(6), 151 (2019)

Rafsanjani, K., Madani, A., Vazin, F.: Effect of iron on yield of corn (*Zea mays* L.) in drought stress. Cercetari Agronomice Moldova. **52**(1), 34–43 (2019)

Rahman, A., Nahar, K., Hasanuzzaman, M., Fujita, M.: Manganese-induced cadmium stress tolerance in rice seedlings: coordinated action of antioxidant defense, glyoxalase system and nutrient homeostasis. C. R. Biol. **339**(11–12), 462–474 (2016)

Rattan, R.K.: Mineral nutrition of plants. In: Rattan, R.K., Katyal, J.C., Dwivedi, B.S., Sarkar, A.K., Bhattacharyya, T., Tarafdar, J.C., Kukal, S.S. (eds.) Soil Science: An Introduction, pp. 499–539. Indian Society of Soil Science, New Delhi (2015)

Reddy, K.J., Munn, L.C., Wang, L.: Chemistry and mineralogy of molybdenum in soils. In: Gupta, U.C. (ed.) Molybdenum in Agriculture. Cambridge University Press, Cambridge (1997)

Rengel, Z., Graham, R.D., Pedler, J.F.: Manganese nutrition and accumulation of phenolics and lignin as related to differential resistance of wheat genotypes to the take-all fungus. Plant Soil. **151**(2), 255–263 (1993)

Rihan, H.Z., Al-Issawi, M., Al Shamari, M., Woldie, W.A., Kiernan, M., Fuller, M.P.: The effect of molybdenum on the molecular control of cold tolerance in cauliflower (*Brassica oleracea* var. botrytis) artificial seeds. Plant Cell Tissue Organ Cult (PCTOC). **118**(2), 215–228 (2014)

Rizwan, M., Ali, S., Adrees, M., Rizvi, H., Zia-ur-Rehman, M., Hannan, F., et al.: Cadmium stress in rice: toxic effects, tolerance mechanisms, and management: a critical review. Environ. Sci. Pol. **23**(18), 17859–17879 (2016)

Rizwan, M., Ali, S., Hussain, A., Ali, Q., Shakoor, M.B., Zia-ur-Rehman, M., et al.: Effect of zinc-lysine on growth, yield and cadmium uptake in wheat (*Triticum aestivum* L.) and health risk assessment. Chemosphere. **187**, 35–42 (2017)

Rizwan, M., Ali, S., Ali, B., Adrees, M., Arshad, M., Hussain, A., et al.: Zinc and iron oxide nanoparticles improved the plant growth and reduced the oxidative stress and cadmium concentration in wheat. Chemosphere. **214**, 269–277 (2019)

Sağlam, A., Yetişsin, F., Demiralay, M., Terzi, R.: Copper stress and responses in plants. In: Plant Metal Interaction, pp. 21–40. Elsevier, Amsterdam (2016)

Sah, S., Singh, N., Singh, R.: Iron acquisition in maize (*Zea mays* L.) using *Pseudomonas* siderophore. 3 Biotech. **7**(2), 121 (2017)

Saha, B., Prabhakar, M., Saha, S., Pal, M., Mandal, J., Singh, R., Singh, A.P., Ranjan, A., Shamim, M., Singh, P.K.: Could maneuvering the methods of zinc and boron application influences yield and agro-morphological traits of brinjal in inceptisols? Int. J. Chem. Stud. **8**(2), 11–15 (2020)

Said-Al Ahl, H.A.H., Mahmoud, A.A.: Effect of zinc and/or iron foliar application on growth and essential oil of sweet basil (*Ocimum basilicum* L.) under salt stress. Ozean J. Appl. Sci. **3**(1), 97–111 (2010)

Saif, S., Tahir, A., Chen, Y.: Green synthesis of iron nanoparticles and their environmental applications and implications. Nanomaterials (Basel). **6**(11), 209 (2016)

Saini, A.K., Singh, R.: Effect of sulphur and iron fertilization on growth and yield of green gram (*Vigna radiata* L.). Int. J. Curr. Microbiol. App. Sci. **6**(6), 1922–1929 (2017)

Sajedi, N.A., Ardakani, M.R., Madani, H., Naderi, A., Miransari, M.: The effects of selenium and other micronutrients on the antioxidant activities and yield of corn (*Zea mays* L.) under drought stress. Physiol. Mol. Biol. Plants. **17**(3), 215–222 (2011)

Salha, B., Chaabane, R.: Effect of molybdenum and nitrogen on *Phaseolus vulgaris* L., *Cicer arietinum* L. and *Lens culinaris* M. seedlings grown under salt stress. Int. J. Biosci. **8**(6), 34–42 (2016). https://doi.org/10.12692/ijb/8.6.34-42

Sarwar, N., Malhi, S.S., Zia, M.H., Naeem, A., Bibi, S., Farid, G.: Role of mineral nutrition in minimizing cadmium accumulation by plants. J. Sci. Food Agric. **90**(6), 925–937 (2010)

Schmidt, S.B., Husted, S.: The biochemical properties of manganese in plants. Plan. Theory. **8**(10), 381 (2019)

Sebastian, A., Prasad, M.N.V.: Modulatory role of mineral nutrients on cadmium accumulation and stress tolerance in *Oryza sativa* L. seedlings. Environ. Sci. Pollut. Res. Int. **23**(2), 1224–1233 (2016)

Shahverdi, M.A., Omidi, H., Tabatabaei, S.J.: Plant growth and steviol glycosides as affected by foliar application of selenium, boron, and iron under NaCl stress in *Stevia rebaudiana* Bertoni. Ind. Crop. Prod. **125**, 408–415 (2018)

Skrzypczak, D., Ligas, B., Mikula, K., Witek-Krowiak, A., Samoraj, M., Moustakas, K., Chojnacka, K.: Valorization of post-extraction biomass residues as carriers of bioavailable micronutrients for plants and livestock. Biomass Convers. Biorefin., 1–16 (2020). https://doi.org/10.1007/s13399-019-00586-z

Soleymani, A., Shahrajabian, M.H.: The effects of Fe, Mn and Zn foliar application on yield, ash and protein percentage of forage sorghum in climatic condition of esfahan. Int. J. Biol. **4**(3), 92–96 (2012)

Soliman, A.S., El-feky, S.A., Darwish, E.: Alleviation of salt stress on *Moringa peregrina* using foliar application of nanofertilizers. J. Hortic. For. **7**(2), 36–47 (2015)

Soni, J., Kushwaha, H.S.: Effect of foliar spray of zinc and iron on productivity of mungbean [*Vigna radiata* (L.) Wilczeck]. Int. J. Pharmacogn. Phytochem. Res. **9**(1), 108–111 (2020)

Stepien, A., Wojtkowiak, K.: Effect of foliar application of Cu, Zn, and Mn on yield and quality indicators of winter wheat grain. Chil. J. Agric. Res. **76**, 219–226 (2016)

Taiz, L., Zeiger, E.: Plant Physiology, 2nd edn. Sinauer Associates Publishers, Sunderland (2002). https://doi.org/10.1071/PP9840361

Tangle Agro. Tangle Agro products catalog 2019. www.tangelagro.com (2019)

Teakle, N.L., Tyerman, S.D.: Mechanisms of Cl^- transport contributing to salt tolerance. Plant Cell Environ. **33**, 566–589 (2010)

TNAU. Plantation crops: coconut: physiological disorders. http://agritech.tnau.ac.in (2015)

Umena, Y., Kawakami, K., Shen, J.R., Kamiya, N.: Crystal structure of oxygen-evolving photosystem II at a resolution of 1.9 Å. Nature. **473**(7345), 55–60 (2011)

Valenciano, J.B., Boto, J.A., Marcelo, V.: Chickpea (*Cicer arietinum* L.) response to zinc, boron and molybdenum application under field conditions. N. Z. J. Crop. Hortic. Sci. **39**(4), 217–229 (2011)

Van Ha, C., Van Nguyen, D., Nguyen, H.M., Le, N.T., Nguyen, K.H., Le, H.M., et al.: Copper nanoparticle application enhances plant growth and grain yield in maize under drought stress conditions. BioRxiv. (2020). https://doi.org/10.1101/2020.02.24.963132

Vieira, R.F., Vieira, C., Cardoso, E.J.B.N., Mosquim, P.R.: Foliar application of molybdenum in common bean. II. Nitrogenase and nitrate reductase activities in a soil of low fertility. J. Plant Nutr. **21**(10), 2141–2151 (1998)

Wang, S., Zhang, X., Liu, K., Fei, P., Chen, J., Li, X., Ning, P., Chen, Y., Shi, J., Tian, X.: Improving zinc concentration and bioavailability of wheat grain through combined foliar applications of zinc and pesticides. J. Agron. **111**(3), 1478–1487 (2019)

Waraich, E.A., Ahmad, R., Ashraf, M.Y.: Role of mineral nutrition in alleviation of drought stress in plants. Aust. J. Crop Sci. **5**(6), 764 (2011)

Weisany, W., Sohrabi, Y., Heidari, G., Siosemardeh, A., Ghassemi-Golezani, K.: Changes in antioxidant enzymes activity and plant performance by salinity stress and zinc application in soybean (*Glycine max* L.). Plant OMICS. **5**(2), 60 (2012)

Weisany, W., Sohrabi, Y., Heidari, G., Siosemardeh, A., Badakhshan, H.: Effects of zinc application on growth, absorption and distribution of mineral nutrients under salinity stress in soybean (*Glycine max* L.). J. Plant Nutr. **37**(14), 2255–2269 (2014)

White, P.J., Broadley, M.R.: Chloride in soils and its uptake and movement within the plant: a review. Ann. Bot. **88**, 967–988 (2001)

Williams, R.J.P., Frausto da Silva, J.J.R.: The involvement of molybdenum in life. Biochem. Biophys. Res. Commun. **292**, 293–299 (2002)

Williams, C.M.J., Maier, N.A., Bartlett, L.: Effect of molybdenum foliar sprays on yield, berry size, seed formation, and petiolar nutrient composition of "Merlot" grapevines. J. Plant Nutr. **27**(11), 1891–1916 (2005)

Wu, F., Zhang, G.: Alleviation of cadmium-toxicity by application of zinc and ascorbic acid in barley. J. Plant Nutr. **25**(12), 2745–2761 (2002)

Yuan, L., Wu, L., Yang, C., Lv, Q.: Effects of iron and zinc foliar applications on rice plants and their grain accumulation and grain nutritional quality. J. Sci. Food Agric. **93**(2), 254–261 (2013)

Zayed, B.A., Salem, A.K.M., El Sharkawy, H.M.: Effect of different micronutrient treatments on rice (*Oryza sativa* L.) growth and yield under saline soil conditions. World J. Agric. Sci. **7**(2), 179–184 (2011)

Zhang, Y., Shi, R., Rezaul, K.M., Zhang, F., Zou, C.: Iron and zinc concentrations in grain and flour of winter wheat as affected by foliar application. J. Agric. Food Chem. **58**(23), 12268–12274 (2010)

Zhang, M., Hu, C., Zhao, X., Tan, Q., Sun, X., Cao, A., et al.: Molybdenum improves antioxidant and osmotic-adjustment ability against salt stress in Chinese cabbage (*Brassica campestris* L. ssp. Pekinensis). Plant Soil. **355**(1–2), 375–383 (2012)

Zhang, M., Hu, C., Sun, X., Zhao, X., Tan, Q., Zhang, Y., Li, N.: Molybdenum affects photosynthesis and ionic homeostasis of Chinese cabbage under salinity stress. Commun. Soil Sci. Plant Anal. **45**(20), 2660–2672 (2014)

Zhang, X., Goatley, M., Conner, J., Wilkins, M., Teshler, I., Liu, J., et al.: Copper chlorophyllin impacts on growth and drought stress tolerance of tomato plants. Hortic. Sci. **54**(12), 2195–2201 (2019)

Zhong, H., Ma, G.R.: The influence of chlorine on physiological effects of potato. J. Zhejiang Agric. **5**, 83–88 (1993)

Chapter 7
Zinc as a Versatile Element in Plants: An Overview on Its Uptake, Translocation, Assimilatory Roles, Deficiency and Toxicity Symptoms

Qazi Fariduddin, Mohd Saleem, Tanveer Ahmad Khan, and Shamsul Hayat

Abstract Zinc (Zn) is the second most abundant transition element after iron in living organisms and is categorized as an essential micronutrient to plants. It takes part in various cellular and physiological functions of plants and exerts a positive role in growth, development and yield. Zn serves structural, catalytic and regulatory roles in enormous numbers of proteins and enzymes. Hence, it is important to get the knowledge and understanding about the dynamics and biochemistry of Zn in soil, its uptake, transport and response of plants to Zn deficiency and in achieving the sustainable solutions for the same in plants. Zn deficiency seems to be an extensive and common problem in crops worldwide, resulting in severe losses in yield affecting nutritional quality. Bearing in mind the current state, an attempt has been made to further dissect the uptake, transport, assimilation, role, deficiency, toxicity symptoms of Zn and economic losses triggered in plants.

Keywords ZIP proteins · Carboxylating enzymes · Auxin hormones · Zn toxicity · HMAs

Introduction

Zinc is an essential micronutrient for both plants and animals. In plants, it plays a vital role in various physiological and metabolic processes, thereby causing an optimal plant growth (Rudani and Prajapati 2018). Zn serves structural, catalytic and regulatory roles in a number of proteins and enzymes such as alcohol

Q. Fariduddin (✉) · M. Saleem · T. A. Khan · S. Hayat
Plant Physiology and Biochemistry Section, Department of Botany, Faculty of Life Sciences, Aligarh Muslim University, Aligarh, India

S. T. Khan, A. Malik (eds.), *Microbial Biofertilizers and Micronutrient Availability*,
https://doi.org/10.1007/978-3-030-76609-2_7

dehydrogenase (ADH), carbonic anhydrase (CA), copper-zinc (Cu–Zn) superoxide dismutase (CuZn-SOD), carboxypeptidase, alkaline phosphatase, RNA polymerase, phospholipase, inorganic pyrophosphatases (PPiases) and Zn fingers (IRRI 2000; Broadley et al. 2007; Rudani and Prajapati 2018). It plays an important role in plant growth and several metabolic processes like nitrogen metabolism, synthesis of cytochrome, proteins, nucleotide, tryptophan, and indole acetic acid (IAA), carbohydrate metabolism, maintenance of membrane integrity and peroxidation of lipids (Broadley et al. 2007; Alloway 2008; Hänsch and Mendel 2009; Jerlin et al. 2017). Moreover, Zn also plays a vital role in the development of pollen, defence against diseases, increases level of antioxidants and pigments in plants and tolerance to various abiotic stresses in plants. Zinc is a vital micronutrient needed for optimum yield and growth of crop plants. About 1000 proteins are said to be associated with Zn for proper functioning in *Arabidopsis thaliana* (Krämer et al. 2007).

Deficiency of Zn is one of the primary problems in many countries affecting approximately half of the world population (WHO 2003; Hafeez et al. 2013). Deficiency of Zn not only hampers the plant growth and yield but also adversely affects the animals. Globally, more than 300 crore individuals are adversely affected by the deficiency of Zn and Fe, which mainly results because crops are grown in Zn-deficient soil. Deficiency of Zn is predominant in those communities which are greatly dependent on a repetitive diet of cereal-based food because Fe and Zn are mainly stored in the hull (seed coat) and are lost during the granulation and refining process (Cakmak 2002; Lu et al. 2013). It is expected by the year 2025, that due to a constant decrease in soil fertility, deficiency of Zn may rise from 42% to 63% and about 50% soil in India could be Zn deficient, posing various physiological alterations affecting the yield of crops (Cakmak 2009). Various biochemical and physiological processes are inhibited by the deficiency of Zn causing stunted growth. Deficiency of Zn in soil can be overcome by external supplementation through the soil in the form of zinc sulphate, which may also be converted into inaccessible forms such as Zn(OH) and Zn (OH_2) depending upon the pH of soil (Sunithakumari et al. 2016). Due to the polishing of seeds (grains) and limited availability of Zn in the soil for plants, it is thus important to understand and obtain knowledge about the dynamics and biochemistry of Zn in plants and adopt sustainable measures to overcome this problem. So to address these issues, this chapter focusses on the following objectives:

- How plants uptake Zn from soil and its subsequent translocation within plants?
- Are various cellular and metabolic processes altered due to Zn?
- Toxicity and tolerance of Zn in plants

Uptake of Zn from Soil to Roots and Its Transport

In soil, Zn occurs in several mineralized forms like (Zn-silicate, Zn-sulphides and Zn-carbonates), adsorbed form (soil surface), solution form (Zn ion and $Zn(OH)^+$) or in organic complex forms (complexed with organic colloids). The release of Zn from mineralized form occurs in the form of ZnS (sphalerite), $ZnCO_3$ (smithsonite), $ZnSiO_4$ (willemite) and $ZnFe_2O_2$ (franklinite). Zn is biologically very crucial at +2 oxidation state. Moreover, the concentration of Zn in the agronomic soil ranges from 10 to 300 μg/g of the soil (Barber 1995). On average, the concentration of Zn in cultivated soil is nearly 65 mg/kg of soil (Alloway 2008). Various physiochemical factors determine the amount and availability of Zn to plants, as shown in Fig. 7.1. These factors include the following:

(i) *Content of Zn in the rhizosphere*: The availability of the soluble form of Zn in the soil is directly correlated with the absorption of Zn by plants through various transporter proteins through facilitated diffusion.
(ii) *pH of soil*: It is one of the important factors and regulates the uptake of Zn from soil (Marschner 1993). Generally, acidic soil favours the uptake, whereas alkaline soil decreases the availability of Zn in soil. High pH stimulates the adsorption of Zn to cation exchange sites of soil particles, thereby diminishing the availability of Zn.

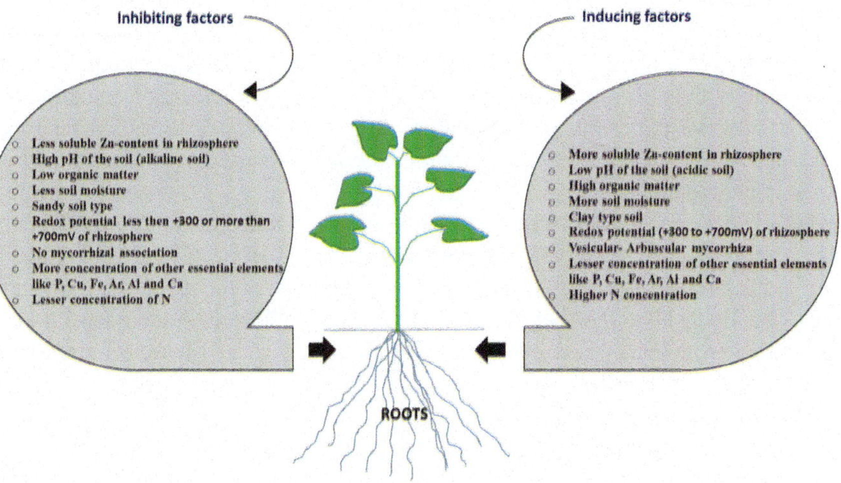

Fig. 7.1 Various physio-chemical factors that affect the uptake and availability of Zn in plants. Factors on left side (inhibiting factors) reduce the solubility and mobility of Zn by enhancing adsorption and precipitation of Zn, limiting its uptake, while factors listed on right side (inducing factors) stimulate the uptake of Zn from soil

(iii) *Organic matter*: High soil organic matter improves the available Zn pool in soil, thereby favouring the growth of plants (Catlett et al. 2002; Obrador et al. 2003). Soil organic matter not only increases the availability but also increases the rate of diffusion (Sharma and Deb 1988).

(iv) *Presence of other elements in soil*:

(a) *Phosphorus*: Increased content or supply of P fertilizers in soil reduces the uptake of Zn. It could be due to some intricate physiological factors of plants (Mousavi et al. 2012).

(b) *Nitrogen*: High level of N increases the availability of Zn to plants.

(c) *Copper*: Cu and Zn both are taken up by the plants via an identical mechanism, so if anyone of them is in excessive quantity it decreases the availability of the other (Mousavi et al. 2012).

(d) *Arsenic*: Low availability of Ar in soil favours the availability of Zn to plants.

(e) *Iron and aluminium*: Higher content of Fe and Al oxides in soil limits the uptake of Zn (Mousavi et al. 2012).

(f) *Calcium*: The higher level of Ca in the form of $CaCO_3$ in soil (calcareous soil) decreases the uptake of Zn because $CaCO_3$ promotes the adsorption and precipitation of Zn, thereby decreasing phytoavailability (Mousavi et al. 2007; Alloway 2008).

(v) *Soil type and moisture*: Out of various types of soils, it was found that light-textured soils (sandy soil) have lesser cation exchange capacity (CEC) and hence lower capacities of Zn adsorption than heavier and fine-textured (clayey) soils; therefore, the deficiency of Zn is more likely to occur in the sandy than clayey soil (Stahl and James 1991). The moisture content of the soil is very important and a high moisture content decreases the availability of Zn (Marschner 1993). Soil with excessive mist decreases the capacity of plants to uptake Zn.

(vi) *Type of species and stage of plant*: Various morphological factors of plant-like root biomass, the height of the plant, number of leaves and leaf area affect the availability of Zn to plants. Besides this, the stage of the plant also determines the Zn level, as the accumulation of Zn varies in flowering and post-flowering stages of plants (Rossi et al. 2004; Samardjieva et al. 2014; Waters and Grusak 2008). Also, there are various Zn hyperaccumulator species, such as *Arabidopsis halleri*, *Arabis paniculata*, *Noccaea caerulescens*, *Noccaea praecox*, *Noccaea fendleri*, *Picris divaricata*, *Potentilla griffithii* and *Sedum plumbizincicola*, which differ in their uptake and accumulation of Zn from non-accumulators (Bert et al. 2003; Milner and Kochian 2008; Huguet et al. 2012; Boyd 2012).

(vii) *Redox potential of rhizosphere (Eh)*: The process of humus formation and mobilization of Zn is greatly influenced by Eh. Optimal Eh for plant growth and development occurs in the range of +300 to +700 mV. Rhizosphere Eh is modified by plant roots, either by association with microbes or by exudation of different chelates and reducing agents (Hartmann et al. 2009). These modification processes affect the uptake, solubility and redox state of Zn in soil. Generally, reduced conditions (Eh $\leq$ +350 mV) in soil decrease the oxygen pool and hence the growth of plants.

(viii) *Vesicular-arbuscular mycorrhiza (VAM)*: VAM generally increases the uptake of P, but there are reports available that this symbiotic association increases the uptake of Zn also (Coccina et al. 2019).

Zn is predominantly taken up by the plants from the soil by the process of facilitated diffusion through roots in the form of free divalent cation (Zn^{2+}) mediated by membrane potential and transporters (Olsen and Palmgren 2014; Tiong et al. 2015). However, at high pH it is also taken up as monovalent cation ($ZnOH^{+}$). Once Zn reaches inside the root, it is transported to the stellar region via epidermis, cortex, endodermis and pericycle tissues sequentially (radial transport). This radial transport of Zn is mediated by either symplastic or apoplastic route. Apoplastic route is followed only up to the endodermal region after that only the symplastic route is followed in radial transport of Zn because of its impermeability through suberized casparian strips (Yang and Jie 2005). Out of these radial transport pathways, symplastic pathway is said to be more selective in controlling the distribution of Zn (Clemens et al. 2002). After crossing the casparian strip barrier, Zn is transported to xylem via pericycle and xylem parenchyma cells, i.e. xylem loading occurs. Loading of Zn to apoplastic xylem from pericycle or xylem parenchyma is mediated by H^{+}-ATPase activity, so xylem loading is an active process (Sondergaard et al. 2004). Mineral ions (Zn^{2+}) in xylem sap are transported as metal complexes with nicotinamide, organic acids or with amino acids asparagine, histidine or some other uncommon amino acids (Takahashi et al. 2003; Olsen and Palmgren 2014). For both intercellular and root-to-shoot translocation of Zn, nicotinamide is said to play an important role as chelator (Olsen and Palmgren 2014; Nishiyama et al. 2012; Clemens et al. 2013). From xylem, Zn reaches the mesophyll cells where xylem unloading and phloem loading occur, but how xylem unloading and phloem loading occur is still a question that needs to be addressed by the scientists. However, different transporter proteins have been revealed to play a role in Zn homeostasis (Lopez-Millan et al. 2004; Chen et al. 2008a; Bashir et al. 2012). These transporter proteins include;

- *ZIP (zinc-, iron-permease family/ZRT-, IRT-like proteins)*: It acts as one of the primary transporters and helps in maintaining the homeostasis of Zn, present in all developmental phases and tissues of a plant (Grotz et al. 1998; Gainza-Cortés et al. 2012; Li et al. 2013). Different types of ZIP transporters are present in plants such as ZIP1, ZIP2, ZIP3, ZIP5, ZIP6, ZIP9, ZIP10 and ZIP12, out of all these proteins ZIP4, ZIP5, ZIP6 and ZIP9 are localized in leaves, whereas the rest are present in shoots. All these ZIP factors are involved in both uptake and translocation of Zn in plant body. It was revealed that ZIP1 and ZIP2 are localized in vacuole and root vasculature, respectively, and are involved in the root-shoot translocation of Zn (Milner et al. 2013); ZIP1, ZIP2, ZIP3, ZIP7 and ZIP9 are present in root cells that help in the uptake of Zn (Tiong et al. 2014); ZIP4, ZIP5 and ZIP8 are involved in the translocation of Zn from root to shoot (Bashir et al. 2013), but ZIP4 is localized in chloroplast (Guerinot 2000). Different views about the localization and functions of these transporters had been proposed, so a detailed investigation about these transporters is a compelling need to unravel a clear picture of Zn uptake and translocation. Till date, 18 different types of ZIP family proteins have been identified in model plant *A. thaliana*. Out of these, 12 are ZIPs (Jain et al. 2013), three are IRTs (iron-regulated transporters) (Vert et al. 2001) and the remaining are ZTP29 (Wang et al. 2010), AtPutZnT (Wang et al. 2008) and AtIAR1 (Lasswell et al. 2000). Broad ZIP proteins are also characterized and classified from other plants (Sivasubramanian et al. 2015).

- *Metal tolerance proteins (MTPs)/cation diffusion facilitator (CDF)*: These proteins help in the storage of Zn in vacuoles (Desbrosses-Fonrouge et al. 2005; Morel et al. 2009). Two MTPs, AtMTP1 and AtMTP3, have been found to be involved in the vacuolar sequestration of Zn (Desbrosses-Fonrouge et al. 2005; Kawachi et al. 2009).
- *Heavy metal adenosine triphosphatases (ATPases) (HMAs)/P1B-type ATPase*: These are involved in the hypertolerance and hyperaccumulation of metals as well as in their transport into vacuoles and root shoot translocation. Various HMAs have been reported like AtHMA1, AtHMA3, AtHMA2 and AtHMA4 in various plants, which play an important role in the homeostasis of Zn. AtHMA1 is present in the inner membrane of chloroplast and helps in the export of Zn from chloroplast, whereas AtHMA3 is present in the tonoplast and helps in the storage of Zn in vacuoles (Morel et al. 2009; Kim et al. 2009). In *A. thaliana*, AtHMA2 and AtHMA4 are involved in xylem loading (Hussain et al. 2004; Sinclair et al. 2007). Homologue of HMA2 (OsHMA2) in rice is involved in the translocation of Zn from root to shoot (Satoh-Nagasawa et al. 2012; Takahashi et al. 2012).
- *Natural resistance-associated macrophage protein 4 (NRAMP4)*: These proteins help in the sequestration and detoxification of Zn and are also involved in the translocation of Zn from root to shoot (Oomen et al. 2009; Verret et al. 2004; Wang et al. 2019b).
- *Yellow stripe like (YSL) protein*: It is a subclass of the oligopeptide transporter (OPT) protein family involved in the transport of chelated heavy metals such as Zn to reproductive organs and leaves (Yen et al. 2001; Waters et al. 2006). Different YSL proteins have been identified in *Arabidopsis* such as AtYSL1, AtYSL3, AtYSL4 and AtYSL6. AtYSL1 and AtYSL3 are essential for the distribution of Zn to the emerging grains (Chu et al. 2010), whereas AtYSL4 and AtYSL6 are localized in chloroplast and act as iron transporters and endomembranous systems (Divol et al. 2013; Chu et al. 2013).
- *Plant cadmium resistance 2 protein (PCR2)*: These proteins increase tolerance to cadmium (Cd^{2+}) ion by restricting its transport. Multiple PCR family proteins have been identified in plant systems such as AtPCR1 and AtPCR2; both of these transporters confer tolerance to Cd and also help in the transport of Zn and Fe (Song et al. 2004, 2010).

- *Vacuolar iron transporter (VIT) family*: Initially, it was thought to be involved in the transport of Fe to vacuoles from cytoplasm, but later on, its role in the transport of other microelements (Mn, Zn) was also identified (Kim et al. 2006; Momonoi et al. 2009). Various VIT proteins are reported from plants such as *AtVIT1* in *Arabidopsis* (expressed in developing seed), OsVIT1 and OsVIT2 (expressed in leaf sheath and blades) from rice and TgVIT1 (expressed in petals of flowers) from *Tulip generiana* (Zhang et al. 2012). These VIT proteins facilitate the transport of Zn^{2+}, Fe^{2+} and Mn^{2+} in vacuoles and also play a role in various physiological processes.

Role of Zn

Zinc plays an important role in both humans and in plants. It is predicted that almost 2800 proteins (10% of human proteins) require Zn for proper organization and functioning (Andreini et al. 2009; Maret and Li 2009), whereas in plants, more than 1200 proteins possess Zn (Krämer and Clemens 2005). In plants, it plays an important role in the regulation of various cellular, physiological and biochemical attributes. Various roles exerted by Zn are as follows:

Zn as a Cofactor

Zn plays an important role as a cofactor (non-proteinaceous part of the enzyme), which stimulates the catalytic activity of enzymes. It is abundant and versatile among all the metallic cofactors, with almost more than 300 enzymes using Zn as the cofactor. Because of the unique chemical properties of Zn, it acts as a good candidate in the form of cofactor. It is the single element that is present in all six classes of enzymes comprising oxidoreductases, transferases, hydrolases, lyases, isomerases and ligases (Sousa et al. 2009). As a cofactor, it acts catalytically (e.g. carbonic anhydrase and carboxypeptidase), co-catalytically, structurally (e.g. alcohol dehydrogenase, CuZn-superoxide dismutase (CuZn-SOD) and proteins involved in replication and transcription) or through protein interface (Auld 2009). When Zn functions catalytically, it gets linked to one molecule of H_2O and three ligands that include histidine, aspartate or glutamate. Structurally, Zn maintains the structure of enzyme by coordinating with four ligands of cysteine. At the co-catalytic sites of enzymes, Zn is linked with either aspartate or histidine residues, and at the protein interface, Zn helps in the bonding of various subunits of proteins (Auld and Bergman 2008; Auld 2009). The activity of various enzymes is modulated by the presence or absence of Zn in plants, which includes alcohol dehydrogenase (ADH), carbonic anhydrases (CAs), CuZn-SOD, alkaline phosphatase, phospholipase, carboxypeptidase, RNA and DNA polymerase, dehydrogenases, aldolases, isomerases and inorganic pyrophosphatases (PPiases) (Castillo-González et al. 2018; Lin and

Kao 1990; Auld and Bergman 2008; Coleman 1992). So being the component of various enzymes, it exerts various cellular, physiological and metabolic changes in plants (Rudani and Prajapati 2018; Solanki et al. 2016).

Role in Photosynthesis via Regulation of Carboxylating Enzymes (CEs) and CA

Zn deficiency affects the photosynthesis as it reduces intercellular CO_2 concentration and stomatal conductance (Cakmak 2008). Zn has been shown to play an important role in the regulation of stomatal aperture by maintaining an elevated content of K in the guard cells (Das and Mukherjee 2014). Zn deficiency triggers the accumulation of saccharides, particularly in phloem sap leaves, which might be associated by decrease in CO_2 fixation. Net photosynthetic rate, stomatal conductance and transpiration rate are strongly reduced by Zn deficiency (Das and Mukherjee 2012; Yao et al. 2019; Liu et al. 2019). Zn plays an important role in the regulation of photosynthesis, due to its presence in carboxylation enzymes (CEs) and carbonic anhydrase (CA) (EC: 4.2.1.1). Green plants convert water (H_2O) and carbon dioxide (CO_2) into sugars, where CO_2 is taken up from atmosphere and is stored in plants as bicarbonate ions in which CA plays an important role. CA contains single Zn ion that catalyses the hydration of CO_2 and converts it into bicarbonate (HCO_3^-), in which Zn^{2+} polarizes the water molecule (lowers the pK_a of water), making it to act more like a OH^- ion, thereby helping the nucleophilic attack of CO_2 to form HCO_3^- in the reaction catalysed by CA, as shown below:

$$CO_2 + H_2O \leftrightarrow HCO_3^- + H^+$$

CA is involved in a wide range of processes and any alteration in its activity directly or indirectly influences the process of photosynthetic CO_2 fixation. CA plays a catalytic role and is present in all organisms, including archaea, blue-green algae (BGA), algae, plants and animals. Almost six different classes of CAs, such as α, β, γ, δ, ε and ζ, are found; out of which α, β and γ are found in plants as well as other organisms, whereas δ, ε and ζ are found only in algae (Moroney et al. 2001; Rudenko et al. 2015). It occurs in chloroplast, cytoplasm or mitochondria of plants, and β-CA (beta carbonic anhydrase) present in cytoplasm is crucial for the fixation of carbon through phosphoenolpyruvate carboxylase (PEPcase) in C4 plants, as shown in Fig. 7.2. Commonly in higher plants, CA triggers the transfer of CO_2 to Rubisco (ribulose-1,5-bisphosphate carboxylase/oxygenase) present in chloroplast and in this case, the CA catalyses the reaction in opposite directions (Rudenko et al. 2015; Zabaleta et al. 2012). In higher plants, β- and γ-CAs (beta- and gamma carbonic anhydrases) appear to assist in the recycling of CO_2 from mitochondria (Zabaleta et al. 2012). In BGA, an efficient carbon-concentrating mechanism (CCM) is established by CA in carboxysomes and less efficient CCM occurs in green algae through the activity of CA in the lumen of chloroplast (Giordano et al. 2005; Price et al.

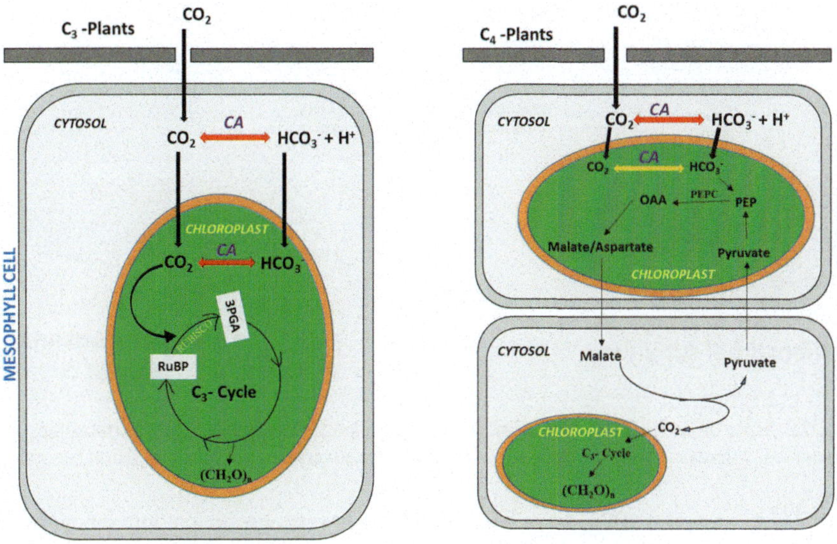

Fig. 7.2 A schematic presentation of role of carbonic anhydrase in mesophyll and bundle sheath cells of C_3 and C_4 plants

2013). CA helps in the fixation of carbon by photosynthesis, as in C_3 plants β-CA elevates the CO_2 concentration in plastids, which act as the substrate for the Rubisco, whereas in C4 and CAM plants, HCO_3^- serves as the substrate for PEPCase and CA also facilitates the delivery of CO_2 via cell membrane and chloroplast in bicarbonate form, as shown in Fig. 7.2 (Tiwari et al. 2005; Marschner 2011). Apart from the CA, Zn has also been shown to play an important role as a carboxylating enzyme, i.e. RUBPCase (ribulose-1,5-bisphosphate carboxylase/oxygenase) (Brookes et al. 1981; Das and Mukherjee 2012).

Role in Transcription and Translation

Zn plays an important role in DNA replication, transcription, translation (central dogma of a cell) and hence gene regulation with the help of Zn-metalloproteins (Zn-dependent proteins) (Coleman 1992; Andreini et al. 2009). Transcription factors (TFs) act as the chief mediator of response from the extranuclear region, bZIP TFs (Zn-dependent) form the largest TF family protein, which is present in all eukaryotes (Andreini et al. 2009). On the basis of the location and number of cysteine and histidine residues, Zn finger TFs have been categorized into different classes such as C_2H_2, C_2C_2, C_2HC, $C_2C_2C_2C_2$ and $C_2HCC_2C_2$ (Laity et al. 2001). C_2H_2 Zn finger type of protein functions as the critical regulator of transcription in plants against a broad range of environmental stimuli like temperature extremes, water stress, salt and light stress (Wang et al. 2019a). In *Arabidopsis*, two types of TFs

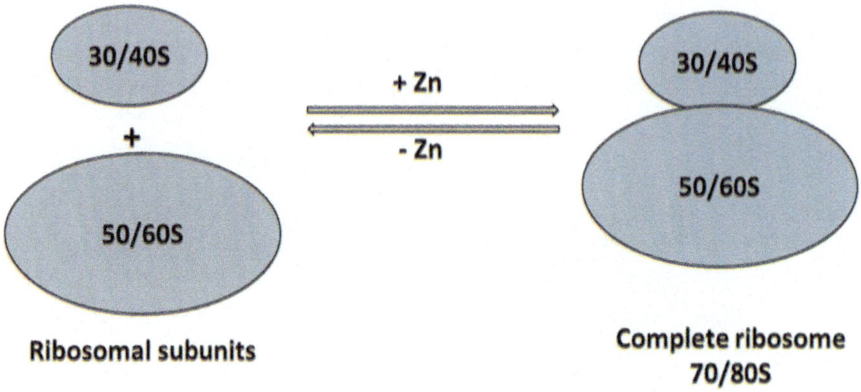

Fig. 7.3 Role of Zn in association of ribosomal subunits. In the presence of Zn, ribosomes are assembled, whereas in its deficiency, subunits dissociate adversely affecting translation process

(AtbZIP19 and AtbZIP23) have been identified in response to Zn deficiency. bZIP TFs regulate the homeostasis of Zn in the cell by binding to the promotor element of DNA, i.e. Zn deficiency response elements (ZDREs) (Assunção et al. 2010; Milner et al. 2013). These TFs are said to control different cellular and physiological processes such as maturation of seeds, senescence, tissue and organ differentiation, hormone and light signalling, and stress responses (Weisshaar et al. 1991; Choi et al. 2000; Jakoby et al. 2002; Liu et al. 2013; Gao et al. 2014; Yin et al. 2017). Zn plays an important role in the expression of genes by either suppressing or activating DNA elements (Kirkby and Romheld 2004). Zn also plays a vital role in translation, as it is an important structural component of the protein factory of the cell, i.e. ribosomes, vital for the association of ribosomal subunits shown in Fig. 7.3, shields the rRNA from ribonuclease enzyme and is critical for the RNA polymerase activity (Mendoza-Cózatl et al. 2006; Elsby and Roberts 2008; Abud et al. 2008). The deficiency of Zn affects protein synthesis and decreases protein content, whereas supplementation of Zn to a Zn-deficient plant resumes the translation process speedily and accumulates amino acids (Patel et al. 2007). Therefore, it is evident that Zn plays a critical role in the flow of information from DNA to proteins, i.e. the central dogma of a cell.

Metabolism of Carbohydrates

Zn plays an important role in the regulation and metabolism of starch and sugars. It was revealed that a Zn-deficient plant had a reduced content of starch, starch grain number as well as activity of starch synthetase enzyme (Kirkby and Romheld 2004; Barker and Eaton 2015). Aldolase enzyme, also known as fructose 1,6-bisphosphate (F1,6BP) aldolase (EC 4.1.2.13), helps in the formation of triose phosphates such as glyceraldehyde 3-phosphate (GAL3P) and dihydroxyacetone phosphate (DHAP) from fructose 1,6-bisphosphate. Two isozymes of aldolase, class I/aldolase A and class II/aldolase B which differ in their action on substrates (fructose 1,6-bisphosphate (F1,6BP) and fructose 1-phosphate (F1P)), have been revealed in higher plants. Aldolase A is localized in the chloroplast and specifically acts on the substrate F1,6BP, while as aldolase B resides in cytosol it acts on both F1,6BP and F1P. Out of both of these classes, it was revealed that aldolase A needs cofactor in the form of divalent cations Zn^{+2}, Ca^{+2} or Fe^{+2} to catalyse the reactions (Uematsu et al. 2012). Aldolase enzyme is involved in glycolysis, Calvin cycle and gluconeogenesis, and due to this factor, Zn-deficient plants accumulate more starch and sugars (Kirkby and Romheld 2004; Barker and Eaton 2015; Cai et al. 2016). Moreover, these enzymes are very critical for growth, photosynthetic capacity, and biomass and regulate the photosynthetic carbon flow via Calvin cycle and increase the regeneration of ribulose-1,5-bisphosphate (RuBP) (Zhu et al. 2007; Uematsu et al. 2012).

Role of Zn in the Synthesis of Auxin Hormone

Auxin is one of the important hormones present ubiquitously in all plants and in various soil- or plant-associated microbes. The actual mechanism by which Zn effects the metabolism of auxin is yet to be revealed; however, it is clear that the most common symptoms of Zn deficiency, i.e. little leaf and stunted growth are associated with the disturbances of IAA metabolism (Hanafy-Ahmed et al. 2012). Zn triggers the biosynthesis of IAA by stimulating the activity of enzyme tryptophan synthetase which helps in the bonding of the indole ring with serine, as shown in Fig. 7.4. Low concentration of IAA in a Zn-deficient plant could possibly be due to either retardation of IAA biosynthesis or induced degradation (Hanafy-Ahmed et al. 2012). Zn-deficient plants possess a low level of IAA due to reduced synthesis of tryptophan, which is said to be the most crucial precursor that controls the biosynthesis of IAA (Wood 2007). There are various contradictory statements regarding the availability of Zn in the biosynthesis of tryptophan (Marschner 2011). Exogenous supplementation of Zn increases the endogenous tryptophan content, gibberellins and growth of plants by increasing the expression of general protein (Marschner 2011; Broadley et al. 2012). Thus, the little leaf and stunted growth of plants due to Zn deficiency could be due to low IAA and gibberellin content.

Role of Zn in Maintenance of Membrane Structure and Integrity

Zn is vital for the integrity and structural maintenance of biomembranes, as it binds with the main components of membranes, i.e. phospholipids and sulphydryl groups (–SH) and forms the tetrahedral complex with the cysteine residues of proteins, thus

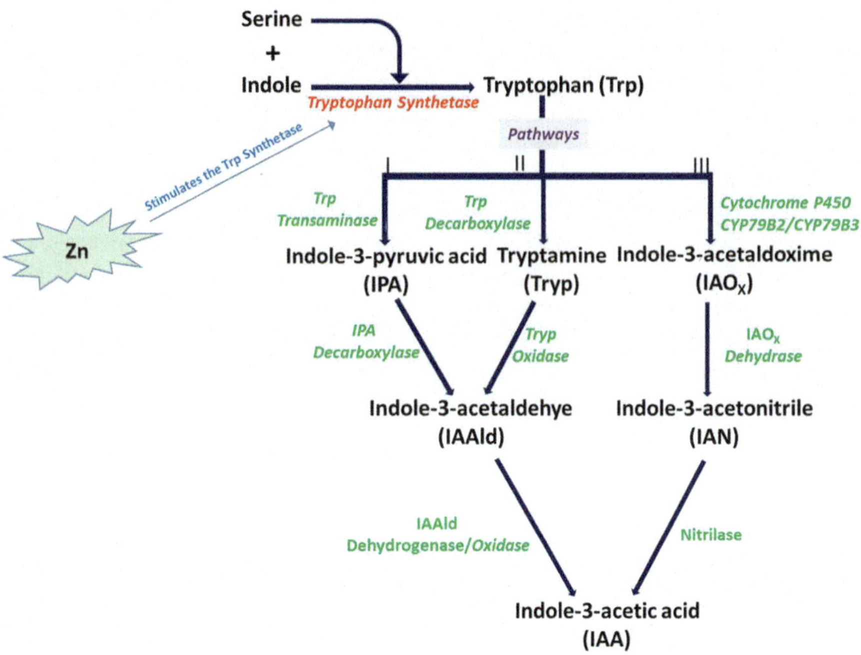

Fig. 7.4 The role of Zn in the synthesis of auxin hormone (IAA). Pathways of IAA biosynthesis operative in plants: (I) indole-3-pyruvic acid (IPA) pathway; (II) tryptamine (Tryp) pathway; (III) indole-3-acetonitrile (IAN) pathway

preventing the membrane proteins and lipids against damage caused by oxidation (Broadley et al. 2012). Zn also controls the oxidation of membrane lipids and proteins through the scavenging system, as it serves as one of the important structural components of an antioxidant enzyme, i.e. copper-zinc superoxide dismutase (CuZn-SOD). Plant cells possess different isoforms of SOD but CuZn-SOD is most abundant where Cu plays a catalytic and Zn plays a structural role (Abreu and Cabelli 2010). Decreased activity of CuZn-SOD and increased membrane permeability have been noted in Zn-deficient plants (Cakmak et al. 1997; Hacisalihoglu et al. 2003). Besides SOD, activities of other antioxidant enzymes such as catalase (CAT) and peroxidase (POX) decrease under Zn-deficient conditions due to reduced protein synthesis and hence lead to the accumulation of various reactive oxygen species (ROS) generating oxidative stress (Cakmak 2000; Chen et al. 2008b). It has been revealed that the permeability of membrane increases under Zn-deficient conditions due to more production of oxygen-free radical ($O_2^{\cdot-}$) by nicotinamide adenine dinucleotide phosphate (NADPH)-dependent $O_2^{\cdot-}$ generating oxidase. Increased level of toxic reactive oxygen species, i.e. $O_2^{\cdot-}$ increases the peroxidation and degradation of membrane lipids and proteins, respectively, and hence electrolyte leakage across membranes (Cakmak and Marschner 1988). The overall summary of the role of Zn in the maintenance of membrane structure and integrity is summarized in Fig. 7.5. The most important deficiency symptoms of Zn include

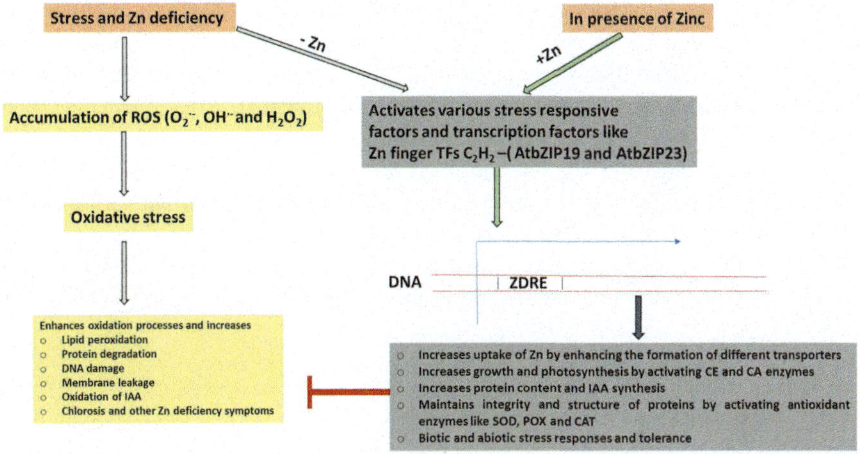

Fig. 7.5 Role of Zn deficiency in various stresses. Zn deficiency and stresses lead to accumulation of ROS like $O_2^{\cdot-}$, $OH^{\cdot-}$ and H_2O_2 resulting in oxidative stress and oxidative damage. However, in the presence or absence of Zn and stress, several TFs like Zn-finger TFs are activated which bind to Zn deficiency responsive element (ZDRE) and leads to the expression of various stress-related genes, providing the tolerance against various biotic and abiotic stresses

chlorosis and necrosis of leaf, higher membrane permeability and retarded shoot growth due to more production of ROS and reduced scavenging system in deficient plants. So, the versatility of Zn in maintaining the structure and permeability of the membrane lies in its potentiality to bind with membrane lipids and proteins as well as trigger the activity of antioxidant enzymatic (Broadley et al. 2007; Upadhyaya et al. 2013).

Role in Stress

Zn plays an important role in the alleviation of both biotic and abiotic stresses (Martos et al. 2016; Faran et al. 2019). Both these stresses pose a serious threat to agriculture globally due to climate change and prevent the plants from attaining their full genetic potential and damage crop (Gill et al. 2015). Zn plays an important role in the alleviation of water logging stress, as this type of stress develops anoxia in soil and alters various metabolic processes of plants. Under stressful conditions, plants cope up with the adverse flooding condition by operating the pentose phosphate pathway instead of glycolysis and develop compounds that provide tolerance to waterlogging. However, in this shifting case the accumulation of malate could be critically lethal to the plants. Accumulation of malate is inhibited by alcohol dehydrogenase (ADH) which reduces acetaldehyde to ethanol (Pardos 2004). This ADH enzyme is present ubiquitously in the cytoplasm of all living organisms and is a dimer where each subunit consists of catalytic and structural domains, both of these units having one Zn atom per domain (Auld and Bergman 2008). So, Zn alleviates water logging conditions with the help of the ADH enzyme. Besides this, current studies have revealed that Zn finger proteins play a key role in the growth, development and tolerance against various abiotic stresses (Liu et al. 2015; Zhai et al. 2016; Wang et al. 2016). There are various types of Zn finger proteins, out of which the C_2H_2 type is extensively studied and plays a key role in plant response to various stresses like drought, salinity, temperature and light stress (Muthamilarasan et al. 2014; Yue et al. 2016). Zn also helps in the alleviation of oxidative stress caused by ROS, through involvement and regulation of an antioxidant system in plants. The overall role of Zn in tolerance to various abiotic stresses is summarized in Fig. 7.5.

Zinc Deficiency and Economic Losses

At present, soil pollution is at its peak, due to inappropriate management of soil resources such as rigorous farming without/minor incorporation of organic matter, replenishment of proper nutrients, narrow crop rotation and many anthropogenic activities, which have severely affected the quality and yield of crops with inadequacy of micronutrients in humans, animals, plants as well as soils globally. This insufficiency of micronutrients in plants and soil harmfully affects the heath and yield of animals and humans (Takkar 2005; Shukla and Tiwari 2014). It is predicted that one third of the soil have less amount of Zn, causing reduced yield and nutritional quality, thus generating economic losses of crops (Alloway 2008; Cakmak 2008; Cakmak et al. 2017; Cakmak and Kutman 2018). Soil with a higher pH, lesser organic matter, calcareous and coarse soil faces acute deficiency of Zn (Shukla et al. 2014). Based on the previous data of Zn deficiency (before 2000), it is estimated that the deficiency of Zn may rise up to 63% by the year 2025 (Singh 2009). However, on the basis of the present trends, it is estimated that due to a continuous research on micronutrients, the Zn deficiency would decrease to 21% by the year 2030, provided the condition if the promotion and support from government in continuous use of Zn fertilizers will continue.

As Zn is an immobile element, deficiency symptoms will first appear on the young leaves. Deficiency of Zn causes several disorders in plants and the most common symptoms include: inhibited growth, reduced petioles and the internodes, little deformed leaves (little leaves) that form symptoms of rosette in the initial stages of dicots and a fan-shaped stem in monocots (Snowball and Robson 1986). Dieback (apex of shoot dies) occurs under severe deficiency of Zn, for example, in South Australian forest plantations (Boardman and McGuire 1990). These above-said symptoms are often combined with necrosis, mottled leaf, necrotic patches, thin stem and patchy appearance (Brennan et al. 1993; Benton 2003). Disorders of Zn deficiency are known by various names like little and mottled leaves in vegetables, white bud in maize, decreased fruit formation in citrus, khaira disease in rice and resetting in rice. In Indian soil, the concentration of Zn ranges from 0.01 to 52.9 mg/kg, which constitutes >1% of the total Zn content. Effect of Zn deficiency on the productivity of crop has been established and confirmed in various crops (Takkar et al. 1989; Singh 2009; Shukla and Behera 2012).

Broadley, there is a strong linkage between the Zn deficiency of humans and status of Zn in the soil (Singh 2009; Shukla et al. 2016). Therefore, it is very important to improve the Zn status of the soil, which will not only improve the plant growth but will also consequently improve the human health due to the consumption of zinc-dense crops and plants. The critical level in most species of crops range between 15 and 50 ppm and 15 ppm which is considered as satisfactory in most cases (Benton 2003).

Toxicity and Tolerance

As we know that excess of everything is bad as the success of every living organism is determined by minimal, maximal and optimal factors or a combination of factors. A higher level of Zn causes phytotoxicity which is rarely seen, but contaminated soil (addition of industrial wastes and sewage sludge) leads to toxicity and the level is about 500 μg of Zn/g of soil (Chaney 1993; Broadley et al. 2007). Toxicity of Zn causes chlorosis, decreased elongation of roots, induced deficiency of other divalent cations having almost same radii elements such as Zn^{2+}, Fe^{2+}, Mg^{2+} and Mn^{2+}, decreased Rubisco activity and inhibition of photosystem II (PSII) (Woolhouse 1983; Sagardoy et al. 2009). It has been revealed that the toxic dose of Zn (200 mg/kg soil) drastically reduces the overall plant growth and other photosynthetic attributes but increases the proline content, electrolyte leakage and activity of antioxidant enzymes (CAT, POX and SOD) in the leaves of *Vigna radiata* (Mir et al. 2015). Repressed photosynthesis is mainly caused by competing with and mimicking other divalent cations causing deficiency of Mg required in the activation of Rubisco and Mn required in the photooxidation of water in thylakoid lumen (Van Assche and Clijsters 1986; Das and Mukherjee 2014). Tolerance of Zn is mainly achieved by compartmentalization of Zn, in which excess Zn is sequestered from cytosol to vacuoles involving various transporters like MTP, thereby leading to detoxification of higher concentration of Zn (Kawachi et al. 2009; Gustin et al. 2009). Besides vacuolar sequestration, complexation of Zn with other organic compounds/ligands like malate, citrate, phytochelatins, oxalate, phosphate, phytate and pectate provides tolerance against Zn phytotoxicity (Sarret et al. 2002; Haydon and Cobbett 2007a, b; Clemens and Peršoh 2009).

Conclusion

Zinc is very essential in several ways for plant growth and various factors affect its uptake and availability such as Zn content in soil, pH of soil, amount of organic matter, presence of other elements and their interaction, soil type and moisture, redox potential of rhizosphere and association of roots with fungi. Various transporters are involved in homeostasis and acquisition of Zn which include ZIP, MTPs, HMAs, NRAMP4, YSL, PCR2 and VIT family of proteins. Zn plays a vital role in regulating various cellular, physiological and metabolic attributes by acting as cofactor and thus affecting the metabolic pathways and metabolism of carbohydrates, photosynthesis, transcription and translation. Apart from this, Zn plays a key role in the synthesis of IAA, structural maintenance and integrity of biomembranes, scavenging of ROS and providing tolerance against biotic and abiotic stresses. Deficiency of Zn negatively affects plant growth, photosynthesis and development and common symptoms include little and malformed leaves, dieback, chlorosis and necrosis. Toxicity symptoms include reduced growth, photosynthesis and uptake of other divalent cations having same radii. Deficiency of Zn not only affects the plants but also animals including humans as there is a win-win scenario, i.e. better nutritional health and better yield. Therefore, the only sustainable solution to meet the challenge of Zn insufficiency in humans globally will come from the improvement of agricultural zone. Hence, effort must be made by plant scientists to address the issues like acquisition of Zn, its homeostasis, transport, biochemistry and assimilation, and to make sure that the bond between agriculture and health becomes closer and stronger in the near future.

References

Abreu, I.A., Cabelli, D.E.: Superoxide dismutases: a review of the metal-associated mechanistic variations. Biochim. Biophys. Acta. **1804**(2), 263–274 (2010)

Abud, Y.C., Guillén, J.N., Madrigal, F.H., Trujillo, M.M.: Effect of nickel, lead and zinc on the development of the root of *Nicotiana tabacum* L. Biológicas. **10**(1), 39–48 (2008)

Alloway, B.J.: Zinc in Soils and Crop Nutrition. International Zinc Association, Brussels (2008)

Andreini, C., Bertini, I., Rosato, A.: Metalloproteomes: a bioinformatic approach. Acc. Chem. Res. **42**(10), 1471–1479 (2009)

Assunção, A.G., et al.: *Arabidopsis thaliana* transcription factors bZIP19 and bZIP23 regulate the adaptation to zinc deficiency. PNAS. **107**(22), 10296–10301 (2010)

Auld, D.S., Bergman, T.: Medium-and short-chain dehydrogenase/reductase gene and protein families: The role of zinc for alcohol dehydrogenase structure and function. Cell Mol. Life Sci. **65**(24), 3961–3970 (2008)

Auld, D.S.: The ins and outs of biological zinc sites. Biometals, **22**(1), 141–148 (2009)

Barber, S.A.: Soil Nutrient Bioavailability, 2nd edn. Wiley, New York (1995)

Barker, A.V., Eaton, T.E.: Zinc. In: Barker, A.V., Pilbeam, D.J. (eds.) Handbook of Plant Nutrition, 2nd edn, pp. 537–564. CRC Press, Boca Raton (2015)

Bashir, K., Ishimaru, Y., Nishizawa, N.K.: Molecular mechanisms of zinc uptake and translocation in rice. Plant Soil. **361**(1–2), 189–201 (2012)

Bashir, K., Takahashi, R., Nakanishi, H., Nishizawa, N.K.: The road to micronutrient biofortification of rice: progress and prospects. Front. Plant Sci. **4**, 15 (2013)

Benton, J.J.: Agronomic Handbook; Management of Crops, Soils and Their Fertility. CRC Press, Boca Raton (2003)

Bert, V., et al.: Genetic basis of Cd tolerance and hyperaccumulation in *Arabidopsis halleri*. Plant Soil. **249**(1), 9–18 (2003)

Boardman, R., McGuire, D.O.: The role of zinc in forestry. II. Zinc deficiency and forest management: effect on yield and silviculture of *Pinus radiata* plantations in South Australia. For Ecol Manage. **37**(1–3), 207–218 (1990)

Boyd, R.S.: Plant defense using toxic inorganic ions: conceptual models of the defensive enhancement and joint effects hypotheses. Plant Sci. **195**, 88–95 (2012)

Brennan, R.F., Armour, J.D., Reuter, J.D.: Diagnosis of zinc deficiency. In: Robson, A.D. (ed.) Zinc in Soils and Plants, pp. 167–181. Kluwer Academic Publisher, Dordecht (1993)

Broadley, M.R., White, P.J., Hammond, J.P., Zelko, I., Lux, A.: Zinc in plants. New Phytol. **173**(4), 677–702 (2007)

Broadley, M., Brown, P., Cakmak, I., Rengel, Z., Zhao, F.: Function of nutrients: micronutrients. In: Marschner, P. (ed.) Marschner's Mineral Nutrition of Higher Plants, 3rd edn, pp. 191–248. Academic Press, San Diego (2012)

Brookes, A., Collins, J.C., Thurman, D.A.: The mechanism of zinc tolerance in grasses. J. Plant Nutr. **3**, 695–705 (1981)

Cai, B., et al.: Genome-wide analysis of the fructose 1,6-bisphosphate aldolase (FBA) gene family and functional characterization of FBA7 in tomato. Plant Physiol. Biochem. **108**, 251–265 (2016)

Cakmak, I.: Tansley Review No. 111 Possible roles of zinc in protecting plant cells from damage by reactive oxygen species. New Phytol. **146**(2), 185–205 (2000)

Cakmak, I.: Plant nutrition research: priorities to meet human needs for food in sustainable ways. Plant Soil. **247**(1), 3–24 (2002)

Cakmak, I.: Enrichment of cereal grains with zinc: agronomic or genetic biofortification? Plant Soil. **302**(1–2), 1–17 (2008)

Cakmak, I.: Enrichment of fertilizers with zinc: an excellent investment for humanity and crop production in India. J. Trace Elem. Med. Biol. **23**(4), 281–289 (2009)

Cakmak, I., Kutman, U.B.: Agronomic biofortification of cereals with zinc: a review. Eur. J. Soil Sci. **69**(1), 172–180 (2018)

Cakmak, I., Marschner, H.: Enhanced superoxide radical production in roots of zinc-deficient plants. J. Exp. Bot. **39**(10), 1449–1460 (1988)

Cakmak, I., et al.: Role of rye chromosomes in improvement of zinc efficiency in wheat and triticale. In: Plant Nutrition for Sustainable Food Production and Environment, pp. 237–241. Springer, Dordrecht (1997)

Cakmak, I., McLaughlin, M.J., White, P.: Zinc for better crop production and human health. Plant Soil. **411**, 1–4 (2017)

Catlett, K.M., Heil, D.M., Lindsay, W.L., Ebinger, M.H.: Soil chemical properties controlling zinc 2+ activity in 18 Colorado soils. Soil Sci. Soc. Am. J. **66**(4), 1182–1189 (2002)

Chaney: Risks associated with use of sewage sludge in agriculture. In: Proceedings of the 15th Federal Convention, vol. 1. Australian Water and Wastewater Association, Queensland, Australia (1993)

Castillo-González, J., Ojeda-Barrios, D., Hernández-Rodríguez, A., González-Franco, A.C., Robles-Hernández, L., López-Ochoa, G.R. Zinc metalloenzymes in plants. Interciencia, **43**(4), 242–248 (2018)

Chen, W.R., Feng, Y., Chao, Y.E.: Genomic analysis and expression pattern of OsZIP1, OsZIP3, and OsZIP4 in two rice (*Oryza sativa* L.) genotypes with different zinc efficiency. Russ. J. Plant Physiol. **55**(3), 400–409 (2008a)

Chen, W., Yang, X., He, Z., Feng, Y., Hu, F.: Differential changes in photosynthetic capacity, 77 K chlorophyll fluorescence and chloroplast ultrastructure between Zn-efficient and Zn-inefficient rice genotypes (*Oryza sativa*) under low zinc stress. Physiol. Plant. **132**(1), 89–101 (2008b)

Choi, H.I., Hong, J.H., Ha, J.O., Kang, J.Y., Kim, S.Y.: ABFs, a family of ABA-responsive element binding factors. J. Biol. Chem. **275**(3), 1723–1730 (2000)

Chu, H.H., et al.: Successful reproduction requires the function of *Arabidopsis* Yellow Stripe-like1 and Yellow Stripe-like3 metal-nicotianamine transporters in both vegetative and reproductive structures. Plant Physiol. **154**, 197–210 (2010)

Chu, H.H., Conte, S.S., Chan Rodriguez, D., Vasques, K., Punshon, T., Salt, D.E., Walker, E.L.: Arabidopsis thaliana Yellow Stripe1-Like4 and Yellow Stripe1-Like6 localize to internal cellular membranes and are involved in metal ion homeostasis. Front. Plant Sci. **4**, 283 (2013)
Clemens, S., Peršoh, D.: Multi-tasking phytochelatin synthases. Plant Sci. **177**(4), 266–271 (2009)
Clemens, S., Palmgren, M.G., Krämer, U.: A long way ahead: understanding and engineering plant metal accumulation. Trends Plant Sci. **7**(7), 309–315 (2002)
Clemens, S., Deinlein, U., Ahmadi, H., Höreth, S., Uraguchi, S.: Nicotianamine is a major player in plant Zn homeostasis. Biometals. **26**(4), 623–632 (2013)
Coccina, A., et al.: The mycorrhizal pathway of zinc uptake contributes to zinc accumulation in barley and wheat grain. BMC Plant Biol. **19**(1), 133 (2019)
Coleman, J.E.: Zinc proteins: enzymes, storage proteins, transcription factors, and replication proteins. Annu. Rev. Biochem. **61**(1), 897–946 (1992)
Das, S.K., Mukherjee, I.: Flubendiamide transport through packed soil columns. Bull. Environ. Contam. Toxicol. **88**(2), 229–233 (2012)
Das, S.K., Mukherjee, I.: Influence of microbial community on degradation of flubendiamide in two Indian soils. Environ. Monit. Assess. **186**(5), 3213–3219 (2014)
Desbrosses-Fonrouge, A.G., et al.: Arabidopsis thaliana MTP1 is a Zn transporter in the vacuolar membrane which mediates Zn detoxification and drives leaf Zn accumulation. FEBS Lett. **579**(19), 4165–4174 (2005)
Divol, F., et al.: The Arabidopsis YELLOW STRIPE LIKE4 and 6 transporters control iron release from the chloroplast. Plant Cell. **25**(3), 1040–1055 (2013)
Elsby, L.M., Roberts, S.G.: Interaction of the TFIIB zinc ribbon with RNA polymerase II. Biochem. Soc. Trans. **36**(Pt 4), 595–598 (2008)
Faran, M., Farooq, M., Rehman, A., Nawaz, A., Saleem, M.K., Ali, N., Siddique, K.H.: High intrinsic seed Zn concentration improves abiotic stress tolerance in wheat. Plant Soil. **437**(1–2), 195–213 (2019)
Gainza-Cortés, F., et al.: Characterization of a putative grapevine Zn transporter, VvZIP3, suggests its involvement in early reproductive development in *Vitis vinifera* L. BMC Plant Biol. **12**(1), 111 (2012)
Gao, M., et al.: Evolutionary and expression analyses of basic zipper transcription factors in the highly homozygous model grape PN40024 (*Vitis vinifera* L.). Plant Mol. Biol. Report. **32**(5), 1085–1102 (2014)
Gill, S.S., et al.: Superoxide dismutase-mentor of abiotic stress tolerance in crop plants. Environ. Sci. Pollut. Res. **22**(14), 10375–10394 (2015)
Giordano, M., Beardall, J., Raven, J.A.: CO_2 concentrating mechanisms in algae: mechanisms, environmental modulation, and evolution. Annu. Rev. Plant Biol. **56**, 99–131 (2005)
Grotz, N., et al.: Identification of a family of zinc transporter genes from *Arabidopsis* that respond to zinc deficiency. PNAS. **95**(12), 7220–7224 (1998)
Guerinot, M.L.: The ZIP family of metal transporters. Biochim. Biophys. Acta. **1465**(1–2), 190–198 (2000)
Gustin, J.L., et al.: MTP1-dependent Zn sequestration into shoot vacuoles suggests dual roles in Zn tolerance and accumulation in Zn-hyperaccumulating plants. Plant J. **57**(6), 1116–1127 (2009)
Hacisalihoglu, G., Hart, J.J., Wang, Y.H., Cakmak, I., Kochian, L.V.: Zinc efficiency is correlated with enhanced expression and activity of zinc-requiring enzymes in wheat. Plant Physiol. **131**(2), 595–602 (2003)
Hafeez, B., Khanif, Y.M., Saleem, M.: Role of zinc in plant nutrition – a review. Am. J. Exp. Agric. **3**(2), 374–391 (2013)
Hanafy-Ahmed, A.H., Khalil, M.K., Abd-Ei-Rahman, A.M., Nadia, A.M.: Effect of zinc, tryptophan and indole acetic acid on growth, yield and chemical composition of Valencia orange trees. J. Appl. Sci. Res. **8**, 901–914 (2012)
Hänsch, R., Mendel, R.R.: Physiological functions of mineral micronutrients (Cu, Zn, Mn, Fe, Ni, Mo, B, Cl). Curr. Opin. Plant Biol. **12**(3), 259–266 (2009)
Hartmann, A., Schmid, M., Van Tuinen, D., Berg, G.: Plant-driven selection of microbes. Plant Soil. **321**(1–2), 235–257 (2009)

Haydon, M.J., Cobbett, C.S.: A novel major facilitator superfamily protein at the tonoplast influences zinc tolerance and accumulation in *Arabidopsis*. Plant Physiol. **143**(4), 1705–1719 (2007a)
Haydon, M.J., Cobbett, C.S.: Transporters of ligands for essential metal ions in plants. New Phytol. **174**(3), 499–506 (2007b)
Huguet, S., et al.: Cd speciation and localization in the hyperaccumulator *Arabidopsis halleri*. Environ. Exp. Bot. **82**, 54–65 (2012)
Hussain, D., et al.: P-type ATPase heavy metal transporters with roles in essential zinc homeostasis in Arabidopsis. Plant Cell. **16**(5), 1327–1339 (2004)
IRRI: Nutritional Disorders and Nutrient Management in Rice. International Rice Research Institute, Manila (2000)
Jain, A., Sinilal, B., Dhandapani, G., Meagher, R.B., Sahi, S.V.: Effects of deficiency and excess of zinc on morphophysiological traits and spatiotemporal regulation of zinc-responsive genes reveal incidence of cross talk between micro-and macronutrients. Environ. Sci. Technol. **47**(10), 5327–5335 (2013)
Jakoby, M., et al.: bZIP transcription factors in Arabidopsis. Trends Plant Sci. **7**(3), 106–111 (2002)
Jerlin, B., Sharmila, S., Kathiresan, K., Kayalvizhi, K.: Zinc solubilizing bacteria from rhizospheric soil of mangroves. J. Microbiol. Biotechnol. **2**(3), 148–155 (2017)
Kawachi, M., et al.: A mutant strain Arabidopsis thaliana that lacks vacuolar membrane zinc transporter MTP1 revealed the latent tolerance to excessive zinc. Plant Cell Physiol. **50**(6), 1156–1170 (2009)
Kim, S.A., et al.: Localization of iron in Arabidopsis seed requires the vacuolar membrane transporter VIT1. Science. **314**(5803), 1295–1298 (2006)
Kim, Y.Y., et al.: AtHMA1 contributes to the detoxification of excess Zn (II) in *Arabidopsis*. Plant J. **58**(5), 737–753 (2009)
Kirkby, E.A., Romheld, V.: Micronutrients in Plant Physiology [Recurso Electrónico]: Functions, Uptake and Mobility. International Fertiliser Society, York (2004)
Krämer, U., Clemens, S.: Functions and homeostasis of zinc, copper, and nickel in plants. In: Molecular Biology of Metal Homeostasis and Detoxification, pp. 215–271. Springer, Berlin, Heidelberg (2005)
Krämer, U., Talke, I.N., Hanikenne, M.: Transition metal transport. FEBS Lett. **581**(12), 2263–2272 (2007)
Laity, J.H., Lee, B.M., Wright, P.E.: Zinc finger proteins: new insights into structural and functional diversity. Curr. Opin. Struct. Biol. **11**(1), 39–46 (2001)
Lasswell, J., Rogg, L.E., Nelson, D.C., Rongey, C., Bartel, B.: Cloning and characterization of IAR1, a gene required for auxin conjugate sensitivity in *Arabidopsis*. Plant Cell. **12**(12), 2395–2408 (2000)
Li, S., et al.: Identification and characterization of the zinc-regulated transporters, iron-regulated transporter-like protein (ZIP) gene family in maize. BMC Plant Biol. **13**(1), 114 (2013)
Lin, M.S., Kao, C.H.: Senescence of rice leaves. XIII changes of Zn2+ dependent acid inorganic pyrophosphatase. J. Plant. Physio. **137**, 41–45 (1990)
Liu, W.X., et al.: *Arabidopsis* Di19 functions as a transcription factor and modulates PR1, PR2, and PR5 expression in response to drought stress. Mol. Plant. **6**(5), 1487–1502 (2013)
Liu, Q., Wang, Z., Xu, X., Zhang, H., Li, C.: Genome-wide analysis of C_2H_2 zinc-finger family transcription factors and their responses to abiotic stresses in poplar (*Populus trichocarpa*). PLoS One. **10**, e0134753 (2015)
Liu, C., et al.: Co-application of molybdenum and zinc increases grain yield and photosynthetic efficiency of wheat leaves. Plant Soil Environ. **65**(10), 508–515 (2019)
Lopez-Millan, A.F., Ellis, D.R., Grusak, M.A.: Identification and characterization of several new members of the ZIP family of metal ion transporters in *Medicago truncatula*. Plant Mol. Biol. **54**(4), 583–596 (2004)
Lu, L., et al.: Analysis of metal element distributions in rice (*Oryza sativa* L.) seeds and relocation during germination based on X-ray fluorescence imaging of Zn, Fe, K, Ca and Mn. PLoS One. **8**(2), e57360 (2013)
Maret, W., Li, Y.: Coordination dynamics of zinc in proteins. Chem. Rev. **109**(10), 4682–4707 (2009)

Marschner, H.: Zinc uptake from soils. In: Zinc in Soils and Plants, pp. 59–77. Springer, Dordrecht (1993)

Marschner, H.: Marschner's Mineral Nutrition of Higher Plants, p. 651. Academic Press, San Diego (2011)

Martos, S., et al.: Zinc triggers signaling mechanisms and defense responses promoting resistance to *Alternaria brassicicola* in *Arabidopsis thaliana*. Plant Sci. **249**, 13–24 (2016)

Mendoza-Cózatl, D.G., Rangel-González, E., Moreno-Sánchez, R.: Simultaneous Cd^{2+}, Zn^{2+}, and Pb^{2+} uptake and accumulation by photosynthetic *Euglena gracilis*. Arch. Environ. Contam. Toxicol. **51**(4), 521–528 (2006)

Milner, M.J., Kochian, L.V.: Investigating heavy-metal hyperaccumulation using *Thlaspi caerulescens* as a model system. Ann. Bot. **102**(1), 3–13 (2008)

Milner, M.J., Seamon, J., Craft, E., Kochian, L.V.: Transport properties of members of the ZIP family in plants and their role in Zn and Mn homeostasis. J. Exp. Bot. **64**(1), 369–381 (2013)

Mir, B.A., Khan, T.A., Fariduddin, Q.: 24-epibrassinolide and spermidine modulate photosynthesis and antioxidant systems in *Vigna radiata* under salt and zinc stress. Int. J. Adv. Res. **3**, 592–608 (2015)

Momonoi, K., et al.: A vacuolar iron transporter in tulip, TgVit1, is responsible for blue coloration in petal cells through iron accumulation. Plant J. **59**(3), 437–447 (2009)

Morel, M., et al.: AtHMA3, a P1B-ATPase allowing cd/Zn/co/Pb vacuolar storage in *Arabidopsis*. Plant Physiol. **149**(2), 894–904 (2009)

Moroney, J.V., Bartlett, S.G., Samuelsson, G.: Carbonic anhydrases in plants and algae. Plant Cell Environ. **24**(2), 141–153 (2001)

Mousavi, S.R., Galavi, M., Ahmadvand, G.: Effect of zinc and manganese foliar application on yield, quality and enrichment on potato (*Solanum tuberosum* L.). Asian J. Plant Sci. **6**, 1256–1260 (2007)

Mousavi, S.R., Galavi, M., Rezaei, M.: The interaction of zinc with other elements in plants: a review. Int. J. Agric. Crop Sci. **4**(24), 1881–1884 (2012)

Muthamilarasan, M., et al.: C_2H_2 type of zinc finger transcription factors in foxtail millet define response to abiotic stresses. Funct. Integr. Genomics. **14**, 531–543 (2014)

Nishiyama, R., Kato, M., Nagata, S., Yanagisawa, S., Yoneyama, T.: Identification of Zn-nicotianamine and Fe–2′-deoxymugineic acid in the phloem sap from rice plants (*Oryza sativa* L.). Plant Cell Physiol. **53**(2), 381–390 (2012)

Obrador, A., Novillo, J., Alvarez, J.M.: Mobility and availability to plants of two zinc sources applied to a calcareous soil. Soil Sci. Soc. Am. J. **67**(2), 564–572 (2003)

Olsen, L.I., Palmgren, M.G.: Many rivers to cross: the journey of zinc from soil to seed. Front. Plant Sci. **5**, 30 (2014)

Oomen, R.J.F.J., et al.: Functional characterization of NRAMP3 and NRAMP4 from the metal hyperaccumulator *Thlaspi caerulescens*. New Phytol. **181**, 637–650 (2009)

Pardos, J.A.: Plant responses to waterlogging. Invest. Agrar. Sist. Recur. For. **13**, 101–107 (2004)

Patel, K., Kumar, A., Durani, S.: Analysis of the structural consensus of the zinc coordination centers of metalloprotein structures. Biochim. Biophys. Acta. **1774**(10), 1247–1253 (2007)

Price, G.D., et al.: The cyanobacterial CCM as a source of genes for improving photosynthetic CO_2 fixation in crop species. J. Exp. Bot. **64**(3), 753–768 (2013)

Rossi, G., Figliolia, A., Socciarelli, S.: Zinc and copper bioaccumulation in *Brassica napus* at flowering and maturation. Eng. Life Sci. **4**(3), 271–275 (2004)

Rudani, K., Prajapati, V.P.K.: The importance of zinc in plant growth: a review. IRJNAS. **5**(2), 38–48 (2018)

Rudenko, N.N., Ignatova, L.K., Fedorchuk, T.P., Ivanov, B.N.: Carbonic anhydrases in photosynthetic cells of higher plants. Biochem. Mosc. **80**(6), 674–687 (2015)

Sagardoy, R., Morales, F., López-Millán, A.F., Abadía, A., Abadía, J.: Effects of zinc toxicity on sugar beet (*Beta vulgaris* L.) plants grown in hydroponics. Plant Biol. **11**(3), 339–350 (2009)

Samardjieva, K.A., et al.: Zinc accumulation and tolerance in *Solanum nigrum* are plant growth dependent. Int. J. Phytoremediation. **17**, 272–279 (2014)

Sarret, G., et al.: Forms of zinc accumulated in the hyperaccumulator *Arabidopsis halleri*. Plant Physiol. **130**(4), 1815–1826 (2002)

Satoh-Nagasawa, N., et al.: Mutations in rice (*Oryza sativa*) heavy metal ATPase 2 (OsHMA2) restrict the translocation of zinc and cadmium. Plant Cell Physiol. **53**(1), 213–224 (2012)
Sharma, K.N., Deb, D.L.: Effect of organic manuring on zinc diffusion in soils of varying texture. J. Indian Soc. Soil Sci. **36**, 219–224 (1988)
Shukla, A.K., Behera, S.K.: Micronutrient fertilisers for higher productivity. Indian J. Fert. **8**(4), 100–117 (2012)
Shukla, A.K., Tiwari, P.K.: Progress report of AICRP on Microand Secondary Nutrients and Pollutant Elements in Soils and Plants, IISS, Bhopal (2014).
Shukla, A.K., Tiwari, P.K., Prakash, C.: Micronutrients deficiencies vis-a-vis food and nutritional security of India. Indian J. Fert. **10**(12), 94–112 (2014)
Shukla, A.K., et al.: Spatial variability of soil micronutrients in the intensively cultivated Trans-Gangetic Plains of India. Soil Tillage Res. **163**, 282–289 (2016)
Sinclair, S.A., Sherson, S.M., Jarvis, R., Camakaris, J., Cobbett, C.S.: The use of the zinc-fluorophore, Zinpyr-1, in the study of zinc homeostasis in *Arabidopsis* roots. New Phytol. **174**(1), 39–45 (2007)
Singh, M.V.: Micronutrient nutritional problems in soils of India and improvement for human and animal health. Indian J. Fert. **5**(4), 11–56 (2009)
Sivasubramanian, R., Mukhi, N., Kaur, J.: Arabidopsis thaliana: a model for plant research. In: Plant Biology and Biotechnology, pp. 1–26. Springer, New Delhi (2015)
Snowball, K., Robson, A.D.: Symptoms of Nutrient Deficiencies: Lupins. University of Western Australia Press, Nedlands (1986)
Solanki, M., Didwania, N., Nandal, V.: Potential of zinc solubilizing bacterial inoculants in fodder crops. IJETSR. **3**(12), 1–4 (2016)
Sondergaard, T.E., Schulz, A., Palmgren, M.G.: Energization of transport processes in plants. Roles of the plasma membrane H^+-ATPase. Plant Physiol. **136**(1), 2475–2482 (2004)
Song, W.Y., et al.: A novel family of cys-rich membrane proteins mediates cadmium resistance in *Arabidopsis*. Plant Physiol. **135**(2), 1027–1039 (2004)
Song, W.Y., et al.: *Arabidopsis* PCR2 is a zinc exporter involved in both zinc extrusion and long-distance zinc transport. Plant Cell. **22**(7), 2237–2252 (2010)
Sousa, S.F., Lopes, A.B., Fernandes, P.A., Ramos, M.J.: The Zinc proteome: a tale of stability and functionality. Dalton T, (38), 7946–7956 (2009), https://doi.org/10.1039/b904404c
Stahl, R.S., James, B.R.: Zinc sorption by B horizons soils as a function of pH. Soil Sci. Soc. Am. J. **55**, 1592–1597 (1991)
Sunithakumari, K., Devi, S.P., Vasandha, S.: Zinc solubilizing bacterial isolates from the agricultural fields of Coimbatore, Tamil Nadu, India. Curr. Sci. **110**(2), 196–205 (2016)
Takahashi, M., et al.: Role of nicotianamine in the intracellular delivery of metals and plant reproductive development. Plant Cell. **15**(6), 1263–1280 (2003)
Takahashi, R., et al.: The OsHMA2 transporter is involved in root-to-shoot translocation of Zn and Cd in rice. Plant Cell Environ. **35**(11), 1948–1957 (2012)
Takkar, P.N.: Correcting micronutrient deficiency to reduce health risks. In: FAI Annual Seminar on Economic Growth through Agriculture and Fertilisers. The Fertiliser Association of India, 1–3 December, 2005, New Delhi (2005)
Takkar, P.N., Chhibba, I.M., Mehta, S.K.: Twenty Years of Coordinated Research on Micronutrients in Soils and Plants. 1967-1987. Indian Institute of Soil Science, Bhopal (1989)
Tiong, J., et al.: Hv ZIP 7 mediates zinc accumulation in barley (*Hordeum vulgare*) at moderately high zinc supply. New Phytol. **201**(1), 131–143 (2014)
Tiong, J., et al.: Increased expression of six ZIP family genes by zinc (Zn) deficiency is associated with enhanced uptake and root-to-shoot translocation of Zn in barley (*Hordeum vulgare*). New Phytol. **207**(4), 1097–1109 (2015)
Tiwari, S., et al.: An Inventory and Assessment of Invasive Alien Plant Species of Nepal. IUCN Nepal, Kathmandu (2005)
Uematsu, K., Suzuki, N., Iwamae, T., Inui, M., Yukawa, H.: Increased fructose 1,6-bisphosphate aldolase in plastids enhances growth and photosynthesis of tobacco plants. J. Exp. Bot. **63**(8), 3001–3009 (2012)

Upadhyaya, H., Dutta, B.K., Panda, S.K.: Zinc modulates drought-induced biochemical damages in tea [*Camellia sinensis* (L) O Kuntze]. J. Agric. Food Chem. **61**(27), 6660–6670 (2013)
Van Assche, F., Clijsters, H.: Inhibition of photosynthesis in *Phaseolus vulgaris* by treatment with toxic concentration of zinc: effect on ribulose-1,5-bisphosphate carboxylase/oxygenase. J. Plant Physiol. **125**(3–4), 355–360 (1986)
Verret, F., et al.: Overexpression of AtHMA4 enhances root-to-shoot translocation of zinc and cadmium and plant metal tolerance. FEBS Lett. **576**(3), 306–312 (2004)
Vert, G., Briat, J.F., Curi, C.: Arabidopsis IRT2 gene encodes a root-periphery iron transporter. Plant J. **26**(2), 181–189 (2001)
Wang, Y., et al.: Transcriptome analyses show changes in gene expression to accompany pollen germination and tube growth in *Arabidopsis*. Plant Physiol. **148**(3), 1201–1211 (2008)
Wang, M., Xu, Q., Yu, J., Yuan, M.: The putative Arabidopsis zinc transporter ZTP29 is involved in the response to salt stress. Plant Mol. Biol. **73**(4–5), 467–479 (2010)
Wang, F., et al.: A novel Cys2/His2 zinc finger protein gene from sweet potato, IbZFP1, is involved in salt and drought tolerance in transgenic *Arabidopsis*. Planta. **243**, 783–797 (2016)
Wang, K., Ding, Y., Cai, C., Chen, Z., Zhu, C.: The role of C_2H_2 zinc finger proteins in plant responses to abiotic stresses. Physiol. Plant. **165**(4), 690–700 (2019a)
Wang, N., et al.: AhNRAMP1 enhances manganese and zinc uptake in plants. Front. Plant Sci. **10**, 415 (2019b)
Waters, B.M., Grusak, M.A.: Whole-plant mineral partitioning throughout the life cycle in *Arabidopsis thaliana* ecotypes Columbia, Landsberg erecta, Cape Verde Islands, and the mutant line ysl1ysl3. New Phytol. **177**(2), 389–405 (2008)
Waters, B.M., et al.: Mutations in Arabidopsis yellow stripe-like1 and yellow stripe-like3 reveal their roles in metal ion homeostasis and loading of metal ions in seeds. Plant Physiol. **141**(4), 1446–1458 (2006)
Weisshaar, B., Armstrong, G.A., Block, A., Silva, O.D.C., Hahlbrock, K.: Light-inducible and constitutively expressed DNA-binding proteins recognizing a plant promoter element with functional relevance in light responsiveness. EMBO J. **10**(7), 1777–1786 (1991)
Wood, B.: Correction of zinc deficiency in pecan by soil banding. HortScience. **42**, 1554–1558 (2007)
Woolhouse, H.W.: Toxicity and tolerance in the responses of plants to metals. In: Physiological Plant Ecology III, pp. 245–300. Springer, Berlin, Heidelberg (1983)
World Health Organization: Summary and Conclusions of the Sixty-First Meeting of the Joint FAO/WHO Expert Committee on Food Additives (JECFA). WHO, Rome (2003)
Yang, H., Jie, Y.: Uptake and transport of calcium in plants. J. Plant Physiol. Mol. Biol. **31**(3), 227–234 (2005)
Yao, L., et al.: Effects of zinc fertilizer on photosynthetic characteristics of "Shine Muscat" grape. In: E3S Web of Conferences, vol. 136, p. 07017. EDP Sciences (2019)
Yen, M.R., Tseng, Y.H., Saier, M.H.: Maize Yellow Stripe1, an iron phytosiderophore uptake transporter, is a member of the oligopeptide transporter (OPT) family. Microbiology. **147**, 2881–2883 (2001)
Yin, M., et al.: The Arabidopsis Cys2/His2 zinc finger transcription factor ZAT18 is a positive regulator of plant tolerance to drought stress. J. Exp. Bot. **68**(11), 2991–3005 (2017)
Yue, X., et al.: ZNF1 encodes a putative C_2H_2 zinc-finger protein essential for appressorium differentiation by the rice blast fungus *Magnaporthe oryzae*. Mol. Plant-Microbe Interact. **29**(1), 22–35 (2016)
Zabaleta, E., Martin, M.V., Braun, H.P.: A basal carbon concentrating mechanism in plants? Plant Sci. **187**, 97–104 (2012)
Zhai, H., et al.: A myo-inositol-1-phosphate synthase gene, IbMIPS1, enhances salt and drought tolerance and stem nematode resistance in transgenic sweet potato. Plant Biotechnol. J. **14**, 592–602 (2016)
Zhang, Y., Xu, Y.H., Yi, H.Y., Gong, J.M.: Vacuolar membrane transporters OsVIT1 and OsVIT2 modulate iron translocation between flag leaves and seeds in rice. Plant J. **72**(3), 400–410 (2012)
Zhu, X.G., De Sturler, E., Long, S.P.: Optimizing the distribution of resources between enzymes of carbon metabolism can dramatically increase photosynthetic rate: a numerical simulation using an evolutionary algorithm. Plant Physiol. **145**(2), 513–526 (2007)

Chapter 8
Transition Metal Homeostasis and Its Role in Plant Growth and Development

Yamshi Arif, Priyanka Singh, Husna Siddiqui, Romana Naaz, and Shamsul Hayat

Abstract The transition metals (Fe, Zn, Cu, and Mn) which are also considered as the micronutrient are essential for plants as they regulate various cellular metabolic activities that promote proper plant growth and development. These transition metals are also involved in photosynthesis, respiration, chlorophyll synthesis, transcription, translation, cell division, and cell elongation. However, their deficiencies and toxicity may hamper the plant growth and metabolism. Therefore, to avoid their toxicity, storage and distribution of metals in several cellular compartments are highly regulated for proper plant metabolism and growth. Alteration in metal composition of cellular organelles impairs many crucial metabolic processes of plant and further affects plant growth and yield trait. Therefore, transition metal homeostasis and cellular compartmentalization are the important processes for proper plant metabolism. This chapter highlights the transition metal homeostasis in several organelles like chloroplast, mitochondria, vacuole, vesicles, and Golgi bodies, and also covers the involvement of transporters and proteins in compartmentalization and maintaining homeostasis for proper developed biofortified crop. This chapter also includes the role of transition metals (Fe, Zn, Cu, and Mn) in plant growth and development which will help us to assess their role for sustainable agriculture and crop improvement.

Keywords Micronutrient · Cellular compartmentalization · Deficiencies · Toxicity · Biofortified · Organelles

Y. Arif · P. Singh · H. Siddiqui · S. Hayat (✉)
Plant Physiology Section, Department of Botany, Faculty of Life Sciences, Aligarh Muslim University, Aligarh, Uttar Pradesh, India

R. Naaz
Department of Applied Science and Humanities, Jamia Millia Islamia, New Delhi, India

S. T. Khan, A. Malik (eds.), *Microbial Biofertilizers and Micronutrient Availability*,
https://doi.org/10.1007/978-3-030-76609-2_8

Introduction

Metals like iron (Fe), zinc (Zn), copper (Cu), and manganese (Mn) are highly beneficial in living organism. In plants, they are micronutrients which are involved in various cellular and metabolic activities. Fe, Zn, Cu, and Mn are also regarded as transition metals, because they have diverse chemical properties which make them unique, such as their redox potential, charge, coordination geometry, and ligand exchange properties (Yruela 2013). Furthermore, transition metals also act as metalloproteins and this makes them beneficial in electron transport (Yruela 2013). Transition metals, such as Fe, Cu, and Mn are redox active and are present in several oxidation states and thus can participate in electron transfer reactions, whereas Zn is present only in stable oxidation state inside the cell (Pilon et al. 2009; Yruela 2013). Due to redox potential, these metals (Cu, Fe, and Mn) are able to catalyze the synthesis of radicals (Hall and Williams 2003). Especially, Cu and Fe are more prone to radical formation inside the cell. The relative binding affinities of these metals (in divalent form) for ligands are Mn < Fe < Zn < Cu. Thus, compared to Mn, Fe, and Zn, Cu has the highest potential to displace metals from their binding site to form proteins and enzymes (Pilon et al. 2009). Additionally, transition metal homeostasis is maintained in the inside of cell to avoid their toxicity.

Transition metals are also micronutrients for plants; thus, they play various cellular functions, such as primary and secondary energy metabolism, cell signaling, gene regulation, hormone signal transduction, and reproduction. Additionally, they also have redox properties. Thus, all these properties make them highly essential for plant growth and development in normal and different environmental stresses. Fe is highly important for plants as it is involved in the photosynthetic electron transport, biosynthesis, and stabilization of chlorophyll, respiration, and other metabolic and biochemical processes (Bashir et al. 2016; Iqbal et al. 2018; Pilon et al. 2009). Deficiency of Fe may lead to chlorosis and necrosis, which further hampers the plant growth and development. Thus, Fe in lower doses is highly essential, whereas higher amount of Fe causes toxicity to the plant, and hence, its regulation inside cellular compartment is highly important (Ravet and Pilon 2013; Sah et al. 2017). Cu also plays an important role in sustaining plant life, it is the cofactor of several enzymes which are involved in photosynthesis, respiration, and ethylene signaling. Furthermore, it serves as an important metal in promoting antioxidant machinery (Li et al. 2018; Sağlam et al. 2016). Cu is also required in micro amount, and higher amount further causes toxicity in plants (Mustafa and AlSharif 2018). Zn is not present in several oxidation states, therefore, does not participate in oxidoreduction reactions (Pilon et al. 2009). It acts as a cofactor for metalloenzyme, thus involved in nucleic acid, protein, and lipid metabolism (Balai et al. 2017). Further, it can also make tetragonal complexes or structures with oxygen, nitrogen, and sulfur; therefore, this makes Zn an important structural, functional, or regulatory cofactor of several enzymes (Mousavi 2011). Zn is also crucial for protein having DNA-binding Zn-finger motifs, protein which participates in DNA and RNA synthesis (such as transcription factor, polymerases), LIM domain, and RING fingers (Bashir et al.

2016). Mn is an important element for plants, metal which participates in the oxygen-evolving complex (OEC) at PSII in chloroplast. Further, it serves as a cofactor for several enzymes such as MnSOD, oxalate oxidase, and several Krebs cycle enzymes. It also plays a crucial role in redox reactions, as it is a part of MnSOD enzyme which is the ROS scavenger (Schmidt and Husted 2019). However, deficiencies and toxicity of transition metals impair the plant growth and development and further hampers the several crucial metabolic processes.

The functioning of several cellular organelles are depends on the activities of transition metals (Fe, Cu, Zn, and Mn). Furthermore, deficiency or toxicity of these metals disturbs the functioning of chloroplast, mitochondria, and vacuole (Bashir et al. 2016). Hence, modulating the transition metal uptake, distribution, homeostasis, and compartmentalization is significantly important for proper cellular functioning, plant growth and development. Additionally, it can cause proper breeding of crop plants with optimal sustainability and adaptability to the changing environmental conditions, and can develop healthy food with proper mineral content (Bashir et al. 2016).

This chapter mainly highlights the proteins and the transporters that are involved in maintaining the transition metal homeostasis and compartmentalization in several cellular organelles like mitochondria, chloroplast, vacuole, endoplasmic reticulum, and Golgi bodies. Furthermore, it also focuses on the modulation of transition metal homeostasis transporters and their role in plant growth, crop improvement, and biofortification.

Cellular Transition Metal Homeostasis

Essential transition metals, such as Fe, Zn, Cu, and Mn, require homeostasis and equilibrium inside the cell and cellular organelles (Bashir et al. 2016). Plants uptake metals from the soil with the help of many transporters and transport them to cytoplasm. Plasma membrane proteins and transporters participate in the cellular metal transport and homeostasis. Plants make use of the various chelating agents (citrate, nicotianamine (NA), ascorbate, deoxymugineic acid (DMA), phenolics, and ascorbate) to solubilize toxic metals in cytosol and also maintain the membrane integrity and diminish ROS production (Bashir et al. 2013; Grillet et al. 2014). The transporters and their families which participate in cellular homeostasis are zinc-regulated transporters (ZRT), iron-regulated transporters (IRT), ZIP protein family, cation diffusion facilitator family (CDF), cation exchange family (CAX), heavy metal ATPase (HMA), and vacuolar iron transport family (VIT). Besides this, some other transporter families are major facilitator super family (MFS), natural resistance-associated macrophage protein (NRAMP), and yellow stripe-like family (YSL) (Bashir et al. 2016). These transporters and proteins play a vital role in cellular homeostasis, metal uptake, and their proper distribution in different parts of the plant (López-Millán et al. 2016).

Chloroplast Transition Metal Homeostasis

Transition metals play an important role in modulating the metal homeostasis and metabolic activities (photosynthesis, chlorophyll synthesis) in chloroplast (López-Millán et al. 2016). During the early stages of plant development, ferretin, an important Fe storage protein which localizes in chloroplast. is Hence, chloroplast serves as the pool for Fe storage and participates in the plant growth (Bashir et al. 2016; Briat et al. 2010). Ferric chelate reductase oxidase 7 (FRO7) participates in the reduction of ferric ion (Fe^{3+}) into ferrous ion (Fe^{2+}) which plays a chief role in Fe homeostasis (Jain et al. 2014). Permease in chloroplasts 1 (PIC1) plays a principal role in Fe transport toward chloroplast, and upregulation and down-regulation in PIC1 expression modulates the development of chloroplast which further regulates the plant growth (Duy et al. 2011). In Arabidopsis, YSL4 and YSL6 transporters perform the foremost function in Fe transport, localization, and homeostasis in chloroplast; they releases chloroplast Fe-NA from the chloroplast which further improves the metal equilibrium in the chloroplast (Divol et al. 2013). Cu influx in chloroplast is mediated by P_{IB}-type ATPase (present on chloroplast envelope), also called HMA6 or PAA1 (Boutigny et al. 2014). Another transporter called HMA1 helps in the influx of Cu^{+}. HMA8, also called PAA2, present in the thylakoid membrane, participates in Cu^{+} import and homeostasis (Catty et al. 2011). HMA1 in chloroplast also participates in Zn transportation and homeostasis and further enhances the photosynthesis (Boutigny et al. 2014). Unlike Zn, Mn also plays an important role in photosynthesis. Transporters involved in Zn and Mn influx and homeostasis are not well documented. Thus, modulating transition metal (Fe, Cu, Mn, and Zn) transport in chloroplast enhances the ion homeostasis, metabolism, photosynthesis, and plant growth and development (Bashir et al. 2016).

Vacuolar Transition Metal Homeostasis

Vacuole serves as the main pool to modulate transition ion homeostasis. In particular, to constrain metal accumulation and toxicity, metals are sequestered from cytosol and accumulated in the vacuole. Further, they are withdrawn from the vacuole when required. Various reports suggest that diverse transporters present in the tonoplast are required for the influx and efflux of metals. In Arabidopsis, several transporters like FPN2 and vacuolar Fe transporter1 (VIT1) accumulate Fe in the vacuole (Kim et al. 2006). Proteins similar to Arabidopsis VIT1-like are called VTL. Additionally, several VTLs like VTL1, VTL2, and VTL5 participate in Fe transportation inside the vacuole from cytosol and maintain the metallic homeostasis (Gollhofer et al. 2014) while VITs (VIT1 and VIT2) in rice are responsible for Mn, Zn, and Fe homeostasis (Yoshida and Negishi 2013). NRAMP (NRAMP3 and NRAMP4) participates in the mobilization of Fe, Mn, and Cd (cadmium) from the vacuole to maintain the metal equilibrium (Pottier et al. 2015). CAX is also involved in metal sequestration and compartmentalization in the vacuole. In particular, CAX2 mobilizes Fe, Cd, and Mn, CAX4 mobilizes Mn and Cd, while CAX5 transports Mn to the vacuole (Socha and Guerinot 2014). YSL4 and YSL6 transporters are involved in the mobilization of Fe in the vacuole and to maintain metal homeostasis. Additionally, YSL transporters are also involved in Mn and Ni homeostasis and compartmentalization (Divol et al. 2013). Metal tolerant proteins (MTP), such as MTP1 and MTP3 in Arabidopsis are responsible for sequestration and compartmentalization of toxic Zn into the vacuole (Arrivault et al. 2006). Furthermore, MTP8 is involved in Mn sequestration and compartmentalization inside the vacuole and the activity of MTP8 activity is enhanced under low Fe and high Mn surroundings. Additionally, HMA3 also participates in the mobilization of Zn into vacuole (Bashir et al. 2016). In Arabidopsis roots, ZIP1 cooperates in remobilizing Mn and Zn from the vacuole toward cytoplasm (Milner et al. 2013). Zn-induced facilitators (ZIF), such as ZIF1 and ZIF2 assist in excess Zn homeostasis and compartmentalization. Moreover, ZIFs are involved in Fe, potassium, and cesium mobilization (Remy et al. 2014). IRTs are involved in Fe, Zn, Mn, and Cd transportation and homeostasis (Ishimaru et al. 2012). The Cu transporter family (COPT/Ctr), consisting of transporters, such as COPT1, COPT2, COPT3, and COPT5, is engaged in the transportation of Cu. Furthermore, COPT5 is involved in pumping out Cu from the vacuole into the cytosol and play a key role in the chloroplastic photosynthetic electron transport (Garcia-Molina et al. 2011; Grotz and Guerinot 2006; Klaumann et al. 2011). HMAs family of transporters is generally present in the plasma membrane and participates in the transportation and homeostasis of metals (Zn, Mn, Fe, and Cd) in the vacuole (Miyadate et al. 2011). Under Fe deficiency, FER-like Fe deficiency-Induced Transcription factor (FIT) modulates the transcription of many genes, such as IREG and MTPs to subsist metal toxicity and to maintain the metal homeostasis (Schmidt and Buckhout 2011). Thus, from the above, it is clear that the several transporters in the vacuole play a vital role in metal homeostasis, sequestration, and compartmentalization which further cope with metal toxicity.

Mitochondrial Transition Metal Homeostasis

Mitochondria, an endosymbiosis originated organelle, participates in many important life sustaining functions. It can modulate several retrograde signaling processes to enhance many responses and also assembles Fe-S cluster (ISC) machinery. ISC plays a chief role in cytosolic and nuclear Fe-S protein assembly, which is completely relied on mitochondria. In Arabidopsis, ATM3 or STA1 transports the glutathione trisulfide complex from mitochondria in the cytosol (Lill et al. 2015; Schaedler et al. 2014). Additionally, deformity in its transportation remarkably affects the Fe-S cluster assembly and also assembly of molybdenum inside the cytoplasm (Teschner et al. 2010). Furthermore, glutathione and Fe-S cluster play a role in many responses of plant during metal accessibility (Teschner et al. 2010). Mitochondrial Fe transport and its homeostasis are mediated by FRO3 and FRO8 proteins (Bashir et al. 2016). In rice, mitochondrial Fe transport (MIT) plays an essential role in Fe export from cytoplasm to mitochondria and participates in metal homoeostasis and its compartmentalization; furthermore, it also helps in the process of plant growth and development (Bashir et al. 2011). There are not much reports which suggest that MIT could help in the transportation of Cu. Inside the mitochondria of mit-2 plants, excess of Cu and Mn is mobilized and an equilibrium is maintained (Bashir et al. 2011). In Arabidopsis, there are two homologs of MIT which participate in maintaining the metal stability in the mitochondria (Jain 2014). In rice, mitochondrial iron-regulated (MIR) proteins show the enhanced translational expression during Fe deficiency, therefore, alter the metal homeostasis (Bashir et al. 2016). Apart from the transporters and proteins present in the mitochondria, frataxin and ferritin in mitochondria also play a crucial role in maintaining the metal homeostasis and evenness throughout the plant cell (Vigani et al. 2013).

Golgi Body and Vesicle Transition Metal Homeostasis

Golgi apparatus and vesicles play an important role in the metal homeostasis which further modulates the cellular processes (Seo et al. 2012). Nucleus, Golgi body, and vesicles together participate in the metal distribution and enhance the cellular activities and protein localization (Roschzttardtz et al. 2011). Nucleus contains specific amount of Fe. However, the pore size of the nucleus allows import and export of metal (Bashir et al. 2016). Metal accumulation is toxic for nuclear activity, so nucleus itself maintains the metal homeostasis (Bashir et al. 2016). In Arabidopsis, AtIRT2 found in vesicles of root epidermis participates in the mobilization of Fe^{2+} and modulates its deficiency (Conte and Walker 2011). Arabidopsis MTP11 helps in the transportation and homeostasis of Mn within the plant organelles (Bashir et al. 2016). In rice, OsYSL6 participates in the detoxification of excess and toxic Mn. Thus, it maintains Mn equilibrium inside the cell. OsYSL5 and OsYSL6 are involved in the mobilization of metal–DMA and metal–NA complex into the internal organelles and compartments (Sasaki et al. 2011). In barley, HvYSL5 is localized to vesicular membrane and participates in the transportation of Fe (Bashir et al. 2016). Thus, all these transporters participate in maintaining the subcellular transition metal homeostasis (Bashir et al. 2016). Rice transporters (OsYSL5 and OsYSL6) are encoded by NAS genes which are localized in the vesicles are different and have different roles than Arabidopsis transporters, that is, AtYSL4 and ATYSL6 (Nozoye et al. 2014). Endomembrane system also contains metal transporters which participate in the mobilization of transition metal inside ER and Golgi apparatus. In barley, MTP8.1 and MTP8.2 localized in Golgi body participate in Mn efflux (Pedas et al. 2014). MTP8.1 and MTP8.2 work in different ways in barley roots and MTP8.1 expression is enhanced during Mn deficiency and toxicity, whereas MTP8.2 expression reduced. Accordingly, excess Mn inside the barley leaves decreases the expression of MTP8.1 and MTP8.2 (Pedas et al. 2014). Furthermore, MTP8.1 and MTP8.2 transporters play a crucial role in loading of Mn inside the Golgi body and transporting or delivering via secretary vesicles (Bashir et al. 2016). In Arabidopsis, HMA7, also called RAN1, helps in Cu transportation which is also responsible for the synthesis of ethylene receptors synthesis (Binder et al. 2010). It is also evaluated that transporters and proteins are involved in transition metal homeostasis inside the Golgi apparatus, ER, and vesicles (Bashir et al. 2016).

Therefore, it can be said that various transporters and proteins located in cellular organelles (mitochondria, chloroplast, vacuole, Golgi bodies, and vesicles) and plasma membrane participate in transition metal homeostasis and compartmentalization (Fig. 8.1)

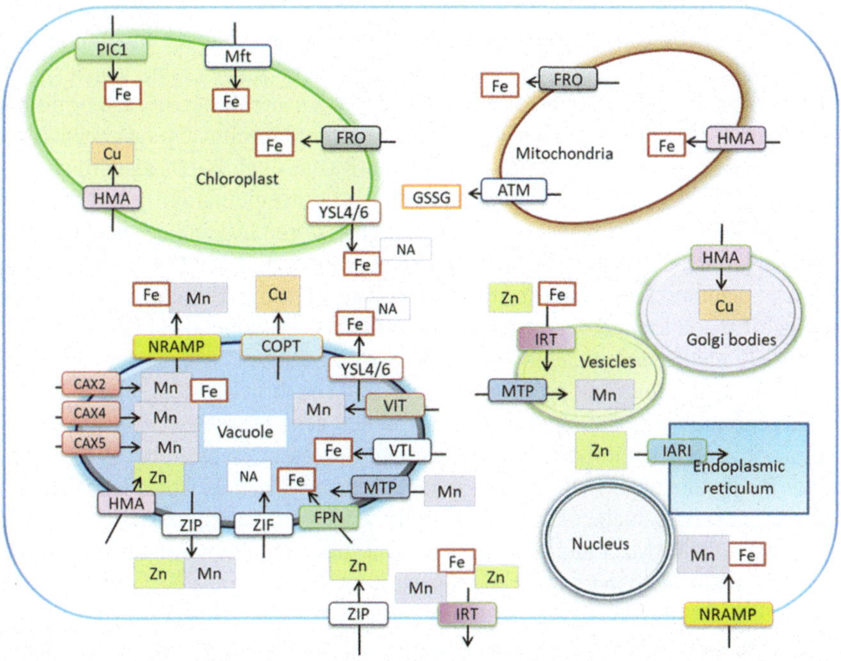

Fig. 8.1 Transition metal (Fe, Zn, Cu, Mn) homeostasis and compartmentalization in cellular organelles. Abbreviation: Fe iron, Zn zinc, Cu copper, Mn manganese, VIT vacuolar iron transporter, NA nicotianamine, FRO ferric chelate reductase oxidase, PIC permease in chloroplast, HMA heavy metal ATPase, NRAMP natural resistance-associated macrophage protein, IRT iron-regulated transporter, CAX cation exchange family, YSL yellow stripe-like family, MTP metal tolerant protein, ZIF Zn-induced facilitator, COPT Cu transporter family, FPN ferroportin, VTL vacuolar iron transporter-like protein

Regulation of Transition Metal Metabolism

Metals compete with each other and the increased content of one metal reduces or alters the transportation of other metal; however, metal accumulation involves complex cellular responses and several signaling processes (Bashir et al. 2016). Cellular responses are enhanced by transporters for mobilization of more than one metal or due to competition among metals. Changes in the content of metals in different cellular organelles affect the genes that participate in metal mobilization and homeostasis, and also affect the cellular metabolism (Vigani et al. 2013). Chloroplast requires large amount of metal for carrying important metabolic process (photosynthesis) which further synthesizes primary and secondary metabolites (López-Millán et al. 2016). Thus, transition metal compartmentalization and their homeostasis modulate activities performed within the chloroplast, nuclear gene expression, and cellular metabolic activities (Bashir et al. 2016). The regulation of PIC1 expression via stress-inducible promoter also plays a crucial role in cellular metabolism, plant growth and development (López-Millán et al. 2016).

Transcriptomic and metabolomic of *mit-2* show that retrograde signaling between organelles like mitochondria and nucleus which regulates gene expression and cellular processes (Vigani et al. 2016). Generally, prominent changes are seen in shoot transcriptomic and metabolomic products of *mit-2* and metabolome which specifically accumulate in shoots are oligosaccharides, pyruvate, fumarate, and orninthine (Vigani et al. 2013). In rice, *opt7-1* (*OsOPT7* mutant) specifically accumulates two times more Fe inside shoots than in wild-type (Bashir et al. 2016). Both *opt7-1* and *mit-2* are different in transcriptomics, because Fe accumulation is modulated in *mit-2* mitochondria (Vigani et al. 2016). The Fe is stored inside the vacuole in *mit-2* mutants, whereas inside chloroplast is *opt7-1* (Vigani et al. 2016). Average amount of metal accumulation enhances the metabolism of plant. Therefore, proper functioning of transition metal transporters helps in transition metal compartmentalization to organelles and maintaining the proper homeostasis and metabolism which further promotes the plant productivity and yield (Bashir et al. 2016).

The cautious modulation of metal transporters gene expression helps in the crop biofortification with important nutrients. The increased expression of IRT1 promotes the uptake of Zn, Mn, Ni, and many other heavy metals which leads to metal toxicity, and under such conditions, vacuolar sequestration plays a vital role (Bashir et al. 2016). NRAMP3 and NRAMP4 are substrate-specific vacuolar metal transporters that also help in toxic metal sequestration (Pottier et al. 2015). In rice, transporter HMA3 eliminates toxic metal from grain (Miyadate et al. 2011). Changes in gene regulation which participates in transition metal homeostasis, accumulation, and compartmentalization are also reported during various environmental stresses (Bashir et al. 2016).

Role of Transition Metal in Plant Growth and Development

The various researches indicate that transition metals are also responsible for the proper growth of the plant and for maintaining proper morphophysiological activities when present in balanced amount. Additionally, they act as cofactor in many enzymes which participate in metabolic processes in plants. Thus, transition metal homeostasis plays an important role in increasing plant growth, yield, and quality. Furthermore, it is critical to maintain proper balance of these metals or nutrients in appropriate amount in plant.

Iron in Plant Growth and Development

Iron is the $3d^6$ element, essential micronutrient, and a transition metal in all living organisms. Fe exists in two active redox metal states that are Fe^{2+} and Fe^{3+}. They participate in plant metabolic processes like oxygen transport and metabolism, cellular proliferation, electron transfer, DNA, and RNA synthesis, and serve as a catalyst in several enzymatic reactions (Sah et al. 2017). Fe is present in very high amount in soil but plants require Fe in very low concentration. The presence of Fe in high amount imbalances the plant metabolism and disturbs ion homeostasis whereas if it exists in very low concentration, then the plant starts showing deficiency symptoms and hampers many metabolic processes such as photosynthesis. For proper plant vitality, growth and development, an appropriate concentration of Fe is needed (Yruela 2013). Furthermore, most of the Fe is present in leaf chloroplast (Pilon et al. 2009) and Fe plays a crucial role during photosynthesis, and also acts as an important metal which helps in electron transfer process. Thus, balanced amount of Fe promotes the photosynthetic machinery, which further induces better plant growth (Ravet and Pilon 2013). Among all transition metals, Fe is a highly essential transition metal which participates in plant growth, development, and metabolic processes, because it is the structural and functional part of several biological molecules, such as heme, cytochrome P450, and hydrogenase (Nakanishi-Masuno et al. 2018). Ferretin is a Fe storage protein which maintains homeostasis and redox potential of Fe in plant chloroplast and carry proper functioning of the metabolic processes involved in growth and development of plant (Ravet and Pilon 2013). Ferretin protein plays the chief role in the synthesis of metalloproteins having Fe-S. Thus, ferretin is the potent Fe donor and helps in the synthesis of photosynthetic apparatus; this further improves the plant metabolism and growth (Yruela 2013). The chloroplast permease protein PIC1 which spans around the membrane plays a crucial role in Fe transport and its homeostasis; subsequently, it enhances plant growth and developmental processes (Duy et al. 2011). Proper Fe distribution in the cell which is under the control of transcription factor and transporter located in the plastids helps to enhance the photosynthesis. Thus, it induces other crucial metabolic processes which increase the morphological traits (Yang et al. 2010). In

Arabidopsis, transporter OPT3 regulates Fe distribution and shoot to root signaling cascade by loading Fe into phloem; this enhances the Fe circulation from xylem to phloem. Thus, Fe partitioning helps in crop development and biofortification strategies (Zhai et al. 2014). Fe is tremendously used as a cofactor of several enzymes, predominantly in light reactions and electron transport. This further enhances the metabolic reactions and biochemical activities which ultimately increase the plant yield (Balk and Schaedler 2014). The Fe in low dose and in balanced form serves as beneficial transition element for proper survival, growth, and development of plant (Palmer and Guerinot 2009). Fe plays a crucial role in plant growth and development but higher doses can be toxic. So, a plant tightly regulates Fe uptake, storage, and translocation (Jeong et al. 2017). NRAMP proteins are membrane-bound transporter present on cell membrane (AtNRAMP1) and tonoplast (AtNRAMP3 and AtNRAMP4) participate in Fe uptake and homeostasis; they also enhance seed germination, photosynthesis, and plant growth (Nakanishi-Masuno et al. 2018). The applications of Fe in oxide form to tomato plants are reported to enhance the growth and physiological attributes (Singh et al. 2019). Fe localizes to distinct layer of seed coat, and thus, enhances germination process and detoxifies developing embryo (Eroglu 2018). Treatment of Fe on mustard enhanced and improved the photosynthate content, redox status, growth, and yield attributes (Rawat et al. 2017). Iron with oxide when applied on plant plays a proficient role in increasing the plant height, nutrient status, growth parameters and yield (Zia-ur-Rehman et al. 2018). Fe in low amount is used for detoxification of cell and crop improvement program (Nikolic and Pavlovic 2018). Low doses of Fe enhance the plant health, productivity, and antioxidant machinery (Tripathi et al. 2018). Apart from this, Fe in lower doses is also beneficial in mitigating the environmental stresses like salinity, drought, and heavy metal in several crops and enhances the proper growth and development of plant by improving physiological and biochemical traits (Rafsanjani et al. 2019; Rizwan et al. 2019; Saif et al. 2016; Soliman et al. 2015). However, Fe in smaller amount plays an important role in sustainable agriculture. Thus, Fe acquisition and homeostasis plays a key role in improving the plant growth, development, and yield.

Zinc in Plant Growth and Development

Zinc is an important micronutrient and second most important transition metal after Fe. It is the only metal which is present in all six classes of enzymes such as hydrolases, oxidoreductases, transferases, isomerases, lyases, and ligases (Prasad et al. 2012). Roots uptake Zn in the divalent form (Zn^{2+}); there it acts as metalloenzyme part or as structural, functional, or regulatory cofactor of several enzymes. Zn has aptness for making up tetragonal complexes or structures with oxygen, nitrogen, and sulfur. Therefore it provides catalytic, activating, and building potency in the enzymes (Mousavi 2011). Zn deficiency in plant is the major constraint for proper plant development and crop production, as it limits plant productivity and yield

(Balai et al. 2017). The plant shows deficiency symptoms, such as stunted plant growth, necrosis, bronzing, formation of rosettes or clusters, inhibition of internode, and reduction in size of root and leaf (Moreira et al. 2018). At a low dose, Zn plays an important role in enhancing the plant growth and yield (Prasad et al. 2012). It is an important transition metal and participates in several plants' metabolic and biochemical processes such as cell division, auxin synthesis, protein synthesis, sexual fertilization, and maintains membrane stability and integrity. Furthermore, it plays a crucial role during photosynthesis, nitrogen metabolism, and chlorophyll synthesis (Mousavi 2011; Said-Al Ahl and Mahmoud 2010). Zn is a part of several enzymes (such as carbonic anhydrase, superoxide dismutase zinc-copper, alcohol dehydrogenase) which participate in enhancing the root growth and development, increasing leaf area, biomass, and yield attributes (Mousavi 2011). Zn plays a potent role in the biosynthesis of indole acetic acid (auxin), an important phytohormone which modulates the plant growth. Zn enhances the plant growth traits and nutrient homeostasis by the mobilization of phosphorus in the soil, as it is a part of the enzyme phosphates (Tarafdar et al. 2014). The transporters of Zn also play a proper role for the distribution and homeostasis of Zn. Transporters (transport proteins) mobilize the Zn to various plant tissues and carry cellular transportation inside the several organelles (mitochondria, nucleus, chloroplast, and vacuole). The Zn^{2+} transporter proteins are zinc–iron permease or Zrt/Irt-like protein (ZIP), metal tolerance proteins (MTPs) or cation diffusion facilitators (CDFs), NRAMP and P-type ATPase or heavy metal ATPases (HMAs), which help in proper compartmentalization and homeostasis of Zn (Moreira et al. 2018). The supplementation of Zn in small amount on soybean plant increased the height, biomass, and branches per plant as well as improved the oil content and grain quality (Choudhary et al. 2014). In wheat, Zn acts as an important micronutrient and helps to improve the growth, biochemical and yield responses (Arif et al. 2017). Foliar application of low doses of Zn on mungbean came out to be beneficial, as it induced seedling growth, chlorophyll content, physiomorphological, and yield attributes. Further, it also modulates the mungbean grain biofortification (Haider et al. 2018). Zn played stimulatory role to enhance the growth, weight, and oil and citral content in mountain pepper and also improved the yield traits to a greater extent (Gao et al. 2019). In peanuts, Zn plays a potent role in promoting germination, growth and development (Prasad et al. 2012). It was reported that Zn application on pearl millet enhances the growth potential of plant by modulating the phosphatase, dehydrogenase, and phytase enzyme activity (Singh et al. 2017; Tarafdar et al. 2014). The application of Zn also enhanced the competence of phosphorous in crop yield and development. In the course of reproductive stage, Zn participates in proper channelization of metabolic processes like photosynthesis as it participates in electron transfer (Balai et al. 2017). Moreover, the application of Zn on several crops is highly beneficial in ameliorating the toxic effect of several abiotic stresses as it increases the plant growth, nutrient uptake and development, and yield attributes even under stress. It also induces the total phenols, total sugars, proline, and antioxidant enzyme (Chrysargyris et al. 2018; Iqbal et al. 2018; Rizwan et al. 2019). Based on the basis of above observations, it can be inferred that Zn, a transition metal, is helpful in increasing the various characteristics of the plant which ultimately leads to increased yields.

Copper in Plant Growth and Development

Copper is a transition metal which participates in several physiological activities of plants such as photosynthesis, respiration, antioxidant machinery, and ethylene signal transduction. Cu acts as an important micronutrient which increase the distribution of carbohydrate and protein metabolism and further helps to enhance the plant growth and developmental processes (Sağlam et al. 2016). Cu is an important redox-active transition metal, its deficiency causes chlorosis followed by necrosis, leaf deformation, and affects reproductive organ (Mustafa and AlSharif 2018). It is the crucial transition metal but when present in excess causes toxicity, where free Cu in excess causes ROS accumulation in plant cell which damages the membrane and further halts the plant growth (Brewer 2010; Li et al. 2018; Shin et al. 2012). Thus, Cu homeostasis should be tightly controlled in organelles and must be intracellularly chelated for maintaining balance and promoting plant growth (Shin et al. 2012). Cu has two oxidation states that are Cu^{1+} and Cu^{2+} and acts as either oxidizing or reducing agent during biochemical reactions. This maintains Cu balance and ignites plant growth and developmental metabolism (Rajput et al. 2018). Copper transport protein (COPT) family such as COPT1 and COPT5 plays an important role in proper plant growth and development. COPT1 accumulates in plant root and diminishes the copper deficiency of plant, whereas, COPT5 is present in the prevacuolar compartment and exports Cu to aerial part (Garcia-Molina et al. 2011; Pilon 2011; Shin et al. 2012). In Arabidopsis, Cu regulates Cu-microRNAs via SPL7 transcription factor. SPL7 regulates Cu homeostasis and helps in proper plant metabolism, growth, and development (Pilon 2017). Antioxidant protein1 (ATX1) in Arabidopsis participates in promoting plant growth and development by inducing its homeostasis (Shin et al. 2012). The sufficient amount of Cu makes the photosynthetic machinery efficient and enhances the plant growth trait (Ravet and Pilon 2013; Sağlam et al. 2016). In *Atriplex halimus*, low dose or balanced amount of Cu acts as an important transition metal in inducing the plant growth, photosynthesis, and nutrient uptake (Mateos-Naranjo et al. 2013). Cu homeostasis and compartmentalization are important events which enhance the cell wall formation and lignin metabolism. Furthermore, it also increases the growth and development of plant reproductive organ (Ryan et al. 2013). In spinach, Cu promotes the germination, antioxidant machinery, and growth parameters (Gong et al. 2019). In *Scopelophilacataractae*, Cu acts as an important microelement and induces the morphophysiological and growth traits. However, high doses of Cu inhibit the growth (Yagura et al. 2019). In cowpea, Cu also increases the growth, yield, and nutrient uptake (Sharif et al. 2019). Homeostatic amount of Cu increased the germination traits and growth in water melon plant (Liu et al. 2016). In *Ocimum basilicum*, Cu at low doses showed proper growth and increased oil content, whereas high doses showed negative effect because of improper Cu compartmentalization (Ghorbanpour et al. 2016). Exogenous application of Cu in proper dose enhanced the plant morphophysiological and biochemical attributes and further induces the plant growth, development, and yield of plant which make the plant tolerant to several environmental stresses like salt, drought, and Cd (Iqbal et al. 2018; Zhang et al. 2019). It is reported that when Cu is present in homeostatic and properly compartmentalized form, it increased the growth of the many plants. However, high doses of Cu causes ion toxicity or stress, and this further impairs plant growth and developmental processes (Ben et al. 2019; Shahid et al. 2016).

Manganese in Plant Growth and Development

Manganese is an essential micronutrient and also a transition metal which has a crucial role in redox reactions, as it is a part of MnSOD enzyme, acts as ROS scavenger (Schmidt and Husted 2019). It also acts as the important metalloenzyme, activator, and cofactor of about 30 enzymes, such as oxalate oxidase (OxOx), catalase, isocitratelyase, malic enzyme, pyruvate carboxylase, and phosphoenol and plays a vital role in oxidoreduction, decarboxylation and hydrolytic reactions. Four atoms of Mn are required in OEC located at PSII and make Mn an important metal which is required for oxygen evolution (DalCorso et al. 2014; Schmidt and Husted 2019). Further, it also participates in the synthesis of proteins, lipids, chlorophyll, and carotenoid component. The deficiency of Mn affects the plant chloroplast to large extent by reducing the thylakoid glycolipids and fatty acids which leads to reduced photosynthesis. Mn exists in a wide range of oxidation states such as from Mn^{1+} to Mn^{7+}; among them, Mn^{2+}, Mn^{3+}, and Mn^{4+} are highly stable and most commonly present in biological system (Socha and Guerinot 2014). Mn^{2+} plays a key role during the entire stages of plant growth development, as it activates a number of different enzymes used in plant metabolic activities (Mills et al. 2008). Its deficiency also leads to chlorosis and pollen infertility. Furthermore, it hampers the plant growth and development. Therefore, Mn is needed for proper plant growth and developmental processes (DalCorso et al. 2014). High concentration of Mn is highly toxic, as it causes necrosis, chlorosis, brown speckles on mature leaves, and reduced plant growth and development. Therefore, for normal plant growth, development and biochemical processes, transport of Mn is highly essential as it maintains cellular homeostasis (Mills et al. 2008). ECA3 (P-type ATPase) plays a key role in Mn homeostasis and for proper plant growth and nutrient status (Mills et al. 2008). Transporters, such as NRAMP, PHO84 (phosphate transporter), ZIP, Calcium-permeable channels, VIT (vacuolar iron transporter), MTP, CDF (cation diffusion facilitator), and CCXs (cation exchanger) play a major role in Mn homeostasis and crop improvement (Shao et al. 2017; Socha and Guerinot 2014). In rice, Mn homeostasis at cellular level via several transporters leads to proper morphophysiological traits (Shao et al. 2017). Iron-regulated transporter 1 (IRT 1) transporter in maize (HvIRT 1) plays a significant role in Mn homeostasis and transport for proper photosynthesis and growth (Long et al. 2018). Mn homeostasis and compartmentalization with the intriguing use of chaperone play a crucial role in proper plant metabolic activity such as elongation, lignin synthesis, development and photosynthesis (Schmidt and Husted 2019). Proper concentration of Mn in plant cell enhances the relative water content, growth, photosynthesis, yield attributes, and quality of the plant (Nozulaidi et al. 2016; Schmidt et al. 2016). Mn application mitigated several environmental stresses by increasing the antioxidant enzyme, leaf area, turgor, growth, and yield of plant (El-Fouly et al. 2011; Hrishikesh et al. 2012). Therefore, Mn transport, homeostasis, and compartmentalization via transporters are highly essential for maintaining the lower and beneficial concentration of Mn inside the cell, which further modulates the plants redox status, photosynthesis, lignin synthesis, growth, and yield.

Conclusions and Future Prospective

Transition metal availability in the rhizosphere is the dynamic process which deals with several biological and physical factors. In particular, transition metal plays a significant role in plant growth and development and acts as a cofactor in many enzymes which participate in metabolic processes in plants. Deficiencies and toxicity of metals further create stress and affect plant growth and metabolism. High uptake of transition metal without compartmentalization leads to cellular accumulation and metal toxicity; this further enhances the ROS amount in the cell and leads to metal stress. Thus, essential transition metal requires homeostasis and equilibrium inside the cell and cellular organelles. Mutation in several genes can also lead to improper distribution of metal and creates metal toxicity. There are several proteins and transporters which are involved in transition metal homeostasis and compartmentalization. Distribution of Fe, Zn, Cu, and Mn in different cell organelles like vacuole, chloroplast, mitochondria, ER, and Golgi helped to enhance the plant growth and metabolism. Metal homeostasis and compartmentalization further enhance sustainable agriculture and crop production with proper nutritional value. The proper distribution and compartmentalization of transition metals (Fe, Cu, Zn, and Mn) serves as the important micronutrient that significantly leads to proper growth, development, and metabolism of plant, as they are crucial in every stage of plant's life cycle. In particular, Fe is an important micronutrient and transition metal that takes part in oxygen transportation, cell division, nucleic acid synthesis, and electron transfer. It serves as one of the most important metals in photosynthetic machinery and also a part of heme, which is used in nitrogen fixation. However, ferretin maintains homeostasis and redox potential of Fe in plant and thus balanced amount of Fe carries proper functioning of metabolic processes involved in growth and development of plant. Cu is also an important transition metal which participates in metabolic processes, such as respiration, photosynthesis, carbohydrate machinery, signal transduction pathway, and antioxidant machinery. Further, it plays a significant role in promoting the germination, growth, and yield. Zn is an important transition metal in plants after Fe and present in all six classes of enzymes. Zn helps in increasing the plant height, nutrient uptake, photosynthesis, and yield parameters. It serves as a crucial metal involved in cell division, fertilization, catalyst of several enzymes, auxin, and protein synthesis. Mn has a significant role in producing oxygen at OEC located at PSII. Besides this, it also serves as an important cofactor for several enzymes like MnSOD, OxOx, catalase, and phosphoenol. Mn participates in promoting plant's growth, chloroplast synthesis, photosynthesis, and antioxidant system and yield traits. Thus, it may be concluded that proper doses of transition metal which can be properly distributed and do not create toxicity plays a vital role in plant growth, metabolism, and crop improvement (Fig. 8.2).

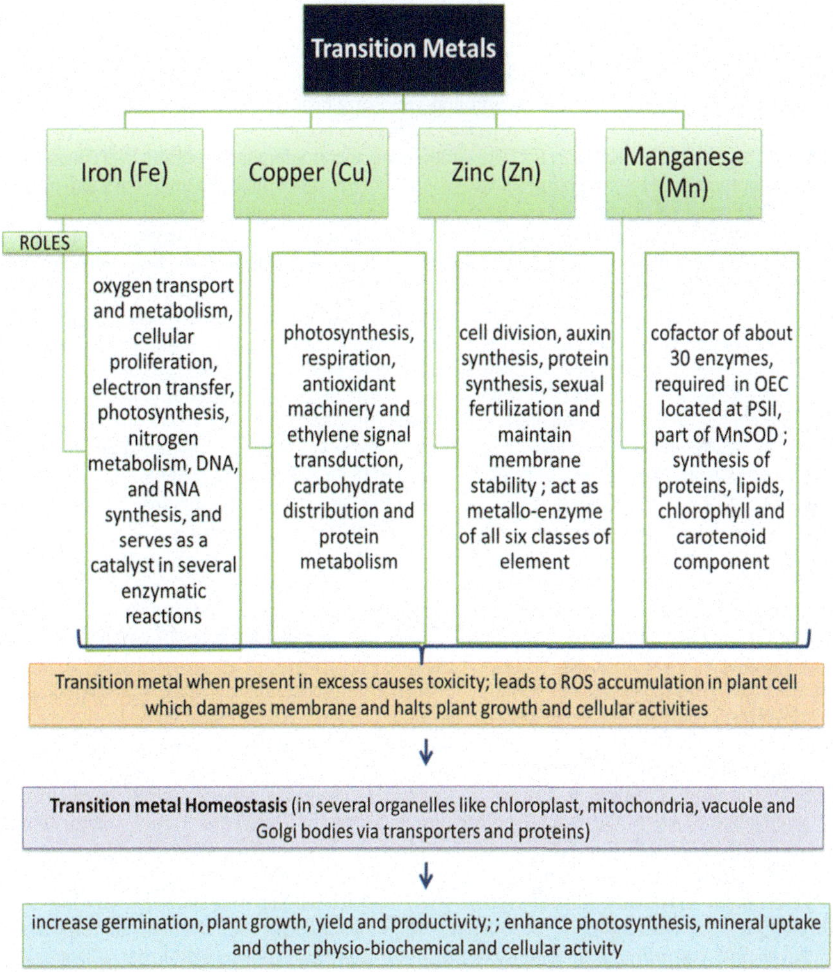

Fig. 8.2 Overview of transition metal homeostasis and their role in plant growth and development

Apart from this, there are other aspects that should be further explored like molecular mechanism and gene involved in the assembly of metals with enzymes as cofactor. A more detailed understanding of transporters and genes involved in homeostasis of transition metals are needed. Furthermore, exploring the transcription factors, regulators, and proteins that can be connected to cellular and organelle homeostasis will also enhance the understanding of the transition metal in a better way.

Acknowledgement Yamshi Arif acknowledge the receipt of Inspire fellowship (IF: 180778) by the Department of Science & Technology, Government of India, New Delhi.

References

Arif, M., Tasneem, M.T., Bashir, F., Yaseen, G., Anwar, A.: Evaluation of different levels of potassium and zinc fertilizer on the growth and yield of wheat. Int. J. Biosens. Bioelectron. **3**(3), 1–5 (2017)

Arrivault, S., Senger, T., Krämer, U.: The Arabidopsis metal tolerance protein AtMTP3 maintains metal homeostasis by mediating Zn exclusion from the shoot under Fe deficiency and Zn oversupply. Plant J. **46**(5), 861–879 (2006)

Balai, K., Sharma, Y., Jajoria, M., Deewan, P., Verma, R.: Effect of phosphorus, and zinc on growth, yield and economics of chickpea (*Cicer aritinum* l.). Int. J. Curr. Microbiol. App. Sci. **6**(3), 1174–1181 (2017)

Balk, J., Schaedler, T.A.: Iron cofactor assembly in plants. Annu. Rev. Plant Biol. **65**, 125–153 (2014)

Bashir, K., Ishimaru, Y., Nishizawa, N.K.: Identification and characterization of the major mitochondrial Fe transporter in rice. Plant Signal. Behav. **6**(10), 1591–1593 (2011)

Bashir, K., Nozoye, T., Ishimaru, Y., Nakanishi, H., Nishizawa, N.K.: Exploiting new tools for iron bio-fortification of rice. Biotechnol. Adv. **31**(8), 1624–1633 (2013)

Bashir, K., Rasheed, S., Kobayashi, T., Seki, M., Nishizawa, N.K.: Regulating subcellular metal homeostasis: the key to crop improvement. Front. Plant Sci. **7**, 1192 (2016)

Ben, M.M., Sakouhi, L., Chaoui, A.: Effect of plant growth regulators, calcium and citric acid on copper toxicity in pea seedlings. J. Plant Nutr. **42**(10), 1230–1242 (2019)

Binder, B.M., Rodríguez, F.I., Bleecker, A.B.: The copper transporter RAN1 is essential for biogenesis of ethylene receptors in Arabidopsis. J. Biol. Chem. **285**(48), 37263–37270 (2010)

Boutigny, S., Sautron, E., Finazzi, G., Rivasseau, C., Frelet-Barrand, A., Pilon, M., et al.: HMA1 and PAA1, two chloroplast-envelope PIB-ATPases, play distinct roles in chloroplast copper homeostasis. J. Exp. Bot. **65**(6), 1529–1540 (2014)

Brewer, G.J.: Risks of copper and iron toxicity during aging in humans. Chem. Res. Toxicol. **23**(2), 319–326 (2010)

Briat, J.F., Duc, C., Ravet, K., Gaymard, F.: Ferritins and iron storage in plants. Biochim. Biophys. Acta. **1800**(8), 806–814 (2010)

Catty, P., Boutigny, S., Miras, R., Joyard, J., Rolland, N., Seigneurin-Berny, D.: Biochemical characterization of AtHMA6/PAA1, a chloroplast envelope Cu (I)-ATPase. J. Biol. Chem. **286**(42), 36188–36197 (2011)

Choudhary, P., Jhajharia, A., Kumar, R.: Influence of sulphur and zinc fertilization on yield, yield components and quality traits of soybean (*Glycine Max* L.). The Bioscan. **9**(1), 137–142 (2014)

Chrysargyris, A., Michailidi, E., Tzortzakis, N.: Physiological and biochemical responses of *Lavandula angustifolia* to salinity under mineral foliar application. Front. Plant Sci. **9**, 489 (2018)

Conte, S.S., Walker, E.L.: Transporters contributing to iron trafficking in plants. Mol. Plant. **4**(3), 464–476 (2011)

DalCorso, G., Manara, A., Piasentin, S., Furini, A.: Nutrient metal elements in plants. Metallomics. **6**(10), 1770–1788 (2014)

Divol, F., Couch, D., Conéjéro, G., Roschzttardtz, H., Mari, S., Curie, C.: The Arabidopsis YELLOW STRIPE LIKE4 and 6 transporters control iron release from the chloroplast. Plant Cell. **25**(3), 1040–1055 (2013)

Duy, D., Stübe, R., Wanner, G., Philippar, K.: The chloroplast permease PIC1 regulates plant growth and development by directing homeostasis and transport of iron. Plant Physiol. **155**(4), 1709–1722 (2011)

El-Fouly, M.M., Mobarak, Z.M., Salama, Z.A.: Micronutrients (Fe, Mn, Zn) foliar spray for increasing salinity tolerance in wheat *Triticum aestivum* L. African J. Plant Sci. **5**(5), 314–322 (2011)

Eroglu, S.: Metal transport in the developing plant seed. In: Advances in Botanical Research, vol. 87, pp. 91–113. Academic Press (2018)

Gao, M., Chen, Y., Wu, L., Wang, Y.: Changes in the profiles of yield, yield component, oil content, and citral content in *Litsea cubeba* (Lour.) persoon following foliar fertilization with zinc and boron. Forests. **10**(1), 59 (2019)

Garcia-Molina, A., Andrés-Colás, N., Perea-García, A., del Valle-Tascón, S., Peñarrubia, L., Puig, S.: The intracellular Arabidopsis COPT5 transport protein is required for photosynthetic electron transport under severe copper deficiency. Plant J. **65**(6), 848–860 (2011)

Ghorbanpour, M., AsgariLajayer, H., Hadian, J.: Influence of copper and zinc on growth, metal accumulation and chemical composition of essential oils in sweet basil (*Ocimum basilicum* L.). J. Med. Plants. **3**(59), 132–144 (2016)

Gollhofer, J., Timofeev, R., Lan, P., Schmidt, W., Buckhout, T.J.: Vacuolar-iron-transporter1-like proteins mediate iron homeostasis in Arabidopsis. PLoS One. **9**(10), e110468 (2014)

Gong, Q., Wang, L., Dai, T., Zhou, J., Kang, Q., Chen, H., et al.: Effects of copper on the growth, antioxidant enzymes and photosynthesis of spinach seedlings. Ecotoxicol. Environ. Saf. **171**, 771–780 (2019)

Grillet, L., Ouerdane, L., Flis, P., Hoang, M.T.T., Isaure, M.P., Lobinski, R., et al.: Ascorbate efflux as a new strategy for iron reduction and transport in plants. J. Biol. Chem. **289**(5), 2515–2525 (2014)

Grotz, N., Guerinot, M.L.: Molecular aspects of Cu, Fe and Zn homeostasis in plants. Biochim. Biophys. Acta. **1763**(7), 595–608 (2006)

Haider, M.U., Farooq, M., Nawaz, A., Hussain, M.: Foliage applied zinc ensures better growth, yield and grain biofortification of mungbean. Int. J. Agric. Biol. **20**(12), 2817–2822 (2018)

Hall, J.L., Williams, L.E.: Transition metal transporters in plants. J. Exp. Bot. **54**(393), 2601–2613 (2003)

Hrishikesh, U., Biman, K.D., Lingaraj, S., Sanjib, K.P.: Comparative effect of Ca, K, Mn and B on post-drought stress recovery in tea [*Camellia sinensis* (L.) O Kuntze]. Am. J. Plant Sci. **3**(4), 18 (2012)

Iqbal, M.N., Rasheed, R., Ashraf, M.Y., Ashraf, M.A., Hussain, I.: Exogenously applied zinc and copper mitigate salinity effect in maize (*Zea mays* L.) by improving key physiological and biochemical attributes. Environ. Sci. Pol. **25**(24), 23883–23896 (2018)

Ishimaru, Y., Bashir, K., Nakanishi, H., Nishizawa, N.K.: OsNRAMP5, a major player for constitutive iron and manganese uptake in rice. Plant Signal. Behav. **7**(7), 763–766 (2012)

Jain, A.: Ferric reductases and transporters that contribute to mitochondrial iron homeostasis (Doctoral dissertation). Retrieved from https://scholarcommons.sc.edu/etd/2971(2014)

Jain, A., Wilson, G.T., Connolly, E.L.: The diverse roles of FRO family metalloreductases in iron and copper homeostasis. Front. Plant Sci. **5**, 100 (2014)

Jeong, J., Merkovich, A., Clyne, M., Connolly, E.L.: Directing iron transport in dicots: regulation of iron acquisition and translocation. Curr. Opin. Plant Biol. **39**, 106–113 (2017)

Kim, S.A., Punshon, T., Lanzirotti, A., Li, L., Alonso, J.M., Ecker, J.R., et al.: Localization of iron in Arabidopsis seed requires the vacuolar membrane transporter VIT1. Science. **314**(5803), 1295–1298 (2006)

Klaumann, S., Nickolaus, S.D., Fürst, S.H., Starck, S., Schneider, S., Ekkehard Neuhaus, H., Trentmann, O.: The tonoplast copper transporter COPT5 acts as an exporter and is required for interorgan allocation of copper in Arabidopsis thaliana. New Phytol. **192**(2), 393–404 (2011)

Li, L., Zhang, K., Gill, R.A., Islam, F., Farooq, M.A., Wang, J., Zhou, W.: Ecotoxicological and interactive effects of copper and chromium on physiochemical, ultrastructural, and molecular profiling in *Brassica napus* L. Biomed. Res. Int. **2018**, 9248123 (2018)

Lill, R., Dutkiewicz, R., Freibert, S.A., Heidenreich, T., Mascarenhas, J., Netz, D.J., et al.: The role of mitochondria and the CIA machinery in the maturation of cytosolic and nuclear iron–sulfur proteins. Eur. J. Cell Biol. **94**(7–9), 280–291 (2015)

Liu, H.Q., Mo, Y.F., Kong, X.Z., Liu, Y., Liu, H.P.: Effects of copper on germination of watermelon seeds and growth of watermelon seedlings. In: Application of Materials Science and Environmental Materials (AMSEM2015) Proceedings of the 3rd International Conference, pp. 108–113 (2016)

Long, L., Persson, D.P., Duan, F., Jørgensen, K., Yuan, L., Schjoerring, J.K., Pedas, P.R.: The iron-regulated transporter 1 plays an essential role in uptake, translocation and grain-loading of manganese, but not iron, in barley. New Phytol. **217**(4), 1640–1653 (2018)

López-Millán, A.F., Duy, D., Philippar, K.: Chloroplast iron transport proteins–function and impact on plant physiology. Front. Plant Sci. **7**, 178 (2016)

Mateos-Naranjo, E., Andrades-Moreno, L., Cambrollé, J., Perez-Martin, A.: Assessing the effect of copper on growth, copper accumulation and physiological responses of grazing species *Atriplex halimus*: ecotoxicological implications. Ecotoxicol. Environ. Saf. **90**, 136–142 (2013)

Mills, R.F., Doherty, M.L., López-Marqués, R.L., Weimar, T., Dupree, P., Palmgren, M.G., et al.: ECA3, a Golgi-localized P2A-type ATPase, plays a crucial role in manganese nutrition in Arabidopsis. Plant Physiol. **146**(1), 116–128 (2008)

Milner, M.J., Seamon, J., Craft, E., Kochian, L.V.: Transport properties of members of the ZIP family in plants and their role in Zn and Mn homeostasis. J. Exp. Bot. **64**(1), 369–381 (2013)

Miyadate, H., Adachi, S., Hiraizumi, A., Tezuka, K., Nakazawa, N., Kawamoto, T., et al.: OsHMA3, a P1B-type of ATPase affects root-to-shoot cadmium translocation in rice by mediating efflux into vacuoles. New Phytol. **189**(1), 190–199 (2011)

Moreira, A., Moraes, L.A., dos Reis, A.R.: The molecular genetics of zinc uptake and utilization efficiency in crop plants. In: Plant Micronutrient Use Efficiency, pp. 87–108. Academic Press (2018)

Mousavi, S.R.: Zinc in crop production and interaction with phosphorus. Aust. J. Basic Appl. Sci. **5**(9), 1503–1509 (2011)

Mustafa, S.K., AlSharif, M.A.: Copper (Cu) an essential redox-active transition metal in living system—a review article. Am. J. Analyt. Chem. **9**(01), 15 (2018)

Nakanishi-Masuno, T., Shitan, N., Sugiyama, A., Takanashi, K., Inaba, S., Kaneko, S., Yazaki, K.: The *Crotalaria juncea* metal transporter CjNRAMP1 has a high Fe uptake activity, even in an environment with high Cd contamination. Int. J. Phytoremediation. **20**(14), 1427–1437 (2018)

Nikolic, M., Pavlovic, J.: Plant responses to iron deficiency and toxicity and iron use efficiency in plants. In: Plant Micronutrient Use Efficiency, pp. 55–69. Academic Press (2018)

Nozoye, T., Nagasaka, S., Bashir, K., Takahashi, M., Kobayashi, T., Nakanishi, H., Nishizawa, N.K.: Nicotianamine synthase 2 localizes to the vesicles of iron-deficient rice roots, and its mutation in the YXX φ or LL motif causes the disruption of vesicle formation or movement in rice. Plant J. **77**(2), 246–260 (2014)

Nozulaidi, N., Nurlnani, M., Khairi, M., Jahan, S.M.D.: Production of corn; effects of manganese application on plant parameters. J. Agric. Res. **1**, 1–7 (2016)

Palmer, C.M., Guerinot, M.L.: Facing the challenges of Cu, Fe and Zn homeostasis in plants. Nat. Chem. Biol. **5**(5), 333 (2009)

Pedas, P., Stokholm, M.S., Hegelund, J.N., Ladegård, A.H., Schjoerring, J.K., Husted, S.: Golgi localized barley MTP8 proteins facilitate Mn transport. PLoS One. **9**(12), e113759 (2014)

Pilon, M.: Moving copper in plants. New Phytol. **192**(2), 305–307 (2011)

Pilon, M.: The copper microRNAs. New Phytol. **213**(3), 1030–1035 (2017)

Pilon, M., Cohu, C.M., Ravet, K., Abdel-Ghany, S.E., Gaymard, F.: Essential transition metal homeostasis in plants. Curr. Opin. Plant Biol. **12**(3), 347–357 (2009)

Pottier, M., Oomen, R., Picco, C., Giraudat, J., Scholz-Starke, J., Richaud, P., et al.: Identification of mutations allowing Natural Resistance Associated Macrophage Proteins (NRAMP) to discriminate against cadmium. Plant J. **83**(4), 625–637 (2015)

Prasad, T.N.V.K.V., Sudhakar, P., Sreenivasulu, Y., Latha, P., Munaswamy, V., Reddy, K.R., et al.: Effect of nanoscale zinc oxide particles on the germination, growth and yield of peanut. J. Plant Nutr. **35**(6), 905–927 (2012)

Rafsanjani, K., Madani, A., Vazin, F.: Effect of iron on yield of corn (Zea mays L.) in drought stress. Cercet. Agron. Mold. **52**(1), 34–43 (2019)

Rajput, V.D., Minkina, T., Suskova, S., Mandzhieva, S., Tsitsuashvili, V., Chapligin, V., Fedorenko, A.: Effects of copper nanoparticles (CuO NPs) on crop plants: a mini review. Bionanoscience. **8**(1), 36–42 (2018)

Ravet, K., Pilon, M.: Copper and iron homeostasis in plants: the challenges of oxidative stress. Antioxid. Redox Signal. **19**(9), 919–932 (2013)

Rawat, M., Nayan, R., Negi, B., Zaidi, M.G.H., Arora, S.: Physio-biochemical basis of iron-sulfide nanoparticle induced growth and seed yield enhancement in B. juncea. Plant Physiol. Biochem. **118**, 274–284 (2017)

Remy, E., Cabrito, T.R., Batista, R.A., Hussein, M.A., Teixeira, M.C., Athanasiadis, A., et al.: Intron retention in the 5′ UTR of the novel ZIF2 transporter enhances translation to promote zinc tolerance in Arabidopsis. PLoS Genet. **10**(5), e1004375 (2014)

Rizwan, M., Ali, S., Ali, B., Adrees, M., Arshad, M., Hussain, A., et al.: Zinc and iron oxide nanoparticles improved the plant growth and reduced the oxidative stress and cadmium concentration in wheat. Chemosphere. **214**, 269–277 (2019)

Roschzttardtz, H., Grillet, L., Isaure, M.P., Conéjéro, G., Ortega, R., Curie, C., Mari, S.: Plant cell nucleolus as a hot spot for iron. J. Biol. Chem. **286**(32), 27863–27866 (2011)

Ryan, B.M., Kirby, J.K., Degryse, F., Harris, H., McLaughlin, M.J., et al.: Copper speciation and isotopic fractionation in plants: uptake and translocation mechanisms. New Phytol. **199**(2), 367–378 (2013)

Sağlam, A., Yetişsin, F., Demiralay, M., Terzi, R.: Copper stress and responses in plants. In: Plant Metal Interaction, pp. 21–40. Elsevier (2016)

Sah, S., Singh, N., Singh, R.: Iron acquisition in maize (*Zea mays* L.) using Pseudomonas siderophore. 3Biotech. **7**(2), 121 (2017)

Said-Al Ahl, H.A.H., Mahmoud, A.A.: Effect of zinc and/or iron foliar application on growth and essential oil of sweet basil (*Ocimum basilicum* L.) under salt stress. Ozean J. Appl. Sci. **3**(1), 97–111 (2010)

Saif, S., Tahir, A., Chen, Y.: Green synthesis of iron nanoparticles and their environmental applications and implications. J. Nanomater. **6**(11), 209 (2016)

Sasaki, A., Yamaji, N., Xia, J., Ma, J.F.: OsYSL6 is involved in the detoxification of excess manganese in rice. Plant Physiol. **157**(4), 1832–1840 (2011)

Schaedler, T.A., Thornton, J.D., Kruse, I., Schwarzländer, M., Meyer, A.J., van Veen, H.W., Balk, J.: A conserved mitochondrial ATP-binding cassette transporter exports glutathione polysulfide for cytosolic metal cofactor assembly. J. Biol. Chem. **289**(34), 23264–23274 (2014)

Schmidt, W., Buckhout, T.J.: A hitchhiker's guide to the Arabidopsis ferrome. Plant Physiol. Biochem. **49**(5), 462–470 (2011)

Schmidt, S.B., Husted, S.: The biochemical properties of manganese in plants. Plan. Theory. **8**(10), 381 (2019)

Schmidt, S.B., Jensen, P.E., Husted, S.: Manganese deficiency in plants: the impact on photosystem II. Trends Plant Sci. **21**(7), 622–632 (2016)

Seo, P.J., Park, J., Park, M.J., Kim, Y.S., Kim, S.G., Jung, J.H., Park, C.M.: A Golgi-localized MATE transporter mediates iron homoeostasis under osmotic stress in Arabidopsis. Biochem. J. **442**(3), 551–561 (2012)

Shahid, M., Shukla, A.K., Bhattacharyya, P., Tripathi, R., Mohanty, S., Kumar, A., et al.: Micronutrients (Fe, Mn, Zn and Cu) balance under long-term application of fertilizer and manure in a tropical rice-rice system. J. Soils Sediments. **16**(3), 737–747 (2016)

Shao, J.F., Yamaji, N., Shen, R.F., Ma, J.F.: The key to Mn homeostasis in plants: regulation of Mn transporters. Trends Plant Sci. **22**(3), 215–224 (2017)

Sharif, A.T., Ali, A.H.M.Z., Rahman, M.K.: Effects of copper and vermicompost on growth and yield of cowpea (*Vigna unguiculata* L.) Walp and nutrient accumulation in its fruits. J. Biodivers. Conserv. Bioresour. Manage. **5**(2), 13–18 (2019)

Shin, L.J., Lo, J.C., Yeh, K.C.: Copper chaperone antioxidant protein1 is essential for copper homeostasis. Plant Physiol. **159**(3), 1099–1110 (2012)

Singh, L., Sharma, P.K., Jajoria, M., Deewan, P., Verma, R.: Effect of phosphorus and zinc application on growth and yield attributes of pearl millet *(Pennisetum glaucum* L.) under rainfed condition. J. Pharmacogn. Phytochem. **6**(1), 388–391 (2017)

Singh, P., Ghosh, D., Manyapu, V., Yadav, M., Majumder, S.: Synergistic impact of iron (iii) oxide nanoparticles and organic waste on growth and development of *Solanum lycopersicum* plants: new paradigm in nanobiofertilizer. Plant Arch. **19**(1), 339–344 (2019)

Socha, A.L., Guerinot, M.L.: Mn-euvering manganese: the role of transporter gene family members in manganese uptake and mobilization in plants. Front. Plant Sci. **5**, 106 (2014)

Soliman, A.S., El-feky, S.A., Darwish, E.: Alleviation of salt stress on *Moringa peregrina* using foliar application of nanofertilizers. J. Hortic. For. **7**(2), 36–47 (2015)

Tarafdar, J.C., Raliya, R., Mahawar, H., Rathore, I.: Development of zinc nanofertilizer to enhance crop production in pearl millet (*Pennisetum americanum*). Agric. Res. **3**(3), 257–262 (2014)

Teschner, J., Lachmann, N., Schulze, J., Geisler, M., Selbach, K., Santamaria-Araujo, J., et al.: A novel role for Arabidopsis mitochondrial ABC transporter ATM3 in molybdenum cofactor biosynthesis. Plant Cell. **22**(2), 468–480 (2010)

Tripathi, D.K., Singh, S., Gaur, S., Singh, S., Yadav, V., Liu, S., et al.: Acquisition and homeostasis of iron in higher plants and their probable role in abiotic stress tolerance. Front. Environ. Sci. **5**, 86 (2018)

Vigani, G., Zocchi, G., Bashir, K., Philippar, K., Briat, J.F.: Signals from chloroplasts and mitochondria for iron homeostasis regulation. Trends Plant Sci. **18**(6), 305–311 (2013)

Vigani, G., Bashir, K., Ishimaru, Y., Lehmann, M., Casiraghi, F.M., Nakanishi, H., et al.: Knocking down mitochondrial iron transporter (MIT) reprograms primary and secondary metabolism in rice plants. J. Exp. Bot. **67**(5), 1357–1368 (2016)

Yagura, R., Imanishi, J., Shibata, S.: Effects of copper ions on the growth and photosynthetic activity of Scopelophilacataractae. Lindbergia. **2019**(1), 1–7 (2019)

Yang, T.J., Lin, W.D., Schmidt, W.: Transcriptional profiling of the Arabidopsis iron deficiency response reveals conserved transition metal homeostasis networks. Plant Physiol. **152**(4), 2130–2141 (2010)

Yoshida, K., Negishi, T.: The identification of a vacuolar iron transporter involved in the blue coloration of cornflower petals. Phytochemistry. **94**, 60–67 (2013)

Yruela, I.: Transition metals in plant photosynthesis. Metallomics. **5**(9), 1090–1109 (2013)

Zhai, Z., Gayomba, S.R., Jung, H.I., Vimalakumari, N.K., Piñeros, M., Craft, E., et al.: OPT3 is a phloem-specific iron transporter that is essential for systemic iron signaling and redistribution of iron and cadmium in Arabidopsis. Plant Cell. **26**(5), 2249–2264 (2014)

Zhang, X., Goatley, M., Conner, J., Wilkins, M., Teshler, I., Liu, J., et al.: Copper chlorophyllin impacts on growth and drought stress tolerance of tomato plants. HortScience. **54**(12), 2195–2201 (2019)

Zia-ur-Rehman, M., Naeem, A., Khalid, H., Rizwan, M., Ali, S., Azhar, M.: Responses of plants to iron oxide nanoparticles. In: Nanomaterials in Plants, Algae, and Microorganisms, pp. 221–238. Academic Press (2018)

Chapter 9
Role of Zinc Homeostasis in Plant Growth

Shams Tabrez Khan, Abdul Malik, and Faizan Ahmad

Abstract Zinc is one of the most important micronutrients required for optimal plant growth and productivity. Paradoxically, a higher concentration of the same can be toxic for the plants, and hence, a constant supply of zinc to maintain an optimal intracellular concentration in plants is required which is referred to as zinc homeostasis. Homeostasis also includes subcellular distribution of zinc and its vesicular storage. To maintain this optimal intracellular concentration, plant cell uses a number of biochemical mechanisms involving a tightly regulated network of dozens of intracellular proteins. Membrane of transport proteins, cation transporters, cation diffusion facilitator, Zn-binding proteins, ligands, members of ZIP family, and metal tolerance proteins are some constituting members of this network. Emerging evidences suggest that these networks of proteins are conserved among land plants. In addition to maintaining the intracellular concentration of zinc, this network coordinates processes like sequestration, zinc uptake from soil, and its transport in the plant ensuring adequate zinc supply to all the cells in the plants. The metal homeostasis network exhibits plasticity to respond to variations in bioavailable soil metal. Due to the chemical similarity between zinc and other divalent cations like iron and cadmium, these metals serve as zinc mimic. The presence of these metals along with zinc therefore affects zinc homeostasis and results in accumulation of unwanted metals in food grains.

Keywords Zinc homeostasis · ZIP proteins · Metal transport

Zinc is called as metal of life, as without zinc-dependent enzymes, metabolic networks of life cannot be sustained. All the six classes of enzymes contain zinc as key component contributing to their structural or catalytic integrity. Folded

S. T. Khan (✉) · A. Malik
Department of Agricultural Microbiology, Faculty of Agricultural Sciences, Aligarh Muslim University, Aligarh, Uttar Pradesh, India

F. Ahmad
Department of Post Harvest Engineering and Technology, Faculty of Agricultural Sciences, Aligarh Muslim University, Aligarh, Uttar Pradesh, India

S. T. Khan, A. Malik (eds.), *Microbial Biofertilizers and Micronutrient Availability*,
https://doi.org/10.1007/978-3-030-76609-2_9

conformations of domains are stabilized by zinc binding, enabling the proteins to interact with other macromolecules and substrate like DNA (Berg and Shi 1996). In addition to its role in metabolic processes, zinc is also involved in signal transduction and cellular control (Sinclair et al. 2018; Kisko et al. 2015). The role of zinc in various plant metabolic processes from photosynthesis, DNA replication, and plant reproduction is discussed in Chap. 1. Though zinc is an essential element, its excess can be even toxic. Therefore, all the living systems, plants, and animals regulate intracellular level of zinc to an optimal concentration, which is called as homeostasis. This involves complicated process of metal or zinc uptake from soil and its transport from roots to other parts of plants. Transport of metal ions like zinc in plants may take place from one cell to another or through intercellular spaces. And all these process like uptake by root epithelial cells, its maintenance in a cell and transport from one cell to another can be regulated to maintain the homeostasis. Importance of zinc in human health is unparalleled, as it plays role in growth, immunity, functioning of nervous system and brain. Its deficiency is affecting almost 30% population of the world with obvious health consequences. Elderly, children, and expecting mothers constitute the most vulnerable groups. Since low dietary intake is the primary reason of zinc deficiency, ensuring adequate supply of dietary zinc will be one of the best ways to fight zinc deficiency in humans. Among various strategies to deal with zinc deficiency one is increasing the zinc content of the plants and cereal grains. It will be of immense importance to understand the mechanism of zinc homeostasis in plants. As it can help in improving the zinc content in plants, especially in cereal grains. Understanding of zinc homeostasis is not only important for improving grain zinc content but also can help in understanding various other zinc-mediated processes like growth, reproduction in plants and events like programmed cell death and survival (Helmersson et al. 2008; Broadley et al. 2007). Proteomic in combination with in silico predictions and confirmation through in vitro experiments can be used for the understanding of zinc homeostasis.

Intracellular Zinc Concentrations

Determining metal concentrations in cells is difficult; furthermore in eukaryotic cells, the concentration of zinc also varies in intracellular compartments. But the Zn^{2+} ion concentrations in *Escherichia coli* and mammalian cells have been determined (Outten and O'Halloran 2001; Dittmer et al. 2009). These studies found that the total Zn^{2+} ion concentration in mammalian cells is typically in the range of 100–500 μM corresponding to 10^8 atoms per cell, while the concentration of "free" or "loosely bound" Zn^{2+} was in the range of pico- to nanomolar. In case of *E. coli* grown on minimal medium, the intracellular zinc concentration was found to be 0.2 mM, which is interestingly 2000 times of the zinc concentration in the ambient medium (Outten and O'Halloran 2001). Concentration of zinc in plant cells has been calculated indirectly using dry weight and zinc concentration ratios (Broadley et al. 2007). As various studies have reported zinc content in dry weight. It is estimated that in leaf epidermal cell, cytoplasm, vacuoles, and cell walls contain 1.8×10^9 atoms, 9.6×10^{10}, and 2.7×10^9 of Zn atoms, respectively. While, in root cortical cells, the cytoplasm, vacuoles, and cell walls contain $\leq 1.5 \times 10^8$, $\leq 3.4 \times 10^8$, and 1.4×10^8 atoms of Zn, respectively. Other studies have also suggested that the vacuoles contain 83–85% of the total cellular zinc while cytoplasm only contains 9–11% of the total zinc (Hacisalihoglu et al. 2003). Besides these reported average values, huge variations from plant to plant have also been reported. At least 15 plant taxa mainly the members of family *Brassicaceae* can accumulate high concentrations of zinc (10,000 mg zinc/g dry weight) without exhibiting any toxicity these plants are also called as hyperaccumulator (Shanmugam et al. 2013; Nkrumah et al. 2018). *Arabidopsis halleri, Arabis paniculata, Noccaea caerulescens, Picris divaricata,* and *Sedum plumbizincicola* are some examples of such plants (BAKER et al. 1994; Bert et al. 2002). Conversely, there are plants that can grow even in the presence of very low concentration of zinc. These two categories of plants may have higher and lower intracellular zinc concentrations and must have developed some mechanism to deal with these concentrations of zinc without affecting their growth.

Zinc Homeostasis in Plants

Zinc homeostasis in plants is a complicated process involving its mobilization and uptake from soil, transport from cell to cell through symplast and through intracellular spaces (apoplast) to reach various tissues of the plants (Fig. 9.1). A whole network of proteins composed of zinc-regulated transporter (ZRT), iron-regulated transporter (ZRT-IRT)-like proteins (ZIPs), zinc-induced facilitator, cation diffusion facilitator (CDF), natural resistance-associated macrophage protein (NRAMP) family members, metal tolerance protein or metal transport protein (MTP), heavy metal ATPases (HMA), and metal-binding compounds is involved in maintaining this homeostasis (Sinclair and Krämer 2012). Investigations have demonstrated that many of these proteins were found to be conserved in many organisms. Details of these proteins are discussed in the review by Sinclair and colleagues and a brief account of these will be included wherever necessary (Sinclair and Krämer 2012). The exact biochemical, molecular, and regulatory process is not yet very well understood, but many studies on the topic are available. One of the most extensively studied plant for various studies is *Arabidopsis* and the mechanism of metal homeostasis has also been reported for this plant (Sinclair and Krämer 2012; Kobae et al. 2004). Interestingly, one species of *Arabidopsis*, *A. halleri* is a metal hyperaccumulator and provides a good model for studying the mechanism of metal homeostasis. It is important to note here that most divalent cations like cadmium and iron act as zinc mimic and the nonspecific behavior of many homeostasis proteins and mechanisms results in interference by these metals.

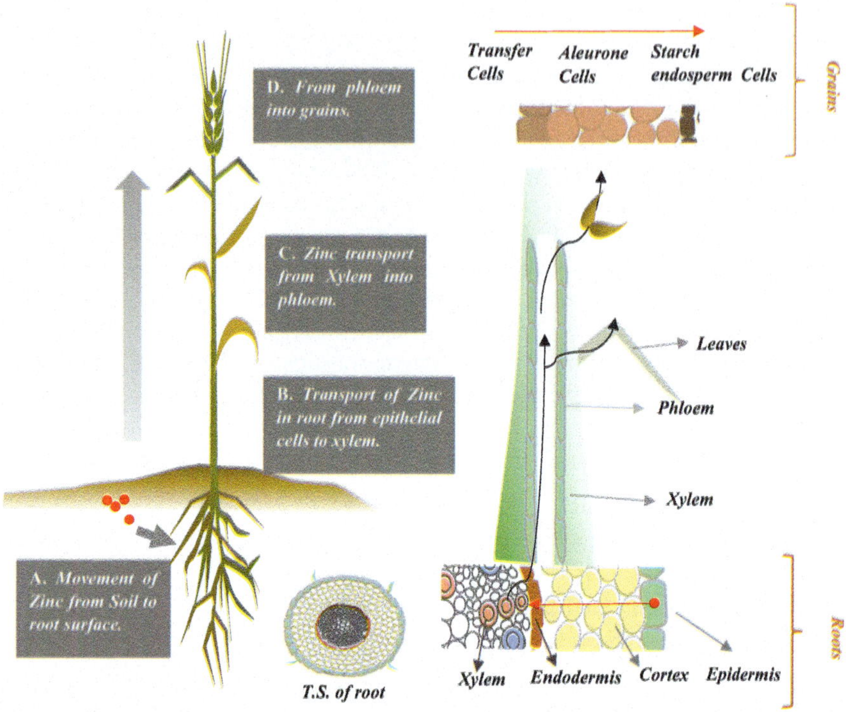

Fig. 9.1 Process of zinc uptake from soil its transport to xylem and other aerial parts of the plants. The transport into the grains through various tissues is also shown

(i) *Zinc Solubilization/Mobilization*

Generally, the concentration of metal ions in the soil is very low and metals are mostly present as mineral fractions. Zinc and other metal ions have the tendency to bind to various inorganic and organic soil components. The availability of these metals depends on a number of soil factors such as fertilization and rain. The first important process involved in plant metal homeostasis is the mobilization of this fixed form of zinc and its acquisition by plants. The main mechanisms of zinc solubilization in the rhizospheric region are acidification and secretion of chelators. Exudates from roots contain a variety of substances such as sugars, amino acids, nucleosides, vitamins, gaseous molecules along with organic acid anions, and phytosiderophores (Dakora and Phillips 2002). Small metal-binding peptides Phytochelatins having repeats of (γ-Glu-Cys)n-Gly have been found in many of the plants studied (Clemens 2001). These compounds form complexes with zinc and transport it to the outer face of the root-cell plasma membrane. The presence of metals and metalloids also induces the production of compounds like phytochelatins (Clemens 2001). Plants acidify the rhizosphere using plasma membrane H^+-ATPases and proton pumps. The increased proton concentration in the rhizospheric soil results in the release of divalent metal ions tightly bound to soil particles through

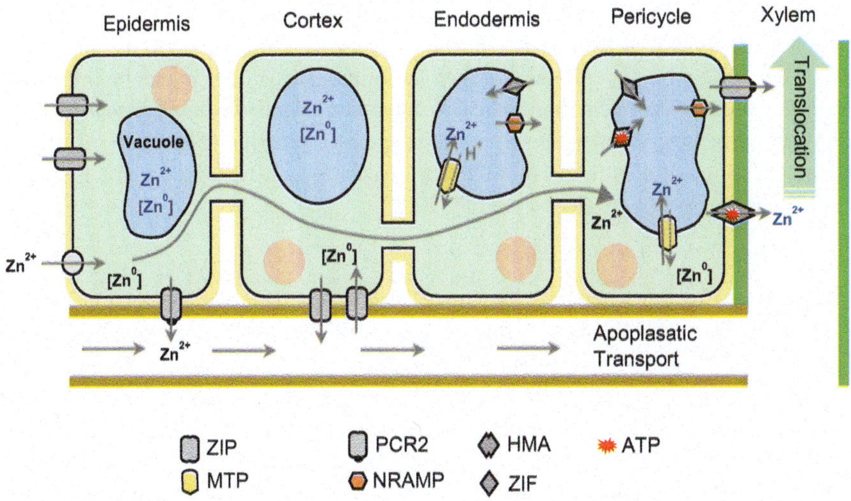

Fig. 9.2 Various proteins involved in zinc uptake, intracellular homeostasis, and xylem loading

cation exchange subsequently increasing zinc uptake by plants (Mench and Martin 1991; Rose et al. 2012). The metal uptake is often nonspecific and the presence or deficiency of one metal influences the uptake of the other metal.

(ii) *Zinc Uptake by Root Epithelial Cells*

Zinc enters into root cell wall free spaces through diffusion and is transported through plasma membrane mainly as free ion (Figs. 9.1 and 9.2). It is to be noted here that the uptake is not specific for zinc only and other metal ions can also be transported across the membrane including toxic metals like cadmium (Palmgren et al. 2008). Uptake of Zn-phytosiderophore or zinc-ligand complex across the plasma membrane is also reported in some plants (von Wiren et al. 1996). The zinc uptake is facilitated through channels or through metal proteins like zinc-regulated transporter proteins of the ZIP family (Guerinot 2000). The process is largely passive and does not require energy due to the negative membrane potential (Olsen and Palmgren 2014; Kochian 1993). ZIP family of proteins having 85 diverse members have been reported in various organisms (Guerinot 2000). These proteins contain transmembrane domains (TM domain) and variable regions between the TM domains. Due to the variation in the length of variable region, the overall length of ZIP proteins varies significantly. A histidine residue is highly conserved in the variable region and is believed to bind to metals for their transport (Ishimaru et al. 2011). In *Arabidopsis*, 18 genes encoding different ZIP proteins have been identified. Although it is now established that ZIP-4 proteins are involved in plant zinc transport. Some reports demonstrate that ZIP family protein iron-regulated transporters 1 and 3 (IRT1 and IRT3) contribute significantly to Zn and Cd uptake by plant roots (Korshunova et al. 1999; Vert et al. 2002). Such transporters are also found in *Oryza sativa*. IRT1 is mainly involved in the uptake of iron and deficiency of iron upregulates the production of IRT1. And upregulation of IRT1 will not only increase the uptake of iron to the cell but will also facilitate uptake of zinc by the cell. Here, it can be understood that how the homeostasis of zinc can be influenced by other metals like iron. Some studies have suggested the involvement of YSL transporter in zinc uptake due to its nonspecific behavior (Murata et al. 2006). However, other studies found that these transporters can transport iron and copper but do not transport zinc (Roberts et al. 2004).

(iii) *Maintenance of Intracellular Homeostasis*

Once the zinc reaches cytoplasm, it is chelated by low molecular weight ligands. Binding of zinc with ligands prevents nonspecific binding to other biomolecules in the cytoplasm and its precipitation. The bound zinc also acts as buffer to maintain low concentration of free zinc ions in the cytoplasm. The zinc-bound molecules may later donate the metal to metalloproteins in apoplast for its transport or efflux (von Wiren et al. 1996; Schaaf et al. 2005). Three main types of chelators found in the plants are phytochelatins, nicotianamine, and metallothionein. Phytochelatins briefly described above are low molecular weight compounds with a characteristic peptide sequence (γGlu-Cys)n-Gly [n = 2–11]) (Schaaf et al. 2005). These phytochelatins chelate zinc through thiol moieties. Nicotianamine is known to bind to metals including iron and zinc (von Wirén et al. 1999). Like phytochelatins, metallothioneins are cysteine-rich metal-binding proteins and can bind to various heavy metals playing important role in metal homeostasis (Zhou and Goldsbrough 1994). Metal requirement in general and zinc requirement in particular of various cellular organelles is different, for example, requirement of chloroplasts and mitochondria is high due to the presence of metallic cofactor requiring electron transport chains in these organelles. The excess zinc in the cells is either immobilized in the apoplast or stored in vacuoles. The transport of zinc from cytoplasm to vacuoles and other membrane-bound organelles will involve various transport proteins to facilitate the movement across the membrane. A number of proteins have been localized in the membrane of vacuoles with possible role in metal transport like cation diffusion facilitator (CDF) proteins. In addition to CDF, studies on *Arabidopsis* show that metal tolerance proteins (MTP) mediate the transport of zinc into the vacuoles. Two such proteins MTP1 and MTP3 have been localized in the vacuolar membrane with possible role in Zn transport (Kobae et al. 2004; Desbrosses-Fonrouge et al. 2005). Proteins belonging to natural resistance-associated macrophage protein (NRAMP) family have also been identified in plants and are found to involve mainly in the transport of iron and manganese. Two members of NRAMP family proteins namely NRAMP3 and NRAMP4 have been found in the vacuolar membrane of *A. thaliana* and are believed to be involved in Zn uptake and transport across vacuolar membrane indirectly (Thomine et al. 2003). As it has been shown that iron transport and homeostasis also influence zinc homeostasis. This zinc is stored in vacuoles as zinc bound to organic molecules and not as free zinc ions. Metals form complex primarily with organic acids such as malate and citrate for storage in the vacuoles of the leaf cells of *Arabidopsis halleri* (Cosio et al. 2004). The most abundant complexes with organic acids in *A. halleri* are Zn-citrate and Zn-malate (Cakmak 2008; Ghori et al. 2019). Histidine is another amino acid which makes complexes with zinc for storage. Since vacuoles are also the site for the degradation of proteins and the metals liberated from the degradation of metalloproteins have to be recaptured and reutilized. Vacuoles serve as transient storage site for these metals. Excess zinc may also be excluded from the cells through plasma membrane efflux. In plants, CDF proteins like TgMtp1 of *Thlaspi goesingense* and HMA2 and HMA4 transporters of *Arabidopsis* are implicated in the exclusion of zinc from the cells. Another exporter protein of zinc in *Arabidopsis* belongs to plant cadmium resistance (PCR) proteins which is involved in the cellular export of various divalent metal cations (Song et al. 2004). PCR2 has been found to be involved in zinc homeostasis indirectly as this protein is mainly involved in the export of zinc and iron (Song et al. 2010). But iron affects zinc homeostasis in several ways which are interrelated.

(iv) *Transport from Root to Other Plant Tissues*

The transport of zinc from roots to aerial parts of the plants depends on the degree of mobilization in the rhizospheric region, the efficiency of its movement from the root epidermal cells into the xylem, and the efficiency of its movement in the xylem vessels. The first step of zinc transport from the root to various plant tissues involves the movement of zinc from root epithelial cells into the xylem called as xylem loading (Palmgren et al. 2008; Hussain et al. 2004). This can involve the movement of zinc through the symplast of interconnected cells connecting root epidermal cells to xylem parenchyma finally reaching vascular tissue of xylem (Fig. 9.2). This movement through the interconnected symplast is called as symplastic Zn flux. Alternatively, zinc can be transported through the intercellular spaces which is called as apoplastic Zn flux (Fig. 9.2). The transport through connected cytoplasm of cells occurs through diffusion, but the transport across cell plasma membrane involving metal transporter may require energy and is carried out actively. Therefore, zinc taken up by root epithelial cells is transported through the cell membrane into the apoplast. Heavy metal transporting ATPases (HMA) transport this zinc across the membrane. Similar mechanism is involved in the transport of Zn from xylem parenchyma into xylem apoplast for xylem loading (Palmgren et al. 2008). Studies on *A. thaliana* have shown the involvement of HMA2 and HMA4 in the transport of zinc across the xylem parenchyma cells into vascular xylem for its transport through the shoot (Hussain et al. 2004). RNAi-based experiments on *A. halleri* have confirmed the involvement of HMA4 in hyperaccumulation of Zn (Hanikenne et al. 2008). The zinc taken up by the root can be transported directly through shoot xylem or can be stored in the vacuoles for remobilization and transported when required by other tissues in the plant cell. Zinc can be transported through xylem and elsewhere in the plants either as zinc ion or as ligand-bound zinc (Schaaf et al. 2005; von Wirén et al. 1999). Excessive zinc can also be stored in the shoot cells mainly in vacuoles. These shoot cell vacuoles may store zinc chelation by low molecular weight ligands which may also occur during its passage through the xylem interfering with its uptake and retention by surrounding cells via Zn^{2+} transporters and Zn retention by metal-binding components of the surrounding cell walls. Xylem is largely composed of dead cells and is surrounded by various types of living cells depending on plant species. Leaf cells, phloem cells, and other cells adjacent to dead xylem are living cells and zinc transport into these cells involves transporters. While its further transport in phloem takes place symplastically (Turgeon and Wolf 2009). Proteins like CDF and MTP are also involved in the transport of zinc through shoot xylem. In *Arabidopsis thaliana,* a Zn transporter MTP1 is found in vacuolar membrane which mediates leaf Zn accumulation and transport across the membrane of vacuoles. Nicotianamine is believed to play central role in the transport of zinc through symplast and phloem (Sinclair and Krämer 2012). Metal ions are transported from leaves into the fruit via the phloem.

(v) *Transport of Zinc to Seed and Zinc Homeostasis in Seed*

From the biofortification point of view also, it is important to understand how the zinc travels from roots to the grains. Since the intake and homeostasis of various metals are interconnected, it is also important that beneficial metals are transported to grains while the toxic metals remain in the roots or nonedible part of the plant (Palmgren et al. 2008). As discussed above, the zinc or other metals travel from roots to dead xylem leaving the symplast of the living cells in a process known as xylem loading. From dead phloem, it is again taken up by the adjacent living cells in the shoot taking these metals to the leaves and other plant tissues. Then, the transport from leaves into the fruit takes place via phloem. During its transport to seed again, the zinc has to leave the symplast and the transport involves specialized metal transport proteins. As the embryo and the endosperm of the seed are not connected symplastically, transport across the membrane is required for metal loading into the grains. This transport has to take place twice once for exiting the supplying phloem cells a process which is called as phloem unloading and a second time for its influx into seed or seed loading or grain filling (Fig. 9.1). The distribution of zinc and other metals within different parts of the seed and its movement from one part to another is poorly understood. The structure of a wheat grains shown in Fig. 9.3 typically consists of testa or pericarp which surrounds the seed and acts as nutrients depository, scutellum, and endosperm. The endosperm supplies nutrient for the developing

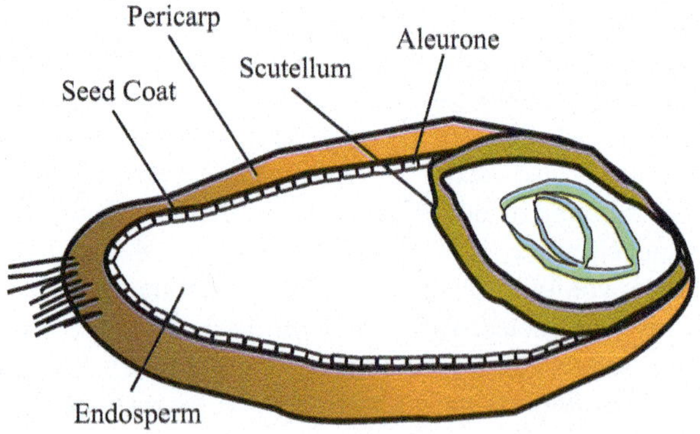

Fig. 9.3 The structure of wheat seed showing location of different tissues

embryo. Aleurone layer surrounds the internal starchy part of the endosperm. The elemental microanalyses of grains have revealed that the highest concentrations of zinc and iron are found in the aleurone and the embryo while the lowest concentrations are in the endosperm (Cakmak 2008; Mazzolini et al. 1985). Zinc is stored in the aleurone and the embryo scutellum in protein storage vacuoles (PSVs). Metallothionein proteins play essential and important role in metal homeostasis in many organisms (Guo et al. 2008). It has been demonstrated that in *A. thaliana,* two metallothionein proteins MT4a and MT4b play important role in the storage of zinc in the seeds (Ren et al. 2012).

Signaling and Regulation of Zinc Homeostasis in Plants

How the cells in plant roots and other tissues sense the change in the zinc status and coordinate the activities of various proteins involved in homeostasis is not well understood. Some studies suggest the role of gene expression in regulating this response. Many of these studies use the heterologous expression of genes in yeast. It was found that genes like *HMA2, NAS,* and *ZIP* are upregulated in zinc deficiency (Ishimaru et al. 2011; Grotz et al. 1998; Pedas et al. 2009). The role of iron-regulated transporter in sensing and regulating zinc homeostasis has also been reported. It was found that transcription level of IRT3 is upregulated in response to Zn deficiency and the levels of the proteins are constitutively high in roots of Zn hyperaccumulators *A. halleri* and *N. caerulescens* (Talke et al. 2006; Becher et al. 2004). In mammalian cells, various response elements like metal responsive elements (MRE) in the promoters of specific genes have been found to respond to the changes in zinc status of the cell (DeMoor et al. 2001; Eide 2006). Binding of Zn to metal responsive elements MTF-1 is known to initiate metallothionein gene transcription and regulate zinc homeostasis in mammalian cell (Andrews 2001). It can be hypothesized that proteins similar to MTF-1 may be involved in the regulation of metal homeostasis in plant. The mechanism in plants therefore is poorly understood requiring future investigations. The work on *A. thaliana* has identified transcription factors bZIP19 and bZIP23 which can bind to zinc deficiency response element (ZDRE) motif (Assunção et al. 2010). In a microarray-based investigation, 23 genes were found to be differentially regulated in response to zinc deficiency, mostly *ZIP* and *NAS*. bZIP19 and bZIP23 orthologs have been identified in poplar, rice, and moss. Furthermore, ZDRE has also been identified in the ZIP gene promoters. Which suggests that ZDRE-mediated mechanism of regulating response to zinc deficiency is widely distributed in vascular plants (Assunção et al. 2010). In an iTRAQ proteomic study, it has been revealed that under the condition of zinc excess, the levels of Zn/H^+-antiporter MTP3 in the vacuolar membrane increase (Arrivault et al. 2006). Another important finding suggests that the signal generated in leaves may influence the expression of various homeostasis proteins in roots, or in other words, the deficiency signals generate in leaves or shoot influence the gene expression in roots to regulate the metal uptake (Grusak and Pezeshgi 1996; Enomoto et al. 2007). In N. tabacum, the application of Fe(II)-EDTA to the leaves or the removal of leaves showing deficiency of iron represses the transcription levels of NtIRT1 and NtFRO1 in the roots. Furthermore, as the homeostasis involves a coordinated activity of a number of proteins including HMA3, IRT1, MTP3, FRO2, and ZIF1, transcriptional levels of the genes coding for these proteins also respond to both deficiency and excess of zinc (Colangelo and Guerinot 2006). The involvement of IRT1 and cadmium-resistant proteins (CRP2) in zinc homeostasis clearly shows that how other metals especially iron influence the zinc homeostasis in plants. Therefore, it is necessary to understand the metal homeostasis in general and how one metal influences the homeostasis of other metal. The understanding is also important, as it can help in selectively improving the uptake of metal with nutritious value and minimizing the uptake of metals that are dangerous for human health.

Toxic Effect of Excess Zinc

As discussed above, plants use a number of mechanisms for maintaining metal/zinc homeostasis. As both low and high concentrations of metals/zinc are deleterious to plant health. Toxicity of high concentrations of zinc and heavy metals is well documented, as these metals are known to induce heavy metal stress and interfere with uptake of other nutrients and enzymatic activities (Yadav 2010; Tkalec et al. 2014; Ghori et al. 2019). Excess of zinc may induce DNA damage in seedlings, decrease the activity of glutathione reductase and superoxide dismutase in roots and seedlings (Tkalec et al. 2014). This may result in inhibition of cell elongation, impaired cell division, wilting and necrosis of the leaves, decline in biomass (Di Baccio et al. 2005; Ruano et al. 1988). It is to be noted that for different plants, the concentration range exhibiting toxicity is different. For example, the toxicity symptoms appear at a concentration of 1000 μM of Zn in *Pisum sativum* while in ryegrass, 50 mM of $ZnSO_4$ causes growth inhibition (Bonnet et al. 2000). Excess of zinc retards various vital metabolic and physiological processes in plants like photosynthesis (Bonnet et al. 2000). Various mechanisms for excess zinc-mediated inhibition of photosynthesis have been documented. Excess of zinc is known to decrease not only the activity of chlorophyll but also the chlorophyll content in plants (Cherif et al. 2010). Decrease in the amount of chlorophyll can be due to decrease in the concentration of Fe-reaching chloroplasts (Bekiaroglou and Karataglis 2002). Due to the high concentration of zinc, Mg^{2+} in the chlorophyll is replaced by Zn^{2+}, impairing the function of light harvesting complex in the chlorophyll (Küpper et al. 1996). Oxygen emission during photosynthesis is also affected as Mn in water splitting complex is also replaced by zinc-inhibiting photolysis (Ralph and Burchett 1998). Zn^{2+} may also inhibit the electron transport during the photosynthesis (Szalontai et al. 1999). Excess of zinc is also known to induce oxidative stress in the plants generating reactive oxygen and nitrogen species such as superoxide radical, nitric oxide, and peroxynitrite (Feigl et al. 2015). Zinc stress stimulates lipoxygenase activities and lipids peroxidation, which activates the plasma membrane-bound NADPH oxidase consequently increasing superoxide anion production (Weckx and Clijsters 2007). Zn^{2+} results in oxidative stress through the stabilization of phenoxyl radicals and the generation of charge transfer complexes. Excess of zinc therefore affects plant growth and development. Inhibition of seed germination due to the higher concentration of zinc is reported for plants like *Cassia angustifolia* and *Vigna unguiculate* (Basha and Selvaraju 2015; Nanda and Agrawal 2016). The development of root and aerial parts of the plants is also reported due to the higher concentration of zinc, which may be due to the inhibition of cell proliferation and elongation (Marichali et al. 2016).

Homeostasis in Zinc Hyperaccumulators

Interestingly at least 15 plant species are known to tolerate high concentration (>10,000 ppm dry weight) of zinc in their aerial parts. When grown with high zinc concentration, *Arabidopsis halleri* and *Noccaea caerulescens* can accumulate extremely high concentrations of Zn of up to 13,620 and 43,710 ppm, respectively (Peer et al. 2006). It is believed that these hyperaccumulators have developed various mechanisms to tolerate high concentrations of zinc. For example, after the zinc uptake by root epithelial cells, root may have mechanism to store excess zinc preventing its movement to other plant tissues. It was observed that the root perendodermis of *N. caerulescens* contains unusual cells with irregularly thickened tangential walls with higher lignin deposition (Zelko et al. 2008). Once the zinc is taken up into the cells symplast, cells must have mechanisms to avoid the toxicity of zinc to other organelles and other intracellular components. Storage of excess zinc in the cell vacuoles is well known. Contrary to this general strategy, hyperaccumulator like *N. caerulescens* is known to store 2.5 times less zinc in root cells vacuole than the sensitive plant species (Lasat et al. 1998). As far as the molecular mechanism of tolerating such high concentration of zinc is concerned. Variation in the expression levels of various proteins involved in zinc homeostasis network plays a significant role as discussed above also (Hanikenne and Nouet 2011; Verbruggen et al. 2009). Transporter and proteins are involved at different stages of zinc uptake and transport. As a first step, zinc is taken up by the plant root, some modification in the root structure and variations in the expression pattern of the transporters have been observed in zinc hyperaccumulators. NcZNT1, a zinc transporter in the plasma membrane of the endodermal cells is constitutively overexpressed in *A. thaliana* increasing the accumulation of Zn and tolerance to its excess (Lin et al. 2016). Similarly, higher levels of other transporters namely IRT3 were found in the roots of *A. halleri* and *N. caerulescens* well-known hyperaccumulators (Talke et al. 2006). The second step in the zinc transport involves the movement of zinc from root to the dead xylem also called as xylem loading. This translocation from the roots to the shoots involves the action of proteins belonging to the heavy metal ATPase "HMA" family. In zinc hyperaccumulator species *A. halleri*, heavy metal ATPases, HMA4 and HMA2, play key role in zinc tolerance and hyperaccumulation (Hanikenne et al. 2008; Milner and Kochian 2008). These HMA are known to promote Zn^{2+} efflux from the root cells for xylem loading (Hussain et al. 2020). Interestingly, the expression of HMA4 was found to be 30-fold higher in the shoots of *H. halleri* compared to *A. thaliana* (Talke et al. 2006). The third step in zinc transport is movement of zinc to other aerial parts of the plant through shoot xylem. Transport of zinc through xylem involves binding with ligands like malate, oxalate, citrate, and succinate. The concentration of these molecules increases in *N. caerulescens* shoots, when the plant is exposed to high concentration of Zn [95]. Finally, the zinc is stored in different plant cells/tissues depending on plant species. As an adaptation strategy, hyperaccumulator plants store small amounts of zinc in mesophyll cells (Cosio et al. 2004). The distribution of zinc in various organelles of the cell and its regulation also are important in the maintenance of homeostasis in hyperaccumulators. Transporter proteins involved in subcellular transport like SaZIP4 of *Sedum alfredii* are key in Zn hyperaccumulation (Yang et al. 2018). The zinc in hyperaccumulators is stored in vacuoles to avoid any damage to other intracellular organelles like chloroplasts. *A. halleri* was also found to contain higher chlorophyll content possibly as a strategy to counter inhibition by high concentration of zinc (Sitko et al. 2017). Studies have demonstrated that when the concentration of zinc is high, HMA1 facilitates Zn^{2+} transport from the chloroplasts to the cytoplasm avoiding damage to the organelle (Kim et al. 2009). Another transport protein known as metal transport protein MTP1 in leaf cell plays a key role in Zn detoxification and tolerance. High levels of this protein in vacuolar membrane in the shoot have been reported in *A. halleri* possibly contributing to Zn accumulation in the shoots (Dräger et al. 2004).

Conclusion

Uptake transport and homeostasis of metals in plants is a very complex and poorly understood mechanism. Further, studies investigating various aspects are required. These studies will provide a basis for improving the micronutrient content of the cereal grains which is emerging as one of the most important issues in relation to human health and well-being. An increased knowledge of biochemical and genetic pathways involved in the homeostasis can help in various ways. It can help in improving the metal uptake from soil even under nutrient-deficient conditions. Transport of metals like zinc from soil to grains also requires further investigations with an aim to improve the process. Furthermore, the uptake of various metals is interrelated and increased uptake of one metal may also result in the accumulation of other undesired metals in the grains as has been described for cadmium which is a dangerous mimic of zinc (Palmgren et al. 2008). With increased knowledge of metal homeostasis and regulatory pathways, it may be possible to devise strategies for selective uptake of desirable micronutrient avoiding metals that are toxic.

Acknowledgments The authors acknowledge the generous financial help provided by the MHRD, Govt. of India, under the Scheme for Promotion of Academic and Research Collaboration (SPARC; P594).

References

Berg, J.M., Shi, Y.: The galvanization of biology: a growing appreciation for the roles of zinc. Science (New York, N.Y.). **271**(5252), 1081–1085 (1996). https://doi.org/10.1126/science.271.5252.1081

Sinclair, S.A., Senger, T., Talke, I.N., Cobbett, C.S., Haydon, M.J., Krämer, U.: Systemic upregulation of MTP2- and HMA2-mediated Zn partitioning to the shoot supplements local Zn deficiency responses. Plant Cell. **30**(10), 2463–2479 (2018). https://doi.org/10.1105/tpc.18.00207

Kisko, M., Bouain, N., Rouached, A., Choudhary, S.P., Rouached, H.: Molecular mechanisms of phosphate and zinc signalling crosstalk in plants: phosphate and zinc loading into root xylem in Arabidopsis. Environ. Exp. Bot. **114**, 57–64 (2015). https://doi.org/10.1016/j.envexpbot.2014.05.013

Helmersson, A., von Arnold, S., Bozhkov, P.V.: The level of free intracellular zinc mediates programmed cell death/cell survival decisions in plant embryos. Plant Physiol. **147**(3), 1158–1167 (2008). https://doi.org/10.1104/pp.108.122598

Broadley, M.R., White, P.J., Hammond, J.P., Zelko, I., Lux, A.: Zinc in plants. New Phytol. **173**(4), 677–702 (2007). https://doi.org/10.1111/j.1469-8137.2007.01996.x

Outten, C.E., O'Halloran, T.V.: Femtomolar sensitivity of metalloregulatory proteins controlling zinc homeostasis. Science (New York, N.Y.). **292**(5526), 2488–2492 (2001). https://doi.org/10.1126/science.1060331

Dittmer, P.J., Miranda, J.G., Gorski, J.A., Palmer, A.E.: Genetically encoded sensors to elucidate spatial distribution of cellular zinc. J. Biol. Chem. **284**(24), 16289–16297 (2009). https://doi.org/10.1074/jbc.M900501200

Hacisalihoglu, G., Hart, J.J., Wang, Y.-H., Cakmak, I., Kochian, L.V.: Zinc efficiency is correlated with enhanced expression and activity of zinc-requiring enzymes in wheat. Plant Physiol. **131**(2), 595–602 (2003). https://doi.org/10.1104/pp.011825

Shanmugam, V., Lo, J.-C., Yeh, K.-C.: Control of Zn uptake in Arabidopsis halleri: a balance between Zn and Fe. Front. Plant Sci. **4**, 281 (2013). https://doi.org/10.3389/fpls.2013.00281

Nkrumah, P.N., Echevarria, G., Erskine, P.D., van der Ent, A.: Contrasting nickel and zinc hyperaccumulation in subspecies of Dichapetalum gelonioides from Southeast Asia. Sci. Rep. **8**(1), 9659 (2018). https://doi.org/10.1038/s41598-018-26859-7

BAKER, A.J.M., REEVES, R.D., HAJAR, A.S.M.: Heavy metal accumulation and tolerance in British populations of the metallophyte Thlaspi caerulescens J. & C. Presl (Brassicaceae). New Phytol. **127**(1), 61–68 (1994). https://doi.org/10.1111/j.1469-8137.1994.tb04259.x

Bert, V., Bonnin, I., Saumitou-Laprade, P., De Laguérie, P., Petit, D.: Do Arabidopsis halleri from nonmetallicolous populations accumulate zinc and cadmium more effectively than those from metallicolous populations? New Phytol. **155**(1), 47–57 (2002). https://doi.org/10.1046/j.1469-8137.2002.00432.x

Sinclair, S.A., Krämer, U.: The zinc homeostasis network of land plants. Biochim. Biophys. Acta. **1823**(9), 1553–1567 (2012). https://doi.org/10.1016/j.bbamcr.2012.05.016

Kobae, Y., Uemura, T., Sato, M.H., Ohnishi, M., Mimura, T., Nakagawa, T., Maeshima, M.: Zinc transporter of Arabidopsis thaliana AtMTP1 is localized to vacuolar membranes and implicated in zinc homeostasis. Plant Cell Physiol. **45**(12), 1749–1758 (2004). https://doi.org/10.1093/pcp/pci015

Dakora, F.D., Phillips, D.A.: Root exudates as mediators of mineral acquisition in low-nutrient environments. Plant Soil. **245**(1), 35–47 (2002). https://doi.org/10.1023/A:1020809400075

Clemens, S.: Molecular mechanisms of plant metal tolerance and homeostasis. Planta. **212**, 475–486 (2001). https://doi.org/10.1007/s004250000458

Mench, M., Martin, E.: Mobilization of cadmium and other metals from two soils by root exudates of Zea mays L., Nicotiana tabacum L. and Nicotiana rustica L. Plant Soil. **132**(2), 187–196 (1991). https://doi.org/10.1007/BF00010399

Rose, T.J., Impa, S.M., Rose, M.T., Pariasca-Tanaka, J., Mori, A., Heuer, S., Johnson-Beebout, S.E., Wissuwa, M.: Enhancing phosphorus and zinc acquisition efficiency in rice: a critical review of root traits and their potential utility in rice breeding. Ann. Bot. **112**(2), 331–345 (2012). https://doi.org/10.1093/aob/mcs217

Palmgren, M.G., Clemens, S., Williams, L.E., Krämer, U., Borg, S., Schjørring, J.K., Sanders, D.: Zinc biofortification of cereals: problems and solutions. Trends Plant Sci. **13**(9), 464–473 (2008). https://doi.org/10.1016/j.tplants.2008.06.005

von Wiren, N., Marschner, H., Romheld, V.: Roots of iron-efficient maize also absorb phytosiderophore-chelated zinc. Plant Physiol. **111**(4), 1119–1125 (1996). https://doi.org/10.1104/pp.111.4.1119

Guerinot, M.L.: The ZIP family of metal transporters. Biochim. Biophys. Acta. **1465**(1–2), 190–198 (2000). https://doi.org/10.1016/s0005-2736(00)00138-3

Olsen, L., Palmgren, M.: Many rivers to cross: the journey of zinc from soil to seed. Front. Plant Sci. **5**, 30 (2014). https://doi.org/10.3389/fpls.2014.00030

Kochian, L.V.: Zinc absorption from hydroponic solutions by plant roots. In: Robson, A.D. (ed.) Zinc in Soils and Plants: Proceedings of the International Symposium on 'Zinc in Soils and Plants' Held at the University of Western Australia, 27–28 September, 1993, pp. 45–57. Springer Netherlands, Dordrecht (1993). https://doi.org/10.1007/978-94-011-0878-2_4

Ishimaru, Y., Bashir, K., Nishizawa, N.K.: Zn uptake and translocation in rice plants. Rice. **4**(1), 21–27 (2011). https://doi.org/10.1007/s12284-011-9061-3

Korshunova, Y.O., Eide, D., Gregg Clark, W., Lou Guerinot, M., Pakrasi, H.B.: The IRT1 protein from Arabidopsis thaliana is a metal transporter with a broad substrate range. Plant Mol. Biol. **40**(1), 37–44 (1999). https://doi.org/10.1023/A:1026438615520

Vert, G., Grotz, N., Dédaldéchamp, F., Gaymard, F., Guerinot, M.L., Briat, J.-F., Curie, C.: IRT1, an Arabidopsis transporter essential for iron uptake from the soil and for plant growth. Plant Cell. **14**(6), 1223–1233 (2002). https://doi.org/10.1105/tpc.001388

Murata, Y., Ma, J.F., Yamaji, N., Ueno, D., Nomoto, K., Iwashita, T.: A specific transporter for iron(III)–phytosiderophore in barley roots. Plant J. **46**(4), 563–572 (2006). https://doi.org/10.1111/j.1365-313X.2006.02714.x

Roberts, L.A., Pierson, A.J., Panaviene, Z., Walker, E.L.: Yellow stripe1. Expanded roles for the maize iron-phytosiderophore transporter. Plant Physiol. **135**(1), 112–120 (2004). https://doi.org/10.1104/pp.103.037572

Schaaf, G., Schikora, A., Häberle, J., Vert, G., Ludewig, U., Briat, J.F., Curie, C., von Wirén, N.: A putative function for the arabidopsis Fe-Phytosiderophore transporter homolog AtYSL2 in Fe and Zn homeostasis. Plant Cell Physiol. **46**(5), 762–774 (2005). https://doi.org/10.1093/pcp/pci081

von Wirén, N., Klair, S., Bansal, S., Briat, J.-F., Khodr, H., Shioiri, T., Leigh, R.A., Hider, R.C.: Nicotianamine chelates both Fe^{III} and Fe^{II}. Implications for metal transport in plants. Plant Physiol. **119**(3), 1107–1114 (1999). https://doi.org/10.1104/pp.119.3.1107

Zhou, J., Goldsbrough, P.B.: Functional homologs of fungal metallothionein genes from Arabidopsis. Plant Cell. **6**(6), 875–884 (1994). https://doi.org/10.1105/tpc.6.6.875

Desbrosses-Fonrouge, A.G., Voigt, K., Schröder, A., Arrivault, S., Thomine, S., Krämer, U.: Arabidopsis thaliana MTP1 is a Zn transporter in the vacuolar membrane which mediates Zn detoxification and drives leaf Zn accumulation. FEBS Lett. **579**(19), 4165–4174 (2005). https://doi.org/10.1016/j.febslet.2005.06.046

Thomine, S., Lelièvre, F., Debarbieux, E., Schroeder, J.I., Barbier-Brygoo, H.: AtNRAMP3, a multispecific vacuolar metal transporter involved in plant responses to iron deficiency. Plant J. **34**(5), 685–695 (2003). https://doi.org/10.1046/j.1365-313x.2003.01760.x

Cosio, C., Martinoia, E., Keller, C.: Hyperaccumulation of cadmium and zinc in Thlaspi caerulescens and Arabidopsis halleri at the leaf cellular level. Plant Physiol. **134**(2), 716–725 (2004). https://doi.org/10.1104/pp.103.031948

Song, W.Y., Martinoia, E., Lee, J., Kim, D., Kim, D.Y., Vogt, E., Shim, D., Choi, K.S., Hwang, I., Lee, Y.: A novel family of cys-rich membrane proteins mediates cadmium resistance in Arabidopsis. Plant Physiol. **135**(2), 1027–1039 (2004). https://doi.org/10.1104/pp.103.037739

Song, W.Y., Choi, K.S., Kim, D.Y., Geisler, M., Park, J., Vincenzetti, V., Schellenberg, M., Kim, S.H., Lim, Y.P., Noh, E.W., Lee, Y., Martinoia, E.: Arabidopsis PCR2 is a zinc exporter involved in both zinc extrusion and long-distance zinc transport. Plant Cell. **22**(7), 2237–2252 (2010). https://doi.org/10.1105/tpc.109.070185

Hussain, D., Haydon, M.J., Wang, Y., Wong, E., Sherson, S.M., Young, J., Camakaris, J., Harper, J.F., Cobbett, C.S.: P-type ATPase heavy metal transporters with roles in essential zinc homeostasis in Arabidopsis. Plant Cell. **16**(5), 1327–1339 (2004). https://doi.org/10.1105/tpc.020487

Hanikenne, M., Talke, I.N., Haydon, M.J., Lanz, C., Nolte, A., Motte, P., Kroymann, J., Weigel, D., Krämer, U.: Evolution of metal hyperaccumulation required cis-regulatory changes and triplication of HMA4. Nature. **453**(7193), 391–395 (2008). https://doi.org/10.1038/nature06877

Turgeon, R., Wolf, S.: Phloem transport: cellular pathways and molecular trafficking. Annu. Rev. Plant Biol. **60**, 207–221 (2009). https://doi.org/10.1146/annurev.arplant.043008.092045

Cakmak, I.: Enrichment of cereal grains with zinc: agronomic or genetic biofortification? Plant Soil. **302**(1), 1–17 (2008). https://doi.org/10.1007/s11104-007-9466-3

Mazzolini, A.P., Pallaghy, C.K., Legge, G.J.F.: Quantitative microanalysis of MN, ZN and other elements in mature wheat seed. New Phytol. **100**(4), 483–509 (1985). https://doi.org/10.1111/j.1469-8137.1985.tb02796.x

Guo, W.-J., Meetam, M., Goldsbrough, P.B.: Examining the specific contributions of individual Arabidopsis metallothioneins to copper distribution and metal tolerance. Plant Physiol. **146**(4), 1697–1706 (2008). https://doi.org/10.1104/pp.108.115782

Ren, Y., Liu, Y., Chen, H., Li, G., Zhang, X., Zhao, J.: Type 4 metallothionein genes are involved in regulating Zn ion accumulation in late embryo and in controlling early seedling growth in Arabidopsis. Plant Cell Environ. **35**(4), 770–789 (2012). https://doi.org/10.1111/j.1365-3040.2011.02450.x

Grotz, N., Fox, T., Connolly, E., Park, W., Guerinot, M.L., Eide, D.: Identification of a family of zinc transporter genes from Arabidopsis that respond to zinc deficiency. Proc. Natl. Acad. Sci. U. S. A. **95**(12), 7220–7224 (1998). https://doi.org/10.1073/pnas.95.12.7220

Pedas, P., Schjoerring, J.K., Husted, S.: Identification and characterization of zinc-starvation-induced ZIP transporters from barley roots. Plant Physiol. Biochem. **47**(5), 377–383 (2009). https://doi.org/10.1016/j.plaphy.2009.01.006

Talke, I.N., Hanikenne, M., Krämer, U.: Zinc-dependent global transcriptional control, transcriptional deregulation, and higher gene copy number for genes in metal homeostasis of the hyperaccumulator *Arabidopsis halleri*. Plant Physiol. **142**(1), 148–167 (2006). https://doi.org/10.1104/pp.105.076232

Becher, M., Talke, I.N., Krall, L., Krämer, U.: Cross-species microarray transcript profiling reveals high constitutive expression of metal homeostasis genes in shoots of the zinc hyperaccumulator Arabidopsis halleri. Plant J. **37**(2), 251–268 (2004). https://doi.org/10.1046/j.1365-313X.2003.01959.x

DeMoor, J.M., Kennette, W.A., Collins, O.M., Koropatnick, J.: Zinc-metallothionein levels are correlated with enhanced glucocorticoid responsiveness in mouse cells exposed to ZnCl2, HgCl2, and heat shock. Toxicol. Sci. **64**(1), 67–76 (2001). https://doi.org/10.1093/toxsci/64.1.67

Eide, D.J.: Zinc transporters and the cellular trafficking of zinc. Biochim. Biophys. Acta. **1763**(7), 711–722 (2006). https://doi.org/10.1016/j.bbamcr.2006.03.005

Andrews, G.K.: Cellular zinc sensors: MTF-1 regulation of gene expression. In: Maret, W. (ed.) Zinc Biochemistry, Physiology, and Homeostasis: Recent Insights and Current Trends, pp. 37–51. Springer Netherlands, Dordrecht (2001). https://doi.org/10.1007/978-94-017-3728-9_3

Assunção, A.G., Herrero, E., Lin, Y.F., Huettel, B., Talukdar, S., Smaczniak, C., Immink, R.G., van Eldik, M., Fiers, M., Schat, H., Aarts, M.G.: Arabidopsis thaliana transcription factors bZIP19 and bZIP23 regulate the adaptation to zinc deficiency. Proc. Natl. Acad. Sci. U. S. A. **107**(22), 10296–10301 (2010). https://doi.org/10.1073/pnas.1004788107

Arrivault, S., Senger, T., Krämer, U.: The Arabidopsis metal tolerance protein AtMTP3 maintains metal homeostasis by mediating Zn exclusion from the shoot under Fe deficiency and Zn oversupply. Plant J. **46**(5), 861–879 (2006). https://doi.org/10.1111/j.1365-313X.2006.02746.x

Grusak, M.A., Pezeshgi, S.: Shoot-to-root signal transmission regulates root Fe(III) reductase activity in the dgl mutant of pea. Plant Physiol. **110**(1), 329–334 (1996). https://doi.org/10.1104/pp.110.1.329

Enomoto, Y., Hodoshima, H., Shimada, H., Shoji, K., Yoshihara, T., Goto, F.: Long-distance signals positively regulate the expression of iron uptake genes in tobacco roots. Planta. **227**(1), 81–89 (2007). https://doi.org/10.1007/s00425-007-0596-x

Colangelo, E.P., Guerinot, M.L.: Put the metal to the petal: metal uptake and transport throughout plants. Curr. Opin. Plant Biol. **9**(3), 322–330 (2006). https://doi.org/10.1016/j.pbi.2006.03.015

Yadav, S.K.: Heavy metals toxicity in plants: an overview on the role of glutathione and phytochelatins in heavy metal stress tolerance of plants. S. Afr. J. Bot. **76**(2), 167–179 (2010). https://doi.org/10.1016/j.sajb.2009.10.007

Tkalec, M., Stefanić, P.P., Cvjetko, P., Sikić, S., Pavlica, M., Balen, B.: The effects of cadmium-zinc interactions on biochemical responses in tobacco seedlings and adult plants. PLoS One. **9**(1), e87582–e87582 (2014). https://doi.org/10.1371/journal.pone.0087582

Ghori, N.H., Ghori, T., Hayat, M.Q., Imadi, S.R., Gul, A., Altay, V., Ozturk, M.: Heavy metal stress and responses in plants. Int. J. Environ. Sci. Technol. **16**(3), 1807–1828 (2019). https://doi.org/10.1007/s13762-019-02215-8

Di Baccio, D., Kopriva, S., Sebastiani, L., Rennenberg, H.: Does glutathione metabolism have a role in the defence of poplar against zinc excess? New Phytol. **167**(1), 73–80 (2005). https://doi.org/10.1111/j.1469-8137.2005.01462.x

Ruano, A., Poschenrieder, C., Barceló, J.: Growth and biomass partitioning in zinc-toxic bush beans. J. Plant Nutr. **11**(5), 577–588 (1988). https://doi.org/10.1080/01904168809363824

Bonnet, M., Camares, O., Veisseire, P.: Effects of zinc and influence of Acremonium lolii on growth parameters, chlorophyll a fluorescence and antioxidant enzyme activities of ryegrass (Lolium perenne L. cv Apollo). J. Exp. Bot. **51**(346), 945–953 (2000). https://doi.org/10.1093/jxb/51.346.945

Cherif, J., Derbel, N., Nakkach, M., Bergmann, H., Jemal, F., Lakhdar, Z.B.: Analysis of in vivo chlorophyll fluorescence spectra to monitor physiological state of tomato plants growing under zinc stress. J. Photochem. Photobiol. B. **101**(3), 332–339 (2010). https://doi.org/10.1016/j.jphotobiol.2010.08.005

Bekiaroglou, P., Karataglis, S.: The effect of lead and zinc on Mentha spicata. J. Agron. Crop Sci. **188**(3), 201–205 (2002). https://doi.org/10.1046/j.1439-037X.2002.00559.x

Küpper, H., Küpper, F., Spiller, M.: Environmental relevance of heavy metal-substituted chlorophylls using the example of water plants. J. Exp. Bot. **47**(2), 259–266 (1996). https://doi.org/10.1093/jxb/47.2.259

Ralph, P.J., Burchett, M.D.: Photosynthetic response of Halophila ovalis to heavy metal stress. Environ. Pollut. **103**(1), 91–101 (1998). https://doi.org/10.1016/S0269-7491(98)00121-3

Szalontai, B., Horváth, L.I., Debreczeny, M., Droppa, M., Horváth, G.: Molecular rearrangements of thylakoids after heavy metal poisoning, as seen by Fourier transform infrared (FTIR) and electron spin resonance (ESR) spectroscopy. Photosynth. Res. **61**(3), 241–252 (1999). https://doi.org/10.1023/A:1006345523919

Feigl, G., Lehotai, N., Molnár, Á., Ördög, A., Rodríguez-Ruiz, M., Palma, J.M., Corpas, F.J., Erdei, L., Kolbert, Z.: Zinc induces distinct changes in the metabolism of reactive oxygen and nitrogen species (ROS and RNS) in the roots of two Brassica species with different sensitivity to zinc stress. Ann. Bot. **116**(4), 613–625 (2015). https://doi.org/10.1093/aob/mcu246

Weckx, J.E.J., Clijsters, H.: Zn phytotoxicity induces oxidative stress in primary leaves of Phaseolus vulgaris. Plant Physiol. Biochem. **35** (2007)

Basha, S.A., Selvaraju, M.: Toxic effect of zinc on growth and nutrient accumulation of cow pea (Vigna unguiculata L.). Int. Lett. Nat. Sci. **43**, 48–53 (2015). https://doi.org/10.18052/www.scipress.com/ILNS.43.48

Nanda, R., Agrawal, V.: Elucidation of zinc and copper induced oxidative stress, DNA damage and activation of defence system during seed germination in Cassia angustifolia Vahl. Environ. Exp. Bot. **125**, 31–41 (2016). https://doi.org/10.1016/j.envexpbot.2016.02.001

Marichali, A., Dallali, S., Ouerghemmi, S., Sebei, H., Casabianca, H., Hosni, K.: Responses of Nigella sativa L. to zinc excess: focus on germination, growth, yield and yield components, lipid and terpene metabolism, and total phenolics and antioxidant activities. J. Agric. Food Chem. **64**(8), 1664–1675 (2016). https://doi.org/10.1021/acs.jafc.6b00274

Peer, W.A., Mahmoudian, M., Freeman, J.L., Lahner, B., Richards, E.L., Reeves, R.D., Murphy, A.S., Salt, D.E.: Assessment of plants from the Brassicaceae family as genetic models for the study of nickel and zinc hyperaccumulation. New Phytol. **172**(2), 248–260 (2006). https://doi.org/10.1111/j.1469-8137.2006.01820.x

Zelko, I., Lux, A., Czibula, K.: Difference in the root structure of hyperaccumulator Thlaspi caerulescens and non-hyperaccumulator Thlaspi arvense. Int. J. Environ. Poll. **33** (2008). https://doi.org/10.1504/IJEP.2008.019387

Lasat, M.M., Baker, A.J.M., Kochian, L.V.: Altered Zn compartmentation in the root symplasm and stimulated Zn absorption into the leaf as mechanisms involved in Zn hyperaccumulation in *Thlaspi caerulescens*. Plant Physiol. **118**(3), 875–883 (1998). https://doi.org/10.1104/pp.118.3.875

Hanikenne, M., Nouet, C.: Metal hyperaccumulation and hypertolerance: a model for plant evolutionary genomics. Curr. Opin. Plant Biol. **14**(3), 252–259 (2011). https://doi.org/10.1016/j.pbi.2011.04.003

Verbruggen, N., Hermans, C., Schat, H.: Molecular mechanisms of metal hyperaccumulation in plants. New Phytol. **181**(4), 759–776 (2009). https://doi.org/10.1111/j.1469-8137.2008.02748.x

Lin, Y.-F., Hassan, Z., Talukdar, S., Schat, H., Aarts, M.G.M.: Expression of the ZNT1 zinc transporter from the metal hyperaccumulator Noccaea caerulescens confers enhanced zinc and cadmium tolerance and accumulation to Arabidopsis thaliana. PLoS One. **11**(3), e0149750 (2016). https://doi.org/10.1371/journal.pone.0149750

Milner, M.J., Kochian, L.V.: Investigating heavy-metal hyperaccumulation using Thlaspi caerulescens as a model system. Ann. Bot. **102**(1), 3–13 (2008). https://doi.org/10.1093/aob/mcn063

Hussain, A., Zahir, Z., Ditta, A., Tahir, M., Ahmad, M., Mumtaz, M., Hayat, K., Hussain, S.: Production and implication of bio-activated organic fertilizer enriched with zinc-solubilizing bacteria to boost up maize (Zea mays L.) production and biofortification under two cropping seasons. Agron. J. **39**, 1–18 (2020). https://doi.org/10.3390/agronomy10010039

Yang, Q., Ma, X., Luo, S., Gao, J., Yang, X., Feng, Y.: SaZIP4, an uptake transporter of Zn/Cd hyperaccumulator Sedum alfredii Hance. Environ. Exp. Bot. **155**, 107–117 (2018). https://doi.org/10.1016/j.envexpbot.2018.06.021

Sitko, K., Rusinowski, S., Kalaji, H.M., Szopiński, M., Małkowski, E.: Photosynthetic efficiency as bioindicator of environmental pressure in *A. halleri*. Plant Physiol. **175**(1), 290–302 (2017). https://doi.org/10.1104/pp.17.00212

Kim, Y.-Y., Choi, H., Segami, S., Cho, H.-T., Martinoia, E., Maeshima, M., Lee, Y.: AtHMA1 contributes to the detoxification of excess Zn(II) in Arabidopsis. Plant J. **58**(5), 737–753 (2009). https://doi.org/10.1111/j.1365-313X.2009.03818.x

Dräger, D.B., Desbrosses-Fonrouge, A.-G., Krach, C., Chardonnens, A.N., Meyer, R.C., Saumitou-Laprade, P., Krämer, U.: Two genes encoding Arabidopsis halleri MTP1 metal transport proteins co-segregate with zinc tolerance and account for high MTP1 transcript levels. Plant J. **39**(3), 425–439 (2004). https://doi.org/10.1111/j.1365-313X.2004.02143.x

Chapter 10
Microbiome and Plant Nutritional Value: Current Knowledge and Management

Alexandre Jousset and Mohammadhossein Ravanbakhsh

Abstract Nutrient deficiencies are a major issue worldwide. Microorganisms can affect plant nutrition and may form the backbone of fortification strategies ensuring high nutritional values of grains and vegetables. However, to date it remains hard to harness this power. We surmise that this difficulty stems from the multiple different mechanisms involved in nutrition. Here, we review the different ways that microorganisms use to impact plant growth. (1) Nutrient mineralization, enzymes, and siderophores. (2) Nutrient mobilization, for instance, arbuscular mycorrhizal fungi (AMF). (3) Manipulation of plant nutrient uptake, by plant hormones or production of physical layer around the roots. (4) Manipulation of plant nutrient translocation. We will review here how these different mechanisms are intertwined in the soil microbiome and how they may serve as multiple leverage points to enhance microbiome function and plant nutrition. Together, these mechanisms may form the base of a sustainable and nutritious agriculture.

Keywords Fortified foods · Plant microbiome and nutrition · Soil degradation · keynonte species · Integrated techniques

The Global Challenge of Hidden Hunger

The worldwide increase in agricultural yield fueled by the green revolution has led to a massive reduction in undernutrition. However, while caloric intake has been steadily rising during the last decades, the problem of malnutrition remains rampant in several regions of the world. Among the many deficiencies, insufficient micronutrient uptake is the source of what has come to be called the "hidden hunger." Despite eating sufficiently, large swaths of the human population are suffering from

A. Jousset (✉) · M. Ravanbakhsh (✉)
Ecology and Biodiversity Group, Institute of Environmental Biology, Utrecht University, Utrecht, The Netherlands
e-mail: A.L.C.Jousset@uu.nl; m.ravanbakhsh@uu.nl

S. T. Khan, A. Malik (eds.), *Microbial Biofertilizers and Micronutrient Availability*,
https://doi.org/10.1007/978-3-030-76609-2_10

health issues associated with deficiencies in elements such as zinc, iron, iodine, selenium, or copper (Bamji et al. 2020). These elements are often referred to as micronutrients or trace elements, as they are required only in tiny amounts in the alimentation, while being toxic at high doses. For instance, iron is required for hemoglobin formation, iodine for proper thyroid function, and zinc for a range of enzymatic processes involved in immunity and neuronal development (Shenkin 2006). Despite many efforts, it is estimated that up to two billion people are suffering from one or more deficiencies. While most cases used to cluster in developing countries, industrialized nations also bear an increasing share in the hidden hunger. Decades of growing societal inequalities have left large swaths of the populations of otherwise prosperous countries unable to afford a proper alimentation. This effect is particularly apparent in Anglo-Saxon countries, which tend to harbor an antisocial stance and boast rules openly tilted against workers. However, the issue is truly global and malnutrition is also worrisomely getting more prevalent in many prosperous economies (Shifler Bowers et al. 2018).

Hidden hunger is particularly pernicious, as it primarily affects children, crippling their physical and intellectual development. At a personal level, malnutrition in childhood may cause lifelong health issues. At a society level, it may affect whole economies and bog them in poverty by hampering social mobility and depriving countries of their potential pool of talents. Hidden hunger may be alleviated, thanks to recent technological advances. In this chapter, we will focus on the central importance of the plant-associated microbiome as the driver of plant nutritional value. Several bacteria and fungi living within and around plants can improve plant micronutrient and vitamin content. After presenting the main types of mechanisms, we will discuss how microorganisms can be implemented in sustainable, affordable, and upscalable strategies to enhance the intake of essential elements by humans.

Loss of Plant Nutritional Value

Plants bring 80% of the worldwide caloric intake in the form of carbohydrates and lipids. In addition, plant-derived aliments are an essential source of proteins, micronutrients, and vitamins. The green revolution has since the 1960s maximized yield and has brought an unprecedented increase of per capita caloric intake. However, this yield increase has been associated with creeping malnutrition issues. For instance, some studies suggest that the nutritional value of several vegetables has dropped sharply since the 1950s, by up to 80% for key minerals such as calcium (Davis et al. 2004). The causes of these reductions are not fully clear, but most likely include a combination of breeding for yield rather than quality and soil exhaustion in response to incomplete fertilization. These early findings have been relativized recently, but a dilution effect does seem to occur in high-yield cultivars (Marles 2017).

Plant Breeding Has Focused on Yield, Not Nutritional Value

Over the past few decades, breeding efforts mainly focus on addressing the quantity of the food by the huge improvement in staple crops. It provided a reasonable amount of food available for the earth's population. However, the increase in the total food supply has not coincided with the increase in micronutrient content, and contributed to cheaper staple food, and depleted soil in terms of soil nutrients. It left the world population with products at a reasonable price, albeit at the cost of low levels of micronutrient. Breeding of new efficient cultivars is restricted more by the passage of time due to reduction in plant genetic pools. Besides, it is tedious, expensive, and economically not viable.

Soil Degradation

Soil degradation is a growing global issue that is not likely to be solved anyway soon (Borrelli et al. 2017). More and more soils now have depleted nutrients due to unsustainable practices, including NPK fertilization without micronutrient supply. Achieving better plant nutrition is further impaired by soil properties, with elements such as Fe and Zn being inherently unavailable for plants in the calcareous soils, despite a huge load of fertilization. Using chelates and nanofertilizers has addressed part of the limitation, but the cost of application is still a barrier.

Existing Solutions

Fortified Foods

Food fortification is commonly used as a strategy to increase the micronutrient uptake in the diet, even when the original food is of low nutritional value or if socio-economic conditions hamper access to a varied and healthy diet. The fortified food is enriched artificially with the vitamin and mineral. While this strategy is widely used in industrialized countries to improve food quality, there is doubt about the advantage of long-term fortification due to its comparatively high costs, difficulties to implement in the production chain, and its indirect contribution to a less diverse alimentation.

Breeding

Breeding new crops with an enhanced nutritional value is in principle an elegant solution to hidden hunger, but it is hampered by a range of technical and regulatory issues. Breeding more nutritious plants is barely achievable with random mutagenesis techniques, due to the tremendous amount of resources and time needed to select and propagate improved mutants. On the other hand, targeted genome-editing techniques, such as CRISPR-CAS9, which may rapidly improve the functional genes involved in nutrition are de facto forbidden in many parts of the world, calling for alternatives.

The Microbiome: The Plant Auxiliary Genetic Pool

The microbiome may be just the toolbox needed to reliably improve crop nutrition. There is a growing awareness that plant-associated microbiomes are a major codeterminant of several plant traits, including nutrient content (Ravanbakhsh et al. 2018, 2020). In the next sections, we will present the main mechanisms through which plant-associated bacteria and fungi coregulate plant nutrient uptake and how they may be used to boost nutrient uptake and vitamin synthesis.

Plants are associated with a specific assemblage of bacteria, fungi, archaea, and protists. Together, these organisms are referred to as microbiomes and have long been recognized to play a role in plant development (Berendsen et al. 2012; Berg et al. 2016; Mendes et al. 2011). Recently, microbiome research has moved from a perspective seeking to isolate and characterize as many beneficial microorganisms (the classical plant growth-promoting (PGPR) bacteria concept) (Lugtenberg and Kamilova 2009) and embrace the community as a whole assemblage of interacting and evolving organisms (Chaparro et al. 2013; Saleem et al. 2019). Their associated microorganisms form such intricate associations that they are increasingly seen as a holobiont, a single ecological and evolutionary unit (Vannier et al. 2019). Recent studies have even shown that plant- and microbial-encoded genes are so intertwined that a given phenotype can be reached by either altering the plant- or microbe-encoded genes involved in the target pathway. One of the best examples for this melting of plant and bacterial traits is ethylene signaling, a hormone involved not only in plant stress tolerance but also in nutrient uptake, is co-regulated by plants and microorganisms. Several bacteria parasitize the plant by using ACC, the precursor of ethylene, as a source of nitrogen. Removing the ACC deaminase gene in model one plant and one bacteria holobiont led to the same phenotype as overexpressing plant ACC synthesis in plants. Furthermore, mutations in the plant genome could even be rescued and the wild-type (WT) phenotype is reestablished by introducing a "mirror mutation" in the plant-associated microorganisms. Strangely enough, ACC deaminase producing bacteria are often touted as PGPR, despite the findings above clearly placing them as parasites hampering plant response to stress as well as decreasing nutritional value. This discrepancy suggests that a deep revamping and revalidation of the "plant growth promoting" gene catalog may be needed to efficiently improve microbiomes in the future.

Associated microorganisms can have multidimensional effects on the plant nutritional value. They can provide some function for the plants which plant lack, for example, fixation of nitrogen, solubilization, and mineralization of elements. In addition, the microorganisms can alter the plant's nutritional value by modification of the plant phenotype and hormonal balance. Diazotroph bacteria can fix N_2 and convert it to ammonia, which can be taken up by plants directly. Fungi cannot directly fix N_2 and make it available for the plant; otherwise, they can provide growth-limiting elements (e.g., carbon and phosphorus) for plants and N_2-fixing bacteria. Most of the micronutrients are unavailable due to specific conditions of the soil chemistry, for example, unbalanced pH, or the presence of precipitating ions such as Ca, and Fe-oxides. The microorganisms are able to increase the mineralization of the elements, for example by secreting phosphorus and zinc solubilizing enzyme, siderophores, and organic acids. Several microorganisms are able to produce the siderophore, natural metabolites that chelate the Fe and make it available for the plants. Microorganisms can create a specific environment around the plants and can fix the micronutrient from leaching. The plants will therefore be able to uptake the nutrient over time. Arbuscular mycorrhizal fungi create a complicated mycelial network to fix the phosphorus and transfer it to the plant. The mycelia also facilitate solubilization and transport of P, secreting siderophores, N assimilation, and CO_2 release (Rashid et al. 2016).

The associated microorganism can alter the plant's nutritional value by altering the plant hormones and regulators. The microbiota produces and secretes hormones, responds to host hormones, and regulates the expression levels of host hormones. Several plant hormones are known to alter the plant's nutritional value. For example, microbial IAA activates several pathways connected to nutrient uptake and localization (Etesami et al. 2015). Ethylene is a plant regulator that is important in several aspects of plant growth and development from germination to senescence. Ethylene is also an important hormone that regulates plant nutritional value. Ethylene induces the plant transporter which is important in the uptake and transport of micronutrients. Inducing ethylene by associated microorganisms improves the plant micronutrient uptake at the same magnitude as the plant genome modification (Ravanbakhsh et al. 2020).

Factors Affecting Microbiome Function

Exposure to stress conditions reduces the concentration of nutrients in the plant tissues. For example, drought condition (Bista et al. 2018), salinity (Huergo et al. 2018), and high temperature (Giri et al. 2017) reduce the supply and uptake of nutrients by plants, directly by altering the plant growth and development and indirectly by altering the plant surrounding (e.g., reducing the diffusion of nutrients via reducing the water source). Drought events, for example, reduce the protein and enzyme crucial for the nitrogen and phosphorus uptake and transport (Bista et al. 2018), and reduce the soil microbial activity (Sanaullah et al. 2012) and diversity. It is important because, a higher microbial diversity is important in providing a wide variety of ecosystem services, including the acquisition of more nutrients by plants (El Mujtar et al. 2019). Different species of microorganisms improve plant fertility by increasing the decomposition of the organic matter, improving the soil's physical condition (Rashid et al. 2016), and realizing the nutrient by producing the chelating agent, acidification, and ion exchange (Singh and Prasanna 2020).

Microbiome Management for High Nutritional Value

The microbiome, with its thousands of fungal, bacterial, viral, and protist species, offers a bounty of functional genes that may serve to increase plant nutritional value. However, the very complexity of the system makes it also difficult to manipulate. We will present some of the existing techniques, with their advantages and drawbacks, and conclude with a starter about potential future advancements.

Amendments (Untargeted) Application of organic fertilizers such as manure, compost, or seaweed is an ancestral way of improving soils. The recent development in fermentation techniques has led to better products with more controlled nutrient content and microbial composition. Soil amendments can improve microbiome composition by selecting functional genes, as it is the case for the chitin amendment, which selects for bacteria producing the lytic enzyme chitinase, involved in both plant health and nutrient cycling (Cretoiu et al. 2013). The addition of organic matter also provides additional resources allowing for a more diverse and complex microbial food web to build up, providing niches for phagotrophic species involved in nutrient cycling (Xiong et al. 2018). Besides, nutrients stored in organic material, such as nitrogen as protein or phosphorus as phytate, often need to be degraded enzymatically before becoming available to the plant. This ensures a given stability, reduces leaching but also means that plant nutrition is only guaranteed with the right microorganisms active around them. This results in compost being a cheap, environment-friendly soil improvement but one that alone is often not enough to reliably restore soils.

Microbial Species Introductions

Plants are surrounded by a specific microbiome, whose members contain many functional genes directly relevant to plant nutrition. One of the first attempts to harness this microbiome has been to introduce species with the desired functional genes into the soil, hoping that they will carry out the expected functions.

This strategy has long been attempted and works well under laboratory conditions, in which one plant and one bacterium are placed in interaction. However, except in a very few cases in which the introduced species carry out a unique function, such as host-specific nitrogen-fixing rhizobia, introducing single species is notoriously unreliable under field conditions. We ascribe this limitation to a range of factors. First of them, we may have well been looking for the wrong traits. Research has focused on molecular mechanisms underlying plant-microbe interaction in simplified systems. While this has allowed characterizing many potential functional genes, very little is known about their relevance in a multispecies microbiome. Further, the traits required for bacterial survival are notoriously overlooked in plant-microbe interaction studies. As a result, bacteria may well carry a given functional gene but will, once introduced in soil, die out before having been able to express it (Xiong et al. 2017). Ultimately, the environmental conditions in different soils are so diverse that no single bacterial strain is likely to strive in all of them.

A potential improvement may come from the use of consortia of different bacteria, complementing each other to build up a more reliable impact on the microbiome. Different species sharing a single functional gene but otherwise harboring different ecological preferences may complement each other, allowing for what is referred to as complementarity effects in community ecology. The key underlying process is that any single species will only be able to use a partial subset of the environmental niches present in the rhizosphere. As such, mixtures of two or more species may be able to catch a larger proportion of the available niches and strive beyond what any single species may achieve. In the context of the rhizosphere environment, imagine an example of two species, one adapted to root tips and one to older parts of the root systems. In isolation, each of them will only reach a limited population density. Together, they may thoroughly colonize wide swaths of the root system, thereby increasing the abundance of the functional genes they have been selected to carry (Saleem et al. 2019). A similar perspective could be used by combining species from early and late plant growth stages, since they harbor key differences in life history strategies and may be competitive at different periods of plant growth (Hu et al. 2020). Another approach would be to use consortia to include multiple functional genes involved in plant nutrition. We speculate here that introducing a mixture of bacteria producing auxin and ethylene may ideally improve plant nutrition, since these two hormones are synergistic in their activity (Ravanbakhsh et al. 2020). However, this approach only partially alleviates one of the main constraints placed on microbial introduction: The microbiome tends to be quite resistant to invasion by new species. As such, even the best consortia will still reach a limited abundance in the rhizosphere and ultimately decline (Hu et al. 2016).

Keystone Species

Not all species in the microbiome share the same usefulness. Introducing keystone species may prove particularly useful for microbiome enhancement. Keystone species typically show a low abundance but, as the key stone of a construction, play a disproportionate role for microbiome functionality. Adding such species may thus have a leverage effect on many other species. Recent research has shed light on the presence of such species for microbiome functionality and plant development (Banerjee et al. 2018). However, it remains to date difficult to clearly identify such species at a functional level and put them at work. More research will certainly make keystones more amendable for practical applications.

Call for New Integrated Techniques

One of the main limitations of all these approaches is that they essentially function as unintegrated interventions, with little account for their interactive effects. In order to reliably enhance plant nutritional value, we need to build an integrative framework, possibly using approaches developed for plant health (Zhong et al. 2020). For instance, more nutritive plants may be bred by selecting for their ability to recruit microorganisms conferring a higher nutritional value. Targeted soil amendments may specifically select functional groups enhancing nutrient mineralization and uptake. And smart fertilizers may be applied, using, for instance, nanotechnology to trigger nutrient release on demand when the plant needs them. Altogether, we are confident that targeting the microbiome from different perspectives will enable a robust improvement in plant nutritional value at the service of healthy mankind.

References

Bamji, M.S., Murty, P.V.V.S., Sudhir, P.D.: Nutritionally sensitive agriculture—an approach to reducing hidden hunger. Eur. J. Clin. Nutr. (2020). https://doi.org/10.1038/s41430-020-00760-x

Banerjee, S., Schlaeppi, K., van der Heijden, M.G.A.: Keystone taxa as drivers of microbiome structure and functioning. Nat. Rev. Microbiol. **16**(9), 567–576 (2018)

Berendsen, R.L., Pieterse, C.M.J., Bakker, P.A.H.M.: The rhizosphere microbiome and plant health. Trends Plant Sci. **17**(8), 478–486 (2012)

Berg, G., Rybakova, D., Grube, M., Koberl, M.: The plant microbiome explored: implications for experimental botany. J. Exp. Bot. **67**(4), 995–1002 (2016)

Bista, D.R., Heckathorn, S.A., Jayawardena, D.M., Mishra, S., Boldt, J.K.: Effects of drought on nutrient uptake and the levels of nutrient-uptake proteins in roots of drought-sensitive and -tolerant grasses. Plants **7**(2), 28 (2018). https://doi.org/10.3390/plants7020028

Borrelli, P., Robinson, D.A., Fleischer, L.R., Lugato, E., Ballabio, C., Alewell, C., Meusburger, K., Modugno, S., Schütt, B., Ferro, V., Bagarello, V., Oost, K.V., Montanarella, L., Panagos, P.: An assessment of the global impact of 21st century land use change on soil erosion. Nat. Commun. **8**(1), 2013 (2017)

Chaparro, J.M., Badri, D.V., Bakker, M.G., Sugiyama, A., Manter, D.K., Vivanco, J.M.: Root exudation of phytochemicals in *Arabidopsis* follows specific patterns that are developmentally programmed and correlate with soil microbial functions. PLoS One. **8**(2), e55731 (2013)

Cretoiu, M.S., Korthals, G.W., Visser, J.H.M., van Elsas, J.D.: Chitin amendment increases soil suppressiveness toward plant pathogens and modulates the actinobacterial and oxalobacteraceal communities in an experimental agricultural field. Appl. Environ. Microbiol. **79**(17), 5291–5301 (2013)

Davis, D.R., Epp, M.D., Riordan, H.D.: Changes in USDA food composition data for 43 garden crops, 1950 to 1999. J. Am. Coll. Nutr. **23**(6), 669–682 (2004)

El Mujtar, V., Muñoz, N., Prack Mc Cormick, B., Pulleman, M., Tittonell, P.: Role and management of soil biodiversity for food security and nutrition; where do we stand? Glob. Food Sec. **20**, 132–144 (2019). https://doi.org/10.1016/j.gfs.2019.01.007

Etesami, H., Alikhani, H.A., Hosseini, H.M.: Indole-3-acetic acid (IAA) production trait, a useful screening to select endophytic and rhizosphere competent bacteria for rice growth promoting agents. MethodsX. **2**, 72–78 (2015). https://doi.org/10.1016/j.mex.2015.02.008. PMID: 26150974; PMCID: PMC4487705

Giri, A., Heckathorn, S., Mishra, S., Krause, C.: Heat stress decreases levels of nutrient-uptake and - assimilation proteins in tomato roots. Plants, **6**, 6 (2017). https://doi.org/10.3390/plants6010006

Hu, J., Wei, Z., Friman, V.P., Gu, S.H., Wang, X.F., Eisenhauer, N., Yang, T.J., Ma, J., Shen, Q.R., Xu, Y.C., Jousset, A.: Probiotic diversity enhances rhizosphere microbiome function and plant disease suppression. MBio. **7**(6), e01790–e01716 (2016)

Hu, J., Wei, Z., Kowalchuk, G.A., Xu, Y., Shen, Q., Jousset, A.: Rhizosphere microbiome functional diversity and pathogen invasion resistance build up during plant development. Environ. Microbiol. **22**(12), 5005–5018 (2020)

Huergo, L.F., Rissi, D.V., Elias, A.S., Gonçalves, M.V., Gernet, M.V., Barreto, F., Dahmer, G.W., Reis, R.A., Pedrosa, F.O., Souza, E.M., Monteiro, R.A., Baura, V.A., Balsanelli, E., Cruz, L.M.: Influence of ancient anthropogenic activities on the mangrove soil microbiome. Sci. Total Environ. **645**, 1–9 (2018). https://doi.org/10.1016/j.scitotenv.2018.07.094. Epub 2018 Jul 14. PMID: 30015113

Lugtenberg, B., Kamilova, F.: Plant-growth-promoting rhizobacteria. Annu. Rev. Microbiol. **63**, 541–556 (2009)

Marles, R.J.: Mineral nutrient composition of vegetables, fruits and grains: the context of reports of apparent historical declines. J. Food Compos. Anal. **56**, 93–103 (2017)

Mendes, R., Kruijt, M., de Bruijn, I., Dekkers, E., van der Voort, M., Schneider, J.H.M., Piceno, Y.M., DeSantis, T.Z., Andersen, G.L., Bakker, P.A.H.M., Raaijmakers, J.M.: Deciphering the rhizosphere microbiome for disease-suppressive bacteria. Science. **332**, 1097–1100 (2011)

Rashid, M.I., Mujawar, L.H., Shahzad, T., Almeelbi, T., Ismail, I.M.I., Oves, M.: Bacteria and fungi can contribute to nutrients bioavailability and aggregate formation in degraded soils. Microbiol. Res. **183**, 26–41 (2016), ISSN 0944-5013. https://doi.org/10.1016/j.micres.2015.11.007

Ravanbakhsh, M., Sasidharan, R., Voesenek, L.A.C.J., et al.: Microbial modulation of plant ethylene signaling: ecological and evolutionary consequences. Microbiome. **6**, 52 (2018). https://doi.org/10.1186/s40168-018-0436-1

Ravanbakhsh, M., Kowalchuk, G.A., Jousset, A.: Targeted plant hologenome editing for plant trait enhancement. New Phytol. **229**(2), 1067–1077 (2020)

Saleem, M., Hu, J., Jousset, A.: More than the sum of its parts: microbiome biodiversity as a driver of plant growth and soil health. Annu. Rev. Ecol. Evol. Syst. **50**(1), 145–168 (2019)

Sanaullah, M., Chabbi, A., Rumpel, C., Kuzyakov, Y.: Carbon allocation in grassland communities under drought stress followed by 14C pulse labeling. Soil Biol. Biochem. **55**, 132–139 (2012), ISSN 0038-0717. https://doi.org/10.1016/j.soilbio.2012.06.004

Shenkin, A.: Micronutrients in health and disease. Postgrad. Med. J. **82**(971), 559–567 (2006). https://doi.org/10.1136/pgmj.2006.047670

Shifler Bowers, K., Francis, E., Kraschnewski, J.L.: The dual burden of malnutrition in the United States and the role of non-profit organizations. Prev. Med. Rep. **12**, 294–297 (2018). https://doi.org/10.1016/j.pmedr.2018.10.002

Singh, D., Prasanna, R.: Potential of microbes in the biofortification of Zn and Fe in dietary food grains. A review. Agron. Sustain. Dev. **40**, 15 (2020). https://doi.org/10.1007/s13593-020-00619-2

Vannier, N., Mony, C., Bittebiere, A.-K., Theis, K.R., Rosenberg, E., Vandenkoornhuyse, P.: Clonal plants as meta-holobionts. mSystems. **4**(2), e00213–e00218 (2019)

Xiong, W., Guo, S., Jousset, A., Zhao, Q., Wu, H., Li, R., Kowalchuk, G.A., Shen, Q.: Bio-fertilizer application induces soil suppressiveness against Fusarium wilt disease by reshaping the soil microbiome. Soil Biol. Biochem. **114**, 238–247 (2017)

Xiong, W., Jousset, A., Guo, S., Karlsson, I., Zhao, Q., Wu, H., Kowalchuk, G.A., Shen, Q., Li, R., Geisen, S.: Soil protist communities form a dynamic hub in the soil microbiome. ISME J. **12**(2), 634–638 (2018)

Zhong, W., Friman, V.P., Pommier, T., Geisen, S., Jousset, A., Shen, Q.: Rhizosphere immunity: targeting the underground for sustainable plant health management. Front. Agric. Sci. Eng. **7**(3), 317–328 (2020)

Chapter 11
Zinc-Solubilizing Bacteria: An Option to Increase Zinc Uptake by Plants

Mahvash Haroon, Shams Tabrez Khan, and Abdul Malik

Abstract Zinc is an essential trace element for life as it is required for the structural integrity and catalytic activity of all the classes of enzymes, and the zinc-binding domains are highly conserved among species. In plants, these enzymes are involved in carbohydrate metabolism, auxin metabolism, and in anti-oxidant activity. Low Zn bioavailability decreases crop Zn content, impairs nutritional quality, as well as the overall crop productivity. Consequent consumption of Zn-deficient crops results in its deficiency affecting one-third of the world population. One of the important reasons of the decreased Zn bioavailability to plants is its availability in soil in forms with limited solubility. Zn uptake by plants can be increased using approaches like (a) regular crop rotation and intercropping, (b) conventional breeding, transgenic and genetic engineering, and (c) use of Zn fertilizers such as zinc sulfate or Zn-EDTA. One of the effective approaches to deal with Zn deficiency is the use of zinc-solubilizing bacteria. Zn solubilization by heterotrophs like strains of *Gluconacetobacter* sp., *Acinetobacter* sp., *Burkholderia* sp., *Klebsiella* sp., *Ralstonia* sp., *Bacillus* sp., *Pseudomonas* sp., *Serratia* sp., and Ericoid mycorrhizal fungi (*Oidiodendron maius*) is well-known. This microbe-mediated Zn biofortification of crops provides a practical and cost-effective option to tackle the global Zn malnutrition problem.

Keywords Zn deficiency in plants · Zn uptake mechanism · Zn solubilizing bacteria · Zn fertilizers · Zn bioavailability

M. Haroon (✉) · S. T. Khan · A. Malik
Department of Agricultural Microbiology, Faculty of Agricultural Sciences, Aligarh Muslim University, Aligarh, Uttar Pradesh, India

S. T. Khan, A. Malik (eds.), *Microbial Biofertilizers and Micronutrient Availability*,
https://doi.org/10.1007/978-3-030-76609-2_11

Introduction

Zinc, one of the prime micronutrients, is requisite in trace amounts for proper development of different plant tissues and organs and overall growth of the plant as well (Hafeez et al. 2013). It plays a key role in photosynthesis and sugar transformations (Alloway 2008b), auxin and protein metabolism (Alloway 2004) and functions as an effective anti-oxidant. Zn finger transcription plays a significant role in the growth and maturation of floral structures, flowering, reproduction, and fruit development (Epstein and Bloom 2005). Zn-deficient crops exhibit chlorotic and necrotic spots on leaves, leaf bronzing, leaf rosetting (due to inhibited internodal growth), and malformed and dwarf ("little") leaf with overall stunted plant growth (Alloway 2004). Zn-deficient plants are more susceptible to temperature and fungal infections (Tavallali et al. 2010). Low bioavailability of Zn hampers crop productivity (biomass production) and also significantly reduces Zn percentage in the harvest output (a mature crop and/or its parts such as fruits, seeds, grains, and rhizomes); in consequence, it negatively affects the nutritive value (Cakmak and Hoffland 2012). Consuming a plant-based zinc-deficient diet may lead to detrimental effects on human health and nutrition and about one-third of the population worldwide is known to suffer from such nutritional deficiency (Myers et al. 2015; Hotz and Brown 2004).

Deficiency symptoms manifest when their supply drops below the critical level (2μg/kg), (Saravanan et al. 2011). The global prevalence of Zn deficiency issues in crop plants is not due to total insufficient levels of Zn, but due to lower solubility of soil Zn (Cakmak 2008). In soil, the total Zn content is significantly high and occurs in complexes that are only sparingly soluble, such as smithsonite ($ZnCO_3$), sphalerite (ZnS), zincite (ZnO), franklinite ($ZnFe_2O_4$), wellemite (Zn_2SiO_4), and hopeite ($Zn_3(PO_4)_2{\cdot}4H_2O$). However, plants can take up zinc as a divalent cation, that is, the soluble form (Kabata-Pendias and Pendias 2001). Literature suggests that the concentration of such soluble form of Zn in soils is very low (4.0–270.0μgL^{-1}) compared to the mean total concentration of Zn i.e., 64.0 $mgkg^{-1}$ (Alloway 2009).

Studies suggest that Zn deficiency is rampant in at least one-third of the cultivated soils globally due to low levels of bioavailable Zn (Sillanpaa 1990). In addition, soils prone to intensive cropping and low drainage, acidic, calcareous, saline and sodic soils, and coarse-textured soils susceptible to high weathering manifest Zn deficiency (Singh et al. 2005). Factors such as high available P and Si, upper horizon drying, sub-soil constraints, and often elevated fertilizer application subscribe to Zn deficiency (Alloway 2009).

Different approaches have long been introduced to reduce zinc deficiency. Zinc fertilizers have been used in the form of $ZnSO_4$ (White and Broadley 2005) or Zn-EDTA (Karak et al. 2005), but their application puts economic and ecological pressure and within seven days of fertilizer application they are converted into insoluble complex forms (Rattan and Shukla 1991). Zn is not only a key mineral micronutrient, but a heavy metal as well. Thus, one of the potential issues associated with the use of Zn fertilizer is its accumulation and subsequent toxicity (Alloway

2009; Broadley et al. 2007). To facilitate the absorption of zinc by plants, in different areas, regular crop rotation and cropping and intercropping systems have been used (Gunes et al. 2007; Zuo and Zhang 2009). Conventional breeding (Cakmak et al. 2010), genetic and transgenic approaches are the other methods (Gustin et al. 2009; Mhatre et al. 2011; Tan et al. 2015), but these technologies are slow, costly, and tedious. A safer potential choice to all of these methods is the employment of Zn-mobilizing and -solubilizing microbial inoculants.

Bacteria depending on their metabolism (autotrophic and heterotrophic bacterial processes) and physicochemical conditions of the associated environment influence Zn solubility (Gadd 2010) and are thus entailed in the biogeochemical cycling of Zn. Autotrophic bacteria-mediated metal solubilization (e.g., sulfur/iron oxidizers) has been researched in abundance and also used in the retrieval and renewal of Zn, copper (Cu), and nickel (Ni) from mineral ores and industrial effluents (Schippers et al. 2013). On the other hand, heterotrophic bacteria have also drawn attention for Zn mobilization and solubilization and are being tested as an inoculum in agronomic studies to increase plant Zn accumulation. In order to address the global Zn malnutrition issue, agronomic biofortification of Zn provides a practicable, cost-effective, and realistic option.

The main focus of this chapter is on zinc-solubilizing microorganisms, their role in Zn solubilization, and Zn uptake mechanisms in relation to soil nutrient dynamics, and the potential to apply this knowledge to the management of a sustainable and environmentally friendly agricultural system (Costerousse et al. 2018).

Importance of Zinc

In both eukaryotes and prokaryotes, the role of zinc in their metabolism and physiology has been extensively studied (Hughes and Poole 1989), as amongst the other micronutrients, Zn ions become a part of the enzyme system as co-factor and metal activator of a large number of enzymes (Parisi and Vallee 1969). Zinc is not only mandatory for the essential catalytic functions and metabolism of the enzymes, but it also induces protein folding and provides stability to the protein subdomains. Zinc deficiency is associated with the occurrence of failure in the production of pigments such as melanin, chrisogenin, prodigiosin, and subtilin in fungi and bacteria (Chernavina 1970). Zinc has an important role in the folding of DNA-binding domains in eukaryotic transcription factors, including zinc finger transcription factors, the broad family of hormone receptor proteins, and in the yeast zinc cluster transcription factors as well. Similarly, Zn plays a highly probable role in the zinc-containing accessory proteins associated with nucleic acid replication as well as RNA polymerases (Coleman 1992).

Zinc in Plants

Zinc is indispensable, even though required in minute quantities for proper plant growth and reproduction, as it performs several key physiological pathways and ensures the structural and functional integrity of the biological membranes as well. Thus, in several cellular processes such as protein synthesis, enzyme activation, gene expression and regulation, phytohormone activity, photosynthetic carbon metabolism and indoleacetic acid metabolism (Oktay et al. 1998; Marschner 1997), pollen formation (fertility), seed formation (Marschner 1995a), and disease resistance, (Alloway 2008a; Rengel 2007), Zn has significant roles. Zn deficiency impairs these physiological functions and compromises plant health and productivity, resulting in lower yields (or even crop failure) and low-grade crop products due to reduction in dry matter production of crop plants (Rengel 2007; El-Fouly et al. 2001; Wang and Jin 2005; Price 1970).

Functions of Zinc in Plants

Some of the functions of Zn in plants are listed as follows:

(a) Zinc is the fundamental element needed in all the six classes of enzyme (oxidoreductases, transferases, hydrolases, lyases, isomerases, and ligases). Zinc is required in the proper functioning of a number of enzymes, which suggests that Zn deficiency hampers protein metabolism, carbohydrate metabolism, auxin responses, and reproductive processes as well (Zhang et al. 1991). Zinc is essential for the normal activity and functioning of metalloenzymes which are entailed in protein and nucleic acid metabolism (RNA stabilization).
(b) Zn levels tends to affect the absorption and translocation of water in plants and helps in reducing the deleterious effects of brief intervals of heat and salt stress (Disante et al. 2011; Kasim 2007; Peck and McDonald 2010; Tavallali et al. 2010).
(c) Zinc deficiency makes plants more vulnerable to abrupt alternations due to high temperature and light intensities (Cakmak 2000; Marschner 1995a).
(d) Biochemical pathways in plants are impaired due to Zn deficiency, manifesting symptoms such as shortening of internodes, small and distorted leaves, and interveinal chlorosis (Marschner 1995b).

    - In barley, Zn deficiency can be visualized by stunted growth, leaf chlorosis and necrosis, mid-leaf collapse, and white spots on leaves (Genc et al. 2003; Lombnaes and Singh 2003).
    - Zn deficiency in wheat is significantly visible in seedlings; while in subsequent stages of growth, brownish-white patches and necrotic lesions are

observed all through the leaf blade together with mid-leaf collapse (Cakmak and Braun 2001; Rengel and Graham 1995a, b).

- The deficiency symptoms manifest with the variations in environmental conditions, stages of plant growth, severity of the deficiency, with respect to the concentration of other nutrients in soil as well (Sadeghzadeh 2013).

Distribution of Zinc Deficiency

In different regions of the world, zinc deficiency is prevalent in a variety of soil types; however, heavily weathered soils of tropical regions, high-pH calcareous soils of semi-arid areas, sandy-textured soils, and acidic soils tend to be the most severely impaired in the changing climate scenario owing to their poor Zn content. Zn deficiency is recorded as the commonly occurring micronutrient deficiency worldwide as about 50% of the soils used for the cultivation of cereals are projected to have low bioavailable Zn. Out of this, 25% are observed with acute deficiency showing visible symptoms and in some cases symptoms are not visible but deficiency can be verified through yield responses to Zn amendments.

In semi-arid parts of India, paddy fields of Pakistan, calcareous water-logged paddy soils of China, and highly alkaline soils with low levels of organic matter in Central Anatolia of Turkey have been recorded with low levels of bioavailable Zn.

Southern Australia, Victoria, Queensland, New South Wales, and Western Australia are the parts of Australia reported with low plant-available Zn. Southwest Australia is considered the world's wide-ranging continuous region of Zn deficiency. Due to such wide-range deficiencies, Australia has become a global subject of study on micronutrients' concern (Sadeghzadeh 2013) (Table 11.1).

Table 11.1 Total zinc concentrations of different soil types

Country	Soil type/climatic zone	Mean Zn (mg Zn kg^{-1})	Range of Zn
India	Arid/semi-arid	59	20–89
India	Humid/sub-humid tropics	52	22–74
India	Vertisols ("heavy cracking clay soil" or "swell-shrink soils")	–	69–76
India	Oxisols (coarse textured)	–	24–30
Philippines	Rice soils	–	63–135
Vietnam	Ferralitic soils (strongly weathered heavily leached soil)	102	40–485
Sri Lanka	Patana soils (wet with high organic matter)	75	35–102
Indonesia	Rhyolites and Peats	–	33–174
Thailand		45	5–158

Zinc Uptake Mechanisms from Soil

Zn uptake by plants occurs primarily through absorption of Zn^{2+} from the soil mineral solution, which is an active and complex exercise and is determined by the root surface ion concentration, plant requirement, and efficiency of absorption by the root system. Zn uptake through plant roots occurs via mass flow, diffusion, and root interception mechanisms.

(a) *Mass flow* is the passive transfer of nutrients to the root system from the soil rhizosphere and is carried out by transpiration. It becomes the dominant pathway that delievers Zn towards the root surface, when a sufficiently high amount of Zn is present in the solution passing through the soil.
(b) *Diffusion* plays a significant role in the transportation of Zn, along with P, K, Cu, Fe, Mn, etc., particularly in soils with low levels of bioavailable Zn, as mass flow carries only small concentrations of nutrients needed by the plants. Diffusion operates only in the close proximity of the plant root system, in comparison to mass flow.
(c) *Root interception* is an important mechanism in which the structure and surface area of the roots help in the uptake of the immobile nutrients such as Zn, through the soil profile. Poor root interception might restrict Zn uptake, even in cases of Zn-fertilizer application (Sadeghzadeh 2013).

Factors Associated with Zn Bioavailability

The term "availability" is often used to describe the capability of the plant to take up available forms of macro- and micronutrients from the soil. The total Zn concentration in soil varies between 10 and 300μg g^{-1} with an average of 50μg g^{-1} (Lindsay 1972).

The issue is, due to one or more reasons or factors, only a limited volume of soil Zn is accessible to the crops. The majority of the overall Zn remains fixed in an insoluble or unexchangeable state in the soil and is difficult to make available for cultivation (Stahl and James 1991). Soils of arid and semi-arid segments and mildly acidic, leached soils of warm and tropical climates are typically most vulnerable to Zn deficiency, but crops are not equally prone to Zn deficiency, as some crops might show deficiency symptoms, while others not, under the same soil.

Bioavailability of Zn is governed by factors such as total soil Zn content, soil pH, organic matter, calcareous soil, high levels of other nutrients like phosphorus or other metals in soil, soil temperature and moisture content, root architecture and rhizospheric effects (Alloway 2008a; Hafeez et al. 2013; Kabata-Pendias and Pendias 1992; Lindsay 1972; Takkar and Randhawa 1978) (Fig. 11.1).

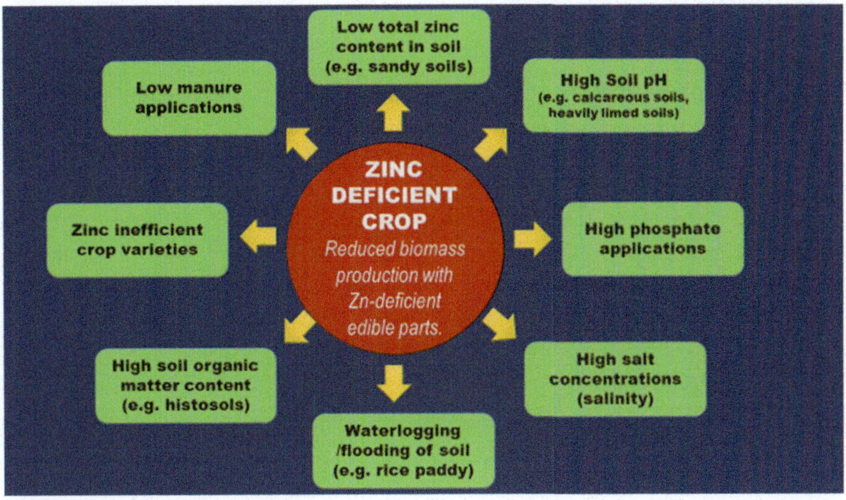

Fig. 11.1 Factors affecting Zn deficiency in crops

Table 11.2 Zinc mineral ores

Mineral Zn	Adsorbed Zn	Complexed Zn
Smithsonite ($ZnCO_3$)	Zn-$CaCO_3$	Manure organic residues
Sphalerite (ZnS)	Zn-$MgCO_3$	
Zincite (ZnO)	Zn-FeO	
Franklinite ($ZnFe_2O_4$)	Zn-MnO	
Wellemite (Zn_2SiO_4)		
Hopeite [$Zn_3(PO_4)_2 \cdot 4H_2O$]		

Total Soil Zn Content

Soils with a total low concentration of Zn are often Zn deficit for crop production. Zn deficiency occurs in soils which are inherently low in total Zn concentrations, such as sandy soil and highly leached acid soils. Parent soil material is the deciding factor of Zn concentrations in its derivative soils. For example, the soils originating from limestone, sandstone, gneisses, and granites are low in total Zn concentrations (Hafeez et al. 2013) (Table 11.2).

Soil pH

Solubility of micronutrients generally decreases, except molybdenum, as soil pH increases, which in turn stimulates adsorption of Zn onto the surface of soil fragments (i.e., metal oxides and clay particles), decreasing its availability to plants (Parker and Walker 1986).

Zn precipitates as $Zn(OH)_2$, $ZnCO_3$ (Saeed and Fox 1977; Shukla and Mittal 1979), and Zn_2SiO_4 under alkaline soils (Parker and Walker 1986). Thus, alkaline soils are more prone to Zn deficiency, in contrast to acidic soils.

Liming of soils increases the alkalinity of the soils, thereby increasing Zn fixing capabilities into unexchangeable forms (Udo et al. 1970), especially in soils having high P concentration (Alloway 2004). Thus, it reduces plant Zn uptake (Shukla and Morris 1967), inducing Zn deficiency (Hafeez et al. 2013; Viets and Prasad 1966).

Soil Organic Matter

Soil organic matter is one of the factors that affects Zn solubility and its availability to the crops. Soils low in organic matter content lead to the commencement of the deficiency symptoms (Alloway 2004), whereas adequate amounts increase the solubility and diffusion rates of Zn. However, soils too high in organic matter content, such as peat and muck, lead to Zn deficiency, as humic substances in the solid state bind Zn^{2+} (Katyal and Randhawa 1983). Humic substances (such as humic and fulvic acids) are the most stable organic compounds having high affinity to metal ions (such as Zn^{2+}) owing to the presence of a number of functional groups (–OH, –COOH, –SH) (Kiekens 1995). Organic compounds (such as hydroxy acids, amino acids, and phosphoric acids) efficaciously chelate Zn^{2+}, thereby increasing solubilization and mobilization of Zn^{2+} in soils (Hafeez et al. 2013; Kabata-Pendias and Pendias 1992).

Soil Texture

Soils having high cation exchange capacity (CEC) have highly receptive sites and thus are able to hold on to Zn better than lighter textured soils (such as sandy soil) (Shukla and Morris 1967). Therefore, in heavier textured clay soil, adsorption of Zn is governed by CEC and pH. Clays such as bentonite and illite have high CEC values than kaolonite, and so they fix Zn more firmly, thus making Zn^{2+} unavailable to crop plants (Ellis 1972; Reddy and Perkins 1974).

Calcareous Soils

Calcareous soils are frequently reported with Zn deficiency; however, the total Zn content in calcareous and noncalcareous soils is comparable (Singh et al. 2005). Calcareous soils at pH greater than 7, along with moderate to high levels of organic matter, are more prone to Zn-deficiency due to high levels of bicarbonate (HCO^{-3}) ions in the soil (Singh et al. 2005).

Phosphate Fertilizers

Crops might be subjected to Zn deficiency under high phosphate concentrations (Alloway 2008b) as a result of high native phosphate levels or due to abundant or prolonged use of phosphate fertilizers (Olsen 1972). Thus, it is referred to as "P-induced Zn deficiency" (Singh et al. 1986).

Soil Moisture

Diffusion is moisture dependent and as a result Zn solubility reduces, lowering Zn uptake, due to the fact that Zn co-precipitates with soluble iron and aluminum under low moisture conditions (Moraghan and Mascagni 1991). The low moisture content of soil influences the rhizospheric nutrient supplementation by restraining diffusion, whereas soil under water-logged conditions transfigures Zn chemistry, where Zn reacts with free sulfide (Mikkelsen and Kuo 1977) and might also change into franklinite ($ZnFe_2O_4$) or amorphous sesquioxide precipitates (Sajwan and Lindsay 1988; Singh and Abrol 1986).

Soil Temperature

The incidence and severity of Zn deficiency is associated with cool and wet atmospheres (Moraghan and Mascagni 1991). A cold rhizospheric temperature reduces rhizospheric colonization of vesicular-arbuscular mycorrhizae (VAM), root architecture, Zn absorption, and translocation into the shoots (Takkar and Walker 1993). Soil temperature is directly proportional to the rate of zinc mineralization (Takkar and Walker 1993). High light intensity and long day-lengths also imparts in plant Zn deficiency (Marschner and Cakmak 1989).

Interaction of Zn with Other Nutrients

Interactions might be interpreted as "an influence, a mutual or reciprocal action of one element upon another in relation to plant growth" (Olsen 1972). Micronutrients and macronutrients happen to make alterations on each other and have potency to effect plant health in an interactive and complementary manner as well. Another aspect is the differential responsiveness of plants to one nutrient in conjunction with the differing ratios of other nutrients introduced simultaneously, that is, the nutrients interact to create an additive effect and the effect is not attributed solely to either of the nutrients (Olsen 1972).

These interactions occurring in soil should be taken into account as these are usually due to the interference of metal ions on the rate of diffusion, bioavailability, and absorption by the root system (Loneragan and Webb 1993). These interactions in soils and also in plants internally may have enhanced or reduced plant development in response to bioavailable Zn (Hafeez et al. 2013; Loneragan and Webb 1993).

Interactions of Zn with Other Macro- and Micronutrients:

Phosphate (P)–Zn Interactions

A high percentage of native available P or prolonged use of P fertilizers might be correlated with Zn deficiency and is referred as "P-induced Zn deficiency." The cause of P-induced Zn deficiency includes the following:

(i) P–Zn interaction: Co-precipitation of P with Zn leads to the formation of $Zn_3(PO_4)_2$, reducing soluble Zn levels in soil.
(ii) Root to shoot translocation rate slows down.
(iii) Growth response to P causes dilution of Zn concentrations in plant apices.
(iv) Due to imbalance in P–Zn ratios, plant cells suffer from metabolic disorders (Olsen 1972).

Nitrogen (N)–Zn Interactions

Fertile soil low in Zn^{2+} and application of N-fertilizers intensifies Zn deficiency by changing the pH (Viets et al. 1957) of the rhizosphere and also by increasing the shoot/root ratio (Loneragan and Webb 1993).

Macronutrients–Zn Interactions

Absorption of Zn^{2+} by plants is retarded by Ca^{2+}, Mg^{2+}, and K^{+}, but is variable according to the effects of their salts based on concentration and soil pH.

Copper–Zn Interactions

Cu^{2+} and Zn^{2+} interact in a number of ways and a strong Cu^{2+} - Zn^{2+} antagonism is observed in soil system (Kausar et al. 1976) as Cu^{2+} strongly and competitively depresses Zn^{2+} absorption and vice versa. In addition, Cu^{2+} actively influences the translocation of Zn^{2+} within plant parts (Hafeez et al. 2013; Loneragan and Webb 1993).

Iron–Zn Interaction

Iron and zinc have a very complex relationship, which varies under different soil conditions. In wheat seedlings, Zn absorption is not affected when the iron level is low (Adriano et al. 1971); however, at higher Zn levels and at levels present in paddy fields, iron suppresses Zn absorption (Giordano et al. 1974). In dicotyledonous species, iron deficit conditions result in acidification of the rhizospheric region, which increases Zn absorption potentially (Marschner and Cakmak 1989). In grasses, production of phytosiderophores (i.e., iron chelators) (Ambler and Brown 1969; Jackson 1962) helps in the mobilization of Zn, thus enhancing Zn uptake from calcareous soils (Hafeez et al. 2013; Treeby et al. 1989; Zhang et al. 1991).

Strategies for Increasing Zinc Content of the Plants

Crop Rotation and Intercropping

An appropriate and practicable cropping system (crop rotation and intercropping) might be able to keep a check on Zn deficiency and provides an efficient and sustainable route to Zn biofortification (Gao et al. 2012).

Crop rotation Microorganisms improve the levels of bioavailable forms of Zn in soil when the crops are grown in rotation, for example, legumes and wheat (Rehman et al. 2018).

Intercropping Intercropping improves the resource utilization such as macro- and micronutrients, available moisture, and solar exposure of the two crop species simultaneously, thus enhancing quality crop output due to spatial and temporal discrepancies of growth factors and also their heterogeneous species-specific physiological reactions owing to ecological stresses. In developing countries like India, Africa, China, Latin America, and Southeast Asia, intercropping is conventional. Intercropping between interspecific species like peanut/maize (Inal et al. 2007), wheat/chickpea (Rehman et al. 2018) and guava/sorghum or maize helps in facilitating zinc and iron content in the seeds, as in the case of wheat and chickpea intercropping, chickpea seeds were recorded to have 2.82 times higher Zn than that in monocropping. Thus, in nutrient-efficient dicot and gramineous crops, intercropping precludes zinc and iron deficiency (Zuo and Zhang 2009).

Genetically Engineered Strains (GESs)

Some of the Genetically Engineered Strains (GESs) have the potential to solubilize and mobilize nutrients to the plant rhizosphere. Thus, generating more GESs with such enhanced solubilization and mobilization capabilities through genetic modifications and DNA technologies is important to ensure an eco-friendly, efficient, and sustainable agricultural practice.

Plant Breeding

Zn deficiency can be tackled through breeding new cereal genotypes, which is the most practical and *cost-effective* approach, although the approach is long-term. *Plant breeding* approach seems to be more efficient but the scale of plant-available Zn reservoir in soils has a considerable impact on the Zn uptake and accumulation in grains by the Zn-efficient (biofortified) cultivars. *Zn-efficient* cultivars of wheat, barley, and rice are available (Genc and McDonald 2004; Gregorio et al. 2000) and are commonly grown in Zn-deficit soils. The production of such Zn accumulators under inadequacy of soil Zn is important in strengthening tolerance to Zn deficiency, increasing crop productivity and improving micronutrient (Zn) concentrations especially in edible plant parts. These Zn-efficient genotypes are reported to have a good deal of Zn-binding chelators, which physiologically makes it available at the cellular level. Furthermore, by placing the limit on the employment of machinery and fertilizers on agricultural fields, biofortified genotypes might be able to mitigate land degradation severity. For the identification and breeding of Zn-efficient varieties, simple, rapid, relevant, and consistent screening procedures are prerequisite, together with an appropriate apprehension of the morphological, physiological, and genetic bases of Zn efficiency of the biocultivars (Sadeghzadeh 2013).

Zn Fertilizers

Zn fertilizers, applied to correct Zn deficiency, are likely to persist in the upper surface in sandy textured soils as well. However, the application of liquid Zn fertilizer to the subsoil (up to 40 cm deep) is better than granular forms in enhancing crop productivity. In addition, developing nations might face problems in affording fertilizers (Rani et al. 2020).

Inorganic Zn Source

Includes ZnO, $ZnCO_3$, $ZnSO_4$, $Zn_3(PO_4)_2$, and $ZnCl_2$. Zinc sulfate heptahydrate ($ZnSO_4.7H_2O$) is the most effective, easily available, and cheapest fertilizer applied through soil or foliar application. ZnO with 80% Zn has very low water solubility especially when applied in granular form. $Zn_3(PO_4)_2$ is efficient for seed coating and is cheaper than ZnO (Mortvedt and Gilkes 1993).

Organic Zn Fertilizers

Include different compounds, majorly Zn-phenolate, Zn-EDTA, and Zn-lignosulfonate. Zn-EDTA enhances the phosphorus uptake and is more efficiently transported to plants as compared to $ZnSO_4$. The activity of biodegradable chelates IDHA [N-(1,2-carboxyethyl)-D, Laspartic acid] and EDDS (ethylene diamine disuccinic acid) or complexes as gluconates and lignosulfonates which are the byproducts from the food and paper industry has shown great results (López-Rayo et al. 2016). Fertilizers such as zincated urea, zincated super and boronated super, when blended with micronutrients have shown long-term effectiveness in increasing the soil fertility and micronutrient deficiency in plants. Due to inefficiency and economic issues, organic fertilizers are less accepted by farmers (Rani et al. 2020).

Advantages

1. Organic fertilizers provide a more balanced nutrient supply by enhancing soil biological activity, thus improving mobilization of both macro- and micronutrients from both organic and inorganic sources and helps in disintegration of toxicants as well.
2. The rate of nutrient release is slow, contributing to a number of micronutrients and also helps in maintaining the reservoir of organic nitrogen and phosphorus in the soil, by fixing P and preventing N leaching.
3. They increase the soil organic matter content, which helps in buffering soil to combat acidity, alkalinity, salinity, pesticides, and heavy metals toxicity.
4. Organic fertilizers improve nutrient exchange, water retention capacity, and also enhances soil aggregation.
5. They enhance the mycorrhizal association, thus improving P reserve.

6. They boost outstretching of the root system due to better arrangement of soil particles (Bokhtiar and Sakurai 2005; Tzen and Chen 2004).

Disadvantages

1. The nutrient composition varies highly; relative to chemical fertilizers, it is high-priced.
2. Comparatively low in nutrient content, it needs to be added in bulk amount to provide ample nourishment to crops.
3. Organic fertilizers might not contain the primary nutrients required by the plants in adequate amounts to sustain maximum crop growth.
4. The nutrients are released at a very slow rate to meet the plant demand, hence deficiency symptoms prevails.
5. Long-term intensive applications on agrarian lands lead to the deposition of salt, nutrients, or heavy metals that have detrimental effects on the living biota (Bokhtiar and Sakurai 2005; Tzen and Chen 2004).

Chemical Fertilizers

Chemical-based fertilizers are thought to be an effective and trustworthy agricultural management measure as they provide essential plant macro- and micronutrients.

Indiscriminate use of agro-chemicals and chemical fertilizers to increase production per unit time per unit area has caused a considerable negative impact on environmental sustainability such as greenhouse effect, soil deterioration, and deficiency of both macro- and micronutrients in the soil system and ultimately in crop plants (Rani et al. 2020) (Table 11.3).

Advantages

1. Nutrients are in soluble form as required by plants; so the effect is direct and fast.
2. They are more competent than organic fertilizers, which makes them more admissible, so often used by agronomists.
3. Smaller doses are enough as these are abundant in nutrients (Bokhtiar and Sakurai 2005; Tzen and Chen 2004).

Disadvantages

1. Over-application may contribute to undesirable effects such as leaching, water pollution, distressing soil micro-flora, crop vulnerability to pathogens, soil acidification or alkalization, or loss of soil fertility—causing irremediable impact on the ecosystem.
2. They enhance soil organic matter decomposition, which contributes to soil structure deterioration.
3. Loss of nutrients from soil via fixation, leaching, or gas emission contributes to decreased fertilizer performance.
4. Oversupply of N results in plant tissue softening, giving rise to plants which are susceptible to pathogens and pests.

Table 11.3 Commonly used zinc fertilizer materials

Compound	Formula	Zinc content (%)
Inorganic compounds		
Zinc sulfate monohydrate	$ZnSO_4{\cdot}H_2O$	36
Zinc sulfate heptahydrate	$ZnSO_4{\cdot}7H_2O$	22
Zinc oxysulfate	$xZnSO_4{\cdot}xZnO$	20–50
Basic zinc sulfate	$ZnSO_4{\cdot}4Zn(OH)_2$	55
Zinc oxide	ZnO	50–80
Zinc carbonate	$ZnCO_3$	50–56
Zinc chloride	$ZnCl_2$	50
Zinc nitrate	$Zn(NO_3)_2{\cdot}3H_2O$	23
Zinc phosphate	$Zn_3(PO_4)_2$	50
Zinc frits	Fritted glass	10–30
Ammoniated zinc sulfate solution	$[Zn(NH_3)_4]SO_4$	10
Organic compounds		
Disodium zinc EDTA	Na_2-Zn-EDTA	8–14
Sodium zinc HEDTA	Na-Zn-HEDTA	6–10
Sodium zinc EDTA	Na-Zn-EDTA	9–13
Zinc polyflavonoid	–	5–10
Zinc lignosulphonate	–	5–8

5. They scale down the plant root–mycorrhizal association and prevent rhizobial symbiotic N fixation due to elevated N fertilization (Bokhtiar and Sakurai 2005; Tzen and Chen 2004).

Microbial Biofertilizers

The need of feasible and practicable agronomic activities with reduced resources and ecological issues is imperative worldwide to meet up with the adequate, cost-beneficial growing human demands.

In this scenario, creative, innovative articulation of technologies is required to ensure enhanced crop productivity through sustainable intensification of agriculture. Thus, the global food security issue fosters reliance in the integration of advanced and efficient technologies, in particular, innovative irrigation systems, integrated fertilization [e.g., modulated release of fertilizers and enhanced efficiency of fertilizers (EEFs)], and biocontrol approaches for nutrients and disease management, along with diverse imaging and sensing technologies which provide crucial data to monitor and secure crop plants. Through the implementation of favorable, significantly positive microbiome and plant–microbe interactions, agricultural microbial biotechnology might provide a promising, alternative sustainable means to amplify agronomic output (Timmusk et al. 2017).

Advances in genomic, post-genomic, biochemistry, ecology, and symbiotic relationships of promising microbes have contributed to the production and commercialization of potential microbe-based products (biofertilizers, biostimulants, biopesticides, etc.) with proven effectiveness in improving agricultural yields and tolerance to environmental physicochemical changes (Anuradha et al. 2015; Lindemann et al. 2016; Mahdi et al. 2010; Umesha et al. 2018).

Presently, microbial biofertilizers are taken into account as one of the main agricultural constituents that enhances agronomic yields and productivity and contributes to sustainable agro-systems, especially bioformulations that have compatible, integrative, and synergic nutrient fertilization effects. It is an intrinsic facet that clusters a range of microbe-dependent bioproducts whose bioactivities are critical for triggering and optimizing the biological metabolism of the complex plant–soil–microbe continuum (Singh et al. 2016). For the development of productive biofertilizers, different groups of soil microbes (especially bacteria and fungi) exhibiting plant growth-promoting traits are employed (Dhaked et al. 2017; Khalid et al. 2009; Lucy et al. 2004; Smith and Read 2008; Vessey 2003).

Microbe-based bioformulation can primarily be categorized into the following types:

(a) *Nitrogen-fixing (NF) bacteria*: Microbes having the ability to fix nitrogen and making it available to the plants either symbiotically (such as *Rhizobium* sp. associated with leguminous plants; *Frankia* with dicotyledonous species; *Azospirillum* sp. with cereal grasses) or nonsymbiotically such as *Cyanobacteria* (or blue-green algae), *Anabaena*, *Nostoc*, and bacterial genera including *Azotobacter, Beijerinckia,* and *Clostridium.* Blue-Green Algae (BGA) and *Azolla* are commonly used to subsidize finances of using N-fertilizers.
(b) *Zinc-solubilizing/mobilizing microorganisms*, such as species of *Bacillus* and *Pseudomonas* have increased the soluble soil Zn content and enhanced the growth of the plants when present in plant rhizosphere either due to natural habitation or due to soil or foliar application of biofertilizer formulations.
(c) *Phosphate-solubilizing/mobilizing microorganisms* such as species of *Azotobacter* and *Azospirillum* have the ability of solubilization and mobilization of soil phosphates, thus making it available to plants for proper growth and development. Arbuscular mycorrhizal fungi also help in the uptake of P and other micronutrients when associated with certain crops or plants.
(d) Some of the microbes might also be involved in providing disease resistance, salt and drought tolerance, etc., to the plants while some others might also act as biopesticides (Pathak and Kumar 2016).

Microbes-Mediated Zn Fortification

Microorganisms are recognized to play an astounding role in solubilization and mobilization of zinc and other minerals such as iron, phosphorus, potassium, and silicates to the plant roots. The root-colonizing, soil-borne microbes proliferate (Kloepper 1994) and assist in nutrient acquisition via solubilization of insoluble zinc complexes into soluble forms, thus making Zn available for plant uptake. Zinc-solubilizing microbes either in pure form or in combination with other biocontrol agents (Glick 2012; Shaikh and Saraf 2017) either in conjugation with inorganic Zn sources or not, help in modulating crop yield and zinc biofortification (Kamran et al. 2017; Khande et al. 2017).

Some opportunistic microbes, for their survival, exploit mineral dissolution and/or mineral formation by utilizing them in a series of chemical and metabolic reactions.

Microbes utilize mineral ions:

1. As an energy source
2. As a terminal acceptor of electrons in respiration
3. As prerequisite trace metal ions
4. To increase competitiveness within the microbial community
5. For the enrichment of their variance and heterogeneity in the microbial community

Microbes either might make use of a single mechanism or multiple mechanisms to ameliorate Zn availability in the rhizosphere and thus boost Zn uptake by plant tissues, one of which is *acidification*. The mobilization of the micronutrients is greatly influenced by the changes in soil pH, such that the availability of Zn^{2+} increases hundred times by the drop of just one unit of pH (Havlin et al. 2005). On this account, plant-available fractions of Zn^{2+} can be increased by lowering the pH of alkaline soil. Zn-solubilizing microbes, owing to the organic acid production and proton extrusion (H^+), lowers the rhizospheric pH, which sequesters zinc cations and improves uptake by plants. (Alexander 1997; Anuradha et al. 2015; Fasim et al. 2002; Wu et al. 2006).

Other mechanisms include siderophore production (Saravanan et al. 2011) and proton extrusion, oxido-reductive system present on the microbial cell membrane, and ligand chelation (Chang et al. 2005; Dhaked et al. 2017; Jones and Darrah 1994; Wakatsuki 1995).

Mycorrhizal fungi are also capable of solubilizing insoluble forms of zinc such as $Zn_3(PO_4)_2$ and ZnO by secreting organic acids (Martino et al. 2003). The rhizospheric soil pH lowers down by the inoculation of arbuscular mycorrhizal fungi and contributes to the release of the Zn^{2+} from soil mineral fractions (Subramanian et al. 2009).

Various plant growth-promoting (PGP) strains are reported to have shown good and flourishing plant growth and together with increased zinc content, when inoculated at different stages of the plant growth, and these include *Pseudomonas, Rhizobium* (Joshi et al. 2013; Naz et al. 2016), and *Bacillus* strains (Hussain et al. 2015; Pawar et al. 2015), *B. aryabhattai* (Ramesh et al. 2014), and *Azospirillum.* Some strains have Zn solubilization abilities on the laboratory scale such as *P. aeruginosa* (Fasim et al. 2002), *P. striata, P. fluorescence, Burkholderia cenocepacia* (Pawar et al. 2015), *Gluconacetobacter diazotrophicus* (Saravanan et al. 2007a, b), *Ralstonia* sp., *Serratia liquefaciens, S. marcescens,* and *B. thuringiensis*

(Abaid-Ullah et al. 2015; Costerousse et al. 2018; Dhaked et al. 2017; Hussain et al. 2018; Kamran et al. 2017; Shaikh and Saraf 2017).

Zinc Solubilizers

(a) Bacteria

- Several bacteria including *Acinetobacter* (Vaid et al. 2014), *Bacillus* sp., *Pseudomonas* sp., *Rhizobium, Azospirillum, Burkholderia cenocepacia* (Li et al. 2010), *Serratia liquefaciens* and *S. marcescens*, *Cyanobacteri*a have been reported to solubilize zinc (Costerousse et al. 2018; Dhaked et al. 2017; Bapiri et al. 2012; (Altomare et al. 1999; Fasim et al. 2002).
- For instance, the Zn-solubilizing ability of soil bacteria *Bacillus* sp. and *Pseudomonas* sp. (Di Simine et al. 1998; Fasim et al. 2002) through organic acid production such as glycolic acid, formic acid, acetic acid, propionic acid, lactic acid, fumaric acid, and succinic acid has been reported (Vazquez et al. 2000).
- The effectiveness of *Pseudomonas* sp. in solubilizing $Zn_3(PO_4)_2$ and ZnO through production of gluconic and 2-keto-gluconic acid has been demonstrated (Di Simine et al. 1998).
- A strain of *P. fluorescens*, solubilized $Zn_3(PO_4)_2$, only in the presence of glucose as a carbon source. Increased concentration of protons (H^+) was observed in the medium, probably as a result of ammonia assimilation and gluconic acid production, which lead to the solubilization (Di Simine et al. 1998).
- Further, *Bacillus* strains solubilize insoluble Zn through production of amino acids, plant hormones, chelating ligands, and organic acids via redox reactions and proton extrusion (Saravanan et al. 2004). Among various *Bacillus* strains, *Bacillus aryabhattai* formed a clear halo greater than 10 mm around the colonies on solid tris-minimal medium amended with different insoluble forms of Zn separately and was highly effective in solubilizing insoluble Zn (Pawar et al. 2015).
- The development of the zone area on ZnO containing modified Bunt and Revira medium (Bunt and Rovira 1955) by *Bacillus* sp. and a comparable halo area for *Acinetobacter* sp. was reported (Goteti et al. 2013; Hussain et al. 2015; Vaid et al. 2014). An investigation was carried out on organic acid production profiling during zinc solubilization by strains (*B. aryabhattai*, *Pseudomonas taiwenensis* and *Bacillus* sp.) (Costerousse et al. 2018).
- In a study, it is explained that solubilization is not observed mainly due to the gluconic acid production alone but in strains like *Curtobacterium oceanosedimentum* and *Streptomyces narbonensis*, solubilization is ascribed to the production of a mixture of six and seven different organic acids, respectively, as they produce marginal quantities of gluconic acid (Costerousse et al. 2018).

(b) *Iron Oxidizers*

Hutchins et al. (1986) reported that *Thiobacillus thioxidans*, *T. ferroxidans*, and facultative thermophilic iron oxidizers solubilized zinc from sulfide ore (sphalerite) (Saravanan et al. 2004).

(c) *Endophytic Diazotroph*

An *in vitro* study reported, *Gluconacetobacter diazotrophicus*, a gram-negative bacteria, as a potential solubilizer for insoluble zinc source especially ZnO, $ZnCO_3$, and $Zn_3(PO_4)_2$. *G. diazotrophicus*. PAl5 solubilizes ZnO more efficiently than $ZnCO_3$ or $Zn_3(PO_4)_2$ through the production of organic acids using glucose and sucrose as the carbon source (Sarathambal et al. 2010; Saravanan et al. 2007a, b). GC-MS (Gas chromatography coupled Mass Spectrometry) analysis shows the presence of a derivative of gluconic acid, that is, 5-ketogluconic acid, as the major organic acid produced by *G. diazotrophicus* PAl5, when cultured with glucose as the sole source of carbon. The production of organic anions might be important in the solubilization of metal compounds.

(d) *Ericoid Mycorrhizal Fungi*

Ericoid mycorrhizal fungi (*Oidiodendron maius*) produce extracellular fungal metabolites, which interact with insoluble metal species and thus, increase the ability of their host plant in micronutrient (Zn) uptake. Organic acid induction (malate and citrate) by the mycorrhizal strains is reckoned for metal solubilization (Fomina et al. 2004; Martino et al. 2003).

(e) *Fungi*

A few fungal genera such as *Aspergillus*, *Trichoderma*, and *Beauveria* substantially solubilize Zn while tolerating high concentrations of Zn in soil (Alloway 2009; Altomare et al. 1999; Anitha et al. 2015; Fomina et al. 2004; van Beelen and Fleuren-Kemilä 1997) (Table 11.4).

Mechanisms of Action of Zinc Solubilization

Different mechanisms employed by zinc solubilizers to improve Zn bioavailability are as follows:

1. *Chelation (via Siderophores production* (Ahmed and Holmström 2014; Brandel et al. 2012; Saravanan et al. 2011) and *chelating ligands* (Hughes and Poole 1989; Johnstone and Nolan 2015; Wakatsuki 1995)
2. *Acidification* (via *organic acid production*)
3. *Cell membrane proton system*
4. *Oxidative dissolution process* (e.g., redox reactions) (Glasauer et al. 2013)
5. *Changes in root architecture* (Chang et al. 2005)

Some of them are discussed in the following text.

Table 11.4 Solubilization of different insoluble forms of Zn by fungi

Fungi	Insoluble Zn sources	Mechanism of solubilization	References
Trichoderma harzianum Rifai	Metallic Zn	Sequestering Zn and increasing the oxidative dissolution process	Altomare et al. (1999)
Ericoid mycorrhizal fungi			
Oidiodendron maius	ZnO and $Zn_3(PO_4)_2$	Sequester Zn by producing malate and citrate	Fomina et al. (2004) and Martino et al. (2003)
Filamentous nonmycorrhizal fungi			
Penicillium luteum	ZnO, $Zn_3(PO_4)_2$	Production of Gluconic acid	Alloway (2009)
Aspergillus niger, A. nomius, A. oryzae	ZnO, $Zn_3(PO_4)_2$	Production of different organic acids, such as citric acid, oxalic acid, etc.	Alloway (2009)
A. terreus	ZnO, $ZnCO_3$, $Zn_3(PO_4)_2$	Decrease in pH	Anitha et al. (2015)
Beauveria caledonica	$Zn_3(PO_4)_2$	Acidolysis	Fomina et al. (2004)

Zinc Chelation

Bioavailable fraction of soil Zn is depressed due to its reactive nature and poor persistency. Zn bioavailability could be improved by the production of Zn-chelating compounds either by the plant root system or by the root-colonizing bacteria (Obrador et al. 2003). These chelators and bacterial metabolites (and siderophores) bind with Zn (Tarkalson et al. 1998), thus reducing Zn^{2+} reactions in soil. These Zn-complexes then move in the direction of plant root, releasing Zn^{2+} in close proximity to the root for absorption, making chelators available to chelate with another Zn^{2+}. Chelation is the dominant phenomenon used by microorganisms to boost Zn bioavailability and absorption by plant roots. Siderophores produced by bacteria help in chelating mineral ions, primarily Fe (Johnstone and Nolan 2015), but are also found to chelate Zn^{2+} (Ahmed and Holmström 2014; Brandel et al. 2012; Hussain et al. 2018).

Bacteria such as *Pseudomonas monteilii*, *Microbacterium saperdae*, and *Enterobacter cancerogenesis* enhance the bioavailability of Zn^{2+} in soil by manufacturing Zn-chelating metallophores (Whiting et al. 2001). It has been reported that biofertilizers containing species of *Pseudomonas*, *Azospirillum lipoferum*, and *Agrobacterium* genera release fixed forms of Zn and these microbes makes it bioavailable for longer time duration by producing chelating agents (Tariq et al. 2007). *Penicillium bilaji* has also been reported to enhance Zn bioavailability via chelation (Hussain et al. 2018; Kucey 1987).

Acidification (Via Organic Acid Production)

Bacteria, through the course of growth and proliferation, have the capability of producing organic acids, due to which pH drops, and nutrients, such as Zn^{2+}, are converted into available forms, ready for plant uptake (Saravanan et al. 2004). The anionic organic acids such as gluconate and ketogluconate chelate Zn^{2+} through carboxyl and hydroxyl groups, in multiple confirmations, thereby enhancing the mineral bioavailability (Welch et al. 2002).

Most of the PGP strains are found to be involved in the production of gluconic acid (Saravanan et al. 2011) or the derivatives of gluconic acid, such as 2- ketogluconic acid (Fasim et al. 2002), 5-ketogluconic acid (Saravanan et al. 2007a, b), and a number of other organic acids such as citric acid, formic acid, glycolic acid, lactic acid, malic acid, oxalic acid, propionic acid, succinic acid, and 2,5-keto-derivatives (Tariq et al. 2007) for Zn solubilization. In a liquid medium, the release of organic acids like caffeic acid, cinnamic acid, chlorogenic acid, ferulic acid, gallic acid, and syringic acid helps in the solubilization of insoluble forms of Zn by *Bacillus* sp. AZ6. The production of organic acids (gluconic acid and its derivatives) was accompanied by the rise in H^+ concentrations in the process of Zn solubilization (Fasim et al. 2002). All these mechanisms help in the enhancement of Zn solubilization and mobilization, uptake, translocation, and accumulation of Zn within the plant parts (Costerousse et al. 2018; Hussain et al. 2018) (Table 11.5).

Cell Membrane Proton System

Protons (H^+) replace Zn cations (Zn^+) due to the presence of same charges at the sorption sites of the metal anions so as to form complexes, resulting in Zn mobilization. The protons might either be directly released by the bacterial cell membranes (e.g., nutrient acquisition or for maintenance of charge neutrality) and/or indirectly from the carboxyl groups of the organic acid produced (at pH higher than pKa) (Costerousse et al. 2018; Glasauer et al. 2013).

Changes in Root Architecture

Due to the native poor bioavailability or low exogenous applications, low concentrations of Zn are present in the rhizosphere, resulting in reduced mobility. For optimum uptake by plants, it should be made available in soluble form. Increased root surface area and improved and enhanced growth of the root system mediate Zn uptake by plants through diffusion (Havlin et al. 2005), even from longer distances. Mycorrhizal fungus has a well-known influence on the root architecture, as it adds up with the root surface and, thus, helps in picking Zn from longer stretches. It has been reported that arbuscular mycorrhizae is capable of picking up Zn from up to a distance of 40 mm from the root surface, and increases the Zn density of the grain by up to 4% (Bürkert and Robson 1994; Subramanian et al. 2009). Enhanced Zn

Table 11.5 Solubilization of different insoluble forms of Zn via organic acid production

Bacteria	Insoluble Zn sources	Organic acid production	References
Pseudomonas fluorescens	$Zn_3(PO_4)_2$	Production of different organic acids	Di Simine et al. (1998)
Pseudomonas aeruginosa	ZnO, $Zn_3(PO_4)_2$	Gluconic acid and ketogluconic acid	Fasim et al. (2002)
Pseudomonas taiwenensis	ZnO, $Zn_3(PO_4)_2$ $ZnCO_3$	Keto-D-glutarate, citric acid, propionic acid, gluconic acid and oxalic acid	Vidyashree et al. (2018)
Pseudomonas sp. and *Bacillus* sp.	ZnO, ZnS and $ZnCO_3$	Production of different organic acids	Saravanan et al. (2004)
Bacillus cereus	ZnO, $Zn_3(PO_4)_2$	Production of different organic acids	Kumar et al. (2017)
B. aryabhattai	ZnO, $Zn_3(PO_4)_2$, $ZnCO_3$	Malic acid, malonic acid, succinic acid, citric acid, propionic acid, keto-D-glutarate and gluconic acid	Vidyashree et al. (2018)
B. megaterium	ZnO, $Zn_3(PO_4)_2$, $ZnCO_3$	Gluconic acid	Dinesh et al. (2018)
Gluconacetobacter diazotrophicus	ZnO, $ZnCO_3$ or $Zn_3(PO_4)_2$	Gluconic acid and its derivative 5-ketogluconic acid	Saravanan et al. (2007a, b) and Madhaiyan et al. (2004)
Klebsiella sp. *Pseudomonas* sp.	ZnO, $Zn_3(PO_4)_2$	Production of different organic acids	Ramesh et al. (2014)
Acinetobacter calcoaceticus	ZnO, $ZnCO_3$	Production of different organic acids	Goteti et al. (2013)
Acinetobacter sp.	ZnO, $ZnSO_4$	Production of different organic acids	Othman et al. (2017)
Serratia sp.	ZnO, $ZnSO_4$	Production of different organic acids	Othman et al. (2017)

biofortification and increased root length (Subramanian et al. 2009), weight, and volume have been observed with the inoculation of Zn-solubilizing microbes and mycorrhizal fungi (Hussain et al. 2018; Tariq et al. 2007).

Role of ZnSB in Combination with Zinc and Other Amendments

The combination of Zn-solubilizing bioinoculants, either solitarily or in consortium (Cakmak 2009a; Marschner 1993; Obrador et al. 2003), along with different organic amendments, have a great potential in assuring macro- and micronutrients' availability, which are usually overlooked by agronomist (Imran et al. 2014). Organic modifications increase the Zn bioavailability, through chelating agents and pH drop

(Cakmak 2009b; Marschner 1995b; Obrador et al. 2003) and by reducing complexation of the native Zn and also the applied Zn with other metal anions (Imran et al. 2014). Organic matter boosts the carbon percentage of the microbial biomass by increasing the production of polysaccharides (Hodgson 1963), thereby increases Zn mobilization and solubilization (Imran et al. 2014). Organic matter increases the rate of decomposition, thus improving the carbon content of soil, and also influences the soil microbiota, ensuring soil quality and fertility through the process of mineralization. An upsurge in the DTPA-extractable Zn was also observed with the inclusion of soil organic carbon (Sidhu and Sharma 2010). Recent studies have shown that, a substantial rise in the Zn content of maize plant was recorded by the use of Zn solubilizers in combination with organic matter amends (Hussain et al. 2004). It has been stated that bioaugmented zinc increased considerably, thereby decreasing the phytate content of the maize grains (Hussain et al. 2018).

Challenges and Future Perspective of Microbe-Based Agro-Inputs

Growing population endangers the nutritional demand, a worldwide concern which necessitates the implementation of innovative, agronomic-inputs, enabling agricultural crisis. Exploitation of microbial resources, in the form of bioinoculants (bioformulations), has become one of the most promising solutions for intensifying sustainable agriculture practices, especially for small-scale agronomists (Du Jardin 2015).

Innumerous scientific reports and field demonstrations support the agronomic effectiveness of biofortification and supplementation strategies, yet these are significantly not succeeding. Efforts are required to come up with novel strategies to make them competent and proficient and to be used as the standard by cultivators to curb Zn deficiency in crop plants.

A number of Zn-solubilizing microbes have been reported to be highly beneficent for crop productivity and soil health; in contrast, some microbes have demonstrated poor, inconsistent results in field trials (Martinez-Viveros et al. 2010; Bhattacharjee et al. 2008). This inconsistency of results indicates the unpredictability in the potency of the microbial inoculant, which could be associated with biological (crop species, native microbial communities) and abiotic factors (physicochemical conditions of the soil) and management of agronomic practices such as cropping and inter-cropping systems, macro- and micronutrient supplementation (fertilization), irrigation system, and biocontrol strategies (Pereg and McMillan 2015).

The interpretation of the biotic and abiotic interactions could help in the manipulation and designing of novel agro-models; thus, next-generation agriculture would be able to achieve suitable associations between biological, chemical, and biocomputing technologies.

Multi- and inter-disciplinary approaches should be adopted in the development of stable microbial-based formulations, as formulation is one of the most decisive and crucial steps (Bashan et al. 2016). Microbial formulations must possess the capability to withstand harsh storage and transportation set-ups, with limited viability loss of the active ingredients to guarantee extended shelf life (Linquist et al. 2013; Trenkel 2010; Shaviv 2000) and must also be suitable to use along with conventional agro-equipments and other agronomic inputs, particularly mineral fertilizers.

Reports and data, along with in-field demonstrations, must be specifically customized and user-defined for farmers' apprehension and acceptance. The next-generation agro-inputs greatly depend on the adoption and integration of these microbial inoculants and the ascendancy of the next-generation agro-inputs fosters through collaborative approach of the agronomists, research scientists, policy-formulators, etc. This approach could help in moving towards a sustainable, integrated, and productive ecosystem, with optimizing nutrient dynamics in reference to varying climate and persistent threats to crop productivity (Bargaz et al. 2018).

Conclusion

The two main reasons defining the need of such microbial strategies for concurrent Zn-mobilization and solubilization are as follows:

- Zn is one of the inevitable micronutrients, which plays an important role in deciding the fate of crop productivity.
- Microbes are eco-friendly and favor sustainable agricultural productivity.

The cognizance of the tripartite relation and interaction of the crop species, soil, and microbial inoculants is a prerequisite for the management of cropping and intercropping systems, agricultural practices, leading to enhanced crop yield productivity and soil health.

In future, microbes are a target, in all possible multidisciplinary fields, so as to design an innovative crop-microbe biosystem having synergy and reciprocal interactivity, to secure higher sustainable yields and productivity.

Acknowledgement The authors acknowledge the generous financial help provided by the MHRD, Govt. of India, under the Scheme for Promotion of Academic and Research Collaboration (SPARC; P594).

References

Abaid-Ullah, M., Nadeem, M., Hassan, M., Ganter, J., Muhammad, B., Nawaz, K., Shah, A.S., Hafeez, F.Y.: Plant growth promoting rhizobacteria: an alternate way to improve yield and quality of wheat (Triticum aestivum). Int. J. Agric. Biol. **17**(1), 51–60 (2015)

Adriano, D.C., Paulsen, G.M., Murphy, L.S.: Phosphorus-iron and phosphorus-zinc relationships in corn (Zea mays L.) seedlings as affected by mineral nutrition 1. Agron. J. **63**(1), 36–39 (1971)

Ahmed, E., Holmström, S.J.: Siderophores in environmental research: roles and applications. Microb. Biotechnol. **7**(3), 196–208 (2014)

Alexander, M.: Introduction to Soil Microbiology. Wiley, New York (1997)
Alloway, B.: Zinc in Soil and Crop Nutrition. Zinc in Plant Nutrition. International Zinc Association, Bruselas (2004)
Alloway, B.J.: Micronutrients and crop production. In: Micronutrient Deficiencies in Global Crop Production. Springer Science & Business Media (2008a)
Alloway, B.J.: Zinc in Soils and Crop Nutrition, p. 139. IZA and IFA, Brussels, Belgium and Paris (2008b)
Alloway, B.J.: Soil factors associated with zinc deficiency in crops and humans. Environ. Geochem. Health. **31**(5), 537–548 (2009)
Altomare, C., Norvell, W.A., Björkman, T., Harman, G.E.: Solubilization of phosphates and micronutrients by the plant-growth-promoting and biocontrol fungus Trichoderma harzianum Rifai 1295–22. Appl. Environ. Microbiol. **65**(7), 2926–2933 (1999)
Ambler, J.E., Brown, J.C.: Cause of differential susceptibility to zinc deficiency in two varieties of navy beans (Phaseolus vulgaris L.) 1. Agron. J. **61**(1), 41–43 (1969)
Anitha, S., Devi, S.N.P., Kumari, K.S.: Isolation and identification of zinc solubilizing fungal isolates from agricultural fields. Indian J. Agric. Sci. **85**(12), 1638–1642 (2015)
Anuradha, P., Syed, I., Swati, M., Patil, V.D.: Solubilization of insoluble zinc compounds by different microbial isolates in vitro condition. Int. J. Trop. Agric. **33**(2 (Part II)), 865–869 (2015)
Bapiri, A., Asgharzadeh, A., Mujallali, H., Khavazi, K., Pazira, E.: Evaluation of zinc solubilization potential by different strains of fluorescent pseudomonads. J. Appl. Sci. Environ. Manag. **16**(3), 295–298 (2012)
Bargaz, A., Lyamlouli, K., Chtouki, M., Zeroual, Y., Dhiba, D.: Soil microbial resources for improving fertilizers efficiency in an integrated plant nutrient management system. Front. Microbiol. **9**, 1606 (2018)
Bokhtiar, S.M., Sakurai, K.: Effects of organic manure and chemical fertilizer on soil fertility and productivity of plant and ratoon crops of sugarcane. Arch. Agron. Soil Sci. **51**(3), 325–334 (2005)
Brandel, J., Humbert, N., Elhabiri, M., Schalk, I.J., Mislin, G.L., Albrecht-Gary, A.M.: Pyochelin, a siderophore of Pseudomonas aeruginosa: physicochemical characterization of the iron (III), copper (II) and zinc (II) complexes. Dalton Trans. **41**(9), 2820–2834 (2012)
Bunt, J.S., Rovira, A.D.: Microbiological studies of some subantarctic soils. J. Soil Sci. **6**(1), 119–128 (1955)
Bürkert, B., Robson, A.: 65Zn uptake in subterranean clover (*Trifolium subterraneum* L.) by three vesicular-arbuscular mycorrhizal fungi in a root-free sandy soil. Soil Biol. Biochem. **26**(9), 1117–1124 (1994)
Broadley, M.R., White, P.J., Hammond, J.P., Zelko, I., Lux, A.: Zinc in plants. New Phytol. **173**(4), 677–702 (2007)
Bhattacharjee, R.B., Singh, A., Mukhopadhyay, S.N.: Use of nitrogen-fixing bacteria as biofertiliser for non-legumes: prospects and challenges. Appl. Microbiol. Biotechnol. **80**(2), 199–209 (2008)
Bashan, Y., Kloepper, J.W., de-Bashan, L.E., Nannipieri, P.: A need for disclosure of the identity of microorganisms, constituents, and application methods when reporting tests with microbe-based or pesticide-based products. Biol. Fertil. Soils. **52**(3), 283 (2016)
Cakmak, I.: Tansley Review No. 111: possible roles of zinc in protecting plant cells from damage by reactive oxygen species. New Phytol. **146**(2), 185–205 (2000)
Cakmak, I.: Enrichment of cereal grains with zinc: agronomic or genetic biofortification? Plant Soil. **302**(1–2), 1–17 (2008)
Cakmak, I.: Enrichment of fertilizers with zinc: an excellent investment for humanity and crop production in India. J. Trace Elem. Med. Biol. **23**(4), 281–289 (2009a)
Cakmak, I.: Agronomic Approaches in Biofortification of Food Crops with Micronutrients. eScholarship, University of California (2009b)
Cakmak, I., Braun, H.J.: Genotypic Variation for Zinc Efficiency, pp. 183–199. Application of physiology in wheat breeding. CIMMYT, Mexico (2001)

Cakmak, I., Hoffland, E.: Zinc for the improvement of crop production and human health. Plant Soil. **361**(1–2), 1–2 (2012)

Cakmak, I., Pfeiffer, W.H., McClafferty, B.: Biofortification of durum wheat with zinc and iron. Cereal Chem. **87**(1), 10–20 (2010)

Chang, H.B., Lin, C.W., Huang, H.J.: Zinc-induced cell death in rice (Oryza sativa L.) roots. Plant Growth Regul. **46**(3), 261–266 (2005)

Chernavina, P.: Importance of trace elements in pigment production of microbes. Mol. Biol. **6**, 340–355 (1970)

Coleman, J.E.: Zinc proteins: enzymes, storage proteins, transcription factors, and replication proteins. Annu. Rev. Biochem. **61**, 897–946 (1992)

Costerousse, B., Schönholzer-Mauclaire, L., Frossard, E., Thonar, C.: Identification of heterotrophic zinc mobilization processes among bacterial strains isolated from wheat rhizosphere (Triticum aestivum L.). Appl. Environ. Microbiol. **84**(1), e01715–e01717 (2018)

Dhaked, B.S., Triveni, S., Reddy, R.S., Padmaja, G.: Isolation and screening of potassium and zinc solubilizing bacteria from different rhizosphere soil. Int. J. Curr. Microbiol. App. Sci. **6**(8), 1271–1281 (2017)

Di Simine, C.D., Sayer, J.A., Gadd, G.M.: Solubilization of zinc phosphate by a strain of Pseudomonas fluorescens isolated from a forest soil. Biol. Fertil. Soils. **28**(1), 87–94 (1998)

Disante, K.B., Fuentes, D., Cortina, J.: Response to drought of Zn-stressed *Quercus suber* L. seedlings. Environ. Exp. Bot. **70**(2–3), 96–103 (2011)

Dinesh, R., Srinivasan, V., Hamza, S., Sarathambal, C., Gowda, S.A., Ganeshamurthy, A.N., Gupta, S.B., Nair, V.A., Subila, K.P., Lijina, A., Divya, V.C.: Isolation and characterization of potential Zn solubilizing bacteria from soil and its effects on soil Zn release rates, soil available Zn and plant Zn content. Geoderma, **321**, 173–186 (2018)

Du Jardin, P.: Plant biostimulants: definition, concept, main categories and regulation. Scientia Horticulturae, **196**, 3–14 (2015)

Ellis, B.G.: Adsorption reactions of micronutrients in soil. In: Micronutrients in Agriculture, pp. 59–78 (1972)

Epstein, E., Bloom, A.J.: Mineral Nutrition of Plants: Principles and Perspectives, 2nd edn. Sinauer Assoc, Sunderland (2005)

El-Fouly, M.M., Nofal, O.A., Mobarak, Z.M.: Effects of soil treatment with iron, manganese and zinc on growth and micronutrient uptake of sunflower plants grown in high-pH soil. J Agron Crop Sci J Agron Crop Sci. **186**(4), 245–251 (2001)

Fasim, F., Ahmed, N., Parsons, R., Gadd, G.M.: Solubilization of zinc salts by a bacterium isolated from the air environment of a tannery. FEMS Microbiol. Lett. **213**(1), 1–6 (2002)

Fomina, M., Alexander, I.J., Hillier, S., Gadd, G.M.: Zinc phosphate and pyromorphite solubilization by soil plant-symbiotic fungi. Geomicrobiol J. **21**(5), 351–366 (2004)

Gadd, G.M.: Metals, minerals and microbes: geomicrobiology and bioremediation. Microbiology. **156**(3), 609–643 (2010)

Gao, X., Hoffland, E., Stomph, T., Grant, C.A., Zou, C., Zhang, F.: Improving zinc bioavailability in transition from flooded to aerobic rice. A review. Agron. Sustain. Dev. **32**(2), 465–478 (2012)

Genc, Y., McDonald, G.K.: The potential of synthetic hexaploid wheats to improve zinc efficiency in modern bread wheat. Plant Soil. **262**(1–2), 23–32 (2004)

Genc, Y., Shepherd, K.W., McDonald, G.K., Graham, R.D., Leon, J.: Inheritance of tolerance to zinc deficiency in barley. Plant Breed. **122**(3), 283–284 (2003)

Giordano, P.M., Noggle, J.C., Mortvedt, J.J.: Zinc uptake by rice, as affected by metabolic inhibitors and competing cations. Plant Soil. **41**(3), 637–646 (1974)

Glasauer, S.M., Beveridge, T.J., Burford, E.P., Harper, F.A., Gadd, G.M.: Metals and Metalloids, Transformation by Microorganisms. Wiley-Interscience, Hoboken (2013)

Glick, B.R.: Plant growth-promoting bacteria: mechanisms and applications. Scientifica. **2012**, 963401 (2012)

Goteti, P.K., Emmanuel, L.D.A., Desai, S., Shaik, M.H.A.: Prospective zinc solubilising bacteria for enhanced nutrient uptake and growth promotion in maize (Zea mays L.). Int. J. Microbiol. **2013**, 1–7 (2013)

Gregorio, G.B., Senadhira, D., Htut, H., Graham, R.D.: Breeding for trace mineral density in rice. Food Nutr. Bull. **21**(4), 382–386 (2000)
Gunes, A., Inal, A., Adak, M.S., Alpaslan, M., Bagci, E.G., Erol, T., Pilbeam, D.J.: Mineral nutrition of wheat, chickpea and lentil as affected by mixed cropping and soil moisture. Nutr. Cycl. Agroecosyst. **78**(1), 83–96 (2007)
Gustin, J.L., Loureiro, M.E., Kim, D., Na, G., Tikhonova, M., Salt, D.E.: MTP1-dependent Zn sequestration into shoot vacuoles suggests dual roles in Zn tolerance and accumulation in Zn-hyperaccumulating plants. Plant J. **57**(6), 1116–1127 (2009)
Hafeez, B., Khanif, Y.M., Saleem, M.: Role of zinc in plant nutrition-a review. J. Exp. Agric. Int. **3**(2), 374–391 (2013)
Havlin, J.L., Beaton, J.D., Tisdale, S.L., Nelson, W.L.: Soil Fertility and Fertilizers: an Introduction to Nutrient Management. Prentice Hall, New Jersey (2005)
Hodgson, J.F.: Chemistry of the micronutrient elements in soils. In: Advances in Agronomy, vol. 15, pp. 119–159. Academic Press (1963)
Hotz, C., Brown, K.H.: Assessment of the risk of zinc deficiency in populations and options for its control. Food Nutr Bull. **25**, S91–S204 (2004)
Hughes, M.N., Poole, R.K.: Metals and Micro-Organisms. Chapman and Hall (1989)
Hussain, D., Haydon, M.J., Wang, Y., Wong, E., Sherson, S.M., Young, J., Camakaris, J., Harper, J.F., Cobbett, C.S.: P-type ATPase heavy metal transporters with roles in essential zinc homeostasis in Arabidopsis. Plant Cell. **16**(5), 1327–1339 (2004)
Hussain, A., Arshad, M., Zahir, Z.A., Asghar, M.: Prospects of zinc solubilizing bacteria for enhancing growth of maize. Pak. J. Agric. Sci. **52**(4), 915–922 (2015)
Hussain, A., Zahir, Z.A., Asghar, H.N., Ahmad, M., Jamil, M., Naveed, M., Akhtar, M.F.U.Z.: Zinc solubilizing bacteria for zinc biofortification in cereals: a step toward sustainable nutritional security. In: Role of Rhizospheric Microbes in Soil, pp. 203–227. Springer, Singapore (2018)
Hutchins, S.R., Davidson, M.S., Brierley, J.A., Brierley, C.L.: Microorganisms in reclamation of metals. Annu. Rev. Microbiol. **40**(1), 311–336 (1986)
Imran, M., Arshad, M., Khalid, A., Kanwal, S., Crowley, D.E.: Perspectives of rhizosphere microflora for improving Zn bioavailability and acquisition by higher plants. Int. J. Agric. Biol. **16**(3), 653–662 (2014)
Inal, A., Gunes, A., Zhang, F., Cakmak, I.: Peanut/maize intercropping induced changes in rhizosphere and nutrient concentrations in shoots. Plant Physiol. Biochem. **45**(5), 350–356 (2007)
Jackson, M.L.: Soil Chemical Analysis, p. 497. Constable and Co. Ltd, London (1962)
Johnstone, T.C., Nolan, E.M.: Beyond iron: non-classical biological functions of bacterial siderophores. Dalton Trans. **44**(14), 6320–6339 (2015)
Jones, D.L., Darrah, P.R.: Role of root derived organic acids in the mobilization of nutrients from the rhizosphere. Plant Soil. **166**(2), 247–257 (1994)
Joshi, D., Negi, G., Vaid, S., Sharma, A.: Enhancement of wheat growth and Zn content in grains by zinc solubilizing bacteria. Int. J. Agric. Environ. Biotechnol. **6**(3), 363–370 (2013)
Kabata-Pendias, A., Pendias, H.: Trace Elements in Soils and Plants. CRC Press, Boca Raton (1992)
Kabata-Pendias, A., Pendias, H.: Trace Elements in Soils and Plants. CRC Press, Boca Raton; London; New-York; Washington (2001)
Kamran, S., Shahid, I., Baig, D.N., Rizwan, M., Malik, K.A., Mehnaz, S.: Contribution of zinc solubilizing bacteria in growth promotion and zinc content of wheat. Front. Microbiol. **8**, 2593 (2017)
Karak, T., Singh, U.K., Das, S., Das, D.K., Kuzyakov, Y.: Comparative efficacy of ZnSO4 and Zn-EDTA application for fertilization of rice (Oryza sativa L.). Arch. Agron. Soil Sci. **51**(3), 253–264 (2005)
Kasim, W.A.: Physiological consequences of structural and ultra-structural changes induced by Zn stress in Phaseolus vulgaris. I. Growth and photosynthetic apparatus. Int. J. Bot. **3**(1), 15–22 (2007)
Katyal, J.C., Randhawa, N.S.: Micronutrients. FAO, fertilizer and plant nutrition. Bulletin. **7** (1983)

Kausar, M.A., Chaudhry, F.M., Rashid, A., Latif, A., Alam, S.M.: Micronutrient availability to cereals from calcareous soils. Plant Soil. **45**(2), 397–410 (1976)
Khalid, A., Arshad, M., Shaharoona, B., Mahmood, T.: Plant growth promoting rhizobacteria and sustainable agriculture. In: Microbial Strategies for Crop Improvement, pp. 133–160. Springer, Berlin, Heidelberg (2009)
Khande, R., Sushil, K.S., Ramesh, A., Mahaveer, P.S.: Zinc solubilizing Bacillus strains that modulate growth, yield and zinc biofortification of soybean and wheat. Rhizosphere. **4**, 126–138 (2017)
Kiekens, L.: Zinc in Heavy Metals. *Soils*. Blackie Academic and Professional, London (1995)
Kloepper, J.W.: Plant growth-promoting rhizobacteria (other systems). Azospirillum/Plant Associations. **187**, 137–166 (1994)
Kucey, R.M.N.: Increased phosphorus uptake by wheat and field beans inoculated with a phosphorus-solubilizing Penicillium bilaji strain and with vesicular-arbuscular mycorrhizal fungi. Appl. Environ. Microbiol. **53**(12), 2699–2703 (1987)
Kumar, P., Srivastava, V.C., Mishra, I.M.: Synthesis of dimethyl carbonate by transesterification reaction using ceria-zinc oxide catalysts prepared with different chelating agents. Appl. Clay Sci. **150**, 275–281 (2017)
Li, W.C., Ye, Z.H., Wong, M.H.: Metal mobilization and production of short-chain organic acids by rhizosphere bacteria associated with a Cd/Zn hyperaccumulating plant, *Sedum alfredii*. Plant Soil. **326**(1–2), 453–467 (2010)
Lindemann, S.R., Bernstein, H.C., Song, H.S., Fredrickson, J.K., Fields, M.W., Shou, W., Johnson, D.R., Beliaev, A.S.: Engineering microbial consortia for controllable outputs. ISME J. **10**(9), 2077–2084 (2016)
Lindsay, W.L.: Zinc in soils and plant nutrition. In: Advances in Agronomy, vol. 24, pp. 147–186. Academic Press (1972)
Linquist, B.A., Liu, L., van Kessel, C., van Groenigen, K.J.: Enhanced efficiency nitrogen fertilizers for rice systems: Meta-analysis of yield and nitrogen uptake. Field. Crops. Res. **154**, 246–254 (2013)
Lombnaes, P., Singh, B.R.: Varietal tolerance to zinc deficiency in wheat and barley grown in chelator-buffered nutrient solution and its effect on uptake of Cu, Fe, and Mn. J. Plant Nutr. Soil Sci. **166**(1), 76–83 (2003)
Loneragan, J.F., Webb, M.J.: Interactions between zinc and other nutrients affecting the growth of plants. In: Zinc in Soils and Plants, pp. 119–134. Springer, Dordrecht (1993)
López-Rayo, S., Nadal, P., Lucena, J.J.: Novel chelating agents for iron, manganese, zinc, and copper mixed fertilisation in high pH soil-less cultures. J. Sci. Food Agric. **96**(4), 1111–1120 (2016)
Lucy, M., Reed, E., Glick, B.R.: Applications of free living plant growth-promoting rhizobacteria. Anton. Leeuw. Int. **86**, 1–25 (2004)
Mahdi, S.S., Hassan, G.I., Samoon, S.A., Rather, H.A., Dar, S.A., Zehra, B.: Bio-fertilizers in organic agriculture. J. Phytol. **2**(10), 42–54 (2010)
Marschner, H.: Zinc uptake from soils. In: Zinc in Soils and Plants, pp. 59–77. Springer, Dordrecht (1993)
Marschner, H.: Adaptation of plants to adverse chemical soil conditions. Mineral nutrition of higher plants (1995a)
Marschner, H.: Mineral Nutrition of Higher Plants, 2nd edn. Academic, Great Britain (1995b)
Marschner, H.: Functions of mineral nutrients, macronutrients. Mineral nutrition of higher plants (1997)
Marschner, H., Cakmak, I.: High light intensity enhances chlorosis and necrosis in leaves of zinc, potassium, and magnesium deficient bean (Phaseolus vulgaris) plants. J. Plant Physiol. **134**(3), 308–315 (1989)
Martino, E., Perotto, S., Parsons, R., Gadd, G.M.: Solubilization of insoluble inorganic zinc compounds by ericoid mycorrhizal fungi derived from heavy metal polluted sites. Soil Biol. Biochem. **35**(1), 133–141 (2003)

Mhatre, M., Srinivas, L., Ganapathi, T.R.: Enhanced iron and zinc accumulation in genetically engineered pineapple plants using soybean ferritin gene. Biol. Trace Elem. Res. **144**(1–3), 1219–1228 (2011)

Mikkelsen, D.S., Kuo, S.: Zinc fertilization and behavior in flooded soils. In: Zinc Fertilization and Behavior in Flooded Soils, vol. 5, (1977)

Moraghan, J.T., Mascagni, H.J.: Environmental and soil factors affecting micronutrient deficiencies and toxicities. In: Micronutrients in Agriculture, vol. 4, pp. 371–425 (1991)

Mortvedt, J.J., Gilkes, R.J.: Zinc Fertilizers. Zinc in Soils and Plants, pp. 33–34. Kluwer Academic Publishers, Dordrecht (1993)

Myers, S.S., Wessells, K.R., Kloog, I., Zanobetti, A., Schwartz, J.: Rising atmospheric CO2 increases global threat of zinc deficiency. The Lancet. Glob. Health. **3**(10), p.e639 (2015)

Madhaiyan, M., Saravanan, V.S., Jovi, D.B.S.S., Lee, H., Thenmozhi, R., Hari, K., Sa, T.: Occurrence of Gluconacetobacter diazotrophicus in tropical and subtropical plants of Western Ghats, India. Microbiol. Res. **159**(3), 233–243 (2004)

Martínez-Viveros, O., Jorquera, M.A., Crowley, D.E., Gajardo, G.M.L.M., Mora, M.L.: Mechanisms and practical considerations involved in plant growth promotion by rhizobacteria. J. Plant. Nutr. Soil Sci. **10**(3), 293–319 (2010)

Naz, I., Ahmad, H., Khokhar, S.N., Khan, K., Shah, A.H.: Impact of zinc solubilizing bacteria on zinc contents of wheat. Am. Euras. J. Agric. Environ. Sci. **16**, 449–454 (2016)

Obrador, A., Novillo, J., Alvarez, J.M.: Mobility and availability to plants of two zinc sources applied to a calcareous soil. Soil Sci. Soc. Am. J. **67**(2), 564–572 (2003)

Olsen, S.R.: Micronutrient interactions. In: Micronutrients in Agriculture, pp. 243–264 (1972)

Oktay, K., Newton, H., Mullan, J., Gosden, R.G.: Development of human primordial follicles to antral stages in SCID/hpg mice stimulated with follicle stimulating hormone. Human reproduction (Oxford, England), **13**(5), 1133–1138 (1998)

Othman, N.M.I., Othman, R., Saud, H.M., Wahab, P.E.M.: Effects of root colonization by zinc-solubilizing bacteria on rice plant (Oryza sativa MR219) growth. Agric. Nat. Resour. **51**(6), 532–537 (2017)

Parisi, B., Vallee, B.L.: Metal enzyme complexes activated by zinc. J. Biol. Chem. **179**, 803–807 (1969)

Parker, M.B., Walker, M.E.: Soil pH and manganese effects on manganese nutrition of peanut 1. Agron. J. **78**(4), 614–620 (1986)

Pathak, D.V., Kumar, M.: Microbial inoculants as biofertilizers and biopesticides. In: Microbial Inoculants in Sustainable Agricultural Productivity, pp. 197–209. Springer, New Delhi (2016)

Pawar, R.C., Choi, D.H., Lee, J.S., Lee, C.S.: Formation of polar surfaces in microstructured ZnO by doping with Cu and applications in photocatalysis using visible light. Mater. Chem. Phys. **151**, 167–180 (2015)

Peck, A.W., McDonald, G.K.: Adequate zinc nutrition alleviates the adverse effects of heat stress in bread wheat. Plant Soil. **337**(1–2), 355–374 (2010)

Price, G.R.: Selection and covariance. Nature, **227**, 520–521 (1970)

Pereg, L., McMillan, M.: Scoping the potential uses of beneficial microorganisms for increasing productivity in cotton cropping systems. Soil Sci Am J. **80**, 349–358 (2015)

Ramesh, A., Sharma, S.K., Sharma, M.P., Yadav, N., Joshi, O.P.: Inoculation of zinc solubilizing Bacillus aryabhattai strains for improved growth, mobilization and biofortification of zinc in soybean and wheat cultivated in Vertisols of central India. Appl. Soil Ecol. **73**, 87–96 (2014)

Rani, N., Kaur, R., Kaur, S.: Zinc solubilizing bacteria to augment soil fertility – a comprehensive review. Int. J. Agric. Sci. Vet. Med. **8**(1), 38–44 (2020)

Rattan, R.K., Shukla, L.M.: Influence of different Zn carriers on the utilization of micronutrients by rice. J. Indian Soc. Soil Sci. **39**, 808–810 (1991)

Reddy, M.R., Perkins, H.F.: Fixation of zinc by clay minerals. Soil Sci. Soc. Am. J. **38**(2), 229–231 (1974)

Rehman, A., Farooq, M., Ozturk, L., Asif, M., Siddique, K.H.: Zinc nutrition in wheat-based cropping systems. Plant Soil. **422**(1–2), 283–315 (2018)

Rengel, Z.: The role of crop residues in improving soil fertility. In: Nutrient Cycling in Terrestrial Ecosystems, pp. 183–214. Springer, Berlin, Heidelberg (2007)

Rengel, Z., Graham, R.D.: Importance of seed Zn content for wheat growth on Zn-deficient soil: II. Grain yield. Plant Soil. **173**(2), 267–274 (1995a)

Rengel, Z., Graham, R.D.: Wheat genotypes differ in Zn efficiency when grown in chelate-buffered nutrient solution. Plant Soil. **176**(2), 307–316 (1995b)

Sadeghzadeh, B.: A review of zinc nutrition and plant breeding. J. Soil Sci. Plant Nutr. **13**(4), 905–927 (2013)

Saeed, M., Fox, R.L.: Relations between suspension pH and zinc solubility in acid and calcareous soils. Soil Sci. **124**(4), 199–204 (1977)

Sajwan, K.S., Lindsay, W.L.: Effect of redox, zinc fertilization and incubation time on DTPA-extractable zinc, iron and manganese. Commun. Soil Sci. Plant Anal. **19**(1), 1–11 (1988)

Sarathambal, C., Thangaraju, M., Paulraj, C., Gomathy, M.: Assessing the zinc solubilization ability of Gluconacetobacter diazotrophicus in maize rhizosphere using labelled 65 Zn compounds. Indian J. Microbiol. **50**(1), 103–109 (2010)

Saravanan, V.S., Subramoniam, S.R., Raj, S.A.: Assessing in vitro solubilization potential of different zinc solubilizing bacterial (ZSB) isolates. Braz. J. Microbiol. **35**(1–2), 121–125 (2004)

Saravanan, V.S., Kalaiarasan, P., Madhaiyan, M., Thangaraju, M.: Solubilization of insoluble zinc compounds by *Gluconacetobacter diazotrophicus* and the detrimental action of zinc ion (Zn2+) and zinc chelates on root knot nematode *Meloidogyne incognita*. Lett. Appl. Microbiol. **44**(3), 235–241 (2007a)

Saravanan, V.S., Madhaiyan, M., Thangaraju, M.: Solubilization of zinc compounds by the diazotrophic, plant growth promoting bacterium *Gluconacetobacter diazotrophicus*. Chemosphere. **66**(9), 1794–1798 (2007b)

Saravanan, V.S., Kumar, M.R., Sa, T.M.: Microbial zinc solubilization and their role on plants. In: Bacteria in Agrobiology: Plant Nutrient Management, pp. 47–63. Springer, Berlin, Heidelberg (2011)

Schippers, A., Hedrich, S., Vasters, J., Drobe, M., Sand, W., Willscher, S.: Biomining: metal recovery from ores with microorganisms. In: Geobiotechnology I, pp. 1–47. Springer, Berlin, Heidelberg (2013)

Shaikh, S.S., Saraf, M.S.: Optimization of growth conditions for zinc solubilizing plant growth associated bacteria and fungi. J. Adv. Res. Biotechnol. **2**(1), 1–9 (2017)

Shukla, U.C., Mittal, S.B.: Characterization of zinc adsorption in some soils of India. Soil Sci. Soc. Am. J. **43**(5), 905–908 (1979)

Shukla, U.C., Morris, H.D.: Relative efficiency of several zinc sources for corn (Zea mays L.) 1. Agron. J. **59**(2), 200–202 (1967)

Shaviv, A.: Advances in controlled release of fertilizers. Adv. Agron. **71**, 1–49 (2000)

Sidhu, G.S., Sharma, B.D.: Diethylenetriaminepentaacetic acid–extractable micronutrients status in soil under a rice–wheat system and their relationship with soil properties in different agroclimatic zones of indo-gangetic plains of India. Commun. Soil Sci. Plant Anal. **41**(1), 29–51 (2010)

Sillanpaa, M.: Micronutrient Assessment at the Country Level: an International Study. Food and Agriculture Organization of the United Nations, Rome (1990)

Singh, M.V., Abrol, I.P.: Transformation and movement of zinc in an alkali soil and their influence on the yield and uptake of zinc by rice and wheat crops. Plant Soil. **94**(3), 445–449 (1986)

Singh, J.P., Karamanos, R.E., Stewart, J.W.B.: Phosphorus-induced zinc deficiency in wheat on residual phosphorus Plots1. Agron. J. **78**(4), 668–675 (1986)

Singh, B., Natesan, S.K.A., Singh, B.K., Usha, K.: Improving zinc efficiency of cereals under zinc deficiency. Curr. Sci. **88**(1), 36–44 (2005)

Singh, D.P., Singh, H.B., Prabha, R. (eds.): Microbial Inoculants in Sustainable Agricultural Productivity, p. 118. Springer, New Delhi (2016)

Smith, S.E., Read, D.J.: Mycorrhizal symbiosis, 3rd edn. Academic Press, New York (2008)

Stahl, R.S., James, B.R.: Zinc sorption by B horizon soils as a function of pH. Soil Sci. Soc. Am. J. **55**(6), 1592–1597 (1991)

Subramanian, K.S., Tenshia, V., Jayalakshmi, K., Ramach, V.: Role of arbuscular mycorrhizal fungus (*Glomus intraradices*) (fungus aided) in zinc nutrition of maize. J. Agric. Biotechnol. Sustainable Dev. **1**(1), 029–038 (2009)

Takkar, P.N., Randhawa, N.S.: Micronutrients in Indian agriculture. Fert. News. **23**(8), 3–26 (1978)

Takkar, P.N., Walker, C.D.: The distribution and correction of zinc deficiency. In: Zinc in Soils and Plants, pp. 151–165. Springer, Dordrecht (1993)

Tan, S., Han, R., Li, P., Yang, G., Li, S., Zhang, P., Wang, W.B., Zhao, W.Z., Yin, L.P.: Over-expression of the MxIRT1 gene increases iron and zinc content in rice seeds. Transgenic Res. **24**(1), 109–122 (2015)

Tariq, M., Hameed, S., Malik, K.A., Hafeez, F.Y.: Plant root associated bacteria for zinc mobilization in rice. Pak. J. Bot. **39**(1), 245 (2007)

Tarkalson, D.D., Jolley, V.D., Robbins, C.W., Terry, R.E.: Mycorrhizal colonization and nutrient uptake of dry bean in manure and compost manure treated subsoil and untreated topsoil and subsoil. J. Plant Nutr. **21**(9), 1867–1878 (1998)

Tavallali, V., Rahemi, M., Eshghi, S., Kholdebarin, B., Ramezanian, A.: Zinc alleviates salt stress and increases antioxidant enzyme activity in the leaves of pistachio (*Pistacia vera* L.'Badami') seedlings. Turk. J. Agric. For. **34**(4), 349–359 (2010)

Timmusk, S., Behers, L., Muthoni, J., Muraya, A., Aronsson, A.C.: Perspectives and challenges of microbial application for crop improvement. Front. Plant Sci. **8**, 49 (2017)

Treeby, M., Marschner, H., Römheld, V.: Mobilization of iron and other micronutrient cations from a calcareous soil by plant-borne, microbial, and synthetic metal chelators. Plant Soil. **114**(2), 217–226 (1989)

Tzen, Z.Y., Chen, J.H.: Evaluation the application rate of animal manure compost in a strongly acidic soil. Soil Environ. **7**, 83–96 (2004)

Trenkel, M.E.: Slow-and controlled-release and stabilized fertilizers: an option for enhancing nutrient use efficiency in agriculture. IFA, International fertilizer industry association. Paris, Francia. Consultado 14 dic. del 2016 (2010)

Udo, E.J., Bohn, H.L., Tucker, T.C.: Zinc adsorption by calcareous soils. Soil Sci. Soc. Am. J. **34**(3), 405–407 (1970)

Umesha, S., Singh, P.K., Singh, R.P.: Microbial biotechnology and sustainable agriculture. In: Biotechnology for Sustainable Agriculture, pp. 185–205. Woodhead Publishing (2018)

van Beelen, P., Fleuren-Kemilä, A.K.: Influence of pH on the toxic effects of zinc, cadmium, and pentachlorophenol on pure cultures of soil microorganisms. Environ. Toxicol. Chem. **16**(2), 146–153 (1997)

Vaid, S.K., Kumar, B., Sharma, A., Shukla, A.K., Srivastava, P.C.: Effect of Zn solubilizing bacteria on growth promotion and Zn nutrition of rice. J. Soil Sci. Plant Nutr. **14**(4), 889–910 (2014)

Vazquez, P., Holguin, G., Puente, M.E., Lopez-Cortes, A., Bashan, Y.: Phosphate-solubilizing microorganisms associated with the rhizosphere of mangroves in a semiarid coastal lagoon. Biol. Fertil. Soils. **30**(5–6), 460–468 (2000)

Vessey, J.K.: Plant growth promoting rhizobacteria as biofertilizers. Plant Soil. **255**(2), 571–586 (2003)

Viets, F.G., Prasad, A.S.: Zinc deficiency in the soil-plant system. In: Zinc Metabolism, pp. 90–127 (1966)

Viets, F.G., Boawn, L.C., Crawford, C.L.: The effect of nitrogen and types of nitrogen carrier on plant uptake of indigenous and applied zinc. Soil Sci. Soc. Am. J. **21**(2), 197–201 (1957)

Vidyashree, D.N., Muthuraju, R., Panneerselvam, P.: Evaluation of zinc solubilizing bacterial (ZSB) strains on growth, yield and quality of tomato (Lycopersicon esculentum). Int. J. Curr. Microbiol. Appl. Sci. **7**(04), p.2018 (2018)

Wakatsuki, T.: Metal oxidoreduction by microbial cells. J. Ind. Microbiol. **14**(2), 169–177 (1995)

Welch, S.A., Taunton, A.E., Banfield, J.F.: Effect of microorganisms and microbial metabolites on apatite dissolution. Geomicrobiol J. **19**(3), 343–367 (2002)

White, P.J., Broadley, M.R.: Biofortifying crops with essential mineral elements. Trends Plant Sci. **10**(12), 586–593 (2005)

Whiting, S.N., de Souza, M.P., Terry, N.: Rhizosphere bacteria mobilize Zn for hyperaccumulation by *Thlaspi caerulescens*. Environ. Sci. Technol. **35**(15), 3144–3150 (2001)

Wu, S.C., Cheung, K.C., Luo, Y.M., Wong, M.H.: Effects of inoculation of plant growth-promoting rhizobacteria on metal uptake by Brassica juncea. Environ. Pollut. **140**(1), 124–135 (2006)

Wang, H., Jin, J.Y.: Photosynthetic rate, chlorophyll fluorescence parameters, and lipid peroxidation of maize leaves as affected by zinc deficiency. Photosynthetica, **43**(4), 591–596 (2005)

Zhang, F.S., Römheld, V., Marschner, H.: Diurnal rhythm of release of phytosiderophores and uptake rate of zinc in iron-deficient wheat. Soil Sci. Plant Nutr. **37**(4), 671–678 (1991)

Zuo, Y., Zhang, F.: Iron and zinc biofortification strategies in dicot plants by intercropping with gramineous species. A review. Agron. Sustain. Dev. **29**(1), 63–71 (2009)

Chapter 12
Microbial Biofertilizers and Micronutrients Bioavailability: Approaches to Deal with Zinc Deficiencies

Jamjam Waqeel and Shams Tabrez Khan

Abstract Micronutrients like boron, iron, manganese, zinc, and copper, although required in small quantities, are essential for plant health and growth. Excessive use of chemical fertilizers to meet the food demand is resulting in nutrient imbalances in soil, plants, and humans leading to environmental and health problems. Use of biofertilizers is an effective, economical, and sustainable approach to tackle the problem of nutrient imbalance. Biofertilizers help in maintaining the micronutrients in soil and making them available to plants. Zinc malnutrition is one of the biggest problems, as its biological consequences are severe and some of the estimates by international organizations are alarming. According to the Food and Agriculture Organization (FAO) of United Nations, 50% of the world's cereal soils are deficient in zinc. The World Health Organization estimates that approximately 800,000 people, mainly children under the age of 5, die annually due to the health complications caused by zinc deficiency. Zinc deficiency affects about 2 billion people worldwide. Various approaches are being used to combat the zinc deficiency such as increasing bioavailability of Zn to plants resulting in Zn biofortification of crops, use of supplements, and food fortification. To increase the Zn content of the plants, various strategies are used such as application of chemical fertilizers (soil and foliar spray), use of biofertilizers, and use of genetic engineering and plant breeding approaches. This chapter discusses some of the strategies used for the Zn biofortification of cereal crops.

Keywords Micronutrients · Biofertilizers · Malnutrition · Bioavailability · Biofortification · Genetic engineering

J. Waqeel (✉) · S. T. Khan
Department of Agricultural Microbiology, Faculty of Agricultural Sciences, Aligarh Muslim University, Aligarh, Uttar Pradesh, India

S. T. Khan, A. Malik (eds.), *Microbial Biofertilizers and Micronutrient Availability*,
https://doi.org/10.1007/978-3-030-76609-2_12

Introduction

Macronutrients (N, P, K, Ca, Mg, S, etc.) and micronutrients (Cu, Fe, Mn, Zn, B, Si, Co, V, etc.) are required by plants for their growth, development, and reproduction. The deficiency of these nutrients has consequences that vary with the degree of deficiency and type of nutrient (Ramesh et al. 2004). Micronutrients, although required in very small amounts, are equally necessary for all living organisms including plants and humans (Miller and Welch 2013). Plants as primary producers acquire these nutrients from soil, but due to the extensive agriculture, the concentrations of various nutrients in agricultural soil are depleting. On the contrary, the demand for food is increasing with increasing world population (Santos et al. 2012).

To meet the nutritional requirement of the plants, farmers are dependent on chemical fertilizers (Thomas and Singh 2019). The use of Zn fertilizers is one such approach; some of the commonly used fertilizers include zinc (*heptahydrate*), zinc (*monohydrate*), and chelated zinc (Zn-EDTA) (DAE 2014). However, the use of these fertilizers is associated with inherent problems like pollution and accumulation in soil (Zahra et al. 2010), leaching to water bodies, and toxicity to other living organisms. Alternatives to these fertilizers and other agricultural chemicals are therefore required to protect the environment in a sustainable manner. One such approach is the microbial inoculants called as "biofertilizers" with obvious benefits to agro-ecosystem (Mahajan et al. 2008). The use of such approaches is highly desirable and urgent, as problems like malnutrition and micronutrient deficiency are becoming prevalent. It is estimated that more than three billion people around the world are at the risk of malnutrition and micronutrient deficiency (Monreal et al. 2015). The micronutrient deficiency, also called as "hidden hunger," in humans is due to various reasons, the most important being the low dietary intake (Mazziotta et al. 2016). Depleting levels of certain nutrients in the plants consumed as food such as cereals mainly contribute to the low dietary intake. The micronutrient deficiency is also called "hidden" or "latent" deficiency in soil because its deficiency can reduce quality and yield without the appearance of any obvious symptom (Alloway 2009). Although all the micronutrients are equally important, zinc deficiency is emerging as a global health concern for various reasons. The prevalence of Zn deficiency, together with lack of awareness about it, is complicating the situation. One reason is rampant zinc deficiency in agricultural soil; according to some estimates, it is over 50% of the world's agricultural soil. Nearly one-third of Uttar Pradesh's soils lacked available zinc, while in states such as Gujarat, Assam, Odisha, Telangana, and Andhra Pradesh, every fourth sample was low in Zn content (Shukla et al. 2014). Zn is also known as metal of life and its deficiency in plants has more severe consequences than other micronutrients such as Fe, Mn, and Cu (Clemens 2006; Broadley et al. 2007). The presence of non-bioavailable zinc in soil and various soil factors like high pH (Sherene 2010); high contents of bicarbonate ions, calcite, salts; and high levels of available phosphorus (Alloway 2009) contribute to Zn deficiency in the soil. Hence, they decrease the bioavailability and mobility of zinc in soil, reducing their dissemination to plant root through promoting the incorporation into soil components, and cause a major decline in food production and quality of nutrition (Henriques et al. 2012). Because cereal crops like wheat and rice are the main component of every day meals, when these foods are consumed by humans, their widespread Zn deficiency is reported in Indian human populations (Singh 2009). In order to resolve this situation, there has been a growing demand for developing varieties of grain crops with increased concentrations of Zn to improve grain nutritional quality for human consumption (White and Broadley 2005; Wissuwa et al. 2008). Different strategies are required to improve the Zn status in soil and plants. Seed priming, foliar application (Alloway 2008a), breeding, and molecular biology approaches are being used to make the crops zinc efficient and improve Zn uptake from soil (van de Mortel et al. 2006).

Macronutrients and Micronutrient for Plants

Nutrients have been categorized as macro- and micronutrients according to their required amounts. A nutritionally dense food is a food that contains both macro- and micronutrient, including carbon, hydrogen, oxygen, nitrogen, phosphorus, potassium, calcium, and magnesium. The micronutrients are boron (B), chlorine (Cl), copper (Cu), iron (Fe), manganese (Mn), nickel (Ni), molybdenum (Mo), and zinc (Zn), (Hänsch and Mendel 2009). Like humans, plants also require macro- and micronutrients, and deficiency of these nutrients has varying consequences, even causing plant death (Ramesh et al. 2004). Plants can fix carbon from atmosphere but acquire other nutrients from the soil through roots (Shinozaki et al. 2020). Nitrogen is one of the most important macronutrients in addition to carbon, hydrogen, and oxygen. Combination of nitrogen with other macronutrients such as O, C, H, S, and, P constitutes various organic nitrogenous compounds like proteins, nucleotides, porphyrins, and alkaloids (Campbell 1999). These nitrogen-containing molecules can be classified into two categories, namely, low molecular weight compounds and high molecular weight compounds (Weissert and Kehr 2017). The nitrogen in the form of protein, nucleic acid, and free amino acid constitutes about 80–85%, 10%, and 5% of the total plant N (Kirkby et al. 2009). Nitrogen is stored in cells as assimilatory amino acids, mainly *Glutamine* and *Asparagines* (Temple et al. 1998). Another important macronutrient is potassium (K) required for the growth and development of the plants. It constitutes up to 10% of the total plant dry weight and is involved in several physiological processes such as osmoregulation, enzyme activation, and maintenance of membrane potential, photosynthesis, and carbohydrate metabolism and translocation (Clarkson and Hanson 1980; Holdaway-Clarke and Hepler 2003; Leigh and Wyn 1984). It also regulates turgor pressure and regulates stomata opening and closing and pollen tube growth (Kim et al. 2010). Another macronutrient for plants is calcium (Franco-Navarro et al. 2016; Raven 2017; Wege et al. 2017). Except for few soils, Ca is generally sufficiently present in soil (Pasricha et al. 2002). Ca provides rigidity to cell walls and serves as a core secondary messenger in signaling pathways. It participates in lipid, protein, and carbohydrate metabolism (Reddy and Reddy 2004). Pectate binds Ca^{+2} in the cell wall, which is therefore the most important reservoir of Ca^{+2} in the plant. The vacuole, ER, plastids, and mitochondria contain higher Ca levels than the cytoplasm (Mitra 2015). In plant tissue, calcium is present as Ca^{2+}, or as calcium carbonate, calcium phosphate, and calcium oxalate (Pandey 2018). The osmoregulation of cells is controlled by the calcium oxalate, wherein $Ca^{2+}$is used for neutralizing anions – phosphate, citrate, malate, and/or oxalate – that may accumulate in high concentrations in vacuoles (Marschner 2012). Chloride ions are the most prevalent inorganic anion in plant cells whenever it accumulates to levels of macronutrients (Marschner 2012). Cl^- ion is an important cofactor for the oxygen production by photosystem II in chloroplast. Stabilization of the water splitting system at the PSII oxidizing site Cl^- helps in maintaining the coordination structure of Mn(4), Ca cluster, and facilitates the flow of protons from the water oxidation complex to the thylakoid lumen (Kawakami et al. 2009; Raven 2017). It also regulates the functioning of certain enzymes such as *asparagine synthetase*, *vacuolar proton-pumping ATPase,* and *amylase* activity (Rognes 1980; Churchill and Sze 1984). It plays a crucial role in signal perception and transduction, since a variety of signals (light, sound, and elicitors) induce depolarization of the membrane by inducing anion efflux (Colmenero-Flores et al. 2019). Magnesium (Mg) is another important element and Mg^{2+} ion is the second most abundant cation in plant cells, after K. Being a component of chlorophyll, it also plays a vital role in photosynthesis. Since it is a cofactor with ATP, it also plays an important role in enzyme catalysis and in the synthesis of biomolecules like nucleic acid and protein. As it is required for the assembly of ribosomal subunits during protein synthesis, its role in protein synthesis is unparalleled (Gerendás and Führs 2013; Marschner 1995a). Vacuole is the main storage organelle for Mg^{2+} and thus helps in regulating Mg^{2+} homeostasis in plant cells (Marschner 1995a). Although adequate amounts of Mg^{2+} ions are present in the soil, the uptake by plants is relatively poor (Wanli et al. 2016). This may be due to the strong cation competition in plasmalemma uptake and lack of magnesium transporters. Low pH and cations like K^+, NH^{4+}, Ca^{2+}, and Mn^{2+} inhibit the absorption of Mg^{2+}. Mg^{2+}deficiency results in decreased root growth impacting the absorption of nutrients and water (Marschner 1995a; Maathuis 2009). Along with N, P, and K, S is the fourth major nutrient required for plant growth (Sulphur Institute, 1986). It is a component of amino acids, proteins, sulfated polysaccharides, sulfolipids, and vitamins (Leustek et al. 2000). Ninety percent of total soil S has been found as organic sulfur (Mitra 2015). Sulfate (SO_4^{2-}) is taken directly from the soil and transported to the xylem (Falk et al. 2007). Nutrients that are required in very small quantities by plants are known as micronutrients. Although micronutrients are also known as minor nutrients or trace elements, the most widely accepted term is "micronutrient" (Dubey 2011). These nutrients, although required in very small quantities, play an equally important role in plant growth and yield (Saini et al. 2019). These nutrients are vital for multiple cellular catabolic and anabolic reactions, including processes like antioxidant resistance, protein synthesis, sugar metabolism, auxin metabolism, and in maintaining the fidelity of genetic information (Clemens 2006; Broadley et al. 2007). The importance of maintaining an adequate supply of micronutrient in soil is becoming increasingly clear (Caliskan et al. 2008; Knez and Graham 2013). The adverse consequences of micronutrient deficiency on plants are being realized together with the fact that agricultural soils are becoming increasingly micronutrient deficient (Report IZA 2019). Micronutrient deficiency has significant consequences on the world economy, reducing the GDP of some countries by 2–11% (Stein and Qaim 2007). The economic loss includes reduced agricultural productivity, increased medical costs, and disability-adjusted life years (DALY). The micronutrient deficit loss costs India 1% of GDP. This reflects a loss of Rs. 27,720 crore per year in terms of illness, productivity, and increased cost caused by death and health care (Brahmam 2007). Malnutrition resulted in about 0.5% of total deaths in 2016 (Dandona et al. 2017).

Micronutrients for Human Health

Plants deficient in micronutrients consequently affect human health, as inadequate dietary intake is the primary reason of micronutrient deficiency in humans (FAO et al. 2017, 2018). It is estimated that more than three billion people around the world are at the risk of malnutrition and micronutrient deficiency (Monreal et al. 2015). In a world survey by Food and Agriculture Organization (FAO), it was estimated that 190.7 million (14.5%) people in India were undernourished during 2014–2016 (FAO et al. 2017, 2018). Deficiencies of micronutrient can be defined as a shortage of mineral and vitamins that are required in small quantities for proper growth and development of humans (Ritchie and Roser 2017). Micronutrient deficits are public health problem, with malnutrition impacting main growth with consequences involving proper physical and mental growth retardation of children (Fig. 12.1), disease susceptibility, blindness, and increased competitiveness and future losses

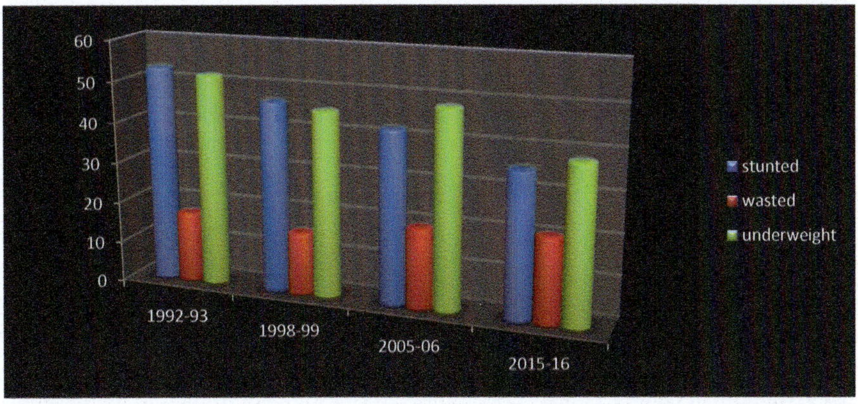

Fig. 12.1 Nutritional deficiency in children under age 5 years. (Source: Different rounds of NFSH 2009a, b; 2018)

(Bailey et al. 2015). Malnutrition associated with micronutrients is also called "hidden hunger" because the effects are not always apparent. It generally refers to the deficiencies of vitamin A, iron, iodine, and zinc, and is emerging as health problem of global concern (Mazziotta et al. 2016). Market analysis indicates that about 90% of farmers believe that they have a crisis because nutrients are not available to the plant via soil (Hodges 2010; Aref 2012). Iron-deficient anemia is a major public health issue that can induce deteriorating clinical results throughout age groups, races, geographic regions, and surgical conditions. However, the body requires about 25 mg of iron a day, often used to create hemoglobin in erythrocytes (Cappellini et al. 2020). Reduced iron consumption may be a direct product of hunger and malnutrition, as is the case for many children and pregnant women in developed countries, or due to vegan or vegetarian iron-poor diets (Camaschella 2015, 2017). Anemia affects one-third of the world's population with the main source being iron-deficiency anemia (IDA) (Kassebaum et al. 2014; GHO 2016). Iron deficiency anemia (IDA) is particularly predominant among preschool infants (<5), reproductive-age adults, and pregnant women with incidence levels of up to 41.7%, 32.8%, and 40.1%, respectively (GHO 2016). In addition, a strong link is found between whole-grain consumption and concentration of manganese. Mn also affects humans at the upper trophic level of the food chain (Huang et al. 2018). Whole grains are known to be a rich source of manganese in human diets (Pan 2018) and healthy drinking water; the WHO suggested manganese concentrations of 0.4 mg/l to prevent toxicity (WHO 2019). Manganese is another important trace metal found in virtually every tissue and necessary for reproduction, digestion, blood sugar control, and homeostasis in humans (Aschner and Aschner 2005). This is one of the omnipresent micronutrients actively involved in bone formation, immune system, cellular energy, and balanced lipid, protein, and carbohydrate metabolism (Erikson et al. 2005). The maternal and child complications were premature rupture of membranes (5%), bleeding (8.8%), hypertension (6.3%), small for gestational age (8.8%), low APGAR score (8.8%), low birth weight (12.6%), stillbirth (5%), and preterm birth (5%) (Iqbal et al. 2020). Copper is the third most abundant important transition metal in humans with the highest levels in the liver and brain (Szerdahelyi and Kasa 1986; Lewińska-Preis et al. 2011). Copper is implicated in a variety of physiological processes including skin pigmentation, maintenance of blood vessel integrity, myelination, iron homeostasis, antioxidant resistance, and neurotransmitter synthesis (Bulcke et al. 2017; Pohanka 2019; Franco-Navarro et al. 2016). Cu deficiency can cause heart hypertrophy with no improvement in diet or bodyweight (Zhou et al. 2009; Majewski et al. 2019). Ceruloplasmin is still the most abundant enzyme containing copper (representing 95% of human plasma copper), being a ferro-oxidase that makes it antioxidant (Sabatucci et al. 2007). A recognized cause of anemia and neutropenia is copper deficiency. Hypocupremia is correlated with excess zinc consumption vying with copper for intestinal absorption, as well as gastric surgery, malabsorption, and excess copper chelation (Siddiqui and Ramlal 2019). In human beings, copper-containing enzymes are not limited to metallothionine, tyrosinase, hemocyanin, ceruloplasmin, and amine oxidase but are also important for the respiratory immune and nervous system. For example, cytochrome c oxidase is important for brain activity, and brain energy production (Llanos and Mercer 2002). Menkes disease is a condition of X chromosome-linked copper deficiency found in 1 out of 250,000 live births (Llanos and Mercer 2002; Tønnesen et al. 1991). Albumin is the most abundant blood protein with a high affinity with copper (600 mg/L of blood plasma), resulting in relatively little "free" ionic copper in the blood compared with other metal-binding protein (Linder 2016). Moco biosynthesis deficiency in humans leads to reduced molybdoenzyme function, leading to inheritable irreversible neurological impairment and sometimes even early childhood mortality (Johnson et al. 1980; Schwarz et al. 2006).

Role of Micronutrients in Plant Health and Productivity

As discussed above, a number of micronutrients are required by plants for their optimal growth and development. These micronutrients include B, Mo. Cl, Cu, Fe, Zn, and Mn, and the amounts in which they are required by the plants are given in Table 12.1. Micronutrients are involved in vital metabolic processes such as cell wall formation, respiration, photosynthesis, chlorophyll synthesis, synthesis of plant hormones, and nitrogen fixation and reduction (Das 2003). The amount of these micronutrients required by the plants varies from plant to plant, soil, and with other factors. Time scale micronutrient deficiencies in the project are shown for the period from 1967 to 1987, from 1987 to 1999, from 1999 to 2011, and from 2011 to 2017 (Fig. 12.2). As micronutrients play an important role in plant physiology and Metabolism, their deficiency has varying consequences on the plant health (Kumar 2002). This may impact different types of plants including agriculturally important crops, medicinal plants, as well as ornamental plants, consequently affecting the economy and human health directly or indirectly.

Table 12.1 Requirement of micronutrients by some cereal grains in subtropical region in India (Roy et al. 2006)

Micronutrients	Wheat (kg/ha)	Maize (g/ha)	Rice (kg/ha)
Iron	1.8	2130	150
Zinc	0.5	380	40
Manganese	0.5	340	675
Copper	0.15	110	18
Molybdenum	–	9	2
Boron	–	240	15

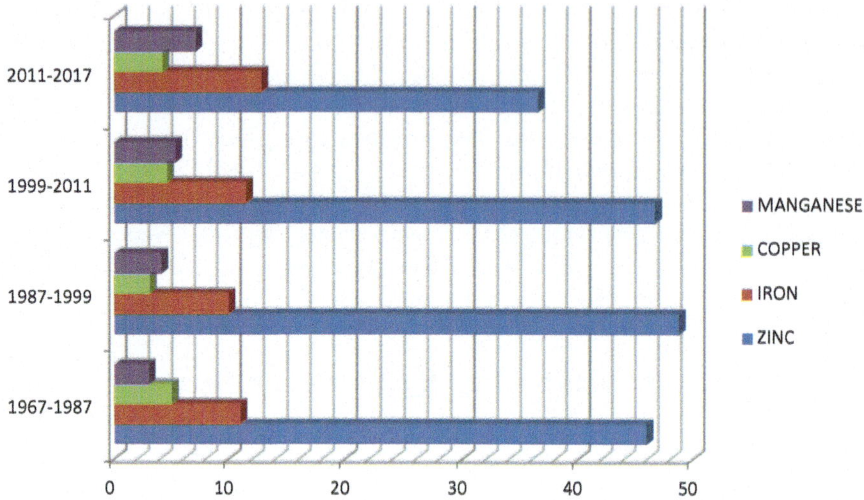

Fig. 12.2 Percentage of micronutrient deficiencies in Indian soil (Shukla and Behra 2019)

Boron

Boron (B) is one of the most important micronutrients required by the plant for their normal growth and development. B requirement of plant varies in a range of 1 μg–1 mg per g per day (Brdar-Jokanović 2020). It is primarily acquired (about 96%) by plant roots as uncharged boric acid (H_3BO_3); however, some quantity of borate anions is also taken up by the plants (Bolaños et al. 2004). Boron is unique in many ways among which its requirement and deficiency concentrations are different for different plant species. Hence, mobility, toxicity, and inter- and intra-species response to insufficient supply vary from plant to plant (Brdar-Jokanović 2020). B seems to be the only micronutrient not directly related to photosynthesis or enzyme activity, yet is correlated with the plant's carbohydrates metabolism and reproduction (Suman et al. 2017). Boric acid readily enters into the cell by penetrating the phospholipids bilayer cell wall (Marschner 1995a; Dordas et al. 2000). Boron deficiency affects root growth more than the shoot growth (Brdar-Jokanović 2020). Importance of B for plants like wheat was reported as early as in 1910 and 1915 (Agulhon 1910). This was later identified as an essential micronutrient (Mazé 1915). Since then, importance of B has been reported for many species of conifers, ferns, phytoplankton, monocotyledons, and dicotyledons (Lewin 1966; Sommer and Lipman 1926). Monocotyledonous plants typically produce less pectin in the cell walls than dicotyledonous plants; thus, they have relatively lower boron requirements and are sensitive to excess boron (Kobayashi et al. 1996; Matoh et al. 1996; Chormova and Fry 2016). Boron deficiency in *Citrus sinensis* causes a decrease in stomatal conductance and CO_2 assimilation and a rise in intercellular carbon dioxide levels (Han et al. 2008). The deficiency also interrupts cell division in apical meristem, resulting in the retarded root growth and sometimes even in the cessation of the root growth. Boron deficiency reduces male fertility by lowering microsporogenesis, germination, and pollen tube elongation (Cheng and Rerkasem 1993). In wheat and barley, boron has inhibited sucrose mobility in phloem, which is a primary product of photosynthesis (Hu et al. 1996; Hu et al. 1997).

Iron

Iron (Fe) is a vital plant micronutrient, which plays a major role in several metabolic pathways such as photosynthesis and respiration (Briat et al. 2015). Soils have different available forms of iron like immediately available iron, available iron, and long-term available iron due to decomposition (Katyal and Deb 1982). Although adequate quantities of Fe are present in Earth's crusts, it is mostly present in the form of peroxides, hydroxides, and oxides, which cannot be easily assimilated by plants (Guerinot and Yi 1994; Romera and Alcǻntara 2004). In India, iron deficiency is directly connected with water scarcity. Absorption of iron from soil is also affected by pH, as alkaline soils have a higher concentration of bicarbonates which block iron uptake into root; acidic soil has a higher concentration of available iron, which may be toxic to the crop (Shukla et al. 2014). Fe deficiency restricts plant development and productivity, and influences crop production and quality (Guerinot and Yi 1994; Jeong and Guerinot 2009). Iron is a key component of chlorophyll synthesis when plants and crops face iron deficiency, leading to interveinal chlorosis in leaves and reduced yield of crops (Briat et al. 2015). Development of roots is necessary for plant life, and plants draw Fe from the soil through the roots. The accumulated Fe should be transmitted by long-distance transporter OsFRDL1, OsYSL, Os IRT1, etc., in a coordinated manner between the different areas of the plant body like leaves, stem, fruits, grains, and flowers (Curie and Briat 2003; Li et al. 2020). Deficiency of iron also stimulates cadmium accumulation, which affects photosynthesis and shoot growth in plants (She et al. 2018). Iron concentration in rice field also decreases with an increase in the concentration of CO_2 in the atmosphere due to human activities (Myers et al. 2014; Zhu et al. 2018). Physiological and molecular studies indicated that many iron chelaters are present in the plant like MAs (Kakei et al. 2009), nicotianamine (Takahashi et al. 2003), and citrate (Takahiro et al. 2009). For chelating Fe^{+3} rice, wheat and maize secrete DMA (*2- deoxymugineic-acid*), while barley secretes *3-epi-hydroxymugineic acid* (Romheld 1990). Plant hormones and signaling molecules play a significant role in regulating Fe deficiency response in rice. Researchers have found that BRs (*Brassinosteroids*), especially EB (*2,4-epibrassinolide,* a *brassinosteroid*), influence the regulation of iron homeostasis both under iron-sufficient and iron-deficient conditions (Li et al. 2020). Ethylene is another hormone, which is involved in controlling Fe status in rice (Lucena et al. 2015; Wu 2011). Other studies have shown that auxin and nitric oxide can influence Fe deficiency in rice (Liu et al. 2015; Buet et al. 2019). Leaf chlorosis caused by iron deficiency in *Arabidopsis* is alleviated by using ABA and iron (Kobayashi et al.

2007, 2009). Fe was shown to be a limiting factor for biomass and seed yield in *Arabidopsis thaliana*, tomato (*Solanum lycopersicum*), spinach (*Spinacia oleracea*), and rice (*Oryza sativa*) (Jin et al. 2009, 2013; Ravet et al. 2009, 2012; Takahashi et al. 2001).

Manganese

Manganese is an essential trace metal, which is required for normal life activities of plants, and it is widely distributed in nature (Doncheva et al. 2009). It plays a major role in enzyme activation, maintaining structure of chloroplast and takes part in photosynthesis by splitting H_2O (González et al. 1998; Mukhopadhyay and Sharma 1991) and is involved in the development of several other compounds necessary for plant metabolism (Arnon and Stout 1939). However, excess Mn in soil inhibits plant growth and is therefore toxic. Mn deficiency in plants can affect plant health and metabolic and physiological functions taking place in plants (Fecht-Christoffers et al. 2003; Delhaize et al. 2007). It is estimated that maximum growth-promoting effects like biomass, total root length, surface area, root volume, root tips, a crossing of *B. papyrifera* are shown by Mn^{+2} at a concentration of 1 mmol/L. Besides, at an Mn concentration of 0–5 mmol / L, all index values of plant roots were higher than those of the control group, indicating that under such conditions, overall plant growth also increases, but at 10 mmol/L, the growth-promoting effect decreases (Huang et al. 2019). Maize is really a greater nutrient-demanding crop, which also needs micronutrients (especially Zn) along with important elements for optimal yield and growth development (Muhammad et al. 2015). Of special concern are Zn (zinc), B (boron), and Mn (manganese); thus, it is important to establish realistic solutions that are willing to comply with integrated nutrient management (Gaind and Singh 2015). Mn (White and Brown 2010) is one of the eight essential micronutrients needed for plant growth as compared to other temperate cereal crop; barley uses Mn more efficiently (Marcar and Graham 1987), and lack of accessible Mn still creates huge barley agronomy problems (Steenbjerg 1935; Goldberg et al. 1983; Reuter et al. 1988). HvIRT1 gene was demonstrated to regulate the Mn deficiency in plants from soil and became an important factor to enhance Mn deficiency via plant breeding in barley (Gherardi and Rengel 2004) like production of phytase exudation (George et al. 2014). Seed priming of wheat with $MnSO_4$ improves the grain yield, Mn content in grain and growth (Muhammad et al. 2015). In maize, seed priming with manganese + zinc + boron (0.1% + 0.5% + 0.01%) improved the grain yield of hybrid maize, and this improvement in yield of grain was attributed to increase in the rate of photosynthesis and efficient uptake of nutrients (Davis and Quick 1998).

Copper

Among all trace metals, Cu is an important micronutrient among trace metals, and plays a crucial role in normal plant growth; even so, at high amounts, Cu can have an extremely toxic impact. The absorbed Cu is normally deposited in the cell wall and in the vacuoles, attaching in hyper accumulator plants with O-rich ligand (Küpper et al. 2009). Copper (Cu) is important as a redox-active cofactor for basic biological processes in plants, including photosynthesis, respiration, protection from oxidative stress, and metabolism in the cell (Burkhead et al. 2009; Puig et al. 2007; Yuan et al. 2010). Copper depletion is also associated with greater oxidative damage by engaging with the superoxide dismutase 1(SOD1) maturation cycle, indicating that analogous molecular toxicity mechanisms could facilitate cell damage in copper dyshomeostasis circumstances (D'Ambrosi and Rossi 2015). In *Arabidopsis thaliana,* roughly one-third of the total Cu in shoots is contained in chloroplasts where Cu is often used as a cofactor for plastocyanin (PC), an electron carrier between the cytochrome (Cyt) b6f complex and the photosystem I (Shikanai et al. 2003; Weigel et al. 2003). Higher levels of N cause a greater deficit in Cu and vice versa. One of SPL7's (SQUAMOSA promoter-binding protein-like7) roles may be to control the N and Cu uptakes. Sufficient N can stimulate plants' growth and result in more Cu requirement (Mermod et al. 2019). Bamboo roots may absorb Cu (II) before decrease, since only Cu(II) was observed in roots (Collin et al. 2014). On the other hand, Cu (II) reduction before absorption is likely occurring in root cells during rice uptake, as exemplified by the slightly lighter Cu isotopes (Jouvin et al. 2012; Ryan et al. 2013). Recently, the consumption/transformation of Cu in situ inside of plant roots was analyzed, and the outcomes presented useful data on both the state of oxidation and the spatial distribution (Ryan et al. 2013; Lu et al. 2017). Feasible reduction of Cu(II) to Cu(I) during root uptake was recommended, as Cu(I) was reported in plant root, like ryegrass root (Zhao et al. 2018), oat root (Ryan et al. 2013), and rice root (Lu et al. 2017), by using bulk X-ray absorption near-edge structure (XANES) spectroscopy, in agreement with the results of Cu isotopic fractionation (Jouvin et al. 2012; Ryan et al. 2013). Plant biomass and yield may be influenced by both Cu excess and shortage (Shaheen et al. 2016; Obiora et al. 2016). As the Cu shortage may stop the photosynthesis process, it reduces plant growth and biomass (Napoli et al. 2019). In contrast, Cu excess can alter plant morphology and growth (Adrees et al. 2015). Various plants have been used for phytoremediation of Cu, some showing low accumulation such as *Brassica nupus* L. *Sagittaria sagittifolia*, *Echinodorus*, *Juncus effusus*, *Acorus calamus*, *Arundina graminifolia*, *B. pilosa*, *Eichhornia crassipes*, *Nymphaea tetragona*, *Pistia lanceolata*, and *P. stratiotes,* and fenugreek plants showed high tolerance to Cu and hence greater metal uptake (Li et al. 2018; Zaheer et al. 2015; Lu et al. 2018; Andreazza et al. 2015). Under varying levels of Cu, however, the concentration and accumulation of Cu in different plant parts were found to be different, and the overall trend was recorded as: root > stem > foliage. (Fatima et al. 2020).

Molybdenum

Mo is likewise an important plant micronutrient. Plants usually suck up Mo as molybdate (MoO_4^{-2}). MoO_4^{-2} is biologically inert and must be integrated to form Moco into the molybdopterin. Moco functions as e-donor or molybdoenzyme acceptors and performs important roles in carbon, nitrogen (N), and sulfur (S) assimilation and biogeochemical cycles (Schwarz and Mendel 2006; Bittner 2014). Mo deficiency is limited in humans, while in crops, Mo deprivation is now an agricultural issue and becoming an agricultural problem for crops grown on acid soils because, in acidic soils, high concentrations of iron oxides/hydroxides are recognized to get strong molybdate adsorption, which deteriorates Mo's bioavailability (Marschner 2012). Mo soil deficiency has been shown to hinder plant growth and productivity in agriculture (Kaiser et al. 2005). Plants with Mo deficiency produce the usual "whiptail" phenotypes, namely, mottled leaf lesions, leaf rolling, and leaf edge wilting (Arnon and Stout 1939). Mo plays an important role in nitrate assimilation, biosynthesis of *abscisic acid*, depletion of *purine*, and *sulfite detoxification* as one of the essential mineral nutrients needed by plants (Schwarz and Mendel 2006; Bittner 2014). Numerous studies have indicated that implementation of Mo enhances the soil availability of phosphorous. Due to the formation of anionic complexes between these elements, such research predominantly shows a synergistic interaction among Mo and P (Barrow et al. 2005; Liu et al. 2010; Nie et al. 2015). There is very less knowledge about use of molybdenum in soil microbial activity (Wen et al. 2018) and changes in soil P relative to soil P-enzyme production in long-term experimental soil rhizosphere (Wei et al. 2017). Legumes are fixing atmospheric nitrogen into available form by their root nodule due to the involvement of Rhizobium bacteria. Molybdenum is an important element which acts as an important cofactor in inducing *Phaseolus vulgaris* nitrogen-fixing activity (Vakhaniya et al. 1990; Reyes et al. 2016). The key effect of application of Mo fertilizer is to increase nodule size and prevent nodule senescence, thus maintaining a longer duration of successful nitrogen fixation (Vieira et al. 1998). Molybdenum is valuable and of significant relevance because of its ability to contribute to the activation of several enzyme systems and physiological activities inside the plant cells. It is a fundamental component of the enzyme nitrate reductase involved in the reduction of NO_3^- to NO_2^- in both microorganisms and higher plants (Lal et al. 2018). Application of molybdenum fertilizers mainly affects the size of root nodule and minimize nodule senescence (Vieira et al. 1998). It really was interesting to note that Mo influenced both the assimilation of carbon and the use of carbohydrates in crop plants. Previous research suggested that the Mo application improved carbon assimilation in several fields and horticultural crops via enhancing aldolase, cytochrome-c oxidase, glycolate oxidase, succinic dehydrogenase, and glycolate dehydrogenase (Agarwala et al. 1978; Agarwala et al. 1986). Experiments have shown that Mo in wheat is influenced by fatty acids during low-temperature stress (Yaneva et al. 1995; Wu et al. 2017). Even so, there is little knowledge available about the role of molybdenum in synthesis, metabolism, concentration, and composition of lipid in drought-tolerant wheat. To further understand the functions of glycerolipids under drought tolerance in winter, researhers employed glycerolipidomics combined with transcriptomics to examine lipid profiles and the expression of associated genes in response to Mo intervention (Wu et al. 2020).

Zinc

Zn is a transition metal essential required for the proper growth and development of all the living organisms. Zn is mainly found in protein or ligand-bound form or a small quantity is found as free Zn ion within the cells. Zn plays a crucial role in metabolism, as it is an important component of various enzymes such as alcohol dehydrogenase and superoxide dismutase (SOD) (Vallee and Auld 1990; Maret 2009). In fact, Zn is the only micronutrient found in all six enzyme classes: *transitions, hydrolases, oxide-reductases, lyases, isomerases, and ligases* (Barak and Helmke 1993). It also acts as a signaling molecule and is involved in the regulation of many proteins like Zn finger domains (Vallee and Auld 1990; Maret 2009). Zn deficiency can therefore disrupt cellular homeostasis, influencing various metabolic processes. Its deficiency in plants is a major problem, as it reduces the quality and quantity of the crop yield (Shinozaki et al. 2020). Zn deficiency in crops, especially in cereals, poses a significant threat to human health. The poor zinc status of cereals is a primary reason for low dietary Zn intake, especially in countries where population is dependent on cereal-based diet (Ritchie and Roser 2017). Zinc has many functions in plants like maintenance of membrane integrity, photosynthesis, gene expression, defense against drought and diseases, and the synthesis of phytohormones. Zinc is a component of *carbonic anhydrase*, which is involved in the photosynthesis, especially in C_4 plants like sugarcane, sorghum, and maize. Its involvement in photosynthesis in C_3 plants (example: soya beans wheat and rice) is, however, uncertain. Other studies also demonstrate that Zn is crucial for the synthesis of IAA precursor tryptophan (Alloway 2004; Nahed and Balbaa 2007). Zinc is crucial for plant reproduction and its inadequate supply affects various flowering, floral growth, anthesis, gametogenesis, fertilization, and seed maturation (Pandey 2010). In reproductive biology, involvement of zinc is significantly occurring as a component of zinc finger proteins like TFIIIA-type zinc finger protein assigned to the regulatory function of reproduction (Kubo et al. 1998; Takatsuji 1999). Zn deficiency is seen as one of Asia's most significant nutritional stresses limiting the production genotypic variations in grain of rice (Ahmad et al. 2012). Importance of Zn as an essential transition metal for plants was first demonstrated in maize, and then in barley and dwarf sunflower (Mazé 1915; Sommer and Lipman 1926). Symptoms of severe Zn deficiency for various crops have been identified (Scaife and Turner 1983; Marschner 1995b; Sharma 2006). Different plant tissues can be used to check the zinc status of the plants. But generally, Zn concentration in young leaves is taken as a good marker of plant Zn status. The average concentration of the Zn in the leaf needed for adequate growth in most crops is approximately 15–20 mg/kg DW

Table 12.2 Functions of micronutrients (FAO 2001)

Micronutrients	Functions
Boron	Help in pollination, cell wall formation, seed development, protein synthesis
Iron	Essential for biosynthesis of chlorophyll, enzyme and RNA metabolism, photosynthesis, oxidation reduction, and respiration
Copper	Essential for vitamin A synthesis, carbohydrate, and nitrogen metabolism
Manganese	Help in activation and regulation of enzyme, translocation of iron, nitrogen metabolism, and chlorophyll synthesis
Molybdenum	Nitrogen fixation in legumes and nitrogen metabolism
Chlorine	Help in opening and closing of stomata, and increases water-holding capacity
Nickel	Essential for seeds to germinate and grow, and responsible for iron absorption
Zinc	Part of many enzyme systems, regulate plant metabolism, utilization of N and P in plants necessary for reproduction, and formation of growth hormones and protein

(Marschner 1995b). Zn content is comparatively higher in nuts, seeds, legumes, and whole-grain cereals while tubers, refined cereals, fruits, and vegetables are considered Zn poor (Brown 2008). Functions of micronutrients in plants are given in Table 12.2.

Zinc: One of the Most Important Micronutrients for Plants and Humans

Role of Zinc in Plant Health

Reduced plant growth can be due to various reasons ranging from unfavorable environmental condition to lack of nutrients required for the growth (Sadras 2005). Both macro- and micronutrient deficiencies are known to limit plant growth significantly (Koerselman and Meuleman 1996; Craine et al. 2008; Tilman et al. 2002). Micronutrient deficiency in plants sometimes has serious consequences on plant health, resulting in decreased plant growth and yield (Van Maarschalkerweerd and Husted 2015; Schmidt et al. 2016). One such important micronutrient is Zn, as its deficiency is known to have serious consequences on plants and humans' health. Above 33% of the world's population has zinc deficiency mainly in countries where the population is dependent on poor cereal-based diet like wheat, maize, and rice (Hotz and Brown 2004). The symptoms of Zn deficiency were first identified in maize by Mazé and were typically referred to as "white bud" (Mazé 1915). Early symptom of serious Zn deficiency is reduced tomato stem elongation (Skoog 1940). Zn deficiency symptoms in tomato include reduced protein and starch synthesis, which can be remedied by resupply of zinc (Hoagland 1944). Other severe Zn deficiency symptoms include little leaves (Dwarf leaves) and root apex necrosis ("dieback"), whereas the mild deficiency causes spatially heterogeneous or interveinal

Fig. 12.3 Brown spot on millets, rice, and yellow stripes on maize and reddish brown on sorghum

chlorosis ("mottle leaf"), development of reddish-brown or bronze tints ("bronzing"), and a range of auxin-like responses such as internode shortening ("rosetting"), epinasty, curling of lamina ("goblet" leaves), and reduction of auxin-like responses such as internode shortening ("rosetting") (Marschner 1995a). Some zinc deficiency symptoms are given in Fig. 12.3 and Table 12.3. Sensitivity to Zn deficiency varies from crop to crop (Alloway 2008a). For example, cereal crops such as maize, sorghum, and rice are highly sensitive to micronutrient deficiency, while crops like wheat, oats, and rye are less sensitive to zinc deficiency (Katyal 2018; Ganeshamurthy et al. 2018; Report IZA 2019). Various factors influence the accumulation of Zn in seeds including the water availability and presence of other divalent cation and ions like iron (Scagel et al. 2012). Phosphorus is also known to interfere with the accumulation of Zn in the seeds (Saha et al. 2013). Zinc-deficient soil has a negative effect on grain yield and quality; it may also increase plant's susceptibility to injury by high temperature and light intensity and fungal disease infection (Ahmad et al. 2012; Cakmak 2000). In various studies, poor soil Zn status is related to poor yield and productivity (Phattarakul et al. 2012). The endosperms (edible portions) of cereal are becoming increasingly low in Zn content. For example, zinc concentration in whole grain of wheat is between 7 and 85 mg/kg (Rengel et al. 1999). Generally, zinc concentration in grain of wheat in cultivated region is in the range of 20–30 mg/kg, and an average value is 28–30 mg/kg (Rengel et al. 1999; Graham et al. 2007). When wheat is grown in zinc-deficient soil, then the concentration of wheat grain is found to be below 20 mg/kg; for example, in Australia, zinc-deficient and zinc-sufficient soils have wheat grain concentration below 10 mg/kg and 20 mg/kg, respectively (Graham et al. 1992; Çakmak et al. 1999, 2010a). Zn is acquired by plants from soil through roots and the process is influenced by several soil and physicochemical properties such as pH, soil composition, organic matter, and carbonate content. It is reported that nearly half of the world's agricultural soils contain decreased amounts of soluble zinc. This problem is compounded by low organic matter and soil moisture, as well as elevated pH and $CaCO_3$ levels, particularly in arid and semi-arid regions (Cakmak 2008; Gonçalves Júnior et al. 2010). Presence of Zn in an insoluble form which is not available to plants also results in Zn deficiency in plants. Plant parts with high growth rate and

Table 12.3 Zn deficiency symptoms of some important cereal crop

Plant species	Symptoms of zinc deficiency	References
Hordeum vulgare	Chlorosis and drying, white spots on leaves, middle leaf becomes collapsed and short leaf tip	Singh et al. (2005)
Triticum aestivum	Adaxial surface of leaves shows brown spot, chlorosis in midrib (mostly young leaves), loss of leaves and lower leaves have brown patches, decreased tillering, inhibited growth (shoot and root), sterile spikelet	Singh et al. (2005)
Brassica nigra	Deep purple color on midrib in young leaves. Plant shows stunted and delayed flowering and maturation	Samant (2009)
Oryza sativa	Brown blotches on leaves and streaks and after that, entirely cover older leaves, growth stunted, and in many cases, it can cause death of plants, reduced yield, and delay in maturity	Neue and Lantin (1994)
Zea mays	Yellow stripes on leaves, white bud formation, and short internodes and red or yellow streak on lower leaves	Alloway (2004)

auxin levels are known to have a high zinc concentration, indicating its requirement for meristematic growth (Haslett et al. 2001; Oguchi et al. 2004). Zinc also controls the synthesis of auxin in plants and its deficiency causes shortening of internodes and distortion of leaves (Skoog 1940; Irshad et al. 2004). Auxin is a group of growth regulator which takes part in cell elongation and cell division (Teale et al. 2006). Zinc deficiency in plants results in low concentration of chlorophyll, and lowers Chl a: b ratio, inhibiting the process of photosystem II (Chen et al. 2008). Zinc is a component of carbonic anhydrase, which regulates RUBISCO activity (Srivastava and Gupta 1996; Storey 2007), a photosynthetic enzyme catalyzing CO_2 diffusion through stomata (Hatch and Slack 1970). The poor supply of this metal in the soil restricts plant growth, decreasing agricultural productivity (Cakmak 2011).

A concentration ranging from 30 to 100 μg zinc g^{-1} dry weight (DW) is considered normal and tolerable by plants. Whereas signs of zinc toxicity are observed when the concentrations exceed 300 μg/g DW (Marschner 1995a). Furthermore, the concentration required by different plant parts and tissues may also be different. For example, meristematic tissues may require a relatively higher concentration due to higher growth rates than mature leaves (Marschner 1995b). Therefore, plants need to regulate an optimal concentration of zinc in different plant tissues, cells, and even in subcellular compartments. This is called zinc homeostasis. Growing crops meant for food in Zn-deficient soil results in poor yields, and low zinc content in food, reducing farmers' profits. The widespread zinc deficiency in soil affects agricultural productivity and yield, requiring steps to improve the soil's zinc status. One approach is the use of zinc fertilizers. But various factors like soil characteristics, irrigation, and rain affect the availability of Zn to the plants (Das and Green 2013).

Role of Zinc in Human Health

Zn is an important micronutrient for humans also and is primarily obtained from diet. Although Zn is obtained from various foods, food containing animal proteins is considered Zn-dense food. Furthermore, phytate and dietary calcium is also known to interfere with Zn absorption in the intestine. Since plant-based diets are known for high phytate content, the low Zn content of plant-based diet along with its high phytate content makes these diet rather unattractive source of Zn (Brown 2008). The zinc deficiency in soil can be directly linked to the Zn deficiency in humans (Report IZA 2019). Zinc status in humans is regulated through a process called Zn homeostasis, which involves a complex interplay of Zn absorption, excretion, transporters, etc. The Zn absorbed in intestine reaches liver via the portal system and then circulatory system. This zinc binds with albumin and represents serum albumin content (FAO/WHO 2004; Joint 2004; IZiNCG 2004; Devergnas et al. 2004). Requirements of Zn for various systems, tissues, cells, and subcellular compartments vary and are maintained through homeostasis (IZiNCG 2004; Hambidge and Krebs 2001; King 2000; Rosen 2009). Similarly, zinc requirements for different age groups also vary and are given in Fig. 12.4. Zn deficiency in humans is caused by a number of factors, the most important being the prolonged inadequate dietary intake. Malabsorption in the intestine also causes its deficiency, which may occur due to various reasons including the genetic disease acrodermatitis enteropathica (Van 1989). Zinc is a very important micronutrient for human health and its deficiency has a number of health consequences in humans. Since Zn cannot be stored in human body, a constant supply of Zn is required through diet. A continuous consumption

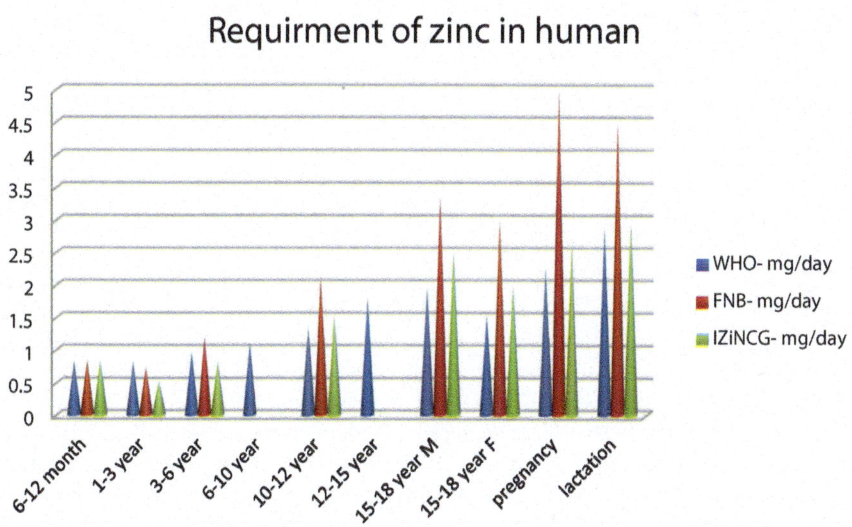

Fig. 12.4 Zinc requirement mg/day in percentage. (Source: Roohani et al. 2013; WHO, FNB, IZiNCG 2004)

of Zn-poor diet results in Zn deficiency in human (Bharadva et al. 2019). In fact, Zn deficiency among humans is emerging as a global health problem, leading to many other health complications and even mortalities (IZiNCG 2004). Its deficiency is known to affect normal growth, brain development, and immunity. Functioning of insulin-like growth factor-1(Cossack 1991; Ninh et al. 1995) that regulates tyrosine phosphorylation and cell growth and cell division is also affected by Zn deficiency (Wilson et al. 2012). Mild zinc deficiency in humans lowers serum testosterone level, decreases natural killer cell activity, T helper cell, and thymulin serum causing oligospermia, hyperammonemia, and hypogeusia (Beck et al. 1997; Prasad et al. 1988). To make the matter worse, the continuously increasing world population has further increased the demand for quality food with adequate quantity of micronutrients (Cakmak et al. 2017; Velu et al. 2014; Detterbeck et al. 2016). This requires efforts at different levels and various strategies have been developed to deal with the problem of Zn deficiency. These strategies include Zn fortification of diet, which can be achieved through dietary modification, diversification, or Zn biofortification of cereal crops (Erenoglu et al. 2011).

Prevalence of Zinc Deficiency

Zn Deficiency in Plants

Soil is one of the most important sources of nutrients in the world containing more than 50% of the total carbon content. Most of the world's food is grown in agricultural soils and only a fraction is obtained from other sources like aquatic sources (Minasny et al. 2017). The continuously increasing anthropogenic activities have resulted in nutrient imbalances in soil, affecting the supply of both macronutrients and micronutrients (Dungait et al. 2012). Micronutrient deficiency of some elements is given in the following order: Zn > B > Mo > Cu > Mn > Fe. In field experiments, deficiency of zinc was found highest in Pakistan and Iraq followed by Nepal, Turkey, Thailand, and Tanzania (Alloway 2008a). Deficiency of phosphorus (P) is one of the most important limiting factors for crop yield worldwide. It is the second most significant reason for reduced soil fertility (Hu et al. 2016; Ye et al. 2018). In a survey, it was estimated that 57 thousand of cultivated soils worldwide do not have sufficient P available in crop production (Wyngaard et al. 2016). However, the most prevailing deficiency of micronutrient in agricultural soil is Zn deficiency. According to some estimates, at least one-third of the world's agricultural soil is deficient in bioavailable Zn (Sillanpää and Vlek 1985). Zinc Nutrient Initiative (ZNI) states that zinc deficiency is the most widespread micronutrient deficiency in the soil, affecting more than half of the world's agricultural soil. Zn deficiency is prevalent in Asia (Turkey, India, China, and Indonesia), Sub-Saharan Africa, and South America's northwest region (Cakmak 2002). Zinc deficiency incidence across Sub-Saharan Africa and South Asia is in the ranges of 15–50%. In the Democratic Republic of

Congo, 54% of the soil samples were found to be Zn deficient (Ritchie and Roser 2017). Micronutrient deficiency in the world is given in Fig. 12.5. In African countries, zinc-deficient crops included maize, rice, wheat, avocados, banana, citrus, nuts, peas, beans, and sugarcane. Most of these Zn-deficient crops were grown in sandy soil, which was highly calcareous (van der Waals and Laker 2008). Zinc deficiency in maize was widespread in Nigeria where the level of zinc is below 1 mg/kg in sandy soil. In countries like Malawi, more than 49% of soil was Zn deficient mainly due to sandy soil texture. In rice, zinc deficiency mainly occurs in Sierra Leone and northeast Nigeria. In Zambian soil, zinc concentration ranges from adequate in some locations to low in other regions (Sillanpää 1982). In Bangladesh, zinc deficiency is a limiting factor for crop production and it was estimated that about 1.75 Mha land for crop mainly wheat and rice is affected by zinc deficiency (Ahsan and Beuter 2000). Zinc deficiency is a major problem in soil and plants, especially rice crop grown in alkaline, waterlogged soils (Gopalan 1992). In Iran, soil mostly used for fruit and cereal crops was found to be Zn deficient, and an estimated 80% of the land is affected by Zn deficiency. It was reported that only wheat agricultural field was affected by up to 40% due to zinc deficiency. In Iran, maize, rice, soya bean, pea, sorghum, and citrus are also highly sensitive to zinc deficiency due to calcareous soil with high pH (7.9–8.5), high application of phosphatic fertilizers, and lack of zinc fertilizers (Alloway 2008b). It is estimated that 70% of agricultural soil in Pakistan is deficient in nitrogen, phosphorus, potassium, and zinc, while around 47% of agricultural soil is Zn-deficient soil (Hamid and Ahmad 2012). Zn deficiency in various crops like rice citrus, cotton, maize, sugar beet, and wheat has been reported in Pakistan (Rashid and Rayan 2008). Many agricultural soils in Afghanistan have calcareous soil texture and high pH of 8–8.5, 8.5–9.0, and 9–9.5 and hence are Zn deficient, which results in Zn deficiency in crops; it is estimated that more than 40% of wheat crop is Zn deficient due to its deficiency in soil (CIMMYT 2000). Predominant zinc deficiency is recognized in Turkey, especially central Anatolia where wheat is major cereal-growing crop. According to FAO, 87% non-irrigated and 13% irrigated soil indicated zinc status, which is very low, that is, 0.62% (Sillanpää 1982). The main reasons for zinc deficiency in Turkey are high pH calcareous soils. Syria is recognized as zinc deficient just because of highly alkaline soil and zinc concentration is found to be 1.21 mg/kg (Sillanpää 1982). This is quite low concentration. Wheat is the crop, which is mainly affected with zinc deficiency (CIMMYT 2000). It is estimated that soil in China is also zinc deficient due to acidic soils of southern and calcareous soil of northern China and 51.1% part of farm land is zinc deficient (Liu 1996; Zou et al. 2008). Rice and maize cereal crops are mostly affected in China due to alkaline paddy field (Zou et al. 2008). Approximately 0.5 lakh ha agricultural rice field is affected in the Philippines (Yoshida et al. 1973) and in Thailand paddy rice soil in the semi-arid area in the northeast are more susceptible toward deficiency of zinc (Katyal and Vlek 1985). It was estimated that 30–50% of soils of Nepal were deficient in zinc (Sillanpää 1990) and 29% of land soils for rice were prone to zinc deficiency (Jensen 2007). Global micronutrient study showed that deficiency of zinc in Iraq is lowest compared to other 30 countries and the widely affected crop was wheat due to lower

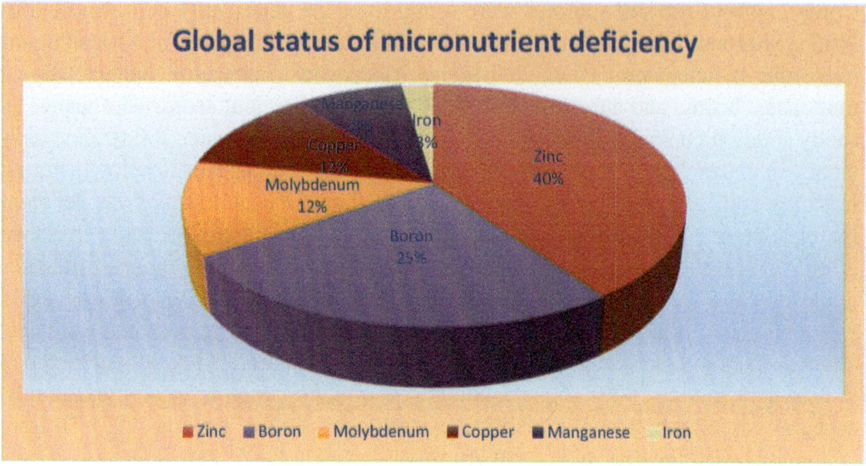

Fig. 12.5 Percentage of micronutrients in world soil. (Source: Graham 2008)

concentration of zinc, which was 0.27 mg/kg in soil. Barley is the main crop grown in Saudi Arabia and 0.1% land area is under cultivation that is highly affected by zinc deficiency due to the presence of saline calcareous and gypsiferous soils (FAO 1972). Rice is zinc deficient in Sri Lanka due to slightly acidic soil, and it occurred in around 10,000 ha of land (Katyal and Vlek 1985). In Australia, different types of soil are found like calcareous sand, black earth, solodised, solodic, solonized, red brown earth, and gray brown. The zinc deficiency is widespread in red brown earth, calcareous sand, red, and black soil (Tiller 1983). In the USA, many soil samples, especially sandy and calcareous soil, have been found to be Zn deficient. The deficiency of zinc is also widespread in maize (Fageria and Stone 2008). Wheat, barley, maize, beans, and flax grown in irrigated, calcareous soils were found to have low Zn content (Alloway 2008b).

Zinc Deficiency in Humans

Various studies have reported the prevalence of Zn deficiency in humans. Different parameters have been used to estimate the levels of Zn deficiency. One of the good indicators can be the food consumed by different populations in different countries. It is estimated that around one-third of the world's population is affected by Zn deficiency, affecting 4–73% of population in different regions. Although serious Zn deficiency is rather uncommon, mild-to-moderate zinc deficiencies are very common (Sandstead 1991). However, old estimates indicate that at least 33% of the world's population is suffering from leading types of micronutrient deficiency (Gorstein et al. 2007). The World Health Organization (WHO) reports that over two billion people worldwide suffer from micronutrient deficiency (Graham et al. 2007).

Many high-income countries, especially across North America, Europe, Central Asia, and Oceania, have zinc deficiency incidences below 5–10% of the total population (Ritchie and Roser 2017). It was reported that 15% of infant death occurs due to diarrhea as compared to AIDS/HIV, which is 2%, and globally, 1.5 million people die due to diarrhea alone (Das and Green 2013). Caggiano et al. (1969) reported the first case of zinc deficiency in the United States in 1969 with symptoms of dwarfism, hypogonadism, hypogammaglobulinemia, giardiasis, strongyloidosis, and schistosomiasis.

Zinc Deficiency Status in India

Zinc Deficiency in Plants

The soil used to cultivate cereals and grains is becoming nutrient deficient resulting in poor nutritional status of these grains (National seminar on millets-2010). Climate change is also affecting the global per capita nutrient availability. The availability of zinc, protein, and iron, decreases by 14.6% 19.5%, and 14.4%, respectively (IFPRI; Science Daily Report 2019). To address the issue of increasing micronutrient deficiency in crops in mid-1960s, the Indian Council of Agricultural Research (ICAR) established the All Bharat coordinated theme of Micronutrients in Soils and Plants throughout 1967 at Punjab Agricultural University (Shukla and Behra 2019). Under the aegis of All India Coordinated Research Project of Micro and Secondary Nutrients and Pollutant Elements in Soils and Plants and other State agencies, the concentrations of total soil status, a large number of soil and plant samples were collected and analyzed for micronutrients, namely Zn, Mn, Fe, Cu, Mo, and B (Fig. 12.6). Of more than 3 lakh soil samples collected from different Indian states, 44% were deficient in Zn with less than 0.6 ppm DTPA-extractable Zn content. The proportion of samples reported to be deficient in Fe, Mn, and Cu is 15%, 6%, and 8%, respectively (Shukla and Behera 2012). Analysis of more than 2.0 lakh soil samples taken from 508 districts of the country during 2011–2017 under the governance of ICAR – Indian Institute of Soil Science, Bhopal – revealed that an average of 36.5%, 12.8%, 7.1%, 4.2%, and 23.2% samples were deficient in zinc, iron, manganese, copper, and boron (Shukla and Behra 2019). Takkar and Nayyar have reported manganese deficiency in wheat crops in the Punjab rice–wheat crop system (Takkar and Nayyar 1981). Copper deficiency is often associated with sandy, calcareous, and eluviated soils rich in organic matter. Boron deficiency in Bihar was observed in highly calcareous soils (Singh 1997; Singh and Shah 1995). Boron concentration varied from 2.6 to 630 mg/kg (Takkar 2011). Boron deficiency is much more prevalent in Bihar and Gujarat (strongly calcareous) and West Bengal, Odisha, and Jharkhand. B deficiency was recognized in India besides the Zn. B availability to plant is governed by soil pH, CaCO3, and organic matter content of the soil. B interactions with other nutrients, plant type, variety, and other environmental factors also influence B's bioavailability (Takkar 2011). In India, 23.2% of soil samples were B deficient while another 21.5% samples were marginally boron deficient (Report IZA 2019). India's iron deficiency increased from 11.0% to 12.8%

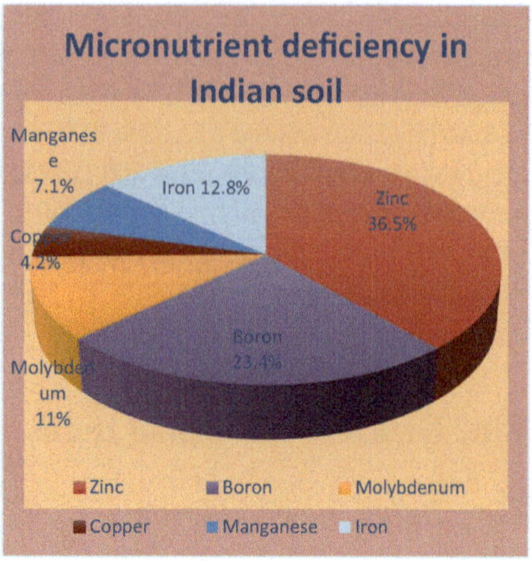

Fig. 12.6 Percentage of micronutrients in India soil. (Source: Shukla et al. 2018)

in 1967–1997 through 2011–2017 mostly in the gray brown soils (25.1%) and old alluvial soils (25%). Manganese deficiency rates raised from 3.5% in 1987 to 7.1% in 2017. Although the level of deficiencies decreases with soil pH, sand content and leaching, the availability of manganese in soil decreases. Most soils have adequate molybdenum levels, but areas of Maharashtra, Kerala, and Odisha have a moderate Mn deficiency (Report IZA 2019). Zn deficiency is among the most prevalent micronutrient deficiency in India, and 36.5% of Indian soils are Zn deficient (Report IZA 2019).

Previously, the prevalence of zinc deficiency was more observed in cereals, especially in the country's rice and wheat belts, but with time, the distribution of zinc deficiency has covered the entire country through crops and crop systems (Shukla and Tiwari 2016). There are many causes for the increasing prevalence of zinc deficiency in India, including significant Zn removals due to high crop yields and crop cultivation systems, overuse of phosphatic fertilizers resulting in Zn deficiency induced by phosphorus, use of highly analyzed fertilizers, lower application of organic manure, and use of poor-quality irrigation water with high $CaCO_3$ content (Shukla and Tiwari 2016). Medium black soils (39.4%) and mixed black and red soils (36.2%) were found to be highly deficient in zinc. Soils which are coarser in texture (sandy / loamy sand), high in pH (>8.5 or alkali / sodic soils) and/or low in organic carbon (<0.4%), or high in CaCO3 (>0.5%) and intensively grown were Zn deficient (Report IZA 2019). Zinc is Asia's most prevalent micronutrient deficiency, as the crops are grown on heavily weathered and leached soils such as tropical red soil (Alloway 2008a). Zinc deficiency in different states of India is given in Fig. 12.7. It ranges from 9.6% in Uttarakhand to as high as 75.3% in Rajasthan (Takkar 2011), while soil sample deficient in DTPA extractable zinc ranged from 8% in Pondicherry to 78% in Maharashtra (Prasad 2013). In Gujarat and Rajasthan, deficiency of zinc is less prevalent. More than 90% of soil has zinc as insoluble Zn, which is not available to the plants, and only 0.1–2.0 mg/kg zinc is in a form available to plants (Singh et al. 2011). Rice and wheat crops grown in calcareous soil show retarded growth due to the Zn deficiency in these soils (Hacisalihoglu and Kochian 2003).

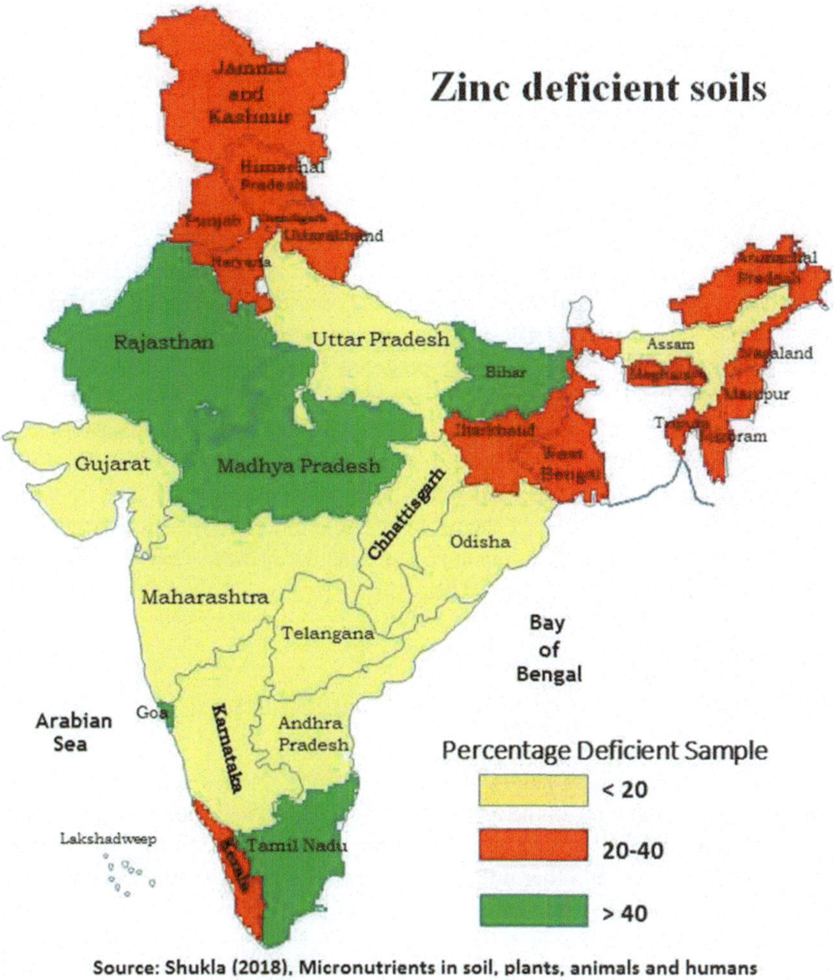

Fig. 12.7 Region of zinc deficiency in India soil. (Source: Zinc deficiency in India Shukla et al. (2018). PwC analysis)

Zinc Deficiency in Human

Surveys of diet conducted by the National Nutrition Monitoring Bureau (NNMB, ICMR) in 9 states of India and by other bodies indicate that cereal–pulse-based Indian diets are deficient in micronutrients, especially iron, calcium, vitamin A, riboflavin, and folic acid. This micronutrient deficiency is also referred to as "hidden hunger." One of the primary reasons for micronutrient deficiency is the poor dietary intake (ICMR 2003). Foods such as pulses, vegetables, particularly green leafy vegetables, are considered Zn-poor (ICMR 2003). As per the Food and Agriculture Organization (FAO) report on State of Food Security and Nutrition in the World, in India, during 2014–2016, 190.7 million (14.5%) people were undernourished (FAO et al. 2017, 2018). In five Indian states, about 43.8% of children under five have zinc deficiency in their serum, that is, <65 μg/dl. The prevalence of Zn deficiency was found in the following sequence: Karnataka (36.2%) < Madhya Pradesh (38.9%) < Gujarat (44.2%) < Uttar Pradesh (48.1%) < Orissa (51.3%) (Kapil and Jain 2011).

Micronutrient deficiencies (such as iodine, magnesium, and vitamin A deficiencies) affect human health and cost between 0.8% and 2.5% of the gross domestic product (Stein and Qaim 2007). India's food scarcity in 2016 resulted in about 0.5% of total deaths. The Asia Pacific regions contribute to more than 60% of the global population with China and India, two of the world's most populated nations. Research has shown that 49.4% of young girls in Delhi suffered from zinc deficiency and in central India, 52% of non-pregnant women suffer from Zn deficiency (Kapil and Jain 2011). The report noted that the highest rate of inadequate zinc intake was related to rice-based diets mainly in the southern and northeastern states of Kerala, Tamil Nadu, Andhra Pradesh, Manipur, and Meghalaya (Times of India 2019). By 2050, India will face the largest burden of Zn deficiency with 50 million people worldwide will be facing the problem. In another study, it is estimated that about 30% of India's population is at risk of zinc deficiency, wherein 400 million men will be affected (India Today 2020).

Strategies to Correct Zinc Deficiency in Plants

Nutrients in soil are added naturally through various sources such as through the addition of organic matter in the form of dead plants, leaves of tree, grass clipping, and other dead organisms (Welch et al. 1991; Sekhon 2003) (http://www.ncagr.gov). Rock outcrops, and other minerals in addition to organic matter also contribute various micronutrients and minerals to the soil (Seaward and Richardson 1989). These minerals are added to the soil through weathering of rocks and through other processes (Wuana and Okieimen 2011). The nutrient status of the soil can also be improved through the addition of fertilizers and other organic sources of nutrients such as manure. Addition of these fertilizers and manures to soil can improve macro- as well as micronutrient status of the soil. In addition to conventionally used fertilizers and manures, crop residues, silage juice, ruined crop and plant remains, wash water, bedding, spent mushroom media, spent soilless media, and spent supplement solutions can also contribute nutrients to the soil (Yanqun et al. 2005). Although a number of fertilizers are commercially available as a source of macronutrient like nitrogen, very few sources of micronutrients are available. Therefore, for correcting micronutrient deficiency, it is important to regulate their concentration in soil and to add these to the soil if required.

Use of Chemical Fertilizers

Due to the continuously rising world population, extensive agricultural practices are becoming increasingly popular to meet the food demand. Among various extensive agricultural practices, one is the use of chemical fertilizers. It is estimated that over 70% of the crop production relies on fertilizers. Most of the industrially produced fertilizers are inorganic fertilizers containing nutrients such as nitrogen, potassium, and magnesium (Kumar and Prakash 2019; Environment 2018). It is estimated that more than 50% of the people feed on crops cultivated in soils amended with synthetic fertilizers (Environment 2018). Extensive use of these fertilizers has immensely improved the world's food productivity (Banerjee et al. 2011; Garai et al. 2014). To meet the food demand and to maintain its quality, it is important to improve the yield and quality of the most commonly used cereals, wheat, rice, and maize, since these cereals are used as staple diet (FAO 2011; Anon 2011). It is important to improve the micronutrient content of crops, as micronutrients can significantly improve both crop productivity and its quality. Among various strategies to attain this target is the use of fertilizers containing micronutrients. In India, the Fertilizer Control Order 1985 targets at including various nutrients in the fertilizers (FCO Amendment 2019). Technically, these micronutrient fertilizers can be grouped as chelated micronutrient mixture fertilizers and micronutrient-fortified fertilizers. Generally, the micronutrient mixture fertilizers contain water-soluble sulfate salts of micronutrients like zinc sulfate and iron sulfate. These fertilizers can be added to the soil using different application methods such as broadcast, positioning of bands, foliar spray, and by irrigation. Other methods include side-dressing or side-banding, foliar application, and soaking or coating of the seeds. Chelated micronutrient fertilizers contain micronutrients fixed to various chelators that help in increasing the bioavailability of the micronutrient to the plants. These chelators include EDTA (ethylenediamine tetra-acetic acid), HEDTA (*hydroxyethylenediaminetriacetic acid*) DTPA (*diethylene triamine pentaacetic acid*), EDDHA (*ethylene diamine di (O-hydroxyphenyl*) acetic acid], and NTA (*nitrilotriacetic acid*) (Report IZA 2019). Report on the use of various micronutrient fertilizer shows that the most widely used micronutrient fertilizer in India was $ZnSO_4$ (1.88.305 t) followed by iron sulfate (21.188 t), boric acid/borax (19.976 t), manganese sulfate (2.740 t), and copper sulfate (1.369 t) during the year 2015–16 (Report FAI, 2016). The application of zinc fertilizers to crops is a simple and cost-effective strategy to correct zinc deficiency in soils. The requirement for Zn varies with crop type, soil type, crop behavior, extent of the Zn deficiency, and with other environmental conditions (Takkar et al. 1989). However, 25 kg/ha $ZnSO_4$heptahydrate, which is equivalent to 5 kg/h zinc, is recommended for soil applications every year or alternate years (Das and Green 2013). Various studies have documented the responses of crops to Zn application from different parts of the world. These studies are based on more than 15,000 field trials from 1967 to 2016. The soil was classified into four categories, namely, non-responsive, marginal or responsive, very responsive, and highly responsive based on the increase in relative economic yield (REY) following the application of Zn fertilizer. Of the 4144 tests performed on cultivated fields during 1967–84, 58% responded to application of Zn fertilizer, while 48% do not show any response (Takkar et al. 1989). It is interesting to note that over the years, the ratio of rice crops responding to Zn application increased to 63%, 72%, and 80% during 1985–2000, 2000–2010, and 2011–2016, respectively (Shukla and Behra 2019). According to an estimate, that use of Cu contribution with Zn and B fertilizers has added 29 Mt. to the present crop production (Shukla and Behera 2011). Of the several Zn sources assessed for their efficacy in different soil-crops, $ZnSO_4{\cdot}7H_2O$ has proved better than or equal to other sources in correcting Zn deficiency (Takkar et al. 1989; Shukla and Behera 2012). In some studies, however, chelated Zn has proved more successful for maize and rice than $ZnSO_4{\cdot}7H_2O$. Many studies have concluded that 2.5–10 kg Zn / ha of $ZnSO_4{\cdot}7H_2O$ is most effective in combating its deficiency and maintaining soil fertility in most crops grown on various Zn-deficient soils. When different methods of fertilizer application were compared, it was observed that soil application, especially under the seed, proved superior to other methods of application. In maintaining high agricultural productivity of rice, sugarcane and vegetables, seedling root dip method with ZnO gave inconclusive results (Shukla and Behra 2019). Among farmers, also seed soaking and foliar application were not popular due to certain constraints and limitations (Shukla and Behra 2019). Soil application of zinc increased the grain yield of durum wheat (*Triticum durum* L.) and bread wheat (*Triticum aestivum* L.) by 20–90% and 60–250%, respectively (Shivay et al. 2008a; Cakmak 2010). Various studies have proved that availability of zinc in soil, its mobilization, and uptake by plants affect the zinc content of the crops and grains. Furthermore, the response to zinc fertilization is also influenced by the zinc efficiency of the plants. Movement of Zn from soil into the plant and its transport through plant vascular tissue ultimately entering grains are also a complex process affecting Zn accumulation in the cereal grains (Patrick and Offler 2001). In *Oryza sativa*, no significant improvement in grain Zn concentration was observed following Zn fertilizer application. However, in wheat, the grain zinc concentration increases from 10 to 60% in response to zinc fertilization (Shivay et al. 2007; Fang et al. 2008; Phattarakul et al. 2009).

Disadvantages of Chemical Fertilizers and Other Agrochemicals

Agrochemicals (excess fertilizer, pesticides, and herbicides, etc.) are a major source of soil and water pollution. Although fertilizer increases the soil's nutritional status, the extensive use of fertilizer poses various environmental threats. These include water eutrophication, soil, and plant nutrient imbalances. Chemical fertilizers are known to contaminate water, soil, and air (Kumar and Prakash 2019). These fertilizers and their byproducts like nitrate and nitrite may enter drinking water and food chain through processes like leaching. Furthermore, excess of these fertilizers has adverse effects on all the living organisms in soil, as nutritional imbalances may adversely affect various food chains in the soil having long-lasting and sometimes irreversible impact (Savci 2012; Sonmez et al. 2007). Since many fertilizers contain small quantities of heavy metals, their long-term use can lead to the accumulation of these heavy metals such as arsenic, and cadmium in soil (Zahra et al. 2010). From soil, these heavy metals can be taken up and accumulated by various crops like vegetables and cereal crops grown in these soil (Sonmez et al. 2007). Abundance of nitrogen results in softening of plant tissue, leading to increased pest susceptibility. Furthermore, pollutants from agricultural fields and setup are often released in very small amounts and their adverse effect becomes clear only after long period of time, resulting in various health issues and diseases like renal disease (Wimalawansa and Wimalawansa 2014). The use of synthetic chemical fertilizers contributes to poor nutritional status of the plants, which consequently affects the human health and that of animals consuming such nutrition-deficient food (Borah et al. 2004). The unregulated use of *superphosphate, phosphate fertilizers, organochlorine, and organophosphate* fertilizers result in the bioaccumulation of pollutants in the soil affecting agricultural productivity and soil fertility (Farnado 2017). The persistence of multiple agrochemicals and their byproducts pose serious risk to soil environment and fertility. These involve the aggregation of various hydrophobic agrochemicals like *endosulfan*, *lindane*, *heptachlor*, and *dithiothreitol* (DDT) and their metabolic intermediates (Jayaraj et al. 2016). A survey in India found that 58% of potable water had organochlorine pesticides far above the standards set by the Environmental Protection Agency (Kole and Bagchi 1995). Studies in Mexico, Vietnam, and Nicaragua's coastal regions show that the population which depends on fishes and other food from aquatic source consume the food contaminated with various agrochemicals like DDT, *lindane*, *hexachlorocyclohexane*, *Endosulfan* (HCHs), and toxaphene (Farnado 2017).

Keeping in view the disadvantages of agrochemicals and excessive use of fertilizers, it is necessary to minimize the use of agrochemicals and to find alternative methods to support extensive agricultural practices using environment-friendly sustainable approaches. Various approaches are being developed to maintain soil fertility such as the use of organic wastes from various sources including the industry and the use of microbial inoculants or biofertilizers (Thomas and Singh 2019).

Biofertilizers

Biofertilizers, unlike chemical fertilizers, pesticides, herbicides, and insecticides, are eco-friendly and cost-effective alternatives that enhance overall fertility of soil in a sustainable manner (Mahdi et al. 2010; Singh et al. 2011). Biofertilizers enhance soil biological activity, nutrient mobilization, improve soil structure and soil nutrient exchange capacity, reduce the loss of nitrogen leaching and phosphorous fixation, lead to better root growth, inhibit certain soil-borne pathogens, increase soil organic matter content and soil water retention, and promote soil aggregates and soil buffering. Biofertilizers are an environmentally friendly and attractive alternative to conventional chemical fertilizer (Anon 2011). The cost of biofertilizers is estimated to be 10% lower than other organic fertilizers, making it an economical alternative (Universidad 2015). The microbial inoculants or "Biofertilizers" are populations of microorganisms such as bacteria, fungi formulated with appropriate carrier material (Boraste et al. 2009). The biofertilizers production technology is easy and costs less than what is required for establishing a facility for the production of chemical fertilizer (Trenkel 1997). Furthermore, being living organism, biofertilizers provide additional nutrients required for the plant growth (Boraste et al. 2009). Various microorganisms belonging to bacteria and fungi in soil possess useful plant growth-promoting activities. These bacteria, fungi, and blue-green algae can be used individually or in combination as biofertilizers (Deshwal and Chaubey 2014). The microbes used as biofertilizers formulations either fix nutrient like nitrogen for plants or mobilize and make the nutrients bioavailable to plants (Anon 2011). It is well reported that many fungal and bacterial species used as biofertilizers show functional relationship with appropriate host plants. They show beneficial effect on host plant which may enhance plant growth (Vessey 2003). Many plants under natural conditions associated with mycorrhizae (Smith and Read 1997) and their mycorrhizal association provide the nutrients and increases the surface area of plant root system (Khan et al. 2000. Mycorrhizae colonization boosts the availability of phosphorus (Bhavikatti and Kadsiddheshwar 2020). Certain bacteria referred to as plant growth-promoting rhizobacteria (PGPR) are known to promote plant growth through various biological processes like nutrients cycling and nutrient mobilization (Glick 1995). In addition to directly stimulating the plant growth through providing nutrients to the plants, these bacteria can promote plant growth indirectly through various other mechanisms. On the basis of colonization PGPR have been classified into rhizospheric and endophytic. In rhizospheric association, PGPR are attached to the surface of root (Backer et al. 2018) for e.g. *Azospirillum* sp., and *Kluyvera ascorbata* sp., and in endophytic association, PGPR occupy the intercellular space like apoplastic space as in *Alnus–Franki* (Huss-Danell 1997), *Parasponia*–rhizobia (Gresshoff et al. 1984, *Azolla–Anabaena* (Peters and Meeks 1989), *Gunnera–Nostoc* (Bergman et al. 1992, and cycads–cyanobacteria (Korzhenevskaya et al. 1999) relationship. It is interesting to note that many PGPR have the ability to fix the atmospheric nitrogen into available form (Hurek et al. 1994). Some of the typically used microorganisms used as biofertilizers are nitrogen-fixing bacteria such as *Rhizobium*, *Azotobacter*, and blue green algae. Cyanobacteria like *Nostoc* and *Anabena* are important nitrogen fixers, especially used for cultivation of rice and wheat (Hashem 2001; Obreht et al. 1993) and it is estimated that BGA alone accounts for 36% of global nitrogen fixation (Kundu and Ladha 1995; Irisarri et al. 2001). The important genera of BGA involved in nitrogen fixation are *Trichodesmium* sp., *Nostoc* sp., and *Anabaena* sp. A number of different organisms can also be associated with specific crops such as *Herbaspirillum* sp. with Maize, *Acetobacter diazotrophicus* with sugarcane and sorghum (Triplett 1996; James et al. 1997; Boddey et al. 2000) *Azoarcus species* with *Leptochloa fusca* (kallar grass) (Malik et al. 1997), *Alcaligenes* sp., *Azospirillum* sp., *Bacillus* sp., *Enterobacter* sp., *Herbaspirillum* sp., *Klebsiella* sp. and *Pseudomonas* sp. with rice and maize (James 2000), and host-specific *Azospirillum* sp. with a variety of annual and perennials plants (Bashan and Holguin 1997). Many studies document an increase in yield following the application of nitrogen biofertilizers. For example, an increase in the growth and crop yield of wheat, corn, sunflower, carrot, oak, sugar beet, tomato, eggplant, pepper, and cotton was observed following the application of *Azospirillum* sp. (Okon 1985; Okon and Labandera-Gonzalez 1994). Phosphate solubilization is another activity of PGPR which increases the availability of phosphorus to the host plants through secretion of organic acids consequently lowering the soil pH (Richardson 2001). A number of organisms have been shown to solubilize phosphorus including *Pseudomonas chlororaphis* and *P. putida* with soybean (Cattelan et al. 1999), *Rhizobium* sp. and *Bradyrhizobium japonicum* with radish (Antoun et al. 1998). *Bacillus* spp., *Pseudomonas* spp., and other soil fungi such as *Aspergillus* spp. *and Penicillium* spp. (Sundara et al. 2002). Other studies demonstrate the production of siderophores by these bacteria like *pyochelin* and *pyoverdine* which promote nutrient mobilization and uptake by plants (Haas and Défago 2005). *Pseudomonas fluorescens* produces iron-chelating siderophores such as *pseudobactin* and *pyoverdine* which bind and absorb ferric ions, making them better iron competitors (Cox and Adams 1985); Leeman et al. 1996). Various other microorganisms are known to produce siderophores include *Trichoderma* sp., *Bacillus* sp., *Streptomyces* sp., *Azospirillum* sp., *Pseudomonas* sp., *Proteus* sp., *Bacillus* sp., *Escherichia* sp., *Xanthomonas* sp., and *Klebsiella* sp. (Mahanty et al. 2017; Maheshwari et al. 2015). A number of microorganisms promote plant growth through the production of plant growth hormones like gibberellins. Gibberellins are a group of phytohormones, which regulate processes of plant development like seed germination, flowering, stem elongation, and senescence of leaves/fruit (Hedden and Phillips 2000). Examples of microorganisms that produce gibberellin include the following genera: *Azospirillum* sp., *Rhizobium* sp., *Bacillus pumilus, Bacillus* sp., *Licheniformis* sp., *Acetobacter diazotrophicus, Herbaspirillumseropedicae* (Atzorn et al. 1988; Gutiérrez-Mañero et al. 2001). Furthermore, some bacteria are known to help plants in tolerating various stresses like soil salinity and aridity. It has been documented that bacteria from desert may help plants tolerate the salinity stress as has been demonstrated for wheat, barley, and cucumbers (Eida et al. 2019). The Basque Institute for Agricultural Research and Development, Neiker-Tecnalia, works to pick such autochthonous bacteria with biofertilizers potential. The strains of bacteria which show positive growth on *A.thaliana* plant shoot biomass under salinity stress were found to be the members of family *Paenibacillaceae, Micrococcaceae, Oxalobacteraceae, Promicromonosporaceae* and *Microbacteriaceae, Enterobacteriaceae,* and *Bacillaceae*. Such families seemed to constitute less than 1% of the overall bacterial endosphere communities (El-Habbasha et al. 2007). *Oxalobacteraceae* and *Promicromonosporaceae* isolates were shown to enhance the fresh weight of the shoot by 193% and 125%, respectively, under salinity stress. These isolates also increase the biomass of the root system (*Oxalobacteraceae* 230%; *Promicromonosporaceae* 283.7%) and contribute to primary root growth (*Oxalobacteraceae* 23.9%; *Promicromonosporaceae*

12.6%). Other studies also demonstrated that bacteria from desert plant roots are capable of promoting plant growth and stress resistance (Mengual et al. 2014; Rolli et al. 2015). Different type of biofertilizers that are available in the market by their trade name are given (Table 12.4).

Table 12.4 Trade name of biofertilizers

Trade name	Microbial fertilizers	Crop plants
Green plus, ROM, biospirillum, azo-S, Spironik, and bio-N	Different strains of *Azospirillum* sp.	Paddy and other crops
Bhoomi Rakshak, Kisaan Azotobacter culture, Azonik, and Bioazoto	Different strains of *Azotobacter*	All crops such as barley, sorghum, maize, wheat, paddy, sunflower, mustard, cotton, sugarcane, grapes, watermelon, papaya, banana, chilly, ladyfinger, cauliflower, linseed, spices tobacco, rapeseed, coconut,, mulberry, fruits, forest plants and plantation crops potato, tomato, onion
Kisaan Rhizobium culture, Rhizo-enrich, Green earth reap, Rhizonik N4, and Biobium,	Rhizobium strains	Oil legumes (soyabeans and groundnut), pulses (peas, cowpea, lentil, gram, arhar, moong, and urd), forest tree legumes (shinsh, subabul, and shisam), fodder, legumes (lucerne and berseem)
MYCORISERISE, get Phos, MycoRhiz, Phosphonive, Kisaan P.S.B. culture, MycoRhiz, reap P, and Biophos	Phosphorus-solubilizing like *Bacillus megaterium, mycorrhizal fungi*	All types of crops
Green earth reap K, Bharpur, potash-cure BioPotash, and BIO-NPK	Potash bacteria like *Bacillus mucilagenosus*	All types of crops
MicroS-109, Sulphonik, Sulf-cure, Siron, S sol B®, and Biosulf	Sulfur-solubilizing microbes	For fiber crops, millets, sugar crops, pulses, oilseeds, plantation crops, forage crops, flowers, aromatic crops, vegetables, fruits, orchards, medicinal crops, ornamental, spices, and cereals
Silica-cure, Silica-109, and bio silica	Silica-solubilizing microbes	For crops like, sugar cane, cereals, onions, legumes, leafy greens, pumpkin, gourd, and cucumber
MicroZ-109, biozinc, zinc-cure, zinc activator, zinc-cure, biofort, Nutrilink, liv zinc, and zinc extra	Zinc-solubilizing microbes	For crops like pulses, pomegranate, citrus, ginger, paddy, and wheat
Sugar-plus	*Gluconacetobacter diazotropicus*	Sugarcane

Source: Singh et al. (2014), India mart (2018), International Panaacea Limited (2018), Biotech International Limited (2018), Biocyclopedia (2018), and National fertilizers limited (2018), ZTM & BPD Unit, ICAR-IARI

Zinc Biofertilizers

Zinc is one of the most important micronutrients for plants, which is often present in soil as insoluble zinc. If this zinc is somehow converted to a soluble form, it will become bioavailable for plant uptake. Various microorganisms have been identified with the ability to solubilize insoluble zinc compounds such as zinc sulfide, zinc oxide, and zinc carbonate. These include *Saccharomyces* sp., *Thiobacillus thiooxidans,* and *Bacillus subtilis* (Rajesha and Ray 2020). Mahdi et al. (2012). found that the application of biofertilizers (*Bacillus* spp.) improves the availability of zinc. Several PGPRs, such as *Rhizobium* sp., *Pseudomonas* sp., *Azospirillum* sp., and *Bacillus* sp., are considered effective zinc solubilizers (Joshi et al. 2013; Naz et al. 2016). These microorganisms employ various methods for the solubilization of Zn. These include production of organic acid, siderophores, and chelated ligand (Saravanan et al. 2011; Wakatsuki 1995; Chang et al. 2005). Different PGPRs have demonstrated an increase in growth and Zn content in plants upon application, which include strains of *Pseudomonas* sp., *Rhizobium* sp. (Joshi et al. 2013; Naz et al. 2016), *Bacillus aryabhattai* (Ramesh et al. 2014), *Bacillus* sp. (Hussain et al. 2015), and *Azospirillum* sp., *Pseudomonas aeruginosa* (Fasim et al. 2002), *Gluconacetobacter diazotrophicus* (Saravanan et al. 2007), *Bacillus* sp., *Pseudomonas striata, Pseudomonas fluorescencia, Burkholderia cenocepacia* (Anuradha et al. 2015), *Serratia liquefaciens, Bacillus thuringiensis,* and *Marcescens* sp. (Abaid-Ullah et al. 2015). Researchers have identified potential zinc-solubilizing bacteria associated with *Zea mays,* which enhances zinc uptake and growth of the plant. *Bacillus* strains were found to promote the growth and yield of soya and wheat through zinc fortification (Khande et al. 2017). Such strains were identified to increase the concentration of Zn in soybean and wheat straw and grains, thus improving the food nutritional status. Vaid et al. (2014) reported an increase in rice productivity, grain yield, and zinc content by 42.7% when inoculated with zinc-solubilizing bacteria. Zinc carbonate was maximally solubilized by bacterial species, compared to other forms of insoluble zinc like zinc oxide (Kamran et al. 2017). Members of the genus *Pseudomonas* have been extensively studied for their PGPR potential. *P. fluorescens* has been shown to produce *indole acetic acid* and solubilization of phosphate and zinc, resulting in increased growth of onion (Reetha et al. 2014; Oteino et al. 2015); Anuradha et al. 2015). Biocontrol properties and HCN production of *Pseudomonas* sp. are also widely reported (Tank and Saraf 2009). Out of five strains, *Pseudomonas fragi, Pantoea dispersa* sp., *Pantoea agglomerans* sp., *Enterobacter cloacae,* and *Rhizobium* sp. show maximum zinc solubilization on $ZnCO_3$ medium and *Pantoea* sp. did not show any solubilization on Zn_3PO_4 medium. All these strains show zinc solubilization on ZnO medium and on Zn_3PO_4 medium, *Pantoea* sp. did not show any solublization (Kamran et al. 2017). The tested strains colonized the *rhizosphere* of sugar cane and wheat increased plant growth (Baig et al. 2012; Verma et al. 2015). When zinc-solubilizing strain like *Pseudomonas fragi, E. cloacae, Rhizobium* sp., were inoculated with wheat plants, they show increased zinc content as well as shoot and root length (Kamran et al. 2017). Strains of *Pseudomonas* sp., *Bacillus* sp., and *Serratia* sp. increased zinc translocation to wheat grains by 7–12% many opposed to plant supplementation of organic zinc (Lefèvre et al. 2014). Inoculation of *Enterobacter ludwigi* strains increased zinc concentration to various degrees in shoots and roots of wheat plants. In wheat shoots, a maximum increase of 82.12% in the Zn concentration was observed ($p < 0.01$) compared to the uninoculated control (Singh et al. 2018). Płociniczak et al. (2016) showed that the metal-tolerant *Brevibacterium casei* strain (MH8a) increased the accumulation of Zn in white willow (*S. alba*) shoots and roots. Other studies also demonstrate that the inoculation of zinc-solubilizing bacteria significantly improves Zn uptake by wheat plant (Chen et al. 2013; Prapagdee et al. 2013). Wheat inoculated with *Azospirillum* sp., and *Pseudomonas* sp. significantly increased zinc content in different parts of the wheat plant at different growth stages (Fasim et al. 2002; Saravanan et al. 2007). Xinxian et al. (2011) documented that endophytic bacteria associated with a perennial herb *Sedum alfredii* exhibit the potential to solubilize $Zn_3(PO_4)_2$ and $ZnCO_3$ in vitro. Spores of Zn-solubilizing *Bacillus* sp. also help the plants to assimilate high quantity of Zn in soybean seeds (Sharma et al. 2011). In similar studies, zinc-solubilizing bacterial strain *Bacillus* sp. (AZ6) also is known to solubilize the insoluble zinc (ZnO), consequently improving the growth of maize plants (Hussain et al. 2015). These studies suggest that it is very important to further isolate and identify such zinc-solubilizing microorganisms and to test their potential as zinc-solubilizing biofertilizers (Fasim et al. 2002; Saravanan et al. 2007).

Zinc Biofortification of Cereal Crops

Deficiency of zinc is a serious problem for both human health and crop production (Duffner et al. 2014). Therefore, zinc biofortification of crops is necessary, which can be achieved through various approaches like breeding programs or agronomic fortification. Figure 12.8 shows the use of different approaches for the biofortification of different crops. For biofortification for improving the micronutrient content

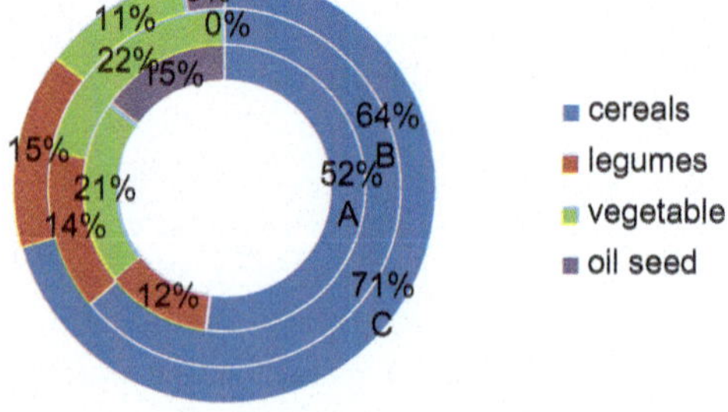

Fig. 12.8 Biofortification in plants (**a**) transgenic, (**b**) breeding, (**c**) agronomy (Garg et al. 2018)

("density") of the food crops, mainly cereals are becoming increasingly essential (Alloway 2008b). The International Rice Research Institute (IRRI) has developed a high iron and zinc rice variety to fight the prevalent zinc deficiency in India. Biofortification of crops with nutrients like iron and zinc is not easy and involves many steps starting from the mobilization from soil into the plant roots and then its transport and accumulation in edible parts like grains (Palmgren et al. 2008). The movement of the nutrients in the plants is also a complex process involving its transport through xylem and phloem. Therefore, it is important to understand the process and challenges involved in the plants' uptake of nutrients and their accumulation in edible parts like grains for successful biofortification of the crops (Waters and Sankaran 2011). The Department of Biotechnology (DBT) and the Government of India's Indian Council for Agricultural Research (ICAR) have also launched funding for biofortification projects aimed at developing biofortified rice cultivars (ICAR 2018a, b). IRRI's International Rice Gene bank holds the world's largest and most extensive array of rice genetic material, with greater than 1.2 lakh unique accessions of rice. This allows IRRI scientists to screen for nutrition-enhancing characteristics in rice and identify the most suitable method of biofortification to improve these characteristics (IRRI report 2018 biofortification). Various other countries like Pakistan and Bangladesh along with India have developed such crops including iron-fortified pearl millet, zinc-fortified wheat, zinc-fortified rice, and golden rice. Some crops which are biofortified are given in Table 12.5.

Agronomic Biofortification

Agronomic fortification refers to the use of zinc fertilizers to meet the Zn requirement of the plants to optimize the uptake and translocation of zinc into the plants and increase the yield (Cakmak 2008). Agronomic fortification is required for soils deficient in Zn, and it has been found that almost half of the global soils are deficient in Zn (Cakmak 2008). Fertilizer application gives better results in the biofortification of wheat or other cereals. The success of fortification also depends on various other factors including edaphic factors such as soil texture and pH (Hirschi 2009; Zhu et al. 2007). Agronomic fortification of zinc and iron can also be carried out through enriching organics such as vermicompost. Various methods of Zn fertilizers application can be used such as foliar application and seed treatment.

Foliar application: Foliar application of Zn fertilizer is one of the most reliable methods for Zn application (Cakmak et al. 2010a). Foliar application is known to enhance rice grain zinc concentration and bioavailability (Fang et al. 2008; Wei et al. 2012; Boonchuay et al. 2013; Jiang et al. 2008; Mabesa et al. 2013; Shivay et al. 2008b; Ram et al. 2016; Mayer et al. 2008). Variations in the duration and solute concentration of foliar spray have also been reported to influence the grain Zn content (Zhang et al. 2010). Various studies document the improvement in the yield and growth following foliar application of zinc. An increase in shoot and root length, stem, leaf, spike and grain dry mass with foliar Se and Zn application has been

Table 12.5 Biofortified crops

Biofortification	Crops	Mineral nutrients	References
Agronomic			
	Rice		
		Iron	Yuan et al. (2013); Wei et al. (2012)
		Zinc	Jiang et al. (2008)
	Wheat		
		Iron	Aciksoz et al. (2011)
		Zinc	Cakmak et al. (2010a, b); Yang et al. (2011)
		P fertilizers + mycorrhiza	Noori et al. (2014)
		Organic and chemical fertilizers (Fe)	Ramzani et al. (2016)
		Bacillus aryabhattai (Zn)	Ramesh et al. (2014)
	Maize		
		Zinc	Alvarez and Rico (2003); Lopez-Valdivia et al. (2002)
		PGPR + cyanobacteria (Zn)	Prasanna et al. (2015)
	Barley		
		Biofertilizers +NPK+ Vermicompost	Maleki et al. (2011)
	Sorghum		
		Mycorrhiza bacteria	Dhawi et al. (2015, 2016)
		Farmyard manure+ biofertilizer	Patidar and Mali (2004)
	Pea		
		Zinc	Poblaciones and Rengel (2016)
	Common bean		
		Zinc	Ibrahim and Ramadan (2015); Ram et al. (2016)
		N, P, K, Cu, Mn, Zn, (chemical +organic fertilizers	Westermann et al. (2011)
	Potato		
		Zinc	Jason and Gardea-Torresdey (2018)
Transgenic			
	Wheat		
		Iron	Xiaoyan et al. (2012); Borg et al. (2012)
	Rice		
		Iron	Takahashi et al. (2001); Lee and An (2009)

(continued)

Table 12.5 (continued)

Biofortification	Crops	Mineral nutrients	References
		Zinc	Lee and An (2009); Masuda et al. (2008)
	Barley		
		Zinc	Ramesh et al. (2004)
	Cassava		
		Iron	Sayre et al. (2011)
	Carrot		
		Calcium	Park et al. (2004); Morris et al. (2008)
	Lettuce		
		Iron	Goto et al. (2000)
Breeding			
	Rice		
		Zinc, iron	HarvestPlus (2010)
		Iron	IRRI (2018), Gregorio et al. (2000)
		Zinc	Gregorio et al. (2000)
	Wheat		
		Zinc	ICAR-IIWBR (2019)
		Zinc and iron	CIAT, CIMMYT, Harvest Plus (2010)
	Sorghum		
		Iron	ICRISAT (2009), HarvestPlus (2010)
		Iron, zinc	Reddy et al. (2005)
	Millets		
		Iron, zinc (peat millet)	ICRISAT (2009), HarvestPlus (2010)
		Iron, zinc	Velu et al. (2007); Rai et al. (2012)
	Lentils		
		Iron, zinc	ICARDA, HarvestPlus (2010)
	Beans		
		Iron, zinc	HarvestPlus (2010)
	Potato		
		Iron, zinc, manganese, and copper	Haynes et al. (2012)

reported (Wu et al. 2020). In multiple field studies, Phattarakul et al. (2012) reported a positive effect of foliar application of Zn content on grain. This experiment recorded an overall 25% increase in Zn concentration in brown rice after foliar application of Zn. An average of 13% improvement in polished rice Zn content was also reported. Zinc foliar application resulted in higher concentrations of grain zinc in bread and durum wheat than in soil application (Zhang et al. 2010;

Khoshgoftarmanesh et al. 2013). Studies revealed notable differences in the accumulation of zinc in grains between various *Triticum* sp. cultivars in soil and foliar Zn application (Cakmak et al. 1997; Khoshgoftarmanesh et al. 2013). In addition, foliar zinc application timing was also found to be critical in improving the zinc content of wheat grains. Foliar zinc application during flowering has demonstrated highest increase in zinc content (Cakmak et al. 2010b).

Seed treatment: Seed treatment is also one of the efficient methods for Zn application because it requires relatively smaller amounts of Zn (Welch and Graham 1999). Seed treatment with Zn-EDTA and fried Zn solutions has resulted in Zn uptake and higher yield compared with banded or broadcast soil application (Kang and Okoro 1976). In general, application of zinc via seed treatment can be divided into seed priming and seed coating (Farooq et al. 2012). Coating of seeds with micronutrients such as zinc before planting is suitable way for enhancing plant growth and production, and for improving nutritional status of the crop for human consumption (Tucker 2003). Zinc seed dressings can effectively supply adequate zinc to a growing crop. The amount of nutrient that can be added to a seed dressing also depends on the seed size. Lesser coating media can hold small seeds than large seeds. In seed processing, the seed is covered by a less or more continuous layer of suspended solids or dissolved containing finely ground liquids or solids (Farooq et al. 2012). For rice, a solution with 0.5% (w/v) Zn was suitable. Slaton et al. (2001) reported better production of dry matter and higher concentration of Zn tissue and grain yields from Zn-primed rice seeds than the soil applications of Zn.

Genetic Biofortification

Zn-dense varieties of crops can also be developed through breeding programs or through genetic engineering approaches (Cakmak 2008). For example, crossing varieties with the ability to accumulate high quantities of zinc in seeds with the varieties that exhibit good growth and productivity can help in developing Zn-efficient varieties (Alloway 2008b). Developing such varieties is regarded as a cost-effective method of alleviating nutritional deficiencies in rural communities of the developing countries where problem of Zn deficiency is prevalent (Mayer et al. 2008). Breeding is possible for cereal crops since there is sufficient genotypic variability in the germplasm (White and Broadley 2009). Various breeding programs around the world are being carried out to develop Zn-dense varieties, especially of cereal crops. The organization CIMMYT used the breeding of wheat species such as *Triticum turgidum* and *Triticum dicoccum* to produce modern hexaploid wheat with higher iron and zinc concentrations (Ortiz-Monasterio et al. 2007). In *Arabidopsis thaliana,* gene expression encoding a Zn transporter protein from the roots of a barley genotype has led to an increase in the concentration of Zn grain (Ramesh et al. 2004). Overexpression of OsIRT1 gene enhanced zinc concentration in GM rice (Lee and An 2009). Overexpression of zinc transporters in transgenic Barley (*Hordeum vulgare*) also improved its zinc content (Ramesh et al. 2004).

Genetic manipulation to reduce plant phytates is another means of tackling the global zinc deficiency (Lonnerdal 2003). Reducing phytate levels using technology like RNAi may improve the bioavailability of Zn as well as other mineral nutrients and thus will enhance the nutritional value of several crop plants (White and Broadley 2009). NAM B1 is a transcription factor, which codes for the mobilization of senescence and nutrition from the vegetative part into the grains (Uauy et al. 2006; Distelfeld et al. 2007). Transgenic rice plants that stimulated the production of nicotianamine more than double in comparison with control line were found to have double concentration of Zn in the grains (Lee et al. 2011; Trijatmiko et al. 2016).

Conclusion

Mineral nutrients, such as Ca, Mg, Mn, B, Fe, Cu, and Zn, are vital to plant growth and development. Growing crops obtain these nutrients generally from the soil. One of the most important micronutrients is Zn and its deficiency causes a significant effect on plant health as documented in numerous studies. The impact of Zn deficiency and responses are different in different plant species. Its requirement for different plant species is also different which depends on plants' age and various other factors. For growing cereal crops such as wheat, rice, and maize, the required zinc is often added in the form of chemical fertilizers. These chemical fertilizers are associated with inherent problems like environmental pollution and toxicity. Therefore, the use of biofertilizers is recommended as the part of sustainable agricultural practices. Furthermore, biofertilizers are cheap, eco-friendly and are known to provide more micronutrients and protect the plant against stress, and pathogens. The application of Zn biofertilizers to the soil can ensure a robust and safer supply of Zn to plants. Since, zinc deficiency is becoming a global problem, a global response and approach are required to tackle the problem. Scientists are seeking to enhance zinc availability through ways like genetic and agronomic biofortification. These efforts should be supported through the active involvement of the government and policymakers. Awareness programs at the global level, especially for farmers, are also required.

Acknowledgments The authors acknowledge the generous financial help provided by the MHRD, Govt. of India, under the Scheme for Promotion of Academic and Research Collaboration (SPARC; P594).

References

Abaid-Ullah, M., Nadeem, M., Hassan, M., Ganter, J., Muhammad, B., Nawaz, K., ... Hafeez, F.Y.: Plant growth promoting rhizobacteria: an alternate way to improve yield and quality of wheat (Triticum aestivum). Int. J. Agric. Biol. **17**(1), (2015)

Aciksoz, S.B., Yazici, A., Ozturk, L., Cakmak, I.: Biofortification of wheat with iron through soil and foliar application of nitrogen and iron fertilizers. Plant Soil. **349**(1–2), 215–225 (2011)

Adrees, M., Ali, S., Rizwan, M., Ibrahim, M., Abbas, F., Farid, M., Zia-ur-Rehman, M., Irshad, M.K., Bharwana, S.A.: The effect of excess copper on growth and physiology of important food crops: a review. Environ. Sci. Pollut. Res. **22**(11), 8148–8162 (2015)

Agarwala, S.C., Sharma, C.P., Farooq, S., Chatterjee, C.: Effect of molybdenum deficiency on the growth and metabolism of corn plants raised in sand culture. Can. J. Bot. **56**(16), 1905–1908 (1978)

Agarwala, S.C., Nautiyal, B.D., Chatterjee, C.: Manganese, copper and molybdenum nutrition of papaya. J. Horticul. Sci. **61**(3), 397–405 (1986)

Agulhon, H.: Presence and utility of boron in plants. Ann. Inst. Pasteur (Paris). **24**, 321–329 (1910)

Ahmad, H.R., Aziz, T., Hussain, S., Akraam, M., Sabir, M., Hanafi, M.M.: Zinc-enriched farm yard manure improves grain yield and grain zinc concentration in rice grown on a saline-sodic soil. Int. J. Agric. Biol. **14**(5) (2012)

Ahsan, E., Beuter, T.: Technical note: Changes in crop production scenario. Bangladesh: Resal. (2000)

Alloway, B.J.: Zinc in Soils and Crop Nutrition. International Zinc Association Communications. IZA Publications, Brussels (2004)

Alloway, B.J. (ed.): Micronutrient deficiencies in global crop production. Springer Science & Business Media. UK. **10**, 978–1007 (2008a)

Alloway, B.J.: Zinc in Soils and Crop Nutrition. IZA and IFA Brussels, Belgium and Paris, France. **2**, (2008b)

Alloway, B.J.: Soil factors associated with zinc deficiency in crops and humans. Environ. Geochem. Health. **31**(5), 537–548 (2009)

Alvarez, J.M., Rico, M.I.: Effects of zinc complexes on the distribution of zinc in calcareous soil and zinc uptake by maize. J. Agric. Food Chem. **51**(19), 5760–5767 (2003)

Andreazza, R., Bortolon, L., Pieniz, S., Bento, F.M., Camargo, F.A.D.O.: Evaluation of two Brazilian indigenous plants for phytostabilization and phytoremediation of copper-contaminated soils. Brazilian J. Biol., (AHEAD) **75**(4), 868–877 (2015)

Anon.: USDA Foreign Agricultural Service Circular Series: Grain 07–11; USDA: Washington, DC (2011)

Antoun, H., Beauchamp, C.J., Goussard, N., Chabot, R., Lalande, R.: Potential of rhizobium and Bradyrhizobium species as plant growth promoting rhizobacteria on non-legumes: effect on radishes (Raphanus sativus L.). In: Molecular Microbial Ecology of the Soil, pp. 57–67. Springer, Dordrecht (1998)

Anuradha, P., Syed, I., Swati, M., Patil, V.D.: Solubilization of insoluble zinc compounds by different microbial isolates in vitro condition. Int. J. Trop. Agric. **33**(2 (Part II)), 865–869 (2015)

Aref, F.: Manganese, iron and copper contents in leaves of maize plants (Zea mays L.) grown with different boron and zinc micronutrients. Afr. J. Biotechnol. **11**(4), 896–903 (2012)

Arnon, D.I., Stout, P.R.: Molybdenum as an essential element for higher plants. Plant Physiol. **14**(3), 599 (1939)

Aschner, J.L., Aschner, M.: Nutritional aspects of manganese homeostasis. Mol. Asp. Med. **26**(4–5), 353–362 (2005)

Atzorn, R., Crozier, A., Wheeler, C.T., Sandberg, G.: Production of gibberellins and indole-3-acetic acid by Rhizobium phaseoli in relation to nodulation of Phaseolus vulgaris roots. Planta. **175**(4), 532–538 (1988)

Backer, R., Rokem, J.S., Ilangumaran, G., Lamont, J., Praslickova, D., Ricci, E., ... Smith, D.L.: Plant growth-promoting rhizobacteria: context, mechanisms of action, and roadmap to commercialization of biostimulants for sustainable agriculture. Front. Plant Sci. **9**, 1473 (2018)

Baig, K.S., Arshad, M., Shaharoona, B., Khalid, A., Ahmed, I.: Comparative effectiveness of Bacillus spp. possessing either dual or single growth-promoting traits for improving phosphorus uptake, growth and yield of wheat (Triticum aestivum L.). Ann. Microbiol. **62**(3), 1109–1119 (2012)

Bailey, R.L., West Jr., K.P., Black, R.E.: The epidemiology of global micronutrient deficiencies. Ann. Nutr. Metab. **66**(Suppl. 2), 22–33 (2015)

Banerjee, A., Datta, J.K., Mondal, N.K., Chanda, T.: Influence of integrated nutrient management on soil properties of old alluvial soil under mustard cropping system. Commun. Soil Sci. Plant Anal. **42**(20), 2473–2492 (2011)

Barak, P., Helmke, P.A.: The chemistry of zinc. Chap 1. In: Robson, A.D. (ed.) Zinc in Soils and Plants. Springer, Dordrecht. **55**, 90–106 (1993)

Barrow, N.J., Cartes, P., Mora, M.L.: Modifications to the Freundlich equation to describe anion sorption over a large range and to describe competition between pairs of ions. Eur. J. Soil Sci. **56**(5), 601–606 (2005)

Bashan, Y., Holguin, G.: Azospirillum–plant relationships: environmental and physiological advances (1990–1996). Canadian J. Microbiol. **43**(2), 103–121 (1997)

Beck, F.W., Kaplan, J., Fine, N., Handschu, W., Prasad, A.S.: Decreased expression of CD73 (ecto-5′-nucleotidase) in the CD8+ subset is associated with zinc deficiency in human patients. J. Lab. Clin. Med. **130**(2), 147–156 (1997)

Bergman, B., Johansson, C., Söderbäck, E.: The Nostoc–Gunnera symbiosis. New Phytol. **122**(3), 379–400 (1992)

Bharadva, K., Mishra, S., Tiwari, S., Yadav, B., Deshmukh, U., Elizabeth, K.E., Banapurmath, C.R.: Prevention of micronutrient deficiencies in young children: consensus statement from infant and young child feeding chapter of Indian Academy of Pediatrics. Indian Pediatr. **56**(7), 577–586 (2019)

Bhavikatti, Ms Bharati Kadsiddheshwar. "Biofertilizers: A Potential approach for Sustainable Agriculture in India." Studies in Indian Place Names 40. **63**, 111–118 (2020)

Biocyclopedia.: https://biocyclopedia.com/index/biotechnology/plant_biotechnology/biofertilizers/biotech_procedures_of_biofertilizer.php (2018)

Biotech International Limited.: https://biotechint.com/biofertilizers.html (2018)

Bittner, F.: Molybdenum metabolism in plants and crosstalk to iron. Front. Plant Sci. **5**, 28 (2014)

Boddey, R.M., Da Silva, L.G., Reis, V., Alves, B.J.R., Urquiaga, S.: Assessment of bacterial nitrogen fixation in grass species. Prokaryotic nitrogen fixation: a model system for the analysis of a biological process., 705–726 (2000)

Boonchuay, P., Cakmak, I., Rerkasem, B., Prom-U-Thai, C.: Effect of different foliar zinc application at different growth stages on seed zinc concentration and its impact on seedling vigor in rice. Soil Sci. Plant Nutr. **59**(2), 180–188 (2013)

Bolaños, L., Lukaszewski, K., Bonilla, I., Blevins, D.: Why boron?. Plant Physiol. Biochem. **42**(11), 907–912 (2004)

Borah, R.C., Talukdar, A., Kataky, J.C.S., Unni, B.G., Modi, M.K., Deka, P.C.: Bioprospecting of commercially important plants. In: Proceedings of the national symposium on "Biochemical approaches for utilization and exploitation of commercially important plants", Jorhat, India, 12-14 Nov. 2003. Indian Society of Agricultural Biochemists (2004)

Boraste, A., Vamsi, K K., Jhadav, A., Khairnar, Y., Gupta, N., Trivedi, S., ... Joshi, B.: Biofertilizers: A novel tool for agriculture. Int. J. Microbiol. Res. **1**(2), 23 (2009)

Borg, S., Brinch-Pedersen, H., Tauris, B., Madsen, L.H., Darbani, B., Noeparvar, S., Holm, P.B.: Wheat ferritins: improving the iron content of the wheat grain. J. Cereal Sci. **56**(2), 204–213 (2012)

Brahmam, S.N.V.: National nutrition monitoring bureau in India-An overview. Indian J. Community Med. **32**(1), 7 (2007)

Brdar-Jokanović, M.: Boron toxicity and deficiency in agricultural plants. Int. J. Mol. Sci. **21**(4), 1424 (2020)

Briat, J.F., Dubos, C., Gaymard, F.: Iron nutrition, biomass production, and plant product quality. Trends Plant Sci. **20**(1), 33–40 (2015)

Broadley, M.R., White, P.J., Hammond, J.P., Zelko, I., Lux, A.: Zinc in plants. New Phytol. **173**(4), 677–702 (2007)

Brown, P.H.: Micronutrient use in agriculture in the United States of America. In: Micronutrient Deficiencies in Global Crop Production, pp. 267–286. Springer, Dordrecht (2008)

Buet, A., Galatro, A., Ramos-Artuso, F., Simontacchi, M.: Nitric oxide and plant mineral nutrition: current knowledge. J. Exp. Bot. **70**(17), 4461–4476 (2019)

Bulcke, F., Dringen, R., Scheiber, I.F.: Neurotoxicity of copper. In: Neurotoxicity of Metals, pp. 313–343. Springer, Cham (2017)

Burkhead, J.L., Gogolin Reynolds, K.A., Abdel-Ghany, S.E., Cohu, C.M., Pilon, M.: Copper homeostasis. New Phytol. **182**(4), 799–816 (2009)

Caggiano, V., Schnitzler, R., Strauss, W., Baker, R.K., Carter, A.C., Josephson, A.S., Wallach, S.: Zinc deficiency in a patient with retarded growth, hypogonadism, hypogammaglobulinemia and chronic infection. Am. J. Med. Sci. **257**, 305–319 (1969)

Cakmak, I.: Tansley Review No. 111. New Phytol. **146**(2), 185–205 (2000)

Cakmak, I.: Plant nutrition research: priorities to meet human needs for food in sustainable ways. Plant Soil. **247**(1), 3–24 (2002)

Cakmak, I.: Enrichment of cereal grains with zinc: agronomic or genetic biofortification? Plant Soil. **302**(1–2), 1–17 (2008)

Cakmak, I.: Biofortification of cereals with zinc and iron through fertilization strategy. In 19th World Congress of Soil Science, Brisbane (2010, August)

Cakmak, I.: Zinc plays critical role in plant growth. 2012-05-20] http://www.zinc.org/crops/resourceserve/zinc_plays_criti-cal_role in_olant-growth 2011

Cakmak, I., Ekiz, H., Yilmaz, A., Torun, B., Köleli, N., Gültekin, I., Alkan, A., Eker, S.: Differential response of rye, triticale, bread and durum wheats to zinc deficiency in calcareous soils. Plant Soil. **188**(1), 1–10 (1997)

Çakmak, İ., Kalaycı, M., Ekiz, H., Braun, H.J., Kılınç, Y., Yılmaz, A.: Zinc deficiency as a practical problem in plant and human nutrition in Turkey: a NATO-science for stability project. Field Crop Res. **60**(1–2), 175–188 (1999)

Cakmak, I., Kalayci, M., Kaya, Y., Torun, A.A., Aydin, N., Wang, Y., Arisoy, Z., Erdem, H.A.M.İ.D.E., Yazici, A., Gokmen, O., Ozturk, L.: Biofortification and localization of zinc in wheat grain. J. Agric. Food Chem. **58**(16), 9092–9102 (2010a)

Cakmak, I., Pfeiffer, W.H., McClafferty, B.: Biofortification of durum wheat with zinc and iron. Cereal Chem. **87**(1), 10–20 (2010b)

Cakmak, I., McLaughlin, M.J., White, P.: Zinc for better crop production and human health. Plant Soil. **411**, 1–4 (2017)

Caliskan, S., Ozkaya, I., Caliskan, M.E., Arslan, M.: The effects of nitrogen and iron fertilization on growth, yield and fertilizer use efficiency of soybean in a Mediterranean-type soil. Field Crop Res. **108**(2), 126–132 (2008)

Camaschella, C.: Iron-deficiency anemia. N. Engl. J. Med. **372**(19), 1832–1843 (2015)

Camaschella, C.: New insights into iron deficiency and iron deficiency anemia. Blood Rev. **31**(4), 225–233 (2017)

Campbell, W.H.: Nitrate reductase structure, function and regulation: bridging the gap between biochemistry and physiology. Annu. Rev. Plant Biol. **50**(1), 277–303 (1999)

Cappellini, M.D., Musallam, K.M., Taher, A.T.: Iron deficiency anaemia revisited. J. Intern. Med. **287**(2), 153–170 (2020)

Cattelan, A.J., Hartel, P.G., Fuhrmann, J.J.: Screening for plant growth–promoting rhizobacteria to promote early soybean growth. Soil Sci. Soc. Am. J. **63**(6), 1670–1680 (1999)

Chang, H.B., Lin, C.W., Huang, H.J.: Zinc-induced cell death in rice (Oryza sativa L.) roots. Plant Growth Regul. **46**(3), 261–266 (2005)

Chen, W., Yang, X., He, Z., Feng, Y., Hu, F.: Differential changes in photosynthetic capacity, 77 K chlorophyll fluorescence and chloroplast ultrastructure between Zn-efficient and Zn-inefficient rice genotypes (Oryza sativa) under low zinc stress. Physiol. Plant. **132**(1), 89–101 (2008)

Chen, Z.J., Sheng, X.F., He, L.Y., Huang, Z., Zhang, W.H.: Effects of root inoculation with bacteria on the growth, Cd uptake and bacterial communities associated with rape grown in Cd-contaminated soil. J. Hazard. Mater. **244**, 709–717 (2013)

Cheng, C., Rerkasem, B.: Effects of boron on pollen viability in wheat. Plant Soil. **155**(1), 313–315 (1993)

Chormova, D., Fry, S.C.: Boron bridging of rhamnogalacturonan-II is promoted in vitro by cationic chaperones, including polyhistidine and wall glycoproteins. New Phytol. **209**(1), 241–251 (2016)

Churchill, K.A., Sze, H.: Anion-sensitive, H+-pumping ATPase of oat roots: direct effects of Cl−, NO3−, and a Disulfonic stilbene. Plant Physiol. **76**(2), 490–497 (1984)

CIMMYT.: International maize and wheat centre 2006.flickr.co

CYMMIT.: http://cimmyt.org?research/wheat/map/developingworld/wheatdevelopingword (2000)

Clarkson, D.T., Hanson, J.B.: The mineral nutrition of higher plants. Annu. Rev. Plant Physiol. **31**(1), 239–298 (1980)

Clemens, S.: Toxic metal accumulation, responses to exposure and mechanisms of tolerance in plants. Biochimie. **88**(11), 1707–1719 (2006)

Collin, B., Doelsch, E., Keller, C., Cazevieille, P., Tella, M., Chaurand, P., Panfili, F., Hazemann, J.L., Meunier, J.D.: Evidence of sulfur-bound reduced copper in bamboo exposed to high silicon and copper concentrations. Environ. Pollut. **187**, 22–30 (2014)

Colmenero-Flores, J.M., Franco-Navarro, J.D., Cubero-Font, P., Peinado-Torrubia, P., Rosales, M.A.: Chloride as a beneficial macronutrient in higher plants: new roles and regulation. Int. J. Mol. Sci. **20**(19), 4686 (2019)

Cossack, Z.T.: Decline in somatomedin-C (insulin-like growth factor-1) with experimentally induced zinc deficiency in human subjects. Clin. Nutr. **10**(5), 284–291 (1991)

Cox, C.D., Adams, P.A.T.R.I.C.I.A.: Siderophore activity of pyoverdin for Pseudomonas aeruginosa. Infect. Immun. **48**(1), 130–138 (1985)

Craine, J.M., Morrow, C., Stock, W.D.: Nutrient concentration ratios and co-limitation in south African grasslands. New Phytol. **179**(3), 829–836 (2008)

Curie, C., Briat, J.F.: Iron transport and signaling in plants. Annu. Rev. Plant Biol. **54**(1), 183–206 (2003)

D'Ambrosi, N., Rossi, L.: Copper at synapse: release, binding and modulation of neurotransmission. Neurochem. Int. **90**, 36–45 (2015)

DAE (Department of Agriculture Extension): List of Import Registration Certificate Holder, pp. 23–74. Dept. Agric., Extn, Farmgate, Dhaka (2014)

Dandona, L., Dandona, R., Kumar, G.A., Shukla, D.K., Paul, V.K., Balakrishnan, K., Prabhakaran, D., Tandon, N., Salvi, S., Dash, A.P., Nandakumar, A.: Nations within a nation: variations in epidemiological transition across the states of India, 1990–2016 in the Global Burden of Disease Study. Lancet. **390**(10111), 2437–2460 (2017)

Das, D.K.: 2001. Micronutrients: Their behaviour in soils and plants. Bangladesh Forest Research Institute. (2003)

Das, S., Green, A.: Importance of zinc in crops and human health. J. SAT Agric. Res. **11**, 1–7 (2013)

Davis, J.G., Quick, J.S.: Nutrient management, cultivar development and selection strategies to optimize water use efficiency. J. Crop. Prod. **1**(2), 221–240 (1998)

Delhaize, E., Gruber, B.D., Pittman, J.K., White, R.G., Leung, H., Miao, Y., Jiang, L., Ryan, P.R., Richardson, A.E.: A role for the AtMTP11 gene of Arabidopsis in manganese transport and tolerance. Plant J. **51**(2), 198–210 (2007)

Deshwal, V.K., Chaubey, A.: Isolation and characterization of Rhizobium leguminosarum from root nodule of Pisum sativum L. J. Acad. Ind. Res. **2**(8), 464–467 (2014)

Detterbeck, A., Pongrac, P., Rensch, S., Reuscher, S., Pečovnik, M., Vavpetič, P., Pelicon, P., Holzheu, S., Krämer, U., Clemens, S.: Spatially resolved analysis of variation in barley (Hordeum vulgare) grain micronutrient accumulation. New Phytol. **211**(4), 1241–1254 (2016)

Devergnas, S., Chimienti, F., Naud, N., Pennequin, A., Coquerel, Y., Chantegrel, J., Favier, A., Seve, M.: Differential regulation of zinc efflux transporters ZnT-1, ZnT-5 and ZnT-7 gene expression by zinc levels: a real-time RT–PCR study. Biochem. Pharmacol. **68**(4), 699–709 (2004)

Dhawi, F., Datta, R., Ramakrishna, W.: Mycorrhiza and PGPB modulate maize biomass, nutrient uptake and metabolic pathways in maize grown in mining-impacted soil. Plant Physiol. Biochem. **97**, 390–399 (2015)

Dhawi, F., Datta, R., Ramakrishna, W.: Mycorrhiza and heavy metal resistant bacteria enhance growth, nutrient uptake and alter metabolic profile of sorghum grown in marginal soil. Chemosphere. **157**, 33–41 (2016)

Distelfeld, A., Cakmak, I., Peleg, Z., Ozturk, L., Yazici, A.M., Budak, H., Saranga, Y., Fahima, T.: Multiple QTL-effects of wheat Gpc-B1 locus on grain protein and micronutrient concentrations. Physiol. Plant. **129**(3), 635–643 (2007)

Doncheva, S.N., Poschenrieder, C., Stoyanova, Z., Georgieva, K., Velichkova, M., Barceló, J.: Silicon amelioration of manganese toxicity in Mn-sensitive and Mn-tolerant maize varieties. Environ. Exp. Bot. **65**(2–3), 189–197 (2009)

Dordas, C., Chrispeels, M.J., Brown, P.H.: Permeability and channel-mediated transport of boric acid across membrane vesicles isolated from squash roots. Plant Physiol. **124**(3), 1349–1362 (2000)

Dubey K.: Plant nutrients and it's role. http://agropedia.iitk.ac.in/content/plant-nutrients-and-its-role. (Verified on Oct. 24, 2017) (2011)

Duffner, A., Hoffland, E., Stomph, T.J., Boonstra, A., Bindraban, P.S.. Eliminating zinc deficiency in rice-based systems. VFRC report, Virtual Fertilizer Research Center; No. 2014/2, pp. 38 (2014)

Dungait, J.A., Cardenas, L.M., Blackwell, M.S., Wu, L., Withers, P.J., Chadwick, D.R., Bol, R., Murray, P.J., Macdonald, A.J., Whitmore, A.P., Goulding, K.W.: Advances in the understanding of nutrient dynamics and management in UK agriculture. Sci. Total Environ. **434**, 39–50 (2012)

Eida, A.A., Alzubaidy, H.S., de Zélicourt, A., Synek, L., Alsharif, W., Lafi, F.F., Hirt, H., Saad, M.M.: Phylogenetically diverse endophytic bacteria from desert plants induce transcriptional changes of tissue-specific ion transporters and salinity stress in Arabidopsis thaliana. Plant Sci. **280**, 228–240 (2019)

El-Habbasha, S.F., Hozayn, M., Khalafallah, M.A.: Integration effect between phosphorus levels and biofertilizers on quality and quantity yield of faba bean (Vicia faba L.) in newly cultivated sandy soils. Res. J. Agric. Biol. Sci. **3**(6), 966–971 (2007)

Environment. ''How do fertilizers affect the environment''. Environmental news Press Centre South Africa, (2018)

Erenoglu, E.B., Kutman, U.B., Ceylan, Y., Yildiz, B., Cakmak, I.: Improved nitrogen nutrition enhances root uptake, root-to-shoot translocation and remobilization of zinc (65Zn) in wheat. New Phytol. **189**(2), 438–448 (2011)

Erikson, K.M., Syversen, T., Aschner, J.L., Aschner, M.: Interactions between excessive manganese exposures and dietary iron-deficiency in neurodegeneration. Environ. Toxicol. Pharmacol. **19**(3), 415–421 (2005)

Fageria, N.K., Stone, L.F.: Micronutrient deficiency problems in South America. In: Micronutrient Deficiencies in Global Crop Production, pp. 245–266. Springer, Dordrecht (2008)

Falk, K.L., Tokuhisa, J.G., Gershenzon, J.: The effect of sulfur nutrition on plant glucosinolate content: Physiology and molecular mechanisms. Plant Biol. **9**(5), 573–581 (2007)

Fang, Y., Wang, L., Xin, Z., Zhao, L., An, X., Hu, Q.: Effect of foliar application of zinc, selenium, and iron fertilizers on nutrients concentration and yield of rice grain in China. J. Agric. Food Chem. **56**(6), 2079–2084 (2008)

FAO: Calcareous Soils. Soils Bulletin No. 21, Food and Agriculture Organization, Rome. www.fao.org/docrep/x5868e/x5858e04.htm (1972)

FAO/International Fertilizer Industry Association (FAO/IFA): 2001 Global estimates of gaseous emissions from agricultural land. Rome. 85 pp.

FAO/WHO.: Expert Consultation on Human Vitamin and Mineral Requirements, Vitamin and mineral requirements in human nutrition: Report of joint FAO/WHO expert consolation. 2nd ed. Bangkok, Thailand: p. 341 (2004)

FAO, FAOSTAT: In http://faostat.fao.org/site/291/default.aspx, Food and Agriculture Organization: 2011.

FAO, IFAD, UNICEF, WFP and WHO.: The State of Food Security and Nutrition in the World 2017: Building Resilience for Peace and Food Security. Rome: Food and Agriculture Organization of the United Nations (FAO) (2017)

FAO, IFAD, UNICEF, WFP and WHO.: The State of Food Security and Nutrition in the World 2018. Building Climate Resilience for Food Security and Nutrition. Rome: Food and Agriculture of the United Nations (FAO) (2018)

Farooq, M., Wahid, A., Siddique, K.H.: Micronutrient application through seed treatments: a review. J. Soil Sci. Plant Nutr. **12**(1), 125–142 (2012)

Fasim, F., Ahmed, N., Parsons, R., Gadd, G.M.: Solubilization of zinc salts by a bacterium isolated from the air environment of a tannery. FEMS Microbiol. Lett. **213**(1), 1–6 (2002)

Fatima, A., Farid, M., Alharby, H.F., Bamagoos, A.A., Rizwan, M., Ali, S.: Efficacy of fenugreek plant for ascorbic acid assisted phytoextraction of copper (Cu); A detailed study of Cu induced morpho-physiological and biochemical alterations. Chemosphere. **251**, 126424 (2020)

FCO Amendment: 2019; Central fertilizer quality control & training institute

Fecht-Christoffers, M.M., Braun, H.P., Lemaitre-Guillier, C., VanDorsselaer, A., Horst, W.J.: Effect of manganese toxicity on the proteome of the leaf apoplast in cowpea. Plant Physiol. **133**(4), 1935–1946 (2003)

Fernando, P.C.: Pesticides, environment, and food safety. Food and Energy Secur. **6**(2), 48–60 (2017)

Franco-Navarro, J.D., Brumós, J., Rosales, M.A., Cubero-Font, P., Talón, M., Colmenero-Flores, J.M.: Chloride regulates leaf cell size and water relations in tobacco plants. J. Exp. Bot. **67**(3), 873–891 (2016)

Gaind, S., Singh, Y.V.: Relative efficiency of fertilization practices to improve productivity and phosphorus balance in rice-wheat cropping system. J. Crop Improv. **29**(1), 23–39 (2015)

Ganeshamurthy, A.N., Raghu-pathi, H.B., Rupa, T.R., Rajendiran, S., Kalaivanan, D.: Micronutrient management in horticultural crops. Indian J. Fertilis. **14**(4), 68–85 (2018)

Garai, T.K., Datta, J.K., Mondal, N.K.: Evaluation of integrated nutrient management on boro rice in alluvial soil and its impacts upon growth, yield attributes, yield and soil nutrient status. Arch. Agron. Soil Sci. **60**(1), 1–14 (2014)

Garg, M., Sharma, N., Sharma, S., Kapoor, P., Kumar, A., Chunduri, V., Arora, P.: Biofortified crops generated by breeding, agronomy, and transgenic approaches are improving lives of millions of people around the world. Front. Nutr. **5**, 12 (2018)

George, T.S., French, A.S., Brown, L.K., Karley, A.J., White, P.J., Ramsay, L., Daniell, T.J.: Genotypic variation in the ability of landraces and commercial cereal varieties to avoid manganese deficiency in soils with limited manganese availability: is there a role for root-exuded phytases? Physiol. Plant. **151**(3), 243–256 (2014)

Gerendas, J., Fuhrs, H.: The significance of magnesium for crop quality. Plant Soil. **368**(1), 101–128 (2013)

Gherardi, M.J., Rengel, Z.: The effect of manganese supply on exudation of carboxylates by roots of lucerne (Medicago sativa). Plant Soil. **260**(1–2), 271–282 (2004)

GHO: Global Health Observatory data repository 2016. http://apps.who.int/gho/data/node.home

Glick, B.R.: The enhancement of plant growth by free-living bacteria. Canadian J. Microbiol. **41**(2), 109–117 (1995)

Goldberg, S.P., Smith, K.A., Holmes, J.C.: The effects of soil compaction, form of nitrogen fertiliser, and fertiliser placement on the availability of manganese to barley. J. Sci. Food Agric. **34**(7), 657–670 (1983)

Gonçalves Júnior, A.C., Nacke, H., Marengoni, N.G., Carvalho, E.A.D., Coelho, G.F.: Productivity and production components of soybean fertilized with different doses of phosphorus, potassium and zinc. Sci. Agrotechnol. **34**(3), 660–666 (2010)

González, A., Steffen, K.L., Lynch, J.P.: Light and excess manganese: implications for oxidative stress in common bean. Plant Physiol. **118**(2), 493–504 (1998)

Gopalan, C. Nutrition in developmental transition in South-East Asia. 141 pp (1992)

Gorstein, J., Sullivan, K.M., Parvanta, I., Begin, F.: Indicators and Methods for Cross-Sectional Surveys of Vitamin and Mineral Status of Populations. The Micronutrient Initiative (Ottawa) and the Centers for Disease Control and Prevention (Atlanta), 53, (2007)

Goto, F., Yoshihara, T., Saiki, H.: Iron accumulation and enhanced growth in transgenic lettuce plants expressing the iron-binding protein ferritin. Theor. Appl. Genet. **100**(5), 658–664 (2000)
Graham, R. D.: Micronutrient deficiencies in crops and their global significance. In Micronutrient deficiencies in global crop production (pp. 41–61). Springer, Dordrecht (2008)
Graham, R.D., Ascher, J.S., Hynes, S.C.: Selecting zinc-efficient cereal genotypes for soils of low zinc status. Plant Soil. **146**(1–2), 241–250 (1992)
Graham, R.D., Welch, R.M., Saunders, D.A., Ortiz-Monasterio, I., Bouis, H.E., Bonierbale, M., De Haan, S., Burgos, G., Thiele, G., Liria, R., Meisner, C.A.: Nutritious subsistence food systems. Adv. Agron. **92**, 1–74 (2007)
Gregorio, G.B., Senadhira, D., Htut, H., Graham, R.D.: Breeding for trace mineral density in rice. Food Nutr. Bull. **21**(4), 382–386 (2000)
Gresshoff, P.M., Newton, S., Mohapatra, S.S., Scott, K.F., Howitt, S., Price, G.D., ... Rolfe, B.G.: Symbiotic nitrogen fixation involving Rhizobium and the non-legume Parasponia. In Advances in Nitrogen Fixation Research (pp. 483–489). Springer, Dordrecht (1984)
Guerinot, M.L., Yi, Y..: Iron: nutritious, noxious, and not readily available. Plant Physiol. **104**(3), 815 (1994)
Gutierrez-Manero, F.J., Ramos-Solano, B., Probanza, A.N., Mehouachi, J.R., Tadeo, F., Talon, M.: The plant-growth-promoting rhizobacteria Bacillus pumilus and Bacillus licheniformis produce high amounts of physiologically active gibberellins. Physiol. Plant. **111**(2), 206–211 (2001)
Haas, D., Défago, G.: Biological control of soil-borne pathogens by fluorescent pseudomonads. Nat. Rev. Microbiol. **3**(4), 307–319 (2005)
Hacisalihoglu, G., Kochian, L.V.: How do some plants tolerate low levels of soil zinc? Mechanisms of zinc efficiency in crop plants. New Phytol. **159**(2), 341–350 (2003)
Hambidge, M., Krebs, N.F.: Interrelationships of key variables of human zinc homeostasis: relevance to dietary zinc requirements. Annu. Rev. Nutr. **21**(1), 429–452 (2001)
Hamid, A., Ahmad, L.: Soil phosphorus fixation chemistry and role of phosphate solubilizing bacteria in enhancing its efficiency for sustainable cropping-A review. J. Pure Appl. Microbiol. **66**, 1905–1911 (2012)
Han, S., Chen, L.S., Jiang, H.X., Smith, B.R., Yang, L.T., Xie, C.Y.: Boron deficiency decreases growth and photosynthesis, and increases starch and hexoses in leaves of citrus seedlings. J. Plant Physiol. **165**(13), 1331–1341 (2008)
Hänsch, R., Mendel, R.R.: Physiological functions of mineral micronutrients (cu, Zn, Mn, Fe, Ni, Mo, B, cl). Curr. Opin. Plant Biol. **12**(3), 259–266 (2009)
HarvestPlus update: 2010. Available from: http://www.harvestplus.org/sites/default/files/HarvestPlus_update_2010_09.02.11%28web%29.pdf. Accessed 2014 January 20
Hashem, M.A.: Problems and prospects of cyanobacterial biofertilizer for rice cultivation. Funct. Plant Biol. **28**(9), 881–888 (2001)
Haslett, B.S., Reid, R.J., Rengel, Z.: Zinc mobility in wheat: uptake and distribution of zinc applied to leaves or roots. Ann. Bot. **87**(3), 379–386 (2001)
Hatch, M.D., Slack, C.R.: Photosynthetic CO2-fixation pathways. Annu. Rev. Plant Physiol. **21**(1), 141–162 (1970)
Haynes, K.G., Yencho, G.C., Clough, M.E., Henninger, M.R., Sterrett, S.B.: Genetic variation for potato tuber micronutrient content and implications for biofortification of potatoes to reduce micronutrient malnutrition. Am. J. Potato Res. **89**(3), 192–198 (2012)
Hedden, P., Phillips, A.L.: Gibberellin metabolism: new insights revealed by the genes. Trends Plant Sci. **5**(12), 523–530 (2000)
Henriques, A.R., Chalfun-Junior, A., Aarts, M.: Strategies to increase zinc deficiency tolerance and homeostasis in plants. Braz. J. Plant Physiol. **24**(1), 3–8 (2012)
Hirschi, K.D.: Nutrient biofortification of food crops. Annual review of nutrition, 29, 401–421 (2009)
Hoagland, D.R.: Lectures on the inorganic nutrition of plants. In: Lectures on the Inorganic Nutrition of Plants (1944)
Hodges, S.C.: Soil Fertility Basics, Soil Science Extension. North Carolina State University, Raleigh. (2010)

Holdaway-Clarke, T.L., Hepler, P.K.: Control of pollen tube growth: role of ion gradients and fluxes. New Phytol. **159**(3), 539–563 (2003)

Hotz, C., & Brown, K. H.; Assessment of the risk of zinc deficiency in populations and options for its control. **25**(1), S91–S204 (2004)

Hu, H., Brown, P.H., Labavitch, J.M.: Species variability in boron requirement is correlated with cell wall pectin. J. Exp. Bot. **47**(2), 227–232 (1996)

Hu, H., Penn, S.G., Lebrilla, C.B., Brown, P.H.: Isolation and characterization of soluble boron complexes in higher plants (the mechanism of phloem mobility of boron). Plant Physiol. **113**(2), 649–655 (1997)

Hu, B., Yang, B., Pang, X., Bao, W., Tian, G.: Responses of soil phosphorus fractions to gap size in a reforested spruce forest. Geoderma. **279**, 61–69 (2016)

Huang, H., Zhao, Y., Xu, Z., Ding, Y., Zhang, W., Wu, L.: Biosorption characteristics of a highly Mn (II)-resistant Ralstonia pickettii strain isolated from Mn ore. PloS One. **13**(8), e0203285 (2018)

Huang, H., Zhao, Y., Xu, Z., Zhang, W., Jiang, K.: Physiological responses of Broussonetia papyrifera to manganese stress, a candidate plant for phytoremediation. Ecotoxicol. Environ. Saf. **181**, 18–25 (2019)

Huss-Danell, K.: Tansley Review No. 93. Actinorhizal symbioses and their N2 fixation. New Phytol. **136**(3), 375–405 (1997)

Hussain, A., Arshad, M., Zahir, Z.A., Asghar, M.: Prospects of zinc solubilizing bacteria for enhancing growth of maize. Pakistan J. Agric. Sci. **52**(4), (2015)

Ibrahim, E.A., Ramadan, W.A.: Effect of zinc foliar spray alone and combined with humic acid or/ and chitosan on growth, nutrient elements content and yield of dry bean (Phaseolus vulgaris L.) plants sown at different dates. Sci. Hortic. **184**, 101–105 (2015)

ICAR-IIWBR. Annual report 2018–2019, ICAR. Indian Institute of Wheat and Barley Research, Kernal, India (2019)

ICMR.: National Nutrition Monitoring Bureau (NNMB), Prevalence of micronutrient deficiencies; Technical report No. 22, National Institute of Nutrition, (2003)

ICRISAT: Innovations for a changing world. Annual Report 2008. In: Patancheru, pp. 16–17. International Crops Research Institute for the Semi-Arid Tropics, Andhr Pradesh, India (2009)

IFPRI. International Food Policy Research Institute. Rising CO2, climate change projected to reduce availability of nutrients worldwide. Science Daily Report (2019)

India Today. 2018: 'Around 50 million people in India to become zinc deficient by 2050, thanks to carbon dioxide increase" (2020)

Indiamart (2018). https://indiamart.com/proddetail/reap-p-17855625762.html

Indian Institute of Rice Research (ICAR): Progress Report-2017, Varietal Improvement All India Coordinated Rice Improvement Programme. Vol. 1. Hyderabad: ICAR- Indian Institute of Rice Research (formerly Directorate of Rice Research) (2018a). Available online at: http://www.icariirr.org/AICRIP/Varietal%20Improvement%20Progress%20Report%202017.pdf (accessed July 2019)

Indian Institute of Rice Research (ICAR): Progress Report-2017, Varietal Improvement All India Coordinated Rice Improvement Programme. Vol. 1. Hyderabad: ICAR- Indian Institute of Rice Research (formerly Directorate of Rice Research) (2018b)

International Panaacea Limited.: http://iplbiologicals.com (2018)

Iqbal, S., Ali, I., Rust, P., Kundi, M., Ekmekcioglu, C.: Selenium, zinc, and manganese status in pregnant women and its relation to maternal and child complications. Nutrients. **12**(3), 725 (2020)

IRRI REPORT: Indian research rice institute. Report biofortification (2018)

Irisarri, P., Gonnet, S., Monza, J.: Cyanobacteria in Uruguayan rice fields: diversity, nitrogen fixing ability and tolerance to herbicides and combined nitrogen. J. Biotechnol. **91**(2–3), 95–103 (2001)

Irshad, M.U.H.A.M.M.A.D., Gill, M.A., Aziz, T.A.R.I.Q., Ahmed, I.: Growth response of cotton cultivars to zinc deficiency stress in chelator-buffered nutrient solution. Pak. J. Bot. **36**(2), 373–380 (2004)

IZiNCG. Brown, K.H., Rivera, J.A., Bhutta, Z., Gibson, R.S., King, J.C., Lönnerdal, B., Ruel, M.T., Sandtröm, B., Wasantwisut, E., Hotz, C., International Zinc Nutrition Consultative

Group (IZiNCG: International Zinc Nutrition Consultative Group (IZiNCG) technical document# 1. Assessment of the risk of zinc deficiency in populations and options for its control. Food Nutr. Bull. **25**(1 Suppl 2), S99–S203 (2004)

James, E.K.: Nitrogen fixation in endophytic and associative symbiosis. Field Crops Res. **65**(2–3), 197–209 (2000)

James, E.K., Olivares, F.L., Baldani, J.I., Dobereiner, J.: Herbaspirillum, an endophytic diazotroph colonizing vascular tissue 3 Sorghum bicolor L. Moench. J. Exp. Bot. **48**(3), 785–798 (1997)

Jason C.W., Gardea-Torresdey, J.: Achieving food security through the very small. Nat. Nanotechnol. **13**(8), 627–629 (2018)

Jayaraj, R., Megha, P., Sreedev, P.: Organochlorine pesticides, their toxic effects on living organisms and their fate in the environment. Interdiscip. Toxicol. **9**(3–4), 90 (2016)

Jensen, K.E.: Boron and zinc deficiency in Nepalese soils: Small-scale variations and the influence of farming system dependent factors. A case study from Sedi Bagar, Kaski District, Nepal (Master's thesis, The University of Bergen) (2007)

Jeong, J., Guerinot, M.L.: Homing in on iron homeostasis in plants. Trends Plant Sci. **14**(5), 280–285 (2009)

Jin, C. W., Du, S.T., Chen, W.W., Li, G.X., Zhang, Y.S., Zheng, S.J.: Elevated carbon dioxide improves plant iron nutrition through enhancing the iron-deficiency-induced responses under iron-limited conditions in tomato. Plant Physiol. **150**(1), 272–280 (2009)

Jin, C. W., Liu, Y., Mao, Q.Q., Wang, Q., Du, S.T.: Mild Fe-deficiency improves biomass production and quality of hydroponic-cultivated spinach plants (Spinacia oleracea L.). Food Chem. **138**(4), 2188–2194 (2013)

Jiang, W., Struik, P.C., Van Keulen, H., Zhao, M., Jin, L.N., Stomph, T.J.: Does increased zinc uptake enhance grain zinc mass concentration in rice? Ann. Appl. Biol. **153**(1), 135–147 (2008)

Johnson, J.L., Waud, W.R., Rajagopalan, K.V., Duran, M., Beemer, F.A., Wadman, S.K.: Inborn errors of molybdenum metabolism: combined deficiencies of sulfite oxidase and xanthine dehydrogenase in a patient lacking the molybdenum cofactor. Proc. Natl. Acad. Sci. **77**(6), 3715–3719 (1980)

Joint, F.A.O/ WHO.: Vitamin and Mineral Requirements in Human Nutrition. Diamond Pocket Books (P) Ltd. WHO, Geneva (2004)

Joshi, D., Negi, G., Vaid, S., Sharma, A.: Enhancement of wheat growth and Zn content in grains by zinc solubilizing bacteria. Int. J. Agric. Environ. Biotechnol. **6**(3), 363–370 (2013)

Jouvin, D., Weiss, D.J., Mason, T.F.M., Bravin, M.N., Louvat, P., Zhao, F., Ferec, F., Hinsinger, P., Benedetti, M.F.: Stable isotopes of Cu and Zn in higher plants: evidence for Cu reduction at the root surface and two conceptual models for isotopic fractionation processes. Environ. Sci. Technol. **46**(5), 2652–2660 (2012)

Kaiser, B.N., Gridley, K.L., Ngaire Brady, J., Phillips, T., Tyerman, S.D.: The role of molybdenum in agricultural plant production. Ann. Bot. **96**(5), 745–754 (2005)

Kakei, Y., Yamaguchi, I., Kobayashi, T., Takahashi, M., Nakanishi, H., Yamakawa, T., Nishizawa, N.K.: A highly sensitive, quick and simple quantification method for nicotianamine and 2′-deoxymugineic acid from minimum samples using LC/ESI-TOF-MS achieves functional analysis of these components in plants. Plant Cell Physiol. **50**(11), 1988–1993 (2009)

Kamran, S., Shahid, I., Baig, D.N., Rizwan, M., Malik, K.A., Mehnaz, S.: Contribution of zinc solubilizing bacteria in growth promotion and zinc content of wheat. Front. Microbiol. **8**, 2593 (2017)

Kang, B.T., Okoro, E.G.: Response of flooded rice grown on a vertisol from northern Nigeria to zinc sources and methods of application. Plant Soil. **44**(1), 15–25 (1976)

Kapil, U., Jain, K.: Magnitude of zinc deficiency amongst under five children in India. Indian J. Pediatr. **78**(9), 1069–1072 (2011)

Kassebaum, N.J., Jasrasaria, R., Naghavi, M., Wulf, S.K., Johns, N., Lozano, R., Regan, M., Weatherall, D., Chou, D.P., Eisele, T.P., Flaxman, S.R.: A systematic analysis of global anemia burden from 1990 to 2010. Blood. **123**(5), 615–624 (2014)

Katyal, J.C.: Micronutrients in Indian agriculture. Indian J. Fertilis. **14**(4), 12–26 (2018)

Katyal, J.C., Deb, D.L.: Review of Soil Research in India, Part I, pp. 146–159. Indian Society of Soil Science, New Delhi (1982)

Katyal, J.C., Vlek, P.L.G.: Micronutrient problems in tropical Asia. In: Micronutrients in Tropical Food Crop Production, pp. 69–94. Springer, Dordrecht (1985)

Kawakami, K., Umena, Y., Kamiya, N., Shen, J.R.: Location of chloride and its possible functions in oxygen-evolving photosystem II revealed by X-ray crystallography. Proc. Natl. Acad. Sci. **106**(21), 8567–8572 (2009)

Khan, A.G., Kuek, C., Chaudhry, T.M., Khoo, C. S., Hayes, W.J.: Role of plants, mycorrhizae and phytochelators in heavy metal contaminated land remediation. Chemosphere. **41**(1–2), 197–207 (2000)

Khande, R., Sharma, S.K., Ramesh, A., Sharma, M.P.: Zinc solubilizing Bacillus strains that modulate growth, yield and zinc biofortification of soybean and wheat. Rhizosphere. **4**, 126–138 (2017)

Khoshgoftarmanesh, A.H., SanaeiOstovar, A., Sadrarhami, A., Chaney, R.: Effect of tire rubber ash and zinc sulfate on yield and grain zinc and cadmium concentrations of different zinc-deficiency tolerance wheat cultivars under field conditions. Eur. J. Agron. **49**, 42–49 (2013)

Kim, T.H., Böhmer, M., Hu, H., Nishimura, N., Schroeder, J.I.: Guard cell signal transduction network: advances in understanding abscisic acid, CO2, and Ca2+ signaling. Annu. Rev. Plant Biol. **61**, 561–591 (2010)

King, J.C.: Determinants of maternal zinc status during pregnancy. Am. J. Clin. Nutr. **71**(5), 1334S–1343S (2000)

Kirkby, E.A., Le Bot, J., Adamowicz, S., Römheld, V.: Nitrogen in physiology. An agronomic perspective and implications for the use of different nitrogen forms. In: IFS Conference, vol. 653, p. 46. The International Fertiliser Society (2009, December)

Knez, M., Graham, R.D.: The impact of micronutrient deficiencies in the impact of micronutrient deficiencies in agricultural soils and crops on the nutritional health of humans. In: Essentials of Medical Geology, pp. 517–533. Springer, Dordrecht (2013)

Kobayashi, M., Matoh, T., Azuma, J.I.: Two chains of rhamnogalacturonan II are cross-linked by borate-diol ester bonds in higher plant cell walls. Plant Physiol. **110**(3), 1017–1020 (1996)

Kobayashi, T., Ogo, Y., Itai, R. N., Nakanishi, H., Takahashi, M., Mori, S., & Nishizawa, N. K.: The transcription factor IDEF1 regulates the response to and tolerance of iron deficiency in plants. Proc. Natl. Acad. Sci. **104**(48), 19150–19155 (2007)

Kobayashi, T., Itai, R. N., Ogo, Y., Kakei, Y., Nakanishi, H., Takahashi, M., Nishizawa, N. K.: The rice transcription factor IDEF1 is essential for the early response to iron deficiency, and induces vegetative expression of late embryogenesis abundant genes. The Plant J. **60**(6), 948–961 (2009)

Koerselman, W., Meuleman, A.F.: The vegetation N: P ratio: a new tool to detect the nature of nutrient limitation. J. App. Ecol. **33**(6), 1441–1450 (1996)

Kole, R.K., Bagchi, M.M.: Pesticide residues in the aquatic environment and their possible ecological hazards. J. Inland Fish Soc. India. **27**(2), 79–89 (1995)

Kubo, K.I., Sakamoto, A., Kobayashi, A., Rybka, Z., Kanno, Y., Nakagawa, H., Nishino, T., Takatsuji, H.: Cys2/His2 zinc-finger protein family of petunia: evolution and general mechanism of target-sequence recognition. Nucleic Acids Res. **26**(2), 608–615 (1998)

Kumar, P.: Managing micronutrient deficiency in ornamental crops. Indian Horticul.(India). **46**(4), 30–31 (2002)

Kumar, R., Prakash, O.: The impact of chemical fertilizers on our environment and ecosystem. Res. Trends Environ. Sci. 69–86 (2019)

Kundu, D.K., Ladha, J.K.: Efficient management of soil and biologically fixed N2 in intensively-cultivated rice fields. Soil Biol. Biochem. **27**(4–5), 431–439 (1995)

Küpper, H., Götz, B., Mijovilovich, A., Küpper, F.C., Meyer-Klaucke, W.: Complexation and toxicity of copper in higher plants. I. Characterization of copper accumulation, speciation, and toxicity in Crassula helmsii as a new copper accumulator. Plant Physiol. **151**(2), 702–714 (2009)

Lal, M., Singh, R.B., Pal, D., Yadav, H.M.S., Pal, A.K., Singh, A.P.: Effect of phosphorus and molybdenum on nutrients content of broad bean (Vicia faba L). J. Pharmacogn. Phytochem. **7**(6), 1560–1562 (2018)

Lee, S., An, G.: Over-expression of OsIRT1 leads to increased iron and zinc accumulations in rice. Plant Cell Environ. **32**(4), 408–416 (2009)

Lee, S., Persson, D.P., Hansen, T.H., Husted, S., Schjoerring, J.K., Kim, Y.S., Jeon, U.S., Kim, Y.K., Kakei, Y., Masuda, H., Nishizawa, N.K.: Bio-available zinc in rice seeds is increased by activation tagging of nicotianamine synthase. Plant Biotechnol. J. **9**(8), 865–873 (2011)

Leeman, M., Den Ouden, F.M., Van Pelt, J.A., Dirkx, F.P.M., Steijl, H., Bakker, P.A.H.M., Schippers, B.: Iron availability affects induction of systemic resistance to Fusarium wilt of radish by Pseudomonas fluorescens. Phytopathology, **86**(2), 149–155 (1996)

Lefèvre, I., VOGEL-MIKUŠ, K. A. T. A. R. I. N. A., Jeromel, L., Vavpetič, P., Planchon, S., Arčon, I., ... Lutts, S.: Differential cadmium and zinc distribution in relation to their physiological impact in the leaves of the accumulating Z ygophyllum fabago L. Plant Cell Environ. **37**(6), 1299–1320 (2014)

Leigh, R.A., Wyn Jones, R.G.: A hypothesis relating critical potassium concentrations for growth to the distribution and functions of this ion in the plant cell. New Phytol. **97**(1), 1–13 (1984)

Leustek, T., Martin, M.N., Bick, J.A., Davies, J.P.: Pathways and regulation of sulfur metabolism revealed through molecular and genetic studies. Annu. Rev. Plant Biol. **51**(1), 141–165 (2000)

Lewin, J.: Boron as a growth requirement for diatoms 1. J. Phycol. **2**(4), 160–163 (1966)

Lewińska-Preis, L., Jabłońska, M., Fabiańska, M.J., Kita, A.: Bioelements and mineral matter in human livers from the highly industrialized region of the upper Silesia Coal Basin (Poland). Environ. Geochem. Health. **33**(6), 595–611 (2011)

Li, L., Zhang, K., Gill, R.A., Islam, F., Farooq, M.A., Wang, J., Zhou, W.: Ecotoxicological and interactive effects of copper and chromium on physiochemical, ultrastructural, and molecular profiling in Brassica napus L. BioMed Res Int. **2018**, 9248123 (2018)

Li, Q., Chen, L., Yang, A.: The molecular mechanisms underlying Iron deficiency responses in Rice. Int. J. Mol. Sci. **21**(1), 43 (2020)

Linder, M.C.: Ceruloplasmin and other copper binding components of blood plasma and their functions: an update. Metallomics. **8**(9), 887–905 (2016)

Liu, Z.: Microelements in Soils of China, p. 188. Jiangsu Science and Technology Publishing House, Nanjing, China (1996)

Liu, H., Hu, C., Hu, X., Nie, Z., Sun, X., Tan, Q., Hu, H.: Interaction of molybdenum and phosphorus supply on uptake and translocation of phosphorus and molybdenum by Brassica napus. J. Plant Nutr. **33**(12), 1751–1760 (2010)

Liu, K., Yue, R., Yuan, C., Liu, J., Zhang, L., Sun, T., ... Shen, C.: Auxin signaling is involved in iron deficiency-induced photosynthetic inhibition and shoot growth defect in rice (Oryza sativa L.). J. Plant Biol. **58**(6), 391–401 (2015)

Llanos, R.M., Mercer, J.F.: The molecular basis of copper homeostasis copper-related disorders. DNA Cell Biol. **21**(4), 259–270 (2002)

Lonnerdal, B.: Genetically modified plants for improved trace element nutrition. J. Nutr. **133**(5), 1490S–1493S (2003)

Lopez-Valdivia, L.M., Fernandez, M.D., Obrador, A., Alvarez, J.M.: Zinc transformations in acidic soil and zinc efficiency on maize by adding six organic zinc complexes. J. Agric. Food Chem. **50**(6), 1455–1460 (2002)

Lu, D., Huang, Q., Deng, C., Zheng, Y.: Phytoremediation of copper pollution by eight aquatic plants. Polish J. Environ. Stud. **27**(1), 175–181 (2018)

Lu, L., Xie, R., Liu, T., Wang, H., Hou, D., Du, Y., He, Z., Yang, X., Sun, H., Tian, S.: Spatial imaging and speciation of Cu in rice (Oryza sativa L.) roots using synchrotron-based X-ray microfluorescence and X-ray absorption spectroscopy. Chemosphere. **175**, 356–364 (2017)

Lucena, C., Romera, F. J., García, M. J., Alcántara, E., Pérez-Vicente, R.: Ethylene participates in the regulation of Fe deficiency responses in Strategy I plants and in rice. Front. Plant Sci. 6, 1056 (2015)

Maathuis, F.J.: Physiological functions of mineral macronutrients. Curr. Opin. Plant Biol. **12**(3), 250–258 (2009)

Mabesa, R.Á., Impa, S.Á., Grewal, D., Johnson-Beebout, S.E.: Contrasting grain-Zn response of biofortification rice (Oryza sativa L.) breeding lines to foliar Zn application. Field Crop Res. **149**, 223–233 (2013)

Mahajan, A., Gupta, R.D., Sharma, R.: Bio-fertilizers-a way to sustainable agriculture. Agrobios Newslett. **6**(9), 36–37 (2008)

Mahanty, T., Bhattacharjee, S., Goswami, M., Bhattacharyya, P., Das, B., Ghosh, A., Tribedi, P.: Biofertilizers: a potential approach for sustainable agriculture development. Environ. Sci. Pollut. Res. **24**(4), 3315–3335 (2017)

Mahdi, S.S., Hassan, G.I., Samoon, S.A., Rather, H.A., Dar, S.A., Zehra, B.: Bio-fertilizers in organic agriculture. J. Phytol. **2**(10), 42–54 (2010)

Mahdi, S.S., Talat, M.A., Hussain Dar, M., Hamid, A., Ahmad, L.: Soil phosphorus fixation chemistry and role of phosphate solubilizing bacteria in enhancing its efficiency for sustainable cropping: a review. J. Pure Appl. Microbiol. **6**(4), 1–7 (2012)

Maheshwari, D.K., Dheeman, S., Agarwal, M.: Phytohormone-producing PGPR for sustainable agriculture. In Bacterial metabolites in sustainable agroecosystem (pp. 159–182). Springer, Cham. (2015)

Majewski, M., Ognik, K., Juśkiewicz, J.: Copper nanoparticles modify the blood plasma antioxidant status and modulate the vascular mechanisms with nitric oxide and prostanoids involved in Wistar rats. Pharmacol. Reports. **71**(3), 509–516 (2019)

Maleki, F.S., Chaichi, M.R., Mazaheri, D., Tavakkol, A.R., Savaghebi, G.: Barley grain mineral analysis as affected by different fertilizing systems and by drought stress. J. Agric. Sci. Technol. **13**(3), 315–326 (2011)

Malik, K.A., Bilal, R., Mehnaz, S., Rasul, G., Mirza, M.S., Ali, S.: Association of nitrogen-fixing, plant-growth-promoting rhizobacteria (PGPR) with kallar grass and rice. In Opportunities for biological nitrogen fixation in rice and other non-legumes (pp. 37–44). Springer, Dordrecht (1997)

Marcar, N.E., Graham, R.D.: Tolerance of wheat, barley, triticale and rye to manganese deficiency during seedling growth. Aust. J. Agric. Res. **38**(3), 501–511 (1987)

Maret, W.: Molecular aspects of human cellular zinc homeostasis: redox control of zinc potentials and zinc signals. Biometals. **22**(1), 149–157 (2009)

Marschner, H.: Mineral Nutrition of Higher Plants, 2nd edn. Academic, Great Britain (1995a)

Marschner, H.: Mineral Nutrition of Higher Plants. Institute of Plant Nutrition, University of Hohenheim. London, UK. USA. VOL 2 (1995b)

Marschner, H.: Marschner's Mineral Nutrition of Higher Plants. Academic pres. Im. Els. London, UK, US. vol. 89, (2012)

Masuda, H., Suzuki, M., Morikawa, K.C., Kobayashi, T., Nakanishi, H., Takahashi, M., Saigusa, M., Mori, S., Nishizawa, N.K.: Increase in iron and zinc concentrations in rice grains via the introduction of barley genes involved in phytosiderophore synthesis. Rice. **1**(1), 100–108 (2008)

Matoh, T., Kawaguchi, S., Kobayashi, M.: Ubiquity of a borate-rhamnogalacturonan II complex in the cell walls of higher plants. Plant Cell Physiol. **37**(5), 636–640 (1996)

Mayer, J.E., Pfeiffer, W.H., Beyer, P.: Biofortified crops to alleviate micronutrient malnutrition. Curr. Opin. Plant Biol. **11**(2), 166–170 (2008)

Mazé, P.: Détermination des éléments minéraux rares nécessaires au développement du maïs. C. R. Hebd. Seances Acad. Sci. **160**, 211–214 (1915)

Mazziotta, M., De Muro, P., Grigoletto, R., Scafetti, F., Lochetti, G.: Measuring food security in three districts of Bangladesh-Satkhira, Netrakona and Rajshahi: a Multidimensional Perspective (2016)

Mengual, C., Schoebitz, M., Azcón, R., Roldán, A.: Microbial inoculants and organic amendment improves plant establishment and soil rehabilitation under semiarid conditions. J. Environ. Manage. **134**, 1–7 (2014)

Mermod, M., Takusagawa, M., Kurata, T., Kamiya, T., Fujiwara, T., Shikanai, T.: SQUAMOSA promoter-binding protein-like 7 mediates copper deficiency response in the presence of high nitrogen in Arabidopsis thaliana. Plant Cell Rep. **38**(7), 835–846 (2019)

Miller, D.D., Welch, R.M.: Food system strategies for preventing micronutrient malnutrition. Food Policy. **42**, 115–128 (2013)

Minasny, B., Malone, B. P., McBratney, A. B., Angers, D. A., Arrouays, D., Chambers, A., ... Winowiecki, L.: Soil carbon 4 per mille. Geoderma. 292, 59–86 (2017)

Mitra, G.N.: Regulation of Nutrient Uptake by Plants, pp. 978–981. Springer, New Delhi (2015)

Monreal, C.M., DeRosa, M., Mallubhotla, S.C., Bindraban, P.S., Dimkpa, C.: The application of nanotechnology for micronutrients in soil-plant systems. VFRC Rep. **3**, 44 (2015)

Morris, J., Hawthorne, K.M., Hotze, T., Abrams, S.A., Hirschi, K.D.: Nutritional impact of elevated calcium transport activity in carrots. Proc. Natl. Acad. Sci. **105**(5), 1431–1435 (2008)

Muhammad, I., Kolla, M., Volker, R., Günter, N.: Impact of nutrient seed priming on germination, seedling development, nutritional status and grain yield of maize. J. Plant Nutr. **38**(12), 1803–1821 (2015)

Mukhopadhyay, M.J., Sharma, A.: Manganese in cell metabolism of higher plants. Bot. Rev. **57**(2), 117–149 (1991)

Myers, S.S., Zanobetti, A., Kloog, I., Huybers, P., Leakey, A.D., Bloom, A.J., Carlisle, E., Dietterich, L.H., Fitzgerald, G., Hasegawa, T., Holbrook, N.M.: Increasing CO 2 threatens human nutrition. Nature. **510**(7503), 139–142 (2014)

Nahed, G.A., Balbaa, L.K.: Influence of tyrosine and zinc on growth, flowering and chemical constituents of Salvia farinacea plants. J. Appl. Sci. Res. **3**(11), 1479–1489 (2007)

Napoli, M., Cecchi, S., Grassi, C., Baldi, A., Zanchi, C.A., Orlandini, S.: Phytoextraction of copper from a contaminated soil using arable and vegetable crops. Chemosphere. **219**, 122–129 (2019)

National Family Health Survey (NFHS-1) 1992–93 [Internet]: Mumbai: International Institute for Population Sciences (IIPS) and Macro International; 2009. http://rchiips.org/nfhs/nfhs1.shtml [cited 2014 Dec 18]

National Family Health Survey (NFHS-2) 1998–99 [Internet]: Mumbai: International Institute for Population Sciences (IIPS) and Macro International; 2009. http://rchiips.org/nfhs/nfhs2.shtml [cited 2014 Dec 18]

National Family Health Survey (NFHS-3) 2005–06 [Internet]: Mumbai: International Institute for Population Sciences (IIPS) and Macro International; 2009. http://rchiips.org/nfhs/nfhs3.shtml [cited 2014 Dec 18]

National Family Health Survey (NFHS-4) 2012–14 [Internet]: Mumbai: International Institute for Population Sciences (IIPS) and Macro International; 2009. http://rchiips.org/nfhs/nfhs4.shtml [cited 2014 Dec 18]

National Family Health Survey, India [Internet]: Mumbai: International Institute for Population Sciences (IIPS) and Macro International; 2009a. http://rchiips.org/nfhs/ [cited 2014 Dec 18]

National Family Health Survey, India [Internet]. Mumbai: International Institute for Population Sciences (IIPS) and Macro International; 2009b. http://rchiips.org/nfhs/ [cited 2014 Dec 18]

National Fertilizers Limited.: http://www.nationalfertilizers.com/index.php?option¼com_content &view¼article&id¼140&Itemid¼156&lang¼en (2018)

Naz, I., Ahmad, H., Khokhar, S.N., Khan, K., Shah, A.H.: Impact of zinc solubilizing bacteria on zinc contents of wheat. Am. Euras. J. Agric. Environ. Sci. **16**, 449–454 (2016)

Neue, H.U., Lantin, R.S.: Micronutrient toxicities and deficiencies in rice. In: Soil Mineral Stresses, pp. 175–200. Springer, Berlin, Heidelberg (1994)

Nie, Z., Li, S., Hu, C., Sun, X., Tan, Q., Liu, H.: Effects of molybdenum and phosphorus fertilizers on cold resistance in winter wheat. J. Plant Nutr. **38**(5), 808–820 (2015)

Ninh, N.X., Thissen, J.P., Maiter, D., Adam, E., Mulumba, N., Ketelslegers, J.M.: Reduced liver insulin-like growth factor-I gene expression in young zinc-deprived rats is associated with a decrease in liver growth hormone (GH) receptors and serum GH-binding protein. J. Endocrinol. **144**(3), 449–456 (1995)

Noori, M., Adibian, M., Sobhkhizi, A., Eyidozehi, K.: Effect of phosphorus fertilizer and mycorrhiza on protein percent, dry weight, weight of 1000 grain in wheat. Int. J. Plant Animal Environ. Sci. **4**(2), 561–564 (2014)

Obiora, S.C., Chukwu, A., Davies, T.C.: Heavy metals and health risk assessment of arable soils and food crops around Pb–Zn mining localities in Enyigba, southeastern Nigeria. J. Afr. Earth Sci. **116**, 182–189 (2016)

Obreht, Z., Kerby, N.W., Gantar, M., Rowell, P.: Effects of root-associated N 2-fixing cyanobacteria on the growth and nitrogen content of wheat (Triticum vulgare L.) seedlings. Biol. Fert. Soils. **15**(1), 68–72 (1993)

Oguchi, K., Tanaka, N., Komatsu, S., Akao, S.: Methylmalonate-semialdehyde dehydrogenase is induced in auxin-stimulated and zinc-stimulated root formation in rice. Plant Cell Rep. **22**(11), 848–858 (2004)

Okon, Y.: Azospirillum as a potential inoculant for agriculture. Trends in Biotechnol. **3**(9), 223–228 (1985)

Okon, Y., Labandera-Gonzalez, C.A.: Agronomic applications of Azospirillum: an evaluation of 20 years worldwide field inoculation. Soil Biol. Biochem. **26**(12), 1591–1601 (1994)

Ortiz-Monasterio, J.I., Palacios-Rojas, N., Meng, E., Pixley, K., Trethowan, R., Pena, R.J.: Enhancing the mineral and vitamin content of wheat and maize through plant breeding. J. Cereal Sci. **46**(3), 293–307 (2007)

Oteino, N., Lally, R.D., Kiwanuka, S., Lloyd, A., Ryan, D., Germaine, K.J., Dowling, D.N.: Plant growth promotion induced by phosphate solubilizing endophytic Pseudomonas isolates. Front. Microbiol. **6**, 745 (2015)

Palmgren, M.G., Clemens, S., Williams, L.E., Kramer, U., Borg, S., Schjorring, J.K., Sanders, D.: Zinc biofortification of cereals: problems and solutions. Trends Plant Sci. **13**(9), 464–473 (2008)

Pan, C.: Manganese metabolism in humans. Front. Biosci. **23**(9), 1655–1679 (2018)

Pandey, N.: Role of micronutrients in reproductive physiology of plants. Plant Stress. **4**(2), 1–13 (2010)

Pandey, N.: Role of plant nutrients in plant growth and physiology. In: Plant Nutrients and Abiotic Stress Tolerance, pp. 51–93. Springer, Singapore (2018)

Park, S., Kim, C.K., Pike, L.M., Smith, R.H., Hirschi, K.D.: Increased calcium in carrots by expression of an Arabidopsis H+/Ca 2+ transporter. Mol. Breed. **14**(3), 275–282 (2004)

Pasricha, N.S., Aulakh, M.S., Vempati, R.K.: Evaluation of available phosphorus soil test methods for peanut in neutral and alkaline soils. Commun. Soil Sci. Plant Anal. **33**(19–20), 3593–3601 (2002)

Patidar, M., Mali, A.L.: Effect of farmyard manure, fertility levels and bio-fertilizers on growth, yield and quality of sorghum (Sorghum bicolor). Indian J. Agronomy. **49**(2), 117–120 (2004)

Patrick, J.W., Offler, C.E.: Compartmentation of transport and transfer events in developing seeds. J. Exp. Bot. **52**(356), 551–564 (2001)

Peters, G.A., Meeks, J.C.: The Azolla-Anabaena symbiosis: basic biology. Annu. Rev. Plant Biol. **40**(1), 193–210 (1989)

Phattarakul, N., Cakmak, I., Boonchuay, P., Wongmo, J., Rerkasem, B.: Role of zinc fertilizers in increasing grain zinc concentration and improving grain yield of rice (2009)

Phattarakul, N.B., Rerkasem, L.J., Li, L.H., Wu, C.Q., Zou, H., Ram, V.S., Sohu, B.S., Kang, H., Surek, M., Kalayci, A., Yazici, F.S., Zhang, I., Cakmak.: Biofortification of rice grain with zinc through zinc fertilization in different countries. Plant Soil. **361**(1–2), 131–141 (2012)

Płociniczak, T., Sinkkonen, A., Romantschuk, M., Sułowicz, S., Piotrowska-Seget, Z.: Rhizospheric bacterial strain Brevibacterium casei MH8a colonizes plant tissues and enhances Cd, Zn, Cu phytoextraction by white mustard. Front. Plant Sci. 7, 101 (2016)

Poblaciones, M.J., Rengel, Z.: Soil and foliar zinc biofortification in field pea (Pisum sativum L.): grain accumulation and bioavailability in raw and cooked grains. Food Chem. **212**, 427–433 (2016)

Pohanka, M.: Copper and copper nanoparticles toxicity and their impact on basic functions in the body. Bratisl. Lek. Listy. **120**, 397–409 (2019)

Prapagdee, B., Chanprasert, M., & Mongkolsuk, S.: Bioaugmentation with cadmium-resistant plant growth-promoting rhizobacteria to assist cadmium phytoextraction by Helianthus annuus. Chemosphere. **92**(6), 659–666 (2013)

Prasad, A.S.: Discovery of human zinc deficiency: its impact on human health and disease. Adv. Nutr. **4**(2), 176–190 (2013)

Prasad, A.S., Meftah, S., Abdallah, J., Kaplan, J., Brewer, G.J., Bach, J.F., Dardenne, M.: Serum thymulin in human zinc deficiency. J. Clin. Invest. **82**(4), 1202–1210 (1988)

Prasanna, R., Bidyarani, N., Babu, S., Hossain, F., Shivay, Y.S., Nain, L.: Cyanobacterial inoculation elicits plant defense response and enhanced Zn mobilization in maize hybrids. Cogent Food Agric. **1**(1), 998507 (2015)

Puig, S., Andrés-Colás, N.U.R.I.A., García-Molina, A.N.T.O.N.I., Penarrubia, L.: Copper and iron homeostasis in Arabidopsis: responses to metal deficiencies, interactions and biotechnological applications. Plant Cell Environ. **30**(3), 271–290 (2007)

Rai, K.N., Govindaraj, M., Rao, A.S.: Genetic enhancement of grain iron and zinc content in pearl millet. Qual. Assur. Saf. Crops Foods. **4**(3), 119–125 (2012)

Rajesha, G., Ray, S.K.: Microbial Bio-fertilizers: A Functional Key Player in Sustainable Agriculture.Promotion of Improved Cultivation Practices in Agri & Allied Sector for Food and Nutritional Security", 37–41 (2020)

Ram, H., Rashid, A., Zhang, W., Duarte, A.Á., Phattarakul, N., Simunji, S., ... Cakmak, I.: Biofortification of wheat, rice and common bean by applying foliar zinc fertilizer along with pesticides in seven countries. Plant Soil. **403**(1), 389–401 (2016)

Ramesh, S.A., Choimes, S., Schachtman, D.P.: Over-expression of an Arabidopsis zinc transporter in Hordeum vulgare increases short-term zinc uptake after zinc deprivation and seed zinc content. Plant Mol. Biol. **54**(3), 373–385 (2004)

Ramesh, A., Sharma, S.K., Sharma, M.P., Yadav, N., Joshi, O.P.: Inoculation of zinc solubilizing Bacillus aryabhattai strains for improved growth, mobilization and biofortification of zinc in soybean and wheat cultivated in Vertisols of Central India. Appl. Soil Ecol. **73**, 87–96 (2014)

Ramzani, P.M.A., Khalid, M., Naveed, M., Ahmad, R., Shahid, M.: Iron biofortification of wheat grains through integrated use of organic and chemical fertilizers in pH affected calcareous soil. Plant Physiol. Biochem. **104**, 284–293 (2016)

Rashid, A., Ryan, J.: Micronutrient constraints to crop production in the Near East. Micronutrient deficiencies in global crop production, 149–180 (2008)

Raven, J.A.: Chloride: essential micronutrient and multifunctional beneficial ion. J. Exp. Bot. **68**(3), 359–367 (2017)

Ravet, K., Touraine, B., Boucherez, J., Briat, J.F., Gaymard, F., Cellier, F.: Ferritins control interaction between iron homeostasis and oxidative stress in Arabidopsis. The Plant J. **57**(3), 400–412 (2009)

Ravet, K., Reyt, G., Arnaud, N., Krouk, G., Djouani, E.B., Boucherez, J., ... Gaymard, F.: Iron and ROS control of the DownSTream mRNA decay pathway is essential for plant fitness. The EMBO J. **31**(1), 175–186 (2012)

Reddy, V.S., Reddy, A.S.: Proteomics of calcium-signaling components in plants. Phytochemistry. **65**(12), 1745–1776 (2004)

Reddy, B.V., Ramesh, S., Longvah, T.: Prospects of breeding for micronutrients and b-carotene-dense sorghums. Int. Sorghum Millets Newslett. **46**, 10–14 (2005)

Rengel, Z., Batten, G.D., Crowley, D.D.: Agronomic approaches for improving the micronutrient density in edible portions of field crops. Field Crop Res. **60**(1–2), 27–40 (1999)

Report FAI, 61St ANNUAL REPORT ON 2015–2016. The Fertiliser Association OF India, New Delhi India 2 (2016)

Reetha, S., Bhuvaneswari, G., Thamizhiniyan, P., Mycin, T.R.: Isolation of indole acetic acid (IAA) producing rhizobacteria of pseudomonas fluorescens and Bacillus subtilis and enhance growth of onion (Allim cepa L.). Int. J. Curr. Microbiol. Appl. Sci. **3**(2), 568–574 (2014)

Report IZA (International Zinc Association).: All for a good harvest: Addressing micronutrient deficiencies (2019)

Reuter, D.J., Alston, A.M., McFarlane, J.D.: Occurrence and correction of manganese deficiency in plants. In: Manganese in Soils and Plants, pp. 205–224. Springer, Dordrecht (1988)
Reyes, J.O., Bendezú, S.G., Joaquín, A.H.: Molybdenum and Cobalt application in bean (Phaseolus vulgaris L.) with two fertilization systems under no-tillage. J. Agric. Sci. Technol. B, **6**(2), 72–77 (2016)
Ritchie, H., Roser, M.: Micronutrient deficiency. Our World in data (2017)
Rognes, S.E.: Anion regulation of lupin asparagine synthetase: chloride activation of the glutamine-utilizing reactions. Phytochemistry. **19**(11), 2287–2293 (1980)
Rolli, E., Marasco, R., Vigani, G., Ettoumi, B., Mapelli, F., Deangelis, M.L., ... Daffonchio, D.: Improved plant resistance to drought is promoted by the root-associated microbiome as a water stress-dependent trait. Environ. Microbiol. **17**(2), 316–331 (2015)
Romera, F.J., Alcántara, E.: Ethylene involvement in the regulation of Fe-deficiency stress responses by strategy I plants. Funct. Plant Biol. **31**(4), 315–328 (2004)
Römheld, V., Marschner, H.: Genotypical differences among graminaceous species in release of phytosiderophores and uptake of iron phytosiderophores. In Genetic Aspects of Plant Mineral Nutrition (pp. 77–83). Springer, Dordrecht, (1990)
Roohani, N., Hurrell, R., Kelishadi, R., Schulin, R.: Zinc and its importance for human health: an integrative review. J. Res. Med. Sci. **18**(2), 144 (2013)
Rosen, C.J.: Primer on the Metabolic Bone Diseases and Disorders of Mineral Metabolism. Wiley Edited by M. J. Favus and the American Society for Bone and Mineral Research. Hagerstown, Maryland. Pp 528, (2009)
Roy, R. N., Finck, A., Blair, G. J., & Tandon, H. L. S.;. Nutrient management guidelines for some major field crops. Plant nutrition for food security. A guide for integrated nutrient management. FAO, Rome, 235–349 (2006)
Ryan, B.M., Kirby, J.K., Degryse, F., Harris, H., McLaughlin, M.J., Scheiderich, K.: Copper speciation and isotopic fractionation in plants: uptake and translocation mechanisms. New Phytol. **199**(2), 367–378 (2013)
Sabatucci, A., Vachette, P., Vasilyev, V.B., Beltramini, M., Sokolov, A., Pulina, M., Salvato, B., Angelucci, C.B., Maccarrone, M., Cozzani, I., Dainese, E.: Structural characterization of the ceruloplasmin: lactoferrin complex in solution. J. Mol. Biol. **371**(4), 1038–1046 (2007)
Sadras, V.O.: A quantitative top-down view of interactions between stresses: theory and analysis of nitrogen–water co-limitation in Mediterranean agro-ecosystems. Aust. J. Agric. Res. **56**(11), 1151–1157 (2005)
Saha, B., Saha, S., Poddar, P., Murmu, S., Singh, A.K.: Uptake of nutrients by wheat as influenced by long-term phosphorus fertilization. The Bioscan. **8**(4), 1331–1335 (2013)
Saini, H., Sourabh, V., Saini, P.: Differential responses of Fe, Zn, B, Cu and Mg on growth and quality attributes of fruit crops. J. Pharm. Phytochem. **8**(5), 1–5 (2019)
Samant, P.K.: Zinc: an essential micronutrient for crop production. Orissa Review, (October), pp. 85–87 (2009)
Sandstead, H.H.: Zinc deficiency: a public health problem?. Am. J. Dis. Child. **145**(8), 853–859 (1991)
Santos, V.B., Araújo, A.S., Leite, L.F., Nunes, L.A., Melo, W.J.: Soil microbial biomass and organic matter fractions during transition from conventional to organic farming systems. Geoderma. **170**, 227–231 (2012)
Saravanan, V.S., Madhaiyan, M., Thangaraju, M.: Solubilization of zinc compounds by the diazotrophic, plant growth promoting bacterium Gluconacetobacter diazotrophicus. Chemosphere. **66**(9), 1794–1798 (2007)
Saravanan, V.S., Kumar, M.R., Sa, T.M.: Microbial zinc solubilization and their role on plants. In: Bacteria in Agrobiology: Plant Nutrient Management, pp. 47, Springer–63, Berlin, Heidelberg (2011)
Savci, S.: Investigation of effect of chemical fertilizers on environment. Apcbee Procedia. **1**, 287–292 (2012)

Sayre, R., Beeching, J.R., Cahoon, E.B., Egesi, C., Fauquet, C., Fellman, J., ... Zhang, P.: The BioCassava plus program: biofortification of cassava for sub-Saharan Africa. Ann. Rev. Plant Biol. **62**, 251–272 (2011)

Scagel, C.F., Bi, G., Fuchigami, L.H., Regan, R.P.: Irrigation frequency alters nutrient uptake in container-grown Rhododendron plants grown with different rates of nitrogen. HortScience. **47**(2), 189–197 (2012)

Scaife, A., Turner, M.: Diagnosis of Mineral Disorders in Plants. Volume 2. Vegetables.144 (5) Her Majesty's Stationery Office, New York (1983)

Schmidt, S.B., Jensen, P.E., Husted, S.: Manganese deficiency in plants: the impact on photosystem II. Trends Plant Sci. **21**(7), 622–632 (2016)

Schwarz, G., Mendel, R.R.: Molybdenum cofactor biosynthesis and molybdenum enzymes. Annu. Rev. Plant Biol. **57**, 623–647 (2006)

Seaward, M. R. D., & Richardson, D. H. S.;. Atmospheric sources of metal pollution and effects on vegetation. Heavy metal tolerance in plants: Evolutionary aspects, UK, USA. **7**, 75–92 (1989)

Sekhon, B.S.: Chelates for micronutrient nutrition among crops. Resonance. **8**(7), 46–53 (2003)

Shaheen, N., Irfan, N.M., Khan, I.N., Islam, S., Islam, M.S., Ahmed, M.K.: Presence of heavy metals in fruits and vegetables: health risk implications in Bangladesh. Chemosphere. **152**, 431–438 (2016)

Sharma, C.P.: Plant micronutrients. CRC Press (2006)

Sharma, S.K.M.P., Ramesh, A., Joshi, O.P.: Characterization of zinc-solubilizing Bacillus isolates and their potential to influence zinc assimilation in soybean seeds, (2011)

She, W., Cui, G., Li, X., Su, X., Jie, Y., Yang, R.: Characterization of cadmium concentration and translocation among ramie cultivars as affected by zinc and iron deficiency. Acta Physiol. Plant. **40**(6), 104 (2018)

Sherene, T.: Mobility and transport of heavy metals in polluted soil environment. Biol. Forum Int. J. **2**(2), 112–121 (2010)

Shikanai, T., Müller-Moulé, P., Munekage, Y., Niyogi, K.K., Pilon, M.: PAA1, a P-type ATPase of Arabidopsis, functions in copper transport in chloroplasts. Plant Cell. **15**(6), 1333–1346 (2003)

Shinozaki, D., Merkulova, E.A., Naya, L., Horie, T., Kanno, Y., Seo, M., Ohsumi, Y., Masclaux-Daubresse, C., Yoshimoto, K.: Autophagy increases zinc bioavailability to avoid light-mediated reactive oxygen species production under zinc deficiency. Plant Physiol. **182**(3), 1284–1296 (2020)

Shivay, Y.S., Kumar, D., Prasad, R.: Effect of zinc-enriched urea on productivity, zinc uptake and efficiency of an aromatic rice–wheat cropping system. Nutr. Cycl. Agroecosyst. **81**(3), 229–243 (2008a)

Shivay, Y.S., Kumar, D., Prasad, R., Ahlawat, I.P.S.: Relative yield and zinc uptake by rice from zinc sulphate and zinc oxide coatings onto urea. Nutr. Cycl. Agroecosyst. **80**(2), 181–188 (2008b)

Shukla, A.K., Behera, S.K.: Zinc management in Indian agriculture: past, present and future. Indian J. Fertiliser. **7**(10), 14–33 (2011)

Shukla, A.K., Behera, S.K.: Micronutrient fertilisers for higher productivity. Indian J. Fertiliser. **8**(4), 100–117 (2012)

Shukla, A.K., Behra, S.K.: All India coordinated research project on micro-and secondary nutrients and pollutant elements in soils and plants: research achievements and future thrusts. Indian J. Fertiliser. **15**(5), 522–543 (2019)

Shukla, A.K., Behera, S.K., Pakhre, A., Chaudhar, S.K.: Micronutrients in soils, plants, animals and humans. Indian Journal of Fertilisers and PwC analysis. ICAR. **14**(4), 30–54 (25 pages), (2018)

Shukla, A.K., Tiwari, P.K.: Micro and secondary nutrients and pollutant elements research in India. Coordinators Report-AICRP on Micro-and Secondary Nutrients and Pollutant Elements in Soils and Plants, ICAR-IISS, Bhopal, pp. 1–196 (2016)

Shukla, A.K., Tiwari, P.K., Prakash, C.: Micronutrients deficiencies vis-a-vis food and nutritional security of India. Indian J. Fert. **10**(12), 94–112 (2014)

Siddiqui, S., Ramlal, R.: "Myelodysplasia" from copper deficiency. Blood J. Am. Soc. Hematol. **133**(8), 883–883 (2019)
Sillanpää, M.: Micronutrients and the nutrient status of soils: a global study. Vol. 48. Food Agric. Org. (1982)
Sillanpää, M., Vlek, P.L.G.: Micronutrients and the agroecology of tropical and Mediterranean regions. In: Micronutrients in Tropical Food Crop Production, pp. 151–167. Springer, Dordrecht (1985)
Sillanpää, M.: Micronutrient assessment at the country level: an international study, pp. 216 (1990)
Singh, M.V.: Micronutrient nutritional problems in soils of India and improvement for human and animal health. Indian J. Fertiliser. **5**(4), 11–56 (2009)
Singh, B., Natesan, S.K.A., Singh, B.K., Usha, K.: Improving zinc efficiency of cereals under zinc deficiency. Curr. Sci. **88**, 36–44 (2005)
Singh, J.S., Pandey, V.C., Singh, D.P.: Efficient soil microorganisms: a new dimension for sustainable agriculture and environmental development. Agric. Ecosyst. Environ. **140**(3–4), 339–353 (2011)
Singh, S., Singh, B.K., Yadav, S.M., Gupta, A.K.: Potential of biofertilizers in crop production in Indian Agriculture. Am. J. Plant Nutrit. Fertiliz. Technol. **4**, 33–40 (2014)
Singh, M.V.: In: Gragory PJ, Philbeam CJ, Walker SH, (eds). Natural resource international—integrated nutrient management on farmers fields: approaches that work. University of Readings, UK: Department of Soil Science, R700/04, (1997)
Singh, M.V., Shah, J.K.: 26th Annual report of 1995–96 for the AICRP of micro- and secondary nutrients and pollutants in soils and plants. Bhopal: Indian Institute of Soil Science (p. 107), (1995)
Singh, R.P., Mishra, S., Jha, P., Raghuvanshi, S., Jha, P.N.: Effect of inoculation of zinc-resistant bacterium Enterobacter ludwigii CDP-14 on growth, biochemical parameters and zinc uptake in wheat (Triticum aestivum L.) plant. Ecol. Eng. **116**, 163–173 (2018)
Skoog, F.: Relationships between zinc and auxin in the growth of higher plants. Am. J. Bot., 939–951 (1940)
Slaton, N.A., Wilson, C.E., Ntamatungiro, S., Norman, R.J., Boothe, D.L.: Evaluation of zinc seed treatments for rice. Agron. J. **93**(1), 152–157 (2001)
Smith, S.E., Read, D.J.: Mycorrhizal symbiosis. Academic press. London, vol. 3, (2010)
Sommer, A.L., Lipman, C.B.: Evidence on the indispensable nature of zinc and boron for higher green plants. Plant Physiol. **1**(3), 231 (1926)
Sonmez, I., Kaplan, M., Sonmez, S.: Investigation of seasonal changes in nitrate contents of soils and irrigation waters in greenhouses Located in Antalya-Demn region. Asian J. Chem. **19**(7), 5639 (2007)
Srivastava, P.C., Gupta, U.C.: Trace Elements in Crop Production. Science Publishers, Inc Citrus Experiment Station, lake Alfred, Fla (1996)
Steenbjerg, F.: The exchangeable manganese in Danish soils and its relation to plant growth. Trans. 3rd Int. Congr. Soil Sci. Oxford. **1**, 198–201 (1935)
Stein, A.J., Qaim, M.: The human and economic cost of hidden hunger. Food Nutr. Bull. **28**(2), 125–134 (2007)
Storey, J.B. Zinc.: In: Barker AV, Pilbeam DJ (eds) Hand-book of plant nutrition. CRC Press, Taylor & FrancisGroup, Boca Raton, pp 411–436, (2007)
Sulphur Institute. Sulphur in Indian agriculture. Dhaka, Bangladesh, India: Sulphur Institute; (Washington, DC) (1986)
Suman, M., Sangma, P.D., Singh, D.: Role of micronutrients (Fe, Zn, B, Cu, Mg, Mn and Mo) in fruit crops. Int. J. Curr. Microbiol. App. Sci. **6**(6), 3240–3250 (2017)
Sundara, B., Natarajan, V., Hari, K.: Influence of phosphorus solubilizing bacteria on the changes in soil available phosphorus and sugarcane and sugar yields. Field Crops Res. **77**(1), 43–49 (2002)
Szerdahelyi, P., Kasa, P.: Histochemical demonstration of copper in normal rat brain and spinal cord. Histochemistry. **85**(4), 341–347 (1986)

Takatsuji, H.: Zinc-finger proteins: the classical zinc finger emerges in contemporary plant science. Plant Mol. Biol. **39**(6), 1073–1078 (1999)

Takahashi, M., Nakanishi, H., Kawasaki, S., Nishizawa, N.K., Mori, S.: Enhanced tolerance of rice to low iron availability in alkaline soils using barley nicotianamine aminotransferase genes. Nat. Biotechnol. **19**(5), 466–469 (2001)

Takahashi, M., Terada, Y., Nakai, I., Nakanishi, H., Yoshimura, E., Mori, S., Nishizawa, N.K.: Role of nicotianamine in the intracellular delivery of metals and plant reproductive development. Plant Cell. **15**(6), 1263–1280 (2003)

Takkar, P.N.: Zinc in human and animal health. Indian J. Fert. **7**, 46–66 (2011)

Takkar, P.N., Chhibba, I.M., Mehta, S.K.: Twenty years of coordinated research on micronutrients in soils and plants, 1967–87 (1989)

Takkar, P.N., Nayyar, V.K.: Zinc deficiency in flood-prone plain soil of Punjab. Prog. Farming. **17**(9), 14–15 (1981)

Takahiro A., Kobayashi, T., Takahashi, M., Nagasaka, S., Usuda, K., Kakei, Y., Ishimaru, Y., Nakanishi, H., Mori, S., Nishizawa, N.K.: OsYSL18 is a rice iron(III)–deoxymugineic acid transporter specifically expressed in reproductive organs and phloem of lamina joints. Plant Mol. Biol. **70**(6), 681–692 (2009)

Tank, N., Saraf, M.: Enhancement of plant growth and decontamination of nickel-spiked soil using PGPR. J. Basic Microbiol. **49**(2), 195–204 (2009)

Teale, W.D., Paponov, I.A., Palme, K.: Auxin in action: signalling, transport and the control of plant growth and development. Nat. Rev. Mol. Cell Biol. **7**(11), 847–859 (2006)

Temple, S.J., Vance, C.P., Gantt, J.S.: Glutamate synthase and nitrogen assimilation. Trends Plant Sci. **3**(2), 51–56 (1998)

Thomas, L., Singh, I.: Microbial biofertilizers: types and applications. In Biofertilizers for sustainable agriculture and environment (pp. 1–19). Springer, Cham. (2019)

Tiller, K. G.. Micronutrients. Soils: An Australian viewpoint. Division of Soils, CSIRO.(Academic Press: Melbourne, Vic.). pp. 365–387 (1983)

Tilman, D., Cassman, K.G., Matson, P.A., Naylor, R., Polasky, S.: Agricultural sustainability and intensive production practices. Nature. **418**(6898), 671–677 (2002)

Times of India: Rising CO2 levels leading to zinc deficiency in diet' http://timesofindia.indiatimes.com/articleshow/68947890.cms?utm_source=contentofinterest&utm_medium=text&utm_campaign=cppst (2019)

Tønnesen, T., Kleijer, W.J., Horn, N.: Incidence of Menkes disease. Hum. Genet. **86**(4), 408–410 (1991)

Trenkel, M.E.: Controlled-release and stabilized fertilizers in agriculture (Vol. 11). Paris: International fertilizer industry association, (1997)

Triplett, E.W.: Diazotrophic endophytes: progress and prospects for nitrogen fixation in monocots. Plant Soil. **186**(1), 29–38 (1996)

Trijatmiko, K.R., Dueñas, C., Tsakirpaloglou, N., Torrizo, L., Arines, F.M., Adeva, C., Balindong, J., Oliva, N., Sapasap, M.V., Borrero, J., Rey, J.: Biofortified indica rice attains iron and zinc nutrition dietary targets in the field. Sci. Rep. **6**, 19792 (2016)

Tucker, G.: Nutritional enhancement of plants. Curr. Opin. Biotechnol. **14**(2), 221–225 (2003)

Uauy, C., Distelfeld, A., Fahima, T., Blechl, A., Dubcovsky, J.: A NAC gene regulating senescence improves grain protein, zinc, and iron content in wheat. Science. **314**(5803), 1298–1301 (2006)

Universidad Politécnica de Madrid. New biofertilizer made from exoskeletons of crustaceans and insects (2015). Retrieved 23 May, 2020

Vaid, S.K., Kumar, B., Sharma, A., Shukla, A.K., Srivastava, P.C.: Effect of Zn solubilizing bacteria on growth promotion and Zn nutrition of rice. J. Soil Sci. Plant Nutr. **14**(4), 889–910 (2014)

Vakhaniya, N.A., Abashidze, N.D., Chokheli, M.D., Nutsubidze, N.N.: Influence of nitrate and molybdate on nitrogen fixation and productivity of kidney beans. Fiziologiya Rastenії. **37**(4), 636–641 (1990)

Vallee, B.L., Auld, D.S.: Zinc coordination, function, and structure of zinc enzymes and other proteins. Biochemistry. **29**(24), 5647–5659 (1990) van de Mortel, J.E., Villanueva, L.A., Schat, H., Kwekkeboom, J., Coughlan, S., Moerland, P.D., van Themaat, E.V.L., Koornneef, M., Aarts, M.G.: Large expression differences in genes for iron and zinc homeostasis, stress response, and lignin biosynthesis distinguish roots of Arabidopsis thaliana and the related metal hyperaccumulator Thlaspi caerulescens. Plant Physiol. **142**(3), 1127–1147 (2006)

van der Waals, J.H., Laker, M.C.: Micronutrient deficiencies in crops in Africa with emphasis on southern Africa. In Micronutrient deficiencies in global crop production (pp. 201–224). Springer, Dordrecht (2008)

van Maarschalkerweerd, M., Husted, S.: Recent developments in fast spectroscopy for plant mineral analysis. Front. Plant Sci. **6**, 169 (2015)

Van Wouwe, J.P.: Clinical and laboratory diagnosis of acrodermatitis enteropathica. Eur. J. Pediatr. **149**(1), 2–8 (1989)

Velu, G., Rai, K.N., Muralidharan, V., Kulkarni, V.N., Longvah, T., Raveendran, T.S.: Prospects of breeding biofortified pearl millet with high grain iron and zinc content. Plant Breed. **126**(2), 182–185 (2007)

Velu, G., Ortiz-Monasterio, I., Cakmak, I., Hao, Y., Singh, R.Á.: Biofortification strategies to increase grain zinc and iron concentrations in wheat. J. Cereal Sci. **59**(3), 365–372 (2014)

Verma, P., Yadav, A.N., Khannam, K.S., Panjiar, N., Kumar, S., Saxena, A.K., Suman, A.: Assessment of genetic diversity and plant growth promoting attributes of psychrotolerant bacteria allied with wheat (Triticum aestivum) from the northern hills zone of India. Ann. Microbiol. **65**(4), 1885–1899 (2015)

Vessey, J.K.: Plant growth promoting rhizobacteria as biofertilizers. Plant Soil. **255**(2), 571–586 (2003)

Vieira, R.F., Cardoso, E.J.B.N., Vieira, C., Cassini, S.T.A.: Foliar application of molybdenum in common beans. I. Nitrogenase and reductase activities in a soil of high fertility. J. Plant Nutr. **21**(1), 169–180 (1998)

Wanli, G., Nazimc, H., Lianga, Z., Dongfeng Yang, D.: Magnesium deficiency in plants: an urgent problem. Crop J. **4**(2), 83–91 (2016)

Wakatsuki, T.: Metal oxidoreduction by microbial cells. J. Ind. Microbiol. **14**(2), 169–177 (1995)

Waters, B.M., Sankaran, R.P.: Moving micronutrients from the soil to the seeds: genes and physiological processes from a biofortification perspective. Plant Sci. **180**(4), 562–574 (2011)

Wege, S., Gilliham, M., Henderson, S.W.: Chloride: not simply a 'cheap osmoticum', but a beneficial plant macronutrient. J. Exp. Bot. **68**(12), 3057–3069 (2017)

Wei, Y., Shohag, M.J.I., Yang, X.: Biofortification and bioavailability of rice grain zinc as affected by different forms of foliar zinc fertilization, e45428, (2012)

Wei, Y., Zhao, Y., Fan, Y., Lu, Q., Li, M., Wei, Q., Zhao, Y., Cao, Z., Wei, Z.: Impact of phosphate-solubilizing bacteria inoculation methods on phosphorus transformation and long-term utilization in composting. Bioresour. Technol. **241**, 134–141 (2017)

Weigel, M., Varotto, C., Pesaresi, P., Finazzi, G., Rappaport, F., Salamini, F., Leister, D.: Plastocyanin is indispensable for photosynthetic electron flow in Arabidopsis thaliana. J. Biol. Chem. **278**(33), 31286–31289 (2003)

Weissert, C., Kehr, J.: Macronutrient sensing and signaling in plants. In: Plant Macronutrient Use Efficiency,University of Hamburg, Hamburg, Germany pp. 45–64. Academic Press (2017)

Welch, R.M., Graham, R.D.: A new paradigm for world agriculture: meeting human needs: productive, sustainable, nutritious. Field Crop Res. **60**(1–2), 1–10 (1999)

Welch, R.M., Allaway, W.H., House, W.A., Kubota, J.: Geographic distribution of trace element problems. Micronutr. Agric. **4**, 31–57 (1991)

Wen, X., Hu, C., Sun, X., Zhao, X., Tan, Q., Liu, P., Xin, J., Qin, S., Wang, P.: Characterization of vegetable nitrogen uptake and soil nitrogen transformation in response to continuous molybdenum application. J. Plant Nutr. Soil Sci. **181**(4), 516–527 (2018)

Westermann, D., Terán, H., Muñoz-Perea, C., Singh, S.: Plant and seed nutrient uptake in common bean in seven organic and conventional production systems. Can. J. Plant Sci. **91**(6), 1089–1099 (2011)

White, P.J., Broadley, M.R.: Biofortifying crops with essential mineral elements. Trends Plant Sci. **10**(12), 586–593 (2005)

White, P.J., Broadley, M.R.: Biofortification of crops with seven mineral elements often lacking in human diets–iron, zinc, copper, calcium, magnesium, selenium and iodine. New Phytol. **182**(1), 49–84 (2009)

White, P.J., Brown, P.H.: Plant nutrition for sustainable development and global health. Ann. Bot. **105**(7), 1073–1080 (2010)

Wilson, M., Hogstrand, C., Maret, W.: Picomolar concentrations of free zinc (II) ions regulate receptor protein-tyrosine phosphatase β activity. J. Biol. Chem. **287**(12), 9322–9326 (2012)

Wimalawansa, S.A., Wimalawansa, S.J.: Impact of changing agricultural practices on human health: Chronic kidney disease of multi-factorial origin in Sri Lanka. Wudpecker J Agric Res. **3**(5), 110–24 (2014)

Wissuwa, M., Ismail, A.M., Graham, R.D.: Rice grain zinc concentrations as affected by genotype, native soil-zinc availability, and zinc fertilization. Plant Soil. **306**(1–2), 37–48 (2008)

World Health Organization (WHO): WHO Guidelines for Drinking Water Quality. https://apps.who.int/iris/bitstream/handle/10665/42852/9241546387.pdf?sequence=1&isAllowed=y (accessed on 23 November 2019)

Wu, J., Wang, C., Zheng, L., Wang, L., Chen, Y., Whelan, J., Shou, H.: Ethylene is involved in the regulation of iron homeostasis by regulating the expression of iron-acquisition-related genes in Oryza sativa. J. Exp. Bot. **62**(2), 667–674 (2011)

Wu, S., Wei, S., Hu, C., Tan, Q., Huang, T., Sun, X.: Molybdenum-induced alteration of fatty acids of thylakoid membranes contributed to low temperature tolerance in wheat. Acta Physiol. Plant. **39**(10), 237 (2017)

Wu, C., Dun, Y., Zhang, Z., Li, M., Wu, G.: Foliar application of selenium and zinc to alleviate wheat (Triticum aestivum L.) cadmium toxicity and uptake from cadmium-contaminated soil. Ecotoxicol. Environ. Saf. **190**, 110091 (2020)

Wuana, R.A., Okieimen, F.E.: Heavy metals in contaminated soils: a review of sources, chemistry, risks and best available strategies for remediation. Isrn Ecol. **2011**, 402647 (2011)

Wyngaard, N., Cabrera, M.L., Jarosch, K.A., Bünemann, E.K.: Phosphorus in the coarse soil fraction is related to soil organic phosphorus mineralization measured by isotopic dilution. Soil Biol. Biochem. **96**, 107–118 (2016)

Xiaoyan, S., Yan, Z., Shubin, W.: Improvement Fe content of wheat (Triticum aestivum) grain by soybean ferritin expression cassette without vector backbone sequence. J. Agric. Biotechnol. **20**(7), 766–773 (2012)

Xinxian, L., Xuemei, C., Yagang, C., Woon-Chung, W.J., Zebin, W., Qitang, W.: Isolation and characterization endophytic bacteria from hyperaccumulator Sedum alfredii Hance and their potential to promote phytoextraction of zinc polluted soil. World J. Microbiol. Biotechnol. **27**(5), 1197–1207 (2011)

Yaneva, I.A., Vunkova-Radeva, R.V., Stefanov, K.L., Tsenov, A.S., Petrova, T.P., Petkov, G.O.: Changes in lipid composition of winter wheat leaves under low temperature stress: effect of molybdenum supply. Biol. Plant. **37**(4), 561 (1995)

Yang, X.W., Tian, X.H., Lu, X.C., Cao, Y.X., Chen, Z.H.: Impacts of phosphorus and zinc levels on phosphorus and zinc nutrition and phytic acid concentration in wheat (Triticum aestivum L.). J. Sci. Food Agric. **91**(13), 2322–2328 (2011)

Yanqun, Z., Yuan, L., Jianjun, C., Haiyan, C., Li, Q. Schvartz, C.: 2005. Hyperaccumulation of Pb, Zn, and Cd in herbaceous grown on lead-zinc mining area in Yunnan, China. Environment International, **31**(5), pp.755–762.

Ye, D., Zhang, X., Li, T., Xu, J., Chen, G.: Phosphorus-acquisition characteristics and rhizosphere properties of wild barley in relation to genotypic differences as dependent on soil phosphorus availability. Plant Soil. **423**(1–2), 503–516 (2018)

Yoshida, S., Ahn, J.S., Forno, D.A.: Occurrence, diagnosis, and correction of zinc deficiency of lowland rice. Soil Sci. Plant Nutr. **19**(2), 83–93 (1973)

Yuan, M., Chu, Z., Li, X., Xu, C., Wang, S.: The bacterial pathogen Xanthomonas oryzae overcomes rice defenses by regulating host copper redistribution. Plant Cell. **22**(9), 3164–3176 (2010)

Yuan, L., Wu, L., Yang, C., Lv, Q.: Effects of iron and zinc foliar applications on rice plants and their grain accumulation and grain nutritional quality. J. Sci. Food Agric. **93**(2), 254–261 (2013)

Zahra, A., Mesdaghinia, A., Nouri, J., Homaee, M., Yunesian, M., Ahmadimoghaddam, M., Mahvi, A.M.: Effect of fertilizer application on soil heavy metal concentration. Environ. Monit. Assess **160**(1–4), 83–89 (2010)

Zaheer, I.E., Ali, S., Rizwan, M., Farid, M., Shakoor, M.B., Gill, R.A., Najeeb, U., Iqbal, N., Ahmad, R.: Citric acid assisted phytoremediation of copper by Brassica napus L. Ecotoxicol. Environ. Saf. **120**, 310–317 (2015)

Zhang, Y., Song, Q., Yan, J., Tang, J., Zhao, R., Zhang, Y., ... Ortiz-Monasterio, I.: Mineral element concentrations in grains of Chinese wheat cultivars. Euphytica. **174**(3), 303–313 (2010)

Zhao, Y.P., Cui, J.L., Chan, T.S., Dong, J.C., Chen, D.L., Li, X.D.: Role of chelant on Cu distribution and speciation in Lolium multiflorum by synchrotron techniques. Sci. Total Environ. **621**, 772–781 (2018)

Zhou, Z., Johnson, W.T., Kang, Y.J.: Regression of copper-deficient heart hypertrophy: reduction in the size of hypertrophic cardiomyocytes. J. Nutr. Biochem. **20**(8), 621–628 (2009)

Zhu, C., Kobayashi, K., Loladze, I., Zhu, J., Jiang, Q., Xu, X., Liu, G., Seneweera, S., Ebi, K.L., Drewnowski, A., Fukagawa, N.K.: Carbon dioxide (CO2) levels this century will alter the protein, micronutrients, and vitamin content of rice grains with potential health consequences for the poorest rice-dependent countries. Sci. Adv. **4**(5), eaaq1012 (2018)

Zhu, C., Naqvi, S., Gomez-Galera, S., Pelacho, A. M., Capell, T., Christou, P.: Transgenic strategies for the nutritional enhancement of plants. Trends Plant Sci. **12**(12), 548–555 (2007)

Zou, C., Gao, X., Shi, R., Fan, X., Zhang, F.: Micronutrient deficiencies in crop production in China. In: Micronutrient deficiencies in global crop production, pp. 127–148. Springer, Dordrecht (2008)

Chapter 13
Development of Microbes-Based Biofertilizer for Zinc Dissolution in Soil

Shaibi Saleem and Shams Tabrez Khan

Abstract Zinc malnutrition is one of the major concerns particularly in the developing countries as most of the population is dependent on zinc-poor cereal-based diet. Low concentration of zinc in cereals and other plants is due to various reasons such as the unavailability of micronutrients in soil which can be linked to extensive use of chemical fertilizers. Poor bioavailability of zinc is due to the presence of insoluble forms of Zn in soil, soil receiving high phosphorus application, physicochemical conditions of soil like alkaline pH, high reactivity of zinc in soil, and water logging. Surveys have revealed that large areas in the world including India are deficient in zinc. Effective, sustainable, and economic strategies are needed to enhance the zinc uptake by plants. The use of chemical fertilizers to supplement the zinc requirement is not effective as it is expensive, leads to chemical pollution, and any such zinc fertilizer added to soil again becomes immobilized in soil and hence becomes unavailable to plants. The utilization of biofertilizers is a well-known approach to overcome the menace of zinc deficiency. Biofertilizer formulations containing zinc solubilizing bacteria (ZSB) can help in the solubilization of zinc and to make it available to crops. However, a systematic and detailed study and review of literature is required for the design of an effective zinc biofertilizer. The parameter that needs to be evaluated include selection of effective ZSB, determination of better carrier material for improving shelf life of ZSB, routine checking of the biofertilizer to satisfy the proper viable count, methods of storage, and production. The success of such biofertilizers will also need wide publicity and awareness programs among farmers. This chapter, therefore, discusses some of the aspects associated with the possible design of zinc biofertilizers.

Keywords Biofertilizer · Plant growth promoting Rhizobia · Zn biofertilizer · Micronutrients · Zn solubilizing microorganisms

S. Saleem (✉) · S. T. Khan
Department of Agricultural Microbiology, Faculty of Agricultural Sciences, Aligarh Muslim University, Aligarh, Uttar Pradesh, India

S. T. Khan, A. Malik (eds.), *Microbial Biofertilizers and Micronutrient Availability*,
https://doi.org/10.1007/978-3-030-76609-2_13

Zinc malnutrition is one of the major concerns particularly in the developing countries as most of the population is dependent on zinc-poor cereal-based diet. Low concentration of zinc in cereals and other plants is due to various reasons such as the unavailability of micronutrients in soil which can be linked to extensive use of chemical fertilizers. Poor bioavailability of zinc is due to the presence of insoluble forms of Zn in soil, soil receiving high phosphorus application, physicochemical conditions of soil like alkaline pH, high reactivity of zinc in soil, and water logging. Surveys have revealed that large areas in the world including India are deficient in zinc. Effective, sustainable, and economic strategies are needed to enhance the zinc uptake by plants. The use of chemical fertilizers to supplement the zinc requirement is not effective as it is expensive, leads to chemical pollution, and any such zinc fertilizer added to soil again becomes immobilized in soil and hence becomes unavailable to plants. The utilization of biofertilizers is a well-known approach to overcome the menace of zinc deficiency. Biofertilizer formulations containing zinc solubilizing bacteria (ZSB) can help in the solubilization of zinc and to make it available to crops. However, a systematic and detailed study and review of literature is required for the design of an effective zinc biofertilizer. The parameter that needs to be evaluated include selection of effective ZSB, determination of better carrier material for improving shelf life of ZSB, routine checking of the biofertilizer to satisfy the proper viable count, methods of storage, and production. The success of such biofertilizers will also need wide publicity and awareness programs among farmers. This chapter, therefore, discusses some of the aspects associated with the possible design of zinc biofertilizers.

Introduction

Zinc is considered as a vital micronutrient required for proper growth and development of all living organisms, including microorganisms, plants, animals, and humans (Hambidge and Krebs 2007; Sommer and Lipman 1926; Wyszkowska et al. 2013; Kidd et al. 1996). Often, food consumed cannot supply all the required micronutrients in adequate amounts (Welch and Graham 2005). One of the most important reasons for micronutrient deficiency is the diet, which is poor in micronutrients. An ever-increasing world population is resulting in food shortage as well as poor food quality (Brahmanand et al. 2013). It is evaluated that agricultural production should rise by 70% by 2050 in order to provide proper food for 9 billion people worldwide. Micronutrient deficiency, especially iron and zinc, affects the health of poor people due to insufficient and poor dietary intake (Ma et al., 2008). The population subgroup, which is at the highest risk, are the women and children in developing and underdeveloped countries (Welch and Graham 2005). Since these groups require comparatively higher amounts of micronutrients due to higher growth rates and other health conditions like pregnancy, it is highly required to increase not only the yield but also improve the nutritional status of food. Therefore, modifications in agricultural practices are required to achieve food security (Welch and Graham

2004). On the contrary, the increasing burden on agricultural lands has resulted in the overuse of agricultural chemicals like chemical fertilizers, pesticides, and insecticides, which are having an adverse effect on the natural environment and are also hazardous to human health. Scientists are searching for alternatives that can reduce the harmful impact on natural resources and the environment.

Zn deficiency is the most important micronutrient deficiency in humans, which is also due to insufficient dietary intake. One of the important reasons for the low Zn level of diet is its deficiency in agricultural soil. The Food and Agriculture Organization of the United Nations (FAO) estimates that 50% of world's soils are zinc deficient (Das and Green 2013). Correcting micronutrient deficiency in agricultural lands can improve dietary intake significantly and will help in improving human health. Increasing the Zinc content in the edible parts of the plants (grains) through Biofortification is one of the most economical and main agronomic strategies to improve the dietary Zinc status. Other strategies include plant breeding and molecular biology-based approaches for improving the Zn status of the crops (Gómez-Galera et al. 2010; Shetty 2011; Ruel and Bouis 1998; Cakmak 2008). However, these approaches have many disadvantages such as cost and social acceptability.

The use of microbial inoculants for mobilizing Zn in soil and improving its accessibility to plants is one of the most attractive alternatives for Zn biofortification of the crops. These strains solubilize Zn using different mechanisms such as the production of organic acids, the production of siderophore, and zinc chelation. Other factors associated with plants or soil, including change in root architecture, low soil pH, also affect the bioavailability of Zn. Many bacterial strains with the potential of solubilizing Zn have been identified and can be formulated as biofertilizers (Mahdi et al. 2010; Kour et al., 2020). The addition of these microbial inoculants to the soil can help in maintaining the required population of such strains to solubilize insoluble Zn and to make it available to plants, which consequently can improve the Zn content in the grains, especially in cereal grains. Biofortification of cereal grains is important as much of the population suffering from Zn deficiency is dependent on a cereals-based diet like wheat, rice, and maize. Zinc biofertilizer can be a safe and economical alternative for zinc biofortification of the crops.

This chapter describes the details of the strains that have been identified to solubilize zinc and how Zn biofertilizers can be produced using these strains. The chapter also discusses the quality control parameters that should be used for the production of such biofertilizers.

Importance of Micronutrients

With the increasing world population, the demand for food is also increasing, and consequently, the burden on agricultural lands is growing. On the other hand, natural resources used for the production and supply of food are not enough (Alexandratos and Bruinsma 2012). To make matters, most of the agricultural practices revolve around the farmer's profitability. It is now being understood that due to the extensive agricultural practices may be the availability of food has slightly improved, but the quality of the food is deteriorating (Huang et al. 2002). One problem which is having obvious health consequences is the micronutrient status of the food. Both plants and humans require micronutrients. These elements are called as "essential trace elements" or micronutrients because they are utilized in minimum amount. Eight micronutrients required by plants are boron (B), copper (Cu), iron (Fe), manganese (Mn), molybdenum (Mo), zinc (Zn), nickel (Ni), and chloride (Cl) (Das and Green 2016). Similarly, humans require at least nine micronutrients [iron (Fe), cobalt (Co), chromium (Cr), copper (Cu), iodine (I), manganese (Mn), selenium (Se), zinc (Zn), and molybdenum (Mo)] (Welch and Graham 2004). As most of the food is becoming micronutrient deficient, inadequate intakes of micronutrients is contributing to stunted growth in children and is also resulting in increment of sickness and death rates among children and in infants (Black 2003; Branca and Ferrari 2002). Micronutrient deficiency also mention as "hidden hunger," is alarming as half of the world's population is at the risk of micronutrient deficiency (Kennedy et al. 2003). Sillanpää in 1990 evaluated that dominant agricultural soils of the world are 49% deficient in zinc (Zn), 31% deficient in boron (B), 15% deficient in molybdenum (Mo), 14% deficient in copper (Cu), 10% deficient in manganese (Mn), and 3% deficient in Iron (Fe) (Graham 2008).

Research has been carried out to manage the growing deficiency of zinc in different parts of the world. Various nutrients are applied routinely to fixed areas of soil and to specific crops for a prolonged time. As a result, the application of micronutrients is expanding in the form of foliage sprays and fertilizers.

Importance of Zn as Micronutrient

Importance of Zn for Humans

Zinc (Zn) is regarded as a vital micronutrient and the second most abundant transition metal (atomic number 30, atomic weight 65.4 u) in living organisms after iron (Fe). Zinc is a critical nutrient required by plants, humans, microorganisms, and other living organisms. It is available in different concentrations and forms in these living organisms. Zn composition in the Earth's crust is 78 mg/kg, which varies in different parent rocks (Nitu et al. 2020). In the magmatic rocks Zn content is 40–120 mg/kg, sedimentary rocks 80–120 mg/kg, sandstones 15–30 mg/kg, and 10–25 mg/kg in dolomites and limestone. Worldwide the soil Zn ranges from 35 to 12,400 mg/kg (Nitu et al. 2020).

During 1934, zinc was considered as an essential micronutrient for experimental animals and in 1961 for humans (Singh & Dubey 2019). The zinc is required by humans in less amount to finish the growth, development, and physiological functions (Hambidge and Krebs 2007). A human adult body normally holds 2–3 g of zinc. It is important for most of the enzymes in the human body (Barak and Helmke 1993); it is also important for the structure of proteins and cell membrane integrity; zinc finger proteins are also known to act as transcription factors; zinc helps in cell signaling (Bhowmik et al. 2010). There are at least 10 Zn transporters and 15 zip transporters in human cells (Bin et al. 2019). If the zinc is lost from biological membranes it results in damage and weakens their functions. The main reason behind zinc shortage includes insufficient dietary consumption, increased demand, malabsorption, increased losses, and impaired utilization (Khalid et al. 2014). Inadequate zinc has also been reported as a reason of many health disorders in human beings. Lack of zinc is the fourth most common deficiency in humans after vitamin A, iron, and iodine (Cakmak 2009; Shivay 2015).

Deficiency of Zn has many obvious effects on human health and is known to weaken the immune system, retard growth, adversely affect the reproductive system, cause skin disorders, and is involved in various types of carcinoma (Prasad 2013). Some studies suggest that 2.7 billion persons suffer from zinc insufficiency globally (Sharma et al. 2013). WHO has reported that about 2 billion (33%) of the world population is affected by inadequate zinc (Singh & Prasad 2014), which results in 450,000 deaths of children under 5 years of age every year (Singh & Prasad 2014). Zn deficiency is responsible for 16% of respiratory disorders, 10% of diarrhea, and 18% of malaria with 800,000 deaths annually in economically weak sections of the society (Katona & Katona-Apte 2008). In developing countries, lack of zinc is the fifth major cause of human diseases and deaths (Hussain et al. 2012).

Importance of Zn for Plants

The essentiality of zinc for plants was established in 1869 (Singh & Dubey 2019). Zinc is needed by plants in very minute amounts (30–100 mg/kg). Zn is essential for growth, development, and reproduction (McCall et al. 2000). It is vital to various metabolic processes in plants including sugar synthesis, photosynthesis and transport, phytohormone synthesis (e.g., auxin), seedling vigor, membrane function, and abiotic stress factors (e.g., drought) (Hajiboland and Amirazad 2010; Gurmani et al. 2012), as it acts as a prosthetic group of about 3000 different proteins in higher plants (Tapiero and Tew 2003). Zn is a constituent of carbonic anhydrase and ribulose 1,5-biphosphate carboxylase (RuBPC), two very important enzymes required for photosynthesis. Therefore, its deficiency can reduce photosynthesis by 50–70%. Zinc also helps in RNA, DNA synthesis, and its regulation protects ribosomal RNA against ribonuclease attack. Zn prevents the peroxidation of lipids and proteins, providing protection against radical oxygen species or ROS (Cakmak 2000). It also helps in maintaining membrane integrity by interacting with phospholipids and sulfhydryl groups. Therefore, deficiency of Zn in plants can adversely affect plant health, agricultural productivity, and the quality of the produce. One of the main reasons for the low Zn status of plants is the deficiency of Zn in agricultural soil.

Table 13.1 Total zinc concentrations, ranges, means and areas of Indian soil

Soil type/climatic zone	Mean zinc	Range of zinc	Area
	(mg Zn kg−1)		
Arid/semi-arid	59	20–89	Western Rajasthan, Haryana, Punjab and Rann of kutch in gujarat
Humid/sub-humid tropics	52	22–74	–
Vertisols	–	69–76	Bombay, Hyderabad and Madya Pradesh states
Oxisols (coarse textured)	–	24–30	Tamil Nadu, Madhya Pradesh, Jharkhand, Odisha, Karnataka and southeast Maharashtra

Modified from: Alloway (2008b), Bhandari et al. (2021), Ahmad (1996), Binkley and Fisher (2019)

The Food and Agriculture Organization (FAO) has determined that compared to other micronutrients, Zn deficiency in agricultural soils is more common, and almost half of the agricultural soils are zinc deficient. Lack of zinc in soils has a negative impact on plant growth, resulting in the low crop yield and lower concentrations of zinc in the edible parts (grains). Various reports have demonstrated a decrease in agricultural productivity and Zn content when grown in Zn insufficient soil (Mori et al. 2016; Noulas et al. 2018).

Zinc in Soils

Several mineral forms of Zn are present in the soil, including $Zn(OH)_2$, smithsonite ($ZnCO_3$), zincite (ZnO), willemite [Zn $(PO_4)_2.4H_2O$], and franklinite (Zn Fe_2O_4). Some of the soils are Zn deficient, while others may have become deficient in Zn due to anthropogenic activities. While still, other soils contain Zn but not in a bioavailable form. Therefore, zinc deficiency in the crops may be either decrease concentration of total zinc in soils or may also be due to its low solubility or bioavailability (Cakmak 2008). Zinc availability and bioavailability are two different things as zinc may be available in sufficient amounts in soil but may not be available in a form suitable for uptake and assimilation by the plants (White & Zasoski, 1999). Zn is absorbed from soil as divalent cation Zn^{2+}. According to Alloway, 2008b, the entire amount of zinc in soils is divided over 5 fractions described as follows:

(i) The water-soluble pool: present in the soil solution.
(ii) Exchangeable pool: ions bound to soil particles by electrical charges.
(iii) Organically bound pool: ions adsorbed, chelated or complexed with organic ligands.
(iv) Pool of zinc sorbed non-exchangeable onto clay minerals and insoluble metallic oxides.

(v) Pool of weathering primary minerals.

The available zinc level in soils is very low as already mentioned in research (4.0–270.0 μg L^{-1}) in link to the mean total Zn level of 64.0 mg kg^{-1} (Alloway 2009). There are many factors which effect the solubilization of the zinc like, pH (Kirk and Bajita 1995; McBride and Blasiak 1979), flooded condition (Hafeez et al. 2013; Alloway 2008a, b), overuse of phosphate fertilizer, high calcite, and iron-induced zinc deficiency (Mandal and Hazra 1997). The solubility of several zinc minerals decreases in the following order: $Zn(OH)_2$ (amorphous) > $ZnCO_3$ (smithsonite) > ZnO (zincite) > Zn $(PO_4)_2.4H_2O)$ (willemite) > soil Zn > Zn Fe_2O_4 (franklinite). All of the $Zn(OH)_2$ minerals, ZnO, and $ZnCO_3$ are about 10^5 times more soluble than soil zinc, which make them satisfactory fertilizer sources of zinc (Alloway 2008b).

Bioavailability of Zn to Plants

Zinc is the essential micronutrient for plant health. It helps to improve plant growth, reproduction, and nutritional status of the crop (Reed and Martens 1996; Mousavi 2011). Zinc was identified as a vital micronutrient for plants by Sommer and Lipman in 1926 for the first time. While, in India, zinc deficiency in plants was first reported by Nene (1966) on paddy and was named as Khaira disease. Since then, various studies have emphasized on the importance of micronutrients for plant growth and quality of the produce (Mousavi 2011). The importance of Zn as an important micronutrient is also well documented (Meena et al. 2017). On the contrary, global surveys have pointed out to the prevalence of Zn deficiency in various agricultural soils (Sillanpää 1982). In similar surveys of agricultural soils in China and India have shown that most of the soils studied are Zn deficient (Zou et al. 2008). In India, analysis of 14,863 soil samples from various parts of the country has shown that 49% of the samples were Zn deficient. While 33% were deficient in B, 12% in Fe, 11% in Mo, and 3% in Cu (Singh, 2008). The critical limit of zinc for some soils and crops was also studied by Shukla and Tiwari (2016) as shown in Table 13.2. Very often Zn is present in soils in insoluble forms, this cannot be assimilated by plants easily. According to Hafeez et al. (2013) and Alloway (2008a, b), there are copious soil factors that affect the availability of Zn to plants (Fig. 13.1).

A. *Soil texture*: The lighter textured soils like sandy soils and histosols contain low levels of Zn, whereas finer texture soils like clay have more Zn than lighter textured soils (Table 13.1) and possess higher cation exchange capacity values. Whereas, other soils like calcareous soils and saline soils have a very low level of zinc for plant uptake due to zinc-sorption. As a result, Zn deficiency is more probably occur in sandy than in clay soils (Hafeez et al. 2013) (Table 13.2).
B. *Formation of insoluble Zn compounds*: Zn may be present in various forms in soil that may and may not be bioavailable for plants. The forms of Zn that are

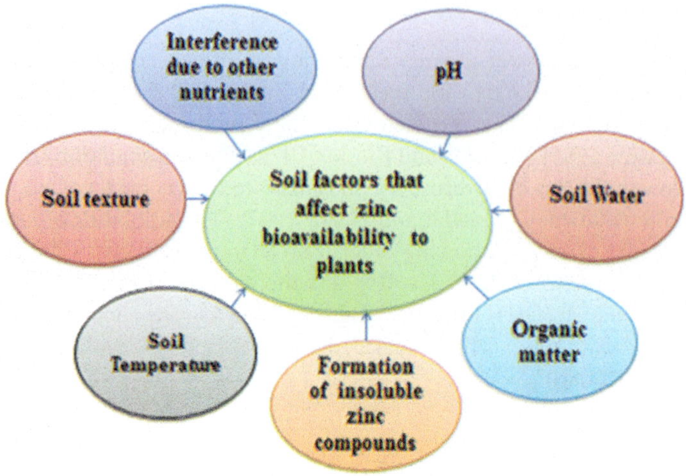

Fig. 13.1 Soil factors that affect zinc bioavailability to plants

Table 13.2 Critical limit of Zinc for some soils and crops at different locations of India

	Zn, mg/kg		
Location	Crop	Soil	Plant
Akola, Maharashtra	Soybean	0.65	24.3
Coimbatore, Tamil Nadu	Maize	0.9	24.8
Pantnagar, Uttarakhand	Lentil	1.2	9.6
Pantnagar, Uttarakhand	Chickpea	1.2	19.2
Pantnagar, Uttarakhand	Maize	0.75	-
Pusa, Bihar	Wheat	0.8	-
Nainital, Uttarakhand	Soybean	1.24	22.3
Odisha, India	Rice	0.8	24.5
Odisha, India	Groundnut	0.7	23
Odisha, India	Potato	0.7	27
West Bengal, India	Rice	0.82	28.5

Modified from: Shukla and Tiwari 2016, Shukla and Behra 2019, Sarangi et al. 2016, Mahata et al. 2013

bioavailable to plants include Zn ions (Zn^{2+} and $ZnOH^{+}$), soluble organic complexes, and labile Zn (easily desorbable Zn) (kiekens 1995).

C. *Soil pH*: *The* availability of Zn is significantly affected by the soil pH. In general, slightly acidic soil pH favors the Zn uptake by plants, while the alkaline pH decreases Zn bioavailability. It has been reported that Zn availability decreases with increasing soil pH (Jeffery and Uren 1983). The bioavailability of Zn increases by 100 times with a decrease of pH by one unit (Zeb et al. 2018). Therefore, soils with alkaline pH are known for low Zn bioavailability.

D. *Soil water*: Flooding is also known to reduce the bioavailability of Zn in soil. Prolonged water logging like in paddy soils adversely affects the zinc bioavailability (Patnaik et al. 2008). Zinc deficiency due to flooding results from Zn reaction with free sulfide (Hafeez et al. 2013). Under the submerged conditions of rice cultivation, Zn is changed into amorphous sesquioxide precipitates or franklinite: $ZnFe_2O_4$ (Hafeez et al. 2013). Thus, a delay in Zn application until after flooding is recommended for rice cultivation.
E. *Interference due to other nutrients*: The presence of certain nutrients like iron and phosphate also decreases the bioavailability of Zn in soil. Studies have demonstrated that the presence of high concentration of iron under anaerobic conditions reduces the zinc bioavailability in soil (Hafeez et al. 2013). Similarly, a high concentration of P also interferes with Zn uptake by roots, as it reacts with Zn in the soil to form insoluble Zn-phosphate complexes (Loneragan and Webb 1993).
F. *Soil temperature*: Low temperatures reduce zinc uptake by plants (Alloway 2008a, b). While warm and moist soils are known to support zinc uptake (Bauer and Lindsay 1965).
G. *Organic matter*: Normally, an *increase* in organic matter contents of a soil increases zinc accessibility. According to Hafeez et al. (2013), if the soil contains extremely high organic matter like in peat soils, this will result in the deficiency of zinc. Because the soil already has humic substances which form chelates with zinc, binding of zinc on solid-state humic substances results in low zinc availability for plant uptake.

Chemical Fertilizers and Their Hidden Dangers

The demand for the food is increasing with the increasing world population, resulting in the shortage of food and decreased nutritive value of the diet. To meet the increased demand for food, farmers are increasingly turning to extensive agricultural practices including the use of chemical fertilizer. The fertilizers are of various types and derived from various sources both natural and industrially produced. The produced fertilizer must contain 5% of one or more of the three nutrients nitrogen (N), phosphorous (P), and potassium (K). Fertilizers also contain minor elements like Zn, Mn, and Fe. The chemical fertilizers help to assist plant growth by enhancing the accessibility of nutrients present in soil or by altering the soil's physical structure (Chandini et al. 2019). Therefore, result in the improved seed and seedling vigor, decreased root uptake and shoot (or grain) accumulation of Cd, increased crop yield, quality parameters, and immediate availability, only when applied in the optimum recommended dose. The plant nutrient requirements are predicted to rise in upcoming years due to constantly growing population (Smith and Paul 1990). According to Alexandratos and Bruinsma (2012) world population is increased by about 2.25 billion by 2012 and expected to reach about 9.15 billion by 2050. The agricultural production and consumption at a global level in 2050 are expected to be 60 percent higher than in 2005/07. Hence, it is predicted that above 70% of the grain yield in upcoming years will be provided by fertilizers. Farmers are using various fertilizers to control the micronutrient deficiency in the soil. As we know, the land used for the production of agriculturally derived food is decreasing in the world due to urbanization (van Vliet et al. 2017). According to van Vliet et al. (2017) 65 Mton of crop production will be replaced by urbanization between 2000 and 2040, resulting in the increased nutrient load per unit area.

Inadequate zinc is the most widespread trace element deficiency in the soil-crop systems and, according to Sharma et al. 2013 about 1/3 of the world's human population (i.e., 2 billion) is at the danger of zinc deficiency. Hence, there is an urgent need to deal with Zn deficiency. Various approaches have been proposed which include crop rotation, plant breeding, and molecular approaches (Humayan Kabir et al. 2014; Rehman et al. 2018; Rawat et al. 2013). One of the most commonly used approaches is the agronomic fortification (zinc fertilizers to increase zinc density) of the crops.

Various zinc fertilizers are being used to correct the zinc deficiency in crops and yield losses in crop plants. The most common sources are ZnO and $ZnSO_4$. $ZnSO_4$ is applied at rates that range from 5 to 25 kg Zn ha^{-1} (Cakmak 2008; Zhao et al. 2014). For example, it is estimated that about 85,000–90,000 tons of zinc sulfate was used by the farmers to improve the zinc status of the plants per year (Devi and Vasandha 2016). The use of zinc fertilizer also called as agronomic fortification is the fast and easiest way to treat the zinc deficiency in the crops. There are three distinct types of compounds used in zinc fertilizers:

A. *Inorganic compounds*: include zinc oxide (ZnO), zinc sulfate ($ZnSO_4$), zinc chloride ($ZnCl_2$), zinc carbonate ($ZnCO_3$), and zinc nitrate $Zn(NO_3)_2$.
B. *Synthetic chelates*: generally, are a combination of a chelating agent such as ethylene di-amine tetra acetic acid (EDTA) with a metal ion. The disodium salt of Zn-EDTA (Na_2Zn-EDTA) is the most commonly used chelated source of zinc. With their high stability, synthetic chelates are eminently suitable for mixing with concentrated fertilizer solutions for soil, fertigation, and hydroponic applications.
C. *Natural organic complexes*: include those which are manufactured by reacting zinc salts with citrates or with organic by products from paper pulp manufacture such as lignosulphonates, phenols, and polyflavonoids.

Various studies have demonstrated the beneficial effect of the use of zinc fertilizers. $ZnSO_4$ and Zn-EDTA has been shown to increase agricultural productivity and Zn content of the produce (White and Broadley 2005). Combined soil and foliar application of $ZnSO_4$ improved grain Zn concentration in wheat (Zhao et al. 2020). Zinc helps to repress fungal diseases of wheat, including Fusarium crown rot (Grewal et al. 1996). Other health benefits of the fertilizer approach for biofortification of wheat with Zn help to compete strongly with cadmium (Cd), which is a

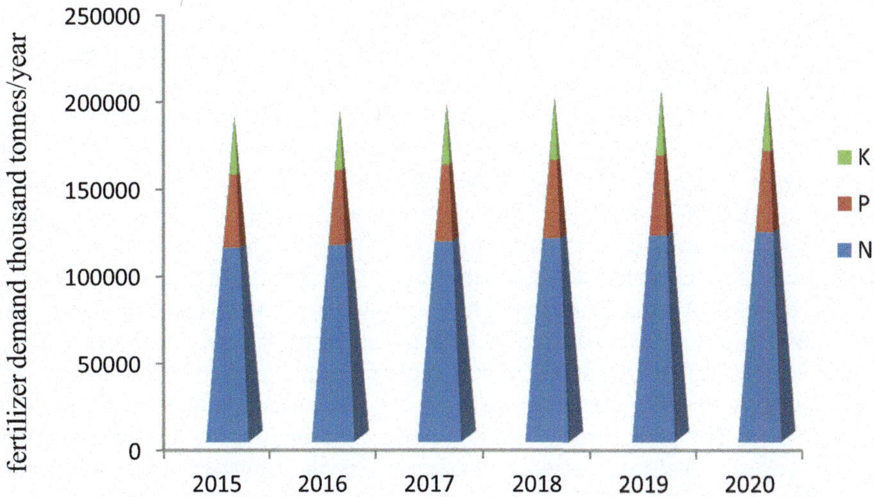

Fig. 13.2 World fertilizer demand of nitrogen, phosphorous, and potassium, 2015–2020 (thousand tons). (Source: FAO 2017)

toxic, non-essential heavy metal that threatens human health (Welch et al. 1999; Reeves and Chaney 2008; Cakmak 2009).

The agriculture systems are using various agricultural chemicals like fertilizers, pesticides, and herbicides to obtain increased yield and profit per unit area, but using high doses of these agricultural chemicals and fertilizers is causing several problems. The consumption of fertilizers is increasing globally and the total supply of (N, P_2 O_5, and K_2O) was expected to reach 273,384 million tons in 2020 (FAO 2017) as shown in Fig. 13.2. China is the largest consumer of fertilizers in the world, consuming about 54,035.856 tons in 2019 (Huang et al. 2008). India is the second largest consumer with a consumption rate of 41,564.0 tons in 2019 (FAI). The overuse of chemical fertilizer is responsible for polluting water resources, environment pollution (soil, air, and water pollution), decreased food quality, release greenhouse gases, decreased fertility of soil, resistance development in different weeds, trace element deficiency in the soil, causes toxicity to different beneficial living organism, low crop production, and eutrophication of water bodies (Youssef and Eissa 2014).

A. *Water pollution*: Nitrogen fertilizers are one of the most widely used chemical fertilizers and are commonly applied to agricultural fields as ammonia, ammonium nitrate, and urea. Even under standard conditions, plants can only use 50% of the total nitrogenous fertilizers applied to soil. Around 2–20% is lost through evaporation, 15–25% react with organic compounds and clay in the soil. While remaining 2–10% enters surface and groundwater. Leaching of nitrogenous fertilizers into drinking water has consequences on environment and human health (Savci 2012). These fertilizers are often converted to nitrate and leaching of nitrate into drinking water is widely reported (Self and Waskom 1992). Nitrate is known to have an adverse effect on human health (Du et al. 2007) and is known to cause diseases like blue baby syndrome. It is a lethal state that occurs when hemoglobin (Fe^{2+}) in an infant's red blood cells is oxidized to methemoglobin (Fe^{3+}). Methemoglobin is unable to transport oxygen hence results in the symptoms of cyanosis. The infants develop blue-gray skin, this condition can cause coma or death if not treated early (Ramanamurthy 2013). According to WHO's guidelines, nitrate toxicity may occur if its concentration exceeds 50 mg NO_3^-/L of drinking water. Typical symptoms of nitrate toxicity include dyspnea, weakness, ataxia, and terminal convulsions. Other diseases include goiter, birth defects, and heart disease (Ward et al. 2018). Another widely reported negative effect of phosphate and nitrogen fertilizer is water eutrophication. Surface waters must contain $\leq$50 μg/liter phosphorus. The increased concentration of phosphorous and nitrogen in the water results in enhanced growth of higher aquatic plants and algae, consequently depleting dissolved oxygen concentration in water affecting its quality. This depleted concentration of oxygen in water reduces the population of all the living species in the aquatic environment that require oxygen like, fish, and aerobic microorganisms, and promote the growth of anaerobic microorganisms and other unwanted species, which deteriorates the water quality often characterized by malodor making the water unfit for drinking (El-Saadony 2015).

B. *Soil pollution:* The overuse of chemical fertilizers can also impact soil health and fertility. Excess of fertilizers is known to cause acidification of soil and decrease in soil organic matter (Rezaeian and Boroon 2018). According to Mahal, the overuse of nitrogen fertilizer decreases the soil organic matter (mineralization) in maize-based agro-ecosystem (Mahal et al. 2019). The two long-term experiments were conducted in central and southern Iowa, USA, from 1999 to 2014, and it was observed that NH^+_4-N fertilizer had a direct suppressive effect on SOM mineralization. The high level of sodium and potassium containing fertilizers shows a negative impact on soil, hence, alters the soil pH and deteriorates soil structure (Savci 2012). The mobile nutrients, like nitrate and nitrogen, are not preserved in soil as the available form for the longer period of time. Nitrate accumulation is a typical physiological characteristic of plants. But the excess of this nitrate nitrogen is poisonous for humans and animals (Savci 2012). Nitrate is converted to nitrite by the existing bacteria in the stomach and the gastrointestinal (GI) tract. When nitrite enters into the blood cells, it converts the hemoglobin into methemoglobin, which result in the symptoms like anoxemia and cyanosis. Research shows that mostly higher concentration of nitrate in our body can be due to drinking water or consuming vegetables with high nitrate. Another concern is that nitrite in the stomach combines with food and creates carcinogenic compounds (Rezaeian and Boroon 2018). Various industrial waste products containing heavy metals are being marketed as micronutrient fertilizers (Mortvedt 1985), which may enter the human food chain and result in the human and environment hazard (Savci 2012).

C. *Air pollution:* Excessive use and production of fertilizers is resulting in the harmful greenhouse gases (Savci 2012). After carbon dioxide and methane, nitrous oxide (N_2O) turns out to be the third most important greenhouse gas. Nitrous oxide also has high global warming potential approximately 310 times greater than carbon dioxide. At the global level, the atmospheric N_2O increases from 0.2% to 0.3% each year. Nitrous oxide leads to ozone destruction and responsible for causing atmospheric holes (Rütting et al. 2018). Excessive use of nitrogenous fertilizers results in air pollution through nitrogen oxides (NO, N_2O, NO_2) emissions (Cooper et al. 2018). Agricultural air pollution occurs basically in the form of ammonia (NH_3). Agriculture-related emission of ammonia originates from sources, for example, manures, slurries, and fertilizer application (Shoji et al. 2001).

D. *Drawbacks of zinc fertilizers:* Agricultural soils are becoming deficient in micronutrient. Studies show widespread micronutrient deficiency in Indian soils, where about 48% of tested soil were found to be micronutrient deficient (Singh 2009). This is further anticipated to rise from 48% to 63% by the year 2025 (Singh 2009). Zinc deficiency is one of the most prevalent forms of micronutrient deficiency. The soluble zinc sulfate ($ZnSO_4$) is commonly used as fertilizer to enhance plant health and agricultural production, may not help in increasing the plant Zn content especially in the long-term use, because about 1–10% of total obtainable zinc is used by the crop and 90% of applied zinc is converted into distinct unavailable zinc pools by precipitation to carbonates or oxides or phosphates (Ma and Uren 1997; Yasrebi et al. 1994). However, it is always an issue to use chemical zinc fertilizer for long period. Insoluble zinc compounds form in ammonium phosphate fertilizers ($ZnNH_4P0_4$); these reaction products are not obtainable for crops, especially on sandy, neutral to alkaline soils under dry conditions (Mortvedt and Gilkes 1993). Zinc fertilizers in the form of zinc sulfate (White and Broadley 2005) or Zn-EDTA (Karak et al. 2005) have been utilized, but their utilization puts an economical and environmental pressure (Rattan and Shukla 1991). Another concern associated with Zn fertilizer application is its accumulation in soil and its toxicity due to the excess accumulation. However, the zinc toxicity has influence on crop plant and inhibits the growth of soil organisms (Broadley et al. 2007; Alloway 2009). Zinc toxicity is common when the crop plant's growth is limited and also to soil polluted by mining and smelting activities, industrial waste, or treated with high-Zn sewage sludge.

Zinc Solubilizing Microorganisms for Zinc Biofertilizers

The use of microbial inoculants to provide nutrition to plants or to mobilize the nutrients in the soil to make these nutrients available to plants is becoming increasingly popular. Such microbial inoculants are called as biofertilizers and have clear advantages over chemical fertilizers. Biofertilizers can be specifically defined as the biologically active microbial inoculants or formulations containing one or more bacteria or fungi, with convenient and economical carrier materials which add, solubilize, and mobilize nutrients to the soil (Mazid and Khan 2015; Javoreková et al. 2015). When microbes are put in an application to soil, seed, and plant surfaces, it colonize the rhizosphere or inside the plant and encourage growth by rising the accessibility of primary nutrients to the host plant (Mazid and Khan 2015). Biofertilizers help to retain the soil habitat abundant in different types of trace elements and macronutrients via nitrogen fixation, zinc solubilization, phosphate and potassium solubilization or mineralization, liberate plant growth regulating substances, production of antibiotics, and biodegradation of organic matter in the soil (Table 13.3).

In addition to various approaches to fortify crops with Zn like the use of chemical Zn fertilizers and development of Zn efficient crops, the use of microbial inoculants to mobilize Zn in the soil making it available to plants is also recommended. Several studies indicated that bacterial, fungal, and actinomycetes are introduced into the soil and approvingly rise zinc uptake by plants (Anitha et al. 2015; Tariq et al. 2007; Shaikh and Saraf 2017). The use of zinc solubilizing soil microorganisms also known as zinc solubilizing bacteria (ZSB) in the form of biofertilizers (Saravanan et al. 2004) is a better alternative due to its effectiveness, economic, and

Table 13.3 Types of biofertilizers

S.No.	Biofertilizers	Example	Reference
1	Nitrogen fixing	*Rhizobium* sp., *Azotobacter* sp., *Azospirillum*	Kaur & Purewal (2019); Sethi and Adhikary (2012)
2	Zinc solubilizing	*B. subtilis, Thiobacillus thioxidans, Pseudomonas*	Khande et al. (2017); Fasim et al. (2002); Sindhu et al. (2019)
3	Phosphate solubilizing	*Achromobacter, Micrococcus, Aerobacter*	Chen et al. (2006); Rodŕiguez and Fraga (1999)
4	Phosphate mobilizing	*Bacillus* spp. *and Aspergillus* sp.	Schneider et al. (2010); Jungk et al. (1993)
5	Plant growth-promoting	*Rhizobium*	Kaur & Purewal (2019)
6	Azolla and blue green algae	*Azolla, Tolypothrix, Nostoc,*	Yao et al. (2018); Chatterjee et al. (2017)
7	Silicon solubilizing microbes	*Thiobacillus thiooxidans,* and *Bacillus globisporus*	Friedrich et al. (1991); Sheng et al. (2008)

eco-friendly nature. These microbes can solubilize and mobilize unavailable zinc in the soil increasing uptake by plants consequently enhancing the plant growth and yield (Ramesh et al. 2014). Various microorganisms in soil like bacteria and fungi can promote plant growth and are generally referred to as plant growth-promoting rhizobacteria (PGPR). Many of these PGP microorganisms are also known to solubilize Zn and to make it bioavailable to plants.

Fungi like *Aspergillus nomius, A. niger*, and *A. oryzae* can solubilize indissoluble zinc compounds ZnO, $Zn_3(PO_4)_2$ and $ZnCO_3$ through the secretion of gluconic acid and its 2- and 2,5-keto-derivatives decreasing the pH (Dhaked et al. 2017) as given in Table 13.4. *Penicillium bilaji, Aspergillus,* are reported to solubilize ZnO, $ZnCO_3$, and ZnS (Nazir et al. 2016). While *A. niger* is known to produce citric and oxalic acid for the solubilization of insoluble ZnO, $Zn_3(PO_4)_2$, and $Ca_3(PO_4)_2$ (White et al. 1997). Furthermore, *A. terreus* (ZSF-9) is involved in solubilization of ZnO, $ZnCO_3$, $Zn_3(PO_4)_2$ through the production of gluconic acid (Anitha et al. 2015). Various genera of bacteria like *Pseudoumonas* and *Bacillus* are also well known for the solubilization of ZnO, $ZnCO_3$, and ZnS increasing bioavailability for plants (Nazir 2017) (Table 13.4). A number of other bacteria

Table 13.4 Grouping of zinc solubilizers with their characteristics

S.No.	Nature of organisms	Characteristics	Example	References
1	Free-living	IAA production, P solubilization, siderophore production, zinc solubilization	*Bacillus megaterium*	Bhatt and Maheshwari (2020)
2	Symbiotic	P solubilization, Zn solubilization	*Azospirillum with VAM*	Volpin and Kapulnik (1994)
4	Bacteria	IAA production, P and Zn solubilization, siderophore production, organic acid production	*Pseudoumonas, Bacillus Thiobacillus thioxidans*	Saravanan et al. (2007); Hussain et al. (2015)
5	Fungi	Organic acid production, inhibit the growth of plant pathogens, zinc solubilization	*Abisidia cylindrospora, Abisidia glauca, Abisidia spinosa, Aspergillus niger, penicillium simplicissimum*	Dhaked et al. (2017); Saravanan et al. (2007); Coles et al. (1999); Saravanan et al. (2011)
6	Arbuscular mycorrhiza	Zinc solubilization, increase root surface area	*Glomus mosseae*	Li and Christie (2001)
7	Ericoid mycorrhiza	Solubilization of zinc	*Oidiodendron maius, Beauveria caledonica, Hymenoscyphus ericae*	Martino et al. (2003); Saravanan et al. (2011)

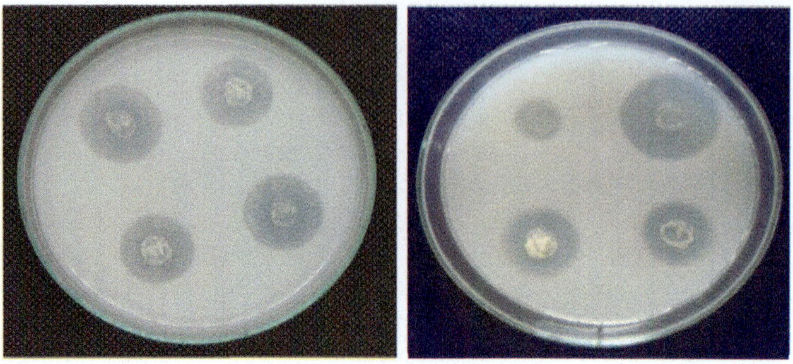

Fig. 13.3 Solubilization of ZnO and $ZnCO_3$ by ZSB (Khan et al., unpublished)

like *Gluconacetobacter diazotrophicus, B. subtilis, Thiobacillus thioxidans, Pseudomonas striata, P. fluoregnence, Burkholderia cenocepacia, Cyanobacteria, Acinetobacter, Serratia liquefaciens*, and *S. marcescens* have also been reported to solubilize insoluble Zn and can be employed as bio-fertilizers (Hussain et al. 2015; Abaid-Ullah et al. 2015; Fasim et al. 2002). The use of zinc solubilizing bacteria (ZSB) along with cheaper insoluble Zn such as ZnO, $ZnCO_3$, and ZnS could be a constructive alternative to high-priced $ZnSO_4$ fertilizers and assist to lower the cost of production of the agricultural commodity therefore can improve agricultural productivity in an economic manner (Mahdi et al. 2010). Use of ZSB is known to promote vegetative growth and productivity of the crops. For example, zinc solubilizing bacteria are reported to improve the development and physiology of maize (Hussain et al. 2015). Its application helps to increase Zn content in different parts of plants like grains. The use of single strain or in combination (multiple-strains) is put in application to the soil to improve zinc concentration in crops, in many cases multiple-strains functions well as compared to isolated strain (Ramesh et al. 2014). According to Yadav and Chandra (2014), the efficient strain for zinc solubilization must have a least 10 mm solubilization zone in guided media having at least 3 mm thickness as shown in Fig. 13.3. Recently many actinobacteria were also demonstrated to have the potential of Zn solubilization. According to Yadav et al. (2018), *Streptomyces* sp. are known to solubilize zinc and also have growth-promoting characters like IAA and siderophore production. Some strains of *Streptomyces* closely related to *S. chattanoogensis, S. rimosus, S. olivaceiscleroticus, S. albofaciens,* and *S. niger* have pathogenic effect against infections caused by soil borne pathogens, namely, *P. capsici* and *S. rolfsii* under in vitro and in plants conditions. *Nocardia* sp. are also known to solubilize different forms of zinc (Mitra et al. 2020).

Mechanisms of Zinc Solubilization by ZSB

The Zn solubilizing bacteria use various mechanisms to solubilize the insoluble form of zinc present in the soil making it bioavailable for plant uptake. Two main mechanisms are production of organic acid and production of chelators (Sindhu et al. 2019) as shown in Fig. 13.5. Few organisms utilize single mechanism, while others may utilize both mechanisms for the solubilization of zinc (Tariq et al. 2007). Microorganisms like VAM induce the change in the root structure, consequently increasing Zn uptake from soil.

A. *Production of organic acid/acidification*: The plant growth-promoting (PGP) bacteria have been accounted to release organic acids and protons, which decreases the pH of the rhizosphere and help in the solubilization of Zn and other metals (Fasim et al. 2002; Wu et al. 2006). Many of ZSB like *Bacillus* sp. and *Pseudomonas* sp. usually produce organic acids which lower soil pH consequently increasing the bioavailability of Zn (Vidyashree et al. 2018). It has been accounted that bioavailability of Zn increases by 100 times with a drop of pH by one unit (Zeb et al. 2018). In a study, it was found that *Pseudomonas fluorescens* secreted 2-ketogluconic acid and gluconic acid for the solubilization of Zn phosphate (Fasim et al. 2002; Saravanan et al. 2011; Di Simine et al. 1998). Microorganisms produce various organic acids like citric, oxalic tartaric, maleic and gluconic acids, etc. Plants take up Zn as (Zn^{2+}) divalent cation. The organic acids released sequester the cations and acidify the micro environment near root for Zn solubilization.
B. *Zinc chelators:* Another mechanism of zinc solubilization is the production of *siderophores* and chelated ligands (Sayer et al. 1997; Wakatsuki 1995; Kamran et al. 2017; Saravanan et al. 2011). Chelators are produced by the plant roots, or by rhizospheric microflora into the soil for the chelation of Zn and other metals, consequently increasing their bioavailability to plants. For example, *Pseudomonas japonica is* a siderophore producing and zinc-solubilizing rhizobacteria (Eshaghi et al. 2019). Chelators present in the soil help to increase the solubility of metal and help to increase the accessibility of trace elements to plants (Mahanty et al. 2017). Chelators bind to the metal ions under high pH conditions and form complex called chelates and help in the availability of several micronutrients like zinc and iron (Rengel et al. 1999). Imran et al., in 2014, have also reported some zinc-chelating compounds that help to elevate the absorption of zinc in the rhizosphere of plants. Some bacteria, for example, *Pseudomonas monteilii, Microbacterium saperdae,* and *Enterobacter cancerogenesis,* have been found to synthesize Zn-chelating metallophores for enhancing plant uptake. Biofertilizer containing the strains of *Pseudomonas* sp. (96–51), *Azospirillum lipoferum* (JCM-1270, ER-20), and *Agrobacterium* sp. (Ca-18) were reported to produce chelating agent ethylene di-amine tetra acetic

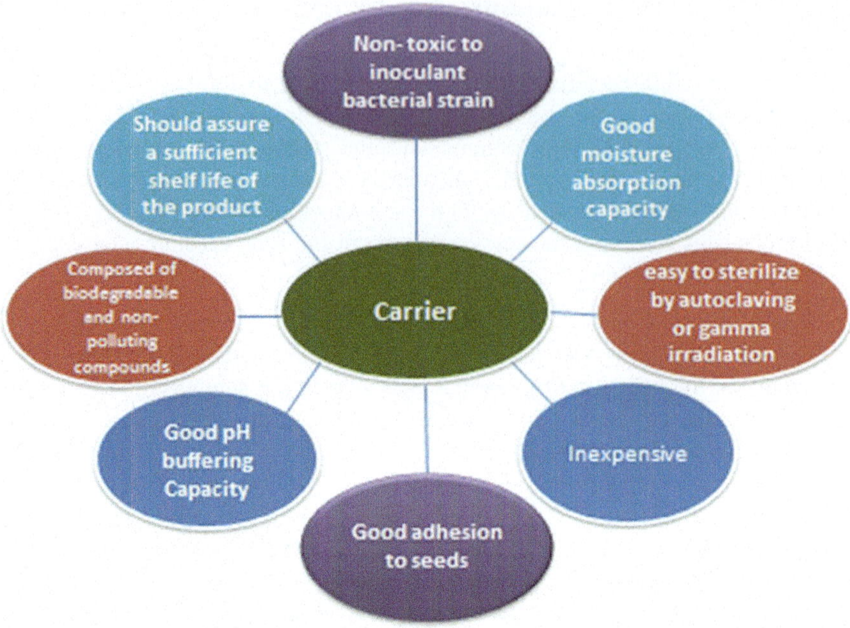

Fig. 13.4 The properties of a superior carrier material (Saif et al. 2021).

acid for releasing fixed insoluble zinc and making it available for prolonged duration (Tariq et al. 2007). Chelators produced by *Penicillium bilaji* increases the accessibility of zinc to plants (Kucey 1988).

C. *Other mechanisms for improving Zn uptake*: As the Zn is consumed by plants through diffusion, its absorption can be increased through increasing the root surface area (Ryan and Angus 2003). Colonization of roots by AM fungi increases the roots surface area which helps plants in nutrients uptake like zinc, copper, manganese, and iron, even from long distances (Adesemoye & Kloepper 2009). Studies have already demonstrated an increase in nutrients uptake due to AM fungi (Fig. 13.5). Elevated Zn assimilation and expansion in vegetative growth of wheat and maize was reported by inoculation of AM fungi and addition of Zn fertilizers in zinc-deficient soils (Kothari et al. 1990).

According to Bürkert and Robson (1994) *arbuscular mycorrhizae* can take zinc from the length of 40 mm from the root surface. Increased uptake of zinc and iron in *Zea mays* due to the AM fungi was also reported (Liu et al. 2000). Infusion of *arbuscular mycorrhizal* in rice roots with fungi was found to promote zinc mobilization and uptake in rice, consequently increasing its growth (Purakayastha and Chhonkar 2001). Likewise, increased uptake of iron, zinc, manganese, and copper was reported in wheat by infusion of *Azospirillum* and *mycorrhizae* in contrast with un-inoculated control plants (Ardakani et al. 2011) (Fig. 13.4).

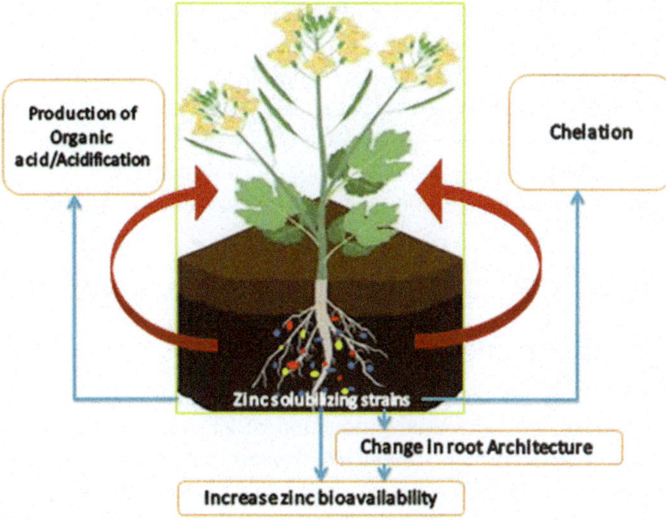

Fig. 13.5 Mechanisms involved for improving zinc bioavailability by zinc solubilizing strains in the rhizosphere of plant (Hussain et al. 2018).

Advantages of Zinc Biofertilizers

Certain microorganisms can fix or mobilize nutrients, or promote plant growth in many other ways and are called as PGPR bacteria. Microbial inoculants based on PGPR are being used in agriculture to promote plant growth and to increase agricultural productivity. These microbial inoculants or biofertilizers are of paramount significance in integrated nutrient management for sustaining agricultural productivity and environmental health (Adesemoye et al. 2009). The utilization of biofertilizers has a definite benefit over chemical fertilizer in addition to increasing soil fertility and bioavailability of nutrients to plants (Mishra and Dash 2014). The plant growth-promoting rhizobacteria and arbuscular mycorrhizal can help to extend the nutrient use efficiency of fertilizers (Mathimaran et al. 2020).

Similarly, the use of Zn biofertilizers is also advantageous over the use of chemical Zn fertilizers. Zinc biofertilizers are ecofriendly, non-toxic, economic, and easy to use (Mazid and Khan 2015; Venkataraman and Shanmugasundaram 1992). Long-term use of zinc biofertilizers is known to maintain the soil fertility and agricultural productivity in a sustainable manner (Mahanty et al. 2017). Other additional benefits of zinc solubilizing bacteria and other biofertilizers include the production of substances like plant growth hormones and mineralization of complex organic molecules in soil (Kamran et al. 2017; Rashid et al. 2016; Hayat et al. 2010). For example, *Azospirillum* is known to secrete gibberellins, ethylene, and auxins (Perrig et al. 2007). *Agrobacterium* and *Rhizobium* are able to solubilize insoluble zinc compound into a soluble form (Khanghahi et al. 2018). Various studies have demonstrated the efficacy of Zn biofertilizers on various crops. Bacterial strains of *Acinetobacter* sp., *Burkholderia* sp., *E. cloacae,* and *Pseudomonas fragi* were shown to improve zinc mobilization in soil consequently increasing the agricultural productivity in crops like rice (Kamran et al. 2017; Vaid et al. 2014). These Zn solubilizing bacteria were also shown to possess additional PGP activities like a strain of *Gluconacetobacter diazotrophicus* showed anti-nematode activity against *Meloidogyne incognita* in addition to zinc solubilization (Saravanan et al. 2007) (Fig. 13.5).

Commercial Production of Zn Biofertilizer

A. *Strain selection*: For industrial production of Zn, biofertilizer based on ZSB would require effective Zn solubilizing bacteria, appropriate growth medium, suitable carrier material, packaging, and quality control facilities as shown in Fig. 13.6. Effectiveness of Zn solubilizing bacteria can be checked under laboratory using basal medium containing insoluble Zn. Generally, strain considered as efficient strain for zinc solubilization must have a minimum 10 mm zone of solubilization in prescribed media (Yadav and Chandra 2014) (Fig. 13.3). It will always be preferable that a candidate strain possesses other PGPR activity in addition to the ability to solubilize Zn. It would also be desirable that the strain has the ability to survive in soil and to compete with other soil microflora. Furthermore, different strains may have different activities for different crops and soils. Therefore, strains meant for different crops and soils may also be different.
B. *Growing selected strain in appropriate medium*: Growth medium should be an economic medium free of toxic substances. For laboratory scale production the pure cultures of ZSB are transferred to sterile medium (~5 ml). And a culture grown to log phase is then transferred to a conical flask containing1000 ml of the sterile medium. Or it can be grown in a laboratory scale fermenter under appropriate conditions. Enough cells should be grown so that while making fertilizer it is to be mixed with sterile carrier. Generally, the population of cells should be up to 10^9 cells/ml.
C. Selection *of carrier and mixing with the cells*: Carrier material allows manageable, high efficacy, and durable storage of biofertilizers. Carrier should be an economic, easily available, buffering, inert, nonpolluting, and easy to sterilize material (Herrmann and Lesueur 2013; Sahu and Brahmaprakash 2016; Malusá et al. 2012) (Fig. 13.4). Different types of carrier materials are used for the industrial production of the biofertilizer. Different soil materials are used to prepare dry inoculants (coal, clays, and peat), organic materials or inert materials (e.g., vermiculite, perlite, and silicates) (Malusá et al. 2012). The liquid biofertilizers contain preferred micro-organisms and chemicals that help in the production of spores and cysts, hence results in the longer duration of biofertilizers (Brahmaprakash and Sahu 2012). Liquid inoculants can be based on broth

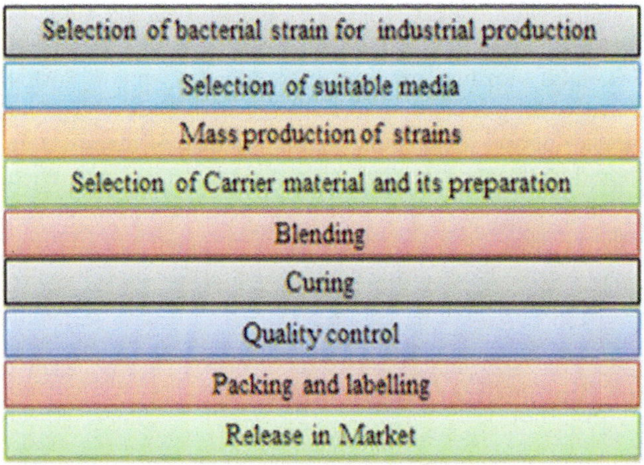

Fig. 13.6 Steps involved in biofertilizer production

cultures, mineral or organic oils, or on oil-in-water suspensions. Generally. powder, beads, or granules are used as solid carriers. The size of granules and beads ranges from 100–200 μm to 3–4 mm in diameter (Malusá et al. 2012). Finally, the presterilized carrier material is mixed with medium containing cells in a ratio of 2:1. The carrier can be sterilized using gamma irradiations (Abd El-Fattah et al. 2013) (Fig. 13.4). When carrier material mixed with broth culture it is suggested to keep it for 24 hours for curing under aseptic conditions. This "curing" (maturation) is known to improve the quality of the product. During this time the microorganisms of interest get acclimatized with the carrier. The biofertilizers can be packed in polythene bags of various weights. These bags can be stored for a week at ambient temperature ranging from 25 to 30 °C. During this period the bacteria may grow to attain the required number. The prepared biofertilizer can be stored appropriately and can be subjected to quality control as shown in (Fig. 13.6).

D. *Limitations and problems associated with biofertilizer production*: Although the use of biofertilizers is the best option of sustainable nutrient management in soil and their advantage over chemical fertilizers is unquestionable. However, still there are some limitations for example, it would require a minimum set up with a well-qualified microbiologist and other skilled staff (Bagyaraj 2004; Debnath et al. 2019). However, it is to be noted that the production of chemical fertilizers would need even more costlier set up with skilled worker. Further research is required to search and validate different strains for different crops, soil and climatic conditions and for their accurate identification and preservation. Development of appropriate carrier materials also needs the research inputs. A wider awareness campaign among farmers makes farmers more confident in using biofertilizers.

E. *Quality control*: The biofertilizers should meet certain quality control parameters as shown in Fig. 13.6, some standards. For example, it contains 10^7 viable cells per gram of fertilizer. The prepared fertilizer should not contain any contaminant even at a dilution of 10^5. The pH of the fertilizer should be in a range of 6.5–7.0 and a moisture percent of 30–40% by weight. Powdered material (particle size) shall pass through 0.15–0.212 mm IS sieve. The inoculum must have an expiry period of 12 months from the date of manufacture. The stored strains should be checked regularly to avoid natural mutants. Each packet containing the biofertilizer should be marked with the information, for example, name of the product, crop for which intended, name and address of manufacturer, type of carrier, batch or manufacture no, expiry date.

Methods of Biofertilizer Application

Various methods are used for the application of biofertilizers, such as seed treatment, seedling root dip, and field application. The choice of method will vary with crop and other factors. Some of the methods are discussed below. In this respect, the carrier should be in the form of fine powder. The standard size of the powder material varies from 75 μm to 0.25 mm (Malusá et al. 2012). To attain compact covering of inoculant on seed surface, use of adhesive such as gum arabic, methylethylcellulose, sucrose solutions, and vegetable oils is generally used (Zaidi et al. 2014).

(A) *Seed inoculation/treatment*: The "seed coating" or "seed inoculation" method is the most widely used method of biofertilizer application. In this method, the fertilizer is mixed with solution containing adhesive material like jaggery to prepare a slurry. Finally, the seeds to be coated are mixed with the slurry under controlled conditions to make a thin coat on the surface of the seed. The seeds are then dried under shade and in conditions to minimize any contamination. The method is recommended for cereals like wheat, rice, maize, and sorghum; oilseeds like groundnut, sunflower, mustard, and safflower; and pulses like cowpea, green gram, black gram, and soybean (Tripti et al. 2015). One packet (200 g) is adequate to treat 10–12 kg seed (Tripti et al. 2015).

(B) *Seedling root treatment*: Another *commonly* used method for the application of biofertilizer is seedling root treatment. This method is recommended for crops like paddy, tobacco, tomato, chilly, onion, cabbage, and cauliflower (Tripti et al. 2015). In this method, the seedling roots of the transplanted crops are dipped in the solution of biofertilizer for 30 minutes. After this treatment, the seedlings are transplanted.

(C) *Direct field application*: In *this* method, the biofertilizer can be added directly to the soil around each sapling. Biofertilizer can be directly added to soil also but for that ten times higher quantities will be required.

Lack of Awareness on Zinc Malnutrition

The awareness regarding the zinc malnutrition among the public and farmers is still lacking, even though an estimated 800,000 people die each year due to inadequate zinc, mostly under the age of five (Das et al. 2016). It was reported that 17.3% of world's population are suffering from zinc deficiency (Wessells and Brown 2012). Some global initiatives by IZiNCG (International Zinc Nutrition Consultative Group) and WHO to highlight the importance of the Zn for human and plant health have already been started. In India, Niti Ayog in 2017 recommends the proper uptake of zinc and other nutrients for proper health of children and women. A bigger outreach program connecting government, health workers, nutrionist, scientists, and farmers is required to achieve the target, where health workers will spread awareness among the high-risk groups, that is, expecting and nursing mothers, children, and elderly. Government should engage in promoting the interest of scientists, farmers, and public for combating the problem of Zn deficiency. Farmers should try to improve the Zn status of crops through various fortification programs. Nutritionist should design and modify diets to increase the Zn content of the diet.

Conclusion

Zinc malnutrition is common among children and is one of the most important issues of the developing world. Among various approaches to increase the Zn content of the crop is the agronomic fortification of Zn. The use of microbial inoculants to improve the Zn status of the crop and cereal grains is one of the easiest, eco-friendly approaches for the mobilization of Zn and making it available to plants. As the use of Zn chemical fertilizers has many inherent problems like a loss in soil fertility, affecting human health, beneficial microbes, and the environment, plant growth-promoting strains are a better alternative. The inoculation of zinc solubilizing bacteria through seed or soil increases plant growth, soil fertility, crop production, and provides protection against plant pathogens. Hence, the use of zinc biofertilizer is a safe and beneficial approach for enhancing crop yields in agriculture. The farmers must be informed about this approach. As we know there is a huge gap between scientists and farmers which needs to be bridged, awareness about the zinc biofertilizer technology should be given to farmers.

Acknowledgments The authors acknowledge the generous financial help provided by the MHRD, Govt. of India, under the Scheme for Promotion of Academic and Research Collaboration (SPARC; P594).

References

Abaid-Ullah, M., Nadeem, M., Hassan, M., Ganter, J., Muhammad, B., Nawaz, K., ...& Hafeez, F. Y.: Plant growth promoting rhizobacteria: an alternate way to improve yield and quality of wheat (Triticum aestivum). Int. J. Agric. Biol. **17**(1), 51–60 (2015)

Abd El-Fattah, D.A., Eweda, W.E., Zayed, M.S., Hassanein, M.K.: Effect of carrier materials, sterilization method, and storage temperature on survival and biological activities of Azotobacter chroococcum inoculant. Ann. Agric. Sci. **58**(2), 111–118 (2013)

Adesemoye, A.O., Kloepper, J.W.: Plant–microbes interactions in enhanced fertilizer-use efficiency. Appl. Microbiol. Biotechnol. **85**(1), 1–12 (2009)

Adesemoye, A.O., Torbert, H.A., Kloepper, J.W.: Plant growth-promoting rhizobacteria allow reduced application rates of chemical fertilizers. Microb. Ecol. **58**(4), 921–929 (2009)

Ahmad, N.: Occurrence and distribution of Vertisols. In Developments in soil science, **24**, pp. 1–41. Elsevier (1996)

Alexandratos, N., Bruinsma, J.: World agriculture towards 2030/2050: the 2012 revision. ESA Working paper No. 12-03. Rome, FAO (2012)

Alloway, B.J.: Micronutrients and crop production: an introduction. In: Micronutrient deficiencies in global crop production, pp. 1–39. Springer, Dordrecht (2008a)

Alloway, B.J.: Zinc in Soils and Crop Nutrition. 2nd Edition, IZA and IFA, Brussels, Belgium and Paris, France (2008b)

Alloway, B.J.: Soil factors associated with zinc deficiency in crops and humans. Environ. Geochem. Health. **31**(5), 537–548 (2009)

Anitha, S., Devi, S.N.P., Kumari, K.S.: Isolation and identification of zinc solubilizing fungal isolates from agricultural fields. Indian J. Agric. Sci. **85**(12), 1638–1642 (2015)

Ardakani, M.R., Mazaheri, D., Mafakheri, S., Moghaddam, A.: Absorption efficiency of N, P, K through triple inoculation of wheat (Triticum aestivum L.) by Azospirillum brasilense, Streptomyces sp., Glomus intraradices and manure application. Physiol. Mol. Biol. Plants. **17**(2), 181–192 (2011)

Bagyaraj, D. J.: Quality control and constraints in biofertilizer production technology. Biofertilizers Technol. 401–408 (2004)

Barak, P., Helmke, P.A.: The chemistry of zinc. In: Zinc in Soils and Plants, pp. 1–13. Springer, Dordrecht (1993)

Bauer, A., Lindsay, W.L.: The effect of soil temperature on the availability of indigenous soil zinc. Soil Sci. Soc. Am. J. **29**(4), 413–416 (1965)

Bhatt, K., Maheshwari, D.K.: Zinc solubilizing bacteria (Bacillus megaterium) with multifarious plant growth promoting activities alleviates growth in Capsicum annuum L. 3 Biotech. **10**(2), 36 (2020)

Bhandari, M. S., Shankhwar, R., Maikhuri, S., Pandey, S., Meena, R. K., Ginwal, H. S., ... & Silveira, L. H. C.: Prediction of ecological and geological niches of Salvadora oleoides in arid zones of India: causes and consequences of global warming. Arabian Journal of Geosciences, **14**(6), 1–18 (2021)

Bhowmik, D., Chiranjib, K.P., Kumar, S.: A potential medicinal importance of zinc in human health and chronic. Int. J. Pharm. **1**(1), 05–11 (2010)

Bin, B.H., Lee, S.H., Bhin, J., Irie, T., Kim, S., Seo, J., et al.: The epithelial zinc transporter ZIP 10 epigenetically regulates human epidermal homeostasis by modulating histone acetyltransferase activity. Br. J. Dermatol. **180**(4), 869–880 (2019)

Binkley, D., & Fisher, R. F.: Ecology and management of forest soils. (Fifth edition, pp.1–435). John Wiley & Sons. USA (2019)

Black, R.: Micronutrient deficiency: an underlying cause of morbidity and mortality. Bull. World Health Organ, **81**(2), 79. (2003). PMID: 12751414, PMCID: PMC2572405

Brahmanand, P.S., Kumar, A., Ghosh, S., Chowdhury, S.R., Singandhupe, R.B., Singh, R., et al.: Challenges to food security in India. Curr. Sci., 841–846 (2013)

Brahmaprakash, G.P., Sahu, P.K.: Biofertilizers for sustainability. J. Indian Institut. Sci. **92**(1), 37–62 (2012)

Branca, F., Ferrari, M.: Impact of micronutrient deficiencies on growth: the stunting syndrome. Ann. Nutr. Metab. **46**(Suppl. 1), 8–17 (2002)

Broadley, M.R., White, P.J., Hammond, J.P., Zelko, I., Lux, A.: Zinc in plants. New Phytol. **173**, 677–702 (2007)

Bürkert, B., Robson, A.: 65Zn uptake in subterranean clover (Trifolium subterraneum L.) by three vesicular-arbuscular mycorrhizal fungi in a root-free sandy soil. Soil Biol. Biochem. **26**(9), 1117–1124 (1994)

Cakmak, I.: Tansley Review No. 111: possible roles of zinc in protecting plant cells from damage by reactive oxygen species. New Phytol. **146**(2), 185–205 (2000)

Cakmak, I.: Enrichment of cereal grains with zinc: agronomic or genetic biofortification? Plant Soil. **302**(1–2), 1–17 (2008)

Cakmak, I.: Enrichment of fertilizers with zinc: an excellent investment for humanity and crop production in India. J. Trace Elem. Med. Biol. **23**(4), 281–289 (2009)

Chandini, K. R., Kumar, R., & Prakash, O.: The Book Research Trends in Environmental Sciences. The Impact of Chemical Fertilizers on Our Environment and Ecosystem. Edition: 2 nd, 69–86 (2019)

Chatterjee, A., Singh, S., Agrawal, C., Yadav, S., Rai, R., & Rai, L. C.: Role of algae as a biofertilizer. In Algal Green Chemistry (pp. 189–200). Elsevier, India (2017)

Chen, Y.P., Rekha, P.D., Arun, A.B., Shen, F.T., Lai, W.A., Young, C.C.: Phosphate solubilizing bacteria from subtropical soil and their tricalcium phosphate solubilizing abilities. Appl. Soil Ecol. **34**(1), 33–41 (2006)

Coles, K.E., David, J.C., Fisher, P.J., Lappin-Scott, H.M., Macnair, M.R.: Solubilisation of zinc compounds by fungi associated with the hyperaccumulator Thlaspi caerulescens. Bot. J. Scotl. **51**(2), 237–247 (1999)

Cooper, J., Reed, E.Y., Hörtenhuber, S., Lindenthal, T., Løes, A.K., Mäder, P., et al.: Phosphorus availability on many organically managed farms in Europe. Nutr. Cycl. Agroecosyst. **110**(2), 227–239 (2018)

Das, S., Green, A.: Importance of zinc in crops and human health. J SAT Agric Res, **11**, 1–7 (2013)

Das, S., Green, A.: Zinc in crops and human health. In: Biofortification of Food Crops, pp. 31–40. Springer, New Delhi (2016)

Debnath, S., Rawat, D., Mukherjee, A. K., Adhikary, S., & Kundu, R.: Applications and constraints of plant beneficial microorganisms in agriculture. In Biostimulants in plant science. IntechOpen, Belgium (2019)

Devi, S.P., Vasandha, S.: Assessment of competence of the Pseudomonas aeruginosa to solubilize insoluble form of zinc under various cultural parameters. Arab. J. Sci. Eng. **41**(6), 2117–2121 (2016)

Dhaked, B.S., Triveni, S., Reddy, R.S., Padmaja, G.: Isolation and screening of potassium and zinc solubilizing bacteria from different rhizosphere soil. Int. J. Curr. Microbiol. App. Sci. **6**(8), 1271–1281 (2017)

Di Simine, C.D., Sayer, J.A., Gadd, G.M.: Solubilization of zinc phosphate by a strain of Pseudomonas fluorescens isolated from a forest soil. Biol. Fertil. Soils. **28**(1), 87–94 (1998)

Du, S.T., Zhang, Y.S., Lin, X.Y.: Accumulation of nitrate in vegetables and its possible implications to human health. Agric. Sci. China. **6**(10), 1246–1255 (2007)

El-Saadony, M.T.A.M.: Applications of Biofertilization and Foliar Feeding in Pea Production (Doctoral dissertation, Faculty of Agriculture, Benha University) (2015)

Eshaghi, E., Nosrati, R., Owlia, P., Malboobi, M.A., Ghaseminejad, P., Ganjali, M.R.: Zinc solubilization characteristics of efficient siderophore-producing soil bacteria. Iranian J. Microbiol. **11**(5), 419 (2019)

FAO.: World Fertilizer Trends and Outlook to 2020: Summary Report (2017)

Fasim, F., Ahmed, N., Parsons, R., Gadd, G.M.: Solubilization of zinc salts by a bacterium isolated from the air environment of a tannery. FEMS Microbiol. Lett. **213**(1), 1–6 (2002)

Friedrich, S., Platonova, N.P., Karavaiko, G.I., Stichel, E., Glombitza, F.: Chemical and microbiological solubilization of silicates. Acta Biotechnol. **11**(3), 187–196 (1991)

Gómez-Galera, S., Rojas, E., Sudhakar, D., Zhu, C., Pelacho, A.M., Capell, T., Christou, P.: Critical evaluation of strategies for mineral fortification of staple food crops. Transgenic Res. **19**(2), 165–180 (2010)

Graham, R.D.: Micronutrient deficiencies in crops and their global significance. In: Micronutrient deficiencies in global crop production, pp. 41–61. Springer, Dordrecht (2008)

Grewal, H.S., Graham, R.D., Rengel, Z.: Genotypic variation in zinc efficiency and resistance to crown rot disease (Fusarium graminearum Schw. Group 1) in wheat. Plant Soil. **186**(2), 219–226 (1996)

Gurmani, A.R., Khan, S.U., Andaleep, R., Waseem, K., Khan, A.: Soil application of zinc improves growth and yield of tomato. Int. J. Agric. Biol. **14**(1), 91–96 (2012)

Hafeez, B., Khanif, Y.M., Saleem, M.: Role of Zinc in plant nutrition-a review. J. Exp. Agric. Int. **50**, 374–391 (2013)

Hajiboland, R., Amirazad, F.: Growth, photosynthesis and antioxidant defense system in Zn-deficient red cabbage plants. Plant Soil Environ. **56**(5), 209–217 (2010)

Hambidge, K.M., Krebs, N.F.: Zinc deficiency: a special challenge. J. Nutr. **137**(4), 1101–1105 (2007)

Hayat, R., Ali, S., Amara, U., Khalid, R., Ahmed, I.: Soil beneficial bacteria and their role in plant growth promotion: a review. Ann. Microbiol. **60**(4), 579–598 (2010)

Herrmann, L., Lesueur, D.: Challenges of formulation and quality of biofertilizers for successful inoculation. Appl. Microbiol. Biotechnol. **97**(20), 8859–8873 (2013)

Huang, J., Pray, C., Rozelle, S.: Enhancing the crops to feed the poor. Nature. **418**(6898), 678–684 (2002)

Huang, J., Hu, R., Cao, J., Rozelle, S.: Training programs and in-the-field guidance to reduce China's overuse of fertilizer without hurting profitability. J. Soil Water Conservat. **63**(5), 165A–167A (2008)

Humayan Kabir, A., Swaraz, A.M., Stangoulis, J.: Zinc-deficiency resistance and biofortification in plants. J. Plant Nutr. Soil Sci. **177**(3), 311–319 (2014)

Hussain, S., Maqsood, M.A., Rengel, Z., Aziz, T.: Biofortification and estimated human bioavailability of zinc in wheat grains as influenced by methods of zinc application. Plant Soil. **361**(1–2), 279–290 (2012)

Hussain, A., Arshad, M., Zahir, Z. A., & Asghar, M.: Prospects of Zinc solubilizing bacteria for enhancing growth of maize. Pak. J. Agric. Sci. **52**(4), 915–922 (2015)

Hussain, A., Zahir, Z. A., Asghar, H. N., Ahmad, M., Jamil, M., Naveed, M., & Akhtar, M. F. U. Z.: Zinc solubilizing bacteria for zinc biofortification in cereals: a step toward sustainable nutritional security. In Role of rhizospheric microbes in soil (pp. 203–227). Springer, Singapore (2018)

Imran, M., Arshad, M., Khalid, A., Kanwal, S., Crowley, D.E.: Perspectives of rhizosphere microflora for improving Zn bioavailability and acquisition by higher plants. Int. J. Agric. Biol. **16**(3), 653–662 (2014)

Javoreková, S., Maková, J., Medo, J., Kovácsová, S., Charousová, I., Horák, J.: Effect of biofertilizers application on microbial diversity and physiological profiling of microorganisms in arable soil. Eurasian J. Soil Sci. **4**(1), 54 (2015)

Jeffery, J.J., Uren, N.C.: Copper and zinc species in the soil solution and the effects of soil pH. Soil Res. **21**(4), 479–488 (1983)

Jungk, A., Seeling, B., Gerke, J.: Mobilization of different phosphate fractions in the rhizosphere. In: Plant Nutrition—From Genetic Engineering to Field Practice, pp. 95–98. Springer, Dordrecht (1993)

Kamran, S., Shahid, I., Baig, D.N., Rizwan, M., Malik, K.A., Mehnaz, S.: Contribution of zinc solubilizing bacteria in growth promotion and zinc content of wheat. Front. Microbiol. **8**, 2593 (2017)

Karak, T., Singh, U.K., Das, S., Das, D.K., Kuzyakov, Y.: Comparative efficacy of ZnSO4 and Zn-EDTA application for fertilization of rice (Oryza sativa L.). Arch. Agron. Soil Sci. **51**(3), 253–264 (2005)

Katona, P., Katona-Apte, J.: The interaction between nutrition and infection. Clin. Infect. Dis. **46**(10), 1582–1588 (2008)

Kaur, P., Purewal, S.S.: Biofertilizers and their role in sustainable agriculture. In: Biofertilizers for Sustainable Agriculture and Environment, pp. 285–300. Springer, Cham (2019)

Kennedy, G., Nantel, G., Shetty, P.: The scourge of "hidden hunger": global dimensions of micronutrient deficiencies. Food Nutrit. Agric. **32**, 8–16 (2003)

Khalid, N., Ahmed, A., Bhatti, M.S., Randhawa, M.A., Ahmad, A., Rafaqat, R.: A question mark on zinc deficiency in 185 million people in Pakistan—possible way out. Crit. Rev. Food Sci. Nutr. **54**(9), 1222–1240 (2014)

Khande, R., Sharma, S.K., Ramesh, A., Sharma, M.P.: Zinc solubilizing Bacillus strains that modulate growth, yield and zinc biofortification of soybean and wheat. Rhizosphere. **4**, 126–138 (2017)

Khanghahi, M.Y., Ricciuti, P., Allegretta, I., Terzano, R., Crecchio, C.: Solubilization of insoluble zinc compounds by zinc solubilizing bacteria (ZSB) and optimization of their growth conditions. Environ. Sci. Pollut. Res. **25**(26), 25862–25868 (2018)

Kidd, M.T., Ferket, P.R., Qureshi, M.A.: Zinc metabolism with special reference to its role in immunity. Worlds Poult. Sci. J. **52**(3), 309–324 (1996)

Kiekens, L.: Zinc in heavy metals. In: Soils. Blackie Academic and Professional, London (1995)

Kirk, G.J.D., Bajita, J.B.: Root-induced iron oxidation, pH changes and zinc solubilization in the rhizosphere of lowland rice. New Phytol. **131**(1), 129–137 (1995)

Kothari, S.K., Marschner, H., George, E.: Effect of VA mycorrhizal fungi and rhizosphere microorganisms on root and shoot morphology, growth and water relations in maize. New Phytol. **116**(2), 303–311 (1990)

Kour, D., Rana, K. L., Yadav, A. N., Yadav, N., Kumar, M., Kumar, V., ... & Saxena, A. K.: Microbial biofertilizers: Bioresources and eco-friendly technologies for agricultural and environmental sustainability. Biocatalysis and Agricultural Biotechnology, **23**, 101487 (2020)

Kucey, R. M. N.: Effect of Penicillium bilaji on the solubility and uptake of P and micronutrients from soil by wheat. Canadian Journal of Soil Science, **68**(2), 261–270 (1988)

Li, X., Christie, P.: Changes in soil solution Zn and pH and uptake of Zn by arbuscular mycorrhizal red clover in Zn-contaminated soil. Chemosphere. **42**(2), 201–207 (2001)

Liu, A., Hamel, C., Hamilton, R.I., Ma, B.L., Smith, D.L.: Acquisition of Cu, Zn, Mn and Fe by mycorrhizal maize (Zea mays L.) grown in soil at different P and micronutrient levels. Mycorrhiza. **9**(6), 331–336 (2000)

Loneragan, J.F., Webb, M.J.: Interactions between zinc and other nutrients affecting the growth of plants. In: Zinc in Soils and Plants, pp. 119–134. Springer, Dordrecht (1993)

Ma, Y.B., Uren, N.C.: The fate and transformations of zinc added to soils. Soil Res. **35**(4), 727–738 (1997)

Ma, G., Jin, Y., Li, Y., Zhai, F., Kok, F.J., Jacobsen, E., Yang, X.: Iron and zinc deficiencies in China: what is a feasible and cost-effective strategy? Public Health Nutr. **11**(6), 632–638 (2008)

Mahal, N.K., Osterholz, W.R., Miguez, F.E., Poffenbarger, H.J., Sawyer, J.E., Olk, D.C., et al.: Nitrogen fertilizer suppresses mineralization of soil organic matter in maize agroecosystems. Front. Ecol. Evol. **7**, 59 (2019)

Mahanty, T., Bhattacharjee, S., Goswami, M., Bhattacharyya, P., Das, B., Ghosh, A., Tribedi, P.: Biofertilizers: a potential approach for sustainable agriculture development. Environ. Sci. Pollut. Res. **24**(4), 3315–3335 (2017)

Mahata, M. K., Debnath, P., & Ghosh, S. K.: Estimation of critical limit of zinc for rice in Terai soils of West Bengal. Journal of the Indian Society of Soil Science, **61**(2), 153–157 (2013)

Mahdi, S. S., Hassan, G. I., Samoon, S. A., Rather, H. A., Dar, S. A., & Zehra, B.: Bio-fertilizers in organic agriculture. J. Phytol. **2**(10), 42–54 (2010)

Malusá, E., Sas-Paszt, L., Ciesielska, J.: Technologies for beneficial microorganisms inocula used as biofertilizers. Sci. World J. **2012**, 491206 (2012)

Mandal, B., Hazra, G.C.: Zinc adsorption in soils as influenced by different soil management practices. Soil Sci. **162**(10), 713–721 (1997)

Martino, E., Perotto, S., Parsons, R., Gadd, G.M.: Solubilization of insoluble inorganic zinc compounds by ericoid mycorrhizal fungi derived from heavy metal polluted sites. Soil Biol. Biochem. **35**(1), 133–141 (2003)

Mathimaran, N., Jegan, S., Thimmegowda, M. N., Prabavathy, V. R., Yuvaraj, P., Kathiravan, R., ... & Mäder, P.: Intercropping transplanted pigeon pea with finger millet: Arbuscular mycorrhizal fungi and plant growth promoting rhizobacteria boost yield while reducing fertilizer input. Frontiers in Sustainable Food Systems, **4**, 88 (2020)

Mazid, M., & Khan, T. A.: Future of bio-fertilizers in Indian agriculture: an overview. Int. J. Agric. Food. Res. **3**(3), 10–23 (2015)

McBride, M.B., Blasiak, J.J.: Zinc and copper solubility as a function of pH in an acid soil. Soil Sci. Soc. Am. J. **43**(5), 866–870 (1979)

McCall, K.A., Huang, C.C., Fierke, C.A.: Function and mechanism of zinc metalloenzymes. J. Nutr. **130**(5), 1437S–1446S (2000)

Meena, V.S., Meena, S.K., Verma, J.P., Kumar, A., Aeron, A., Mishra, P.K., et al.: Plant beneficial rhizospheric microorganism (PBRM) strategies to improve nutrients use efficiency: a review. Ecol. Eng. **107**, 8–32 (2017)

Mishra, P., Dash, D.: Rejuvenation of biofertilizer for sustainable agriculture and economic development. Consilience. **11**, 41–61 (2014)

Mitra, D., Ganeshamurthy, A. N., Sharma, K., Radha, T. K., & Rupa, T. R.: Soil physico-chemical properties and In-Vitrozinc solubilization of rhizo-microbes from the different rhizospheric soil samples of harsh environment. Annals. Food Sci Technol. **21**(1), 226–232 (2020)

Mori, A., Kirk, G.J., Lee, J.S., Morete, M.J., Nanda, A.K., Johnson-Beebout, S.E., Wissuwa, M.: Rice genotype differences in tolerance of zinc-deficient soils: evidence for the importance of root-induced changes in the rhizosphere. Front. Plant Sci. **6**, 1160 (2016)

Mortvedt, J.J.: Plant uptake of heavy metals in zinc fertilizers made from industrial by-products. J. Environ. Qual. **14**(3), 424–427 (1985)

Mortvedt, J.J., Gilkes, R.J.: Zinc fertilizers. In: Zinc in Soils and Plants, pp. 33–44. Springer, Dordrecht (1993)

Mousavi, S.R.: Zinc in crop production and interaction with phosphorus. Aust. J. Basic Appl. Sci. **5**(9), 1503–1509 (2011)

Nazir, Q.: Prespectives of using Zinc impregnated urea for improving yield and Biofortification of Zinc in cereals (Doctoral dissertation, University of Agriculture, Faisalabad) (2017)

Nazir, Q., Hussain, A., Imran, M., Mahmood, S., Ahmad, M., Iqbal, M.M.: Zinc biofortification of cereals through fertilizers: recent advances and future perspectives. Asian J. Agric. Biol. **4**, 140–152 (2016)

Nene, Y. Á.: Symptoms, cause and control of Khaira disease of paddy. Bull Indian Phytopathol Soc, **3**, 97–191 (1966)

Rani, N., Kaur, R., Kaur, S.: Zinc solubilising bacteria to augment soil fertility–a comprehensive review. Int. J. AgriSci. Vet. Med. **8**(1), 38–44 (2020)

Noulas, C., Tziouvalekas, M., Karyotis, T.: Zinc in soils, water and food crops. J. Trace Elem. Med. Biol. **49**, 252–260 (2018)

Patnaik, M.C., Raju, A.S., Raj, G.B.: Effect of soil moisture regimes on zinc availability in a red sandy loam soil of Andhra Pradesh. J. Indian Soc. Soil Sci. **56**(4), 452–453 (2008)

Perrig, D., Boiero, M.L., Masciarelli, O.A., Penna, C., Ruiz, O.A., Cassán, F.D., Luna, M.V.: Plant-growth-promoting compounds produced by two agronomically important strains of Azospirillum brasilense, and implications for inoculant formulation. Appl. Microbiol. Biotechnol. **75**(5), 1143–1150 (2007)

Prasad, A.S.: Discovery of human zinc deficiency: its impact on human health and disease. Adv. Nutr. **4**(2), 176–190 (2013)

Purakayastha, T.J., Chhonkar, P.K.: Influence of vesicular-arbuscular mycorrhizal fungi (Glomus etunicatum L.) on mobilization of zinc in wetland rice (Oryza sativa L.). Biol. Fertil. Soils. **33**(4), 323–327 (2001)

Ramanamurthy, S. V.: Methemoglobinemia: a reappraisal with an Indian perspective. Hematology. **23**(77), 344–349 (2013)

Ramesh, A., Sharma, S.K., Sharma, M.P., Yadav, N., Joshi, O.P.: Inoculation of zinc solubilizing Bacillus aryabhattai strains for improved growth, mobilization and biofortification of zinc in soybean and wheat cultivated in Vertisols of Central India. Appl. Soil Ecol. **73**, 87–96 (2014)

Rashid, M.I., Mujawar, L.H., Shahzad, T., Almeelbi, T., Ismail, I.M., Oves, M.: Bacteria and fungi can contribute to nutrients bioavailability and aggregate formation in degraded soils. Microbiol. Res. **183**, 26–41 (2016)

Rattan, R.K., Shukla, L.M.: Influence of different Zn carriers on the utilization of micronutrients by rice. J. Indian Soc. Soil Sci. **39**, 808–810 (1991)

Rawat, N., Neelam, K., Tiwari, V.K., Dhaliwal, H.S.: Biofortification of cereals to overcome hidden hunger. Plant Breed. **132**(5), 437–445 (2013)

Reed, S.T., Martens, D.C.: Copper and Zinc. Meth. Soil Anal. Part 3 Chem. Meth. **5**, 703–722 (1996)

Reeves, P.G., Chaney, R.L.: Bioavailability as an issue in risk assessment and management of food cadmium: a review. Sci. Total Environ. **398**(1–3), 13–19 (2008)

Rehman, A., Farooq, M., Ozturk, L., Asif, M., Siddique, K.H.: Zinc nutrition in wheat-based cropping systems. Plant Soil. **422**(1–2), 283–315 (2018)

Rengel, Z., Batten, G. D., & Crowley, D. D.: Agronomic approaches for improving the micronutrient density in edible portions of field crops. Field crops research, **60**(1-2), 27–40 (1999)

Rezaeian, S., Boroon, M.R.: A close study on the nitrate fertilizer use and environmental pollution for human health in Iran. Int. J. Agric. Biosyst. Eng. **12**(3), 69–72 (2018)

Rodríguez, H., Fraga, R.: Phosphate solubilizing bacteria and their role in plant growth promotion. Biotechnol. Adv. **17**(4–5), 319–339 (1999)

Ruel, M.T., Bouis, H.E.: Plant breeding: a long-term strategy for the control of zinc deficiency in vulnerable populations. Am. J. Clin. Nutr. **68**(2), 488S–494S (1998)

Rütting, T., Aronsson, H., Delin, S.: Efficient use of nitrogen in agriculture. Nutr. Cycling Agroecosyst. **110**, 1–5 (2018)

Ryan, M.H., Angus, J.F.: Arbuscular mycorrhizae in wheat and field pea crops on a low P soil: increased Zn-uptake but no increase in P-uptake or yield. Plant Soil. **250**(2), 225–239 (2003)

Sahu, P.K., Brahmaprakash, G.P.: Formulations of biofertilizers–approaches and advances. In: Microbial Inoculants in Sustainable Agricultural Productivity, pp. 179–198. Springer, New Delhi (2016)

Saif, S., Abid, Z., Ashiq, M. F., Altaf, M., & Ashraf, R. S.: Biofertilizer Formulations. Biofertilizers: Study and Impact, 211 (2021)

Sarangi, D. R., Jena, D., & Chatterjee, A. K.: Assessment of critical limit of zinc for rice, groundnut and potato in red and laterite soils of Odisha. ORYZA-An International Journal on Rice, **53**(3), 276–281 (2016)

Saravanan, V.S.: An investigation on solubilization of Zinc and certain other insoluble compounds by Gluconacetobacter diazotrophicus (2004)

Saravanan, V.S., Subramoniam, S.R., Raj, S.A.: Assessing in vitro solubilization potential of different zinc solubilizing bacterial (ZSB) isolates. Braz. J. Microbiol. **35**(1–2), 121–125 (2004)

Saravanan, V.S., Madhaiyan, M., Thangaraju, M.: Solubilization of zinc compounds by the diazotrophic, plant growth promoting bacterium Gluconacetobacter diazotrophicus. Chemosphere. **66**(9), 1794–1798 (2007)

Saravanan, V.S., Kumar, M.R., Sa, T.M.: Microbial zinc solubilization and their role on plants. In: Bacteria in Agrobiology: Plant Nutrient Management, pp. 47–63. Springer, Berlin, Heidelberg (2011)

Savci, S.: An agricultural pollutant: chemical fertilizer. Int. J. Environ. Sci. Develop. **3**(1), 73 (2012)

Sayer, J.A., Kierans, M., Gadd, G.M.: Solubilisation of some naturally occurring metal-bearing minerals, limescale and lead phosphate by Aspergillus Niger. FEMS Microbiol. Lett. **154**(1), 29–35 (1997)

Schneider, K.D., Van Straaten, P., De Orduña, R.M., Glasauer, S., Trevors, J., Fallow, D., Smith, P.S.: Comparing phosphorus mobilization strategies using Aspergillus Niger for the mineral dissolution of three phosphate rocks. J. Appl. Microbiol. **108**(1), 366–374 (2010)

Self, J.R., Waskom, R.M.: Nitrates in drinking water. Service in action; no. 0.517 (1992)

Sethi, S.K., Adhikary, S.P.: Azotobacter: a plant growth-promoting rhizobacteria used as biofertilizer. Dyn. Biochem. Process. Biotechnol. Mol. Biol. **6**(1), 68–74 (2012)

Shaikh, S.S., Saraf, M.S.: Optimization of growth conditions for zinc solubilizing plant growth associated bacteria and fungi. J. Adv. Res. Biotechnol. **2**(1), 1–9 (2017)

Sharma, A., Patni, B., Shankhdhar, D., Shankhdhar, S.C.: Zinc–an indispensable micronutrient. Physiol. Mol. Biol. Plants. **19**(1), 11–20 (2013)

Sheng, X.F., Zhao, F., He, L.Y., Qiu, G., Chen, L.: Isolation and characterization of silicate mineral-solubilizing Bacillus globisporus Q12 from the surfaces of weathered feldspar. Can. J. Microbiol. **54**(12), 1064–1068 (2008)

Shetty, P.: Addressing micronutrient malnutrition to achieve nutrition security. Combating micronutrient deficiencies: Food-based approaches, 28 (2011)

Shivay, Y.S.: Zinc deficiency in humans and its amelioration. Indian J. Community Health. **27**(Supp 1) (2015)

Shoji, S., Delgado, J., Mosier, A., Miura, Y.: Use of controlled release fertilizers and nitrification inhibitors to increase nitrogen use efficiency and to conserve air and water quality. Commun. Soil Sci. Plant Anal. **32**(7–8), 1051–1070 (2001)

Shukla, A. K., & Behra, S. K.: All India coordinated research project on micro-and secondary nutrients and pollutant elements in soils and plants: Research achievements and future thrusts. Indian J Fert, **15**(5), 522–543 (2019)

Shukla, A. K., & Tiwari, P. K.: Micro and secondary nutrients and pollutant elements research in India. Coordinators Report-AICRP on Micro-and Secondary Nutrients and Pollutant Elements in Soils and Plants. ICAR-IISS, Bhopal, 1–196 (2016)

Sillanpää, M.: Micronutrients and the nutrient status of soils: a global study (Vol. 48). Food & Agriculture Org, Rome (1982)

Sillanpää, M.: Micronutrient assessment at the country level: an international study. FAO soils bulletin. (63) (1990)

Sindhu, S.S., Sharma, R., Sindhu, S., Phour, M.: Plant nutrient management through inoculation of zinc-solubilizing Bacteria for sustainable agriculture. In: Biofertilizers for Sustainable Agriculture and Environment, pp. 173–201. Springer, Cham (2019)

Singh, M.V.: Micronutrient deficiencies in crops and soils in India. In: Micronutrient deficiencies in global crop production, pp. 93–125. Springer, Dordrecht (2008)

Singh, M.V.: Micronutrient nutritional problems in soils of India and improvement for human and animal health. Indian J. Fertilis. **5**(4), 11–56 (2009)

Singh, S., & Dubey, R. P.: Zinc, and its essentiality for human health. The Pharma. Innov. Journal. **8**(9), 137–139 (2019)

Singh, M.K., Prasad, S.K.: Agronomic aspects of zinc biofortification in rice (Oryza sativa L.). Proc. Nat. Acad. Sci. India Sect. B Biol. Sci. **84**(3), 613–623 (2014)

Smith, J. L., & Paul, E. A.: The significance of soil microbial biomass estimations. Soil Biochem. **6**, 357–396 (1990)

Sommer, A.L., Lipman, C.B.: Evidence on the indispensable nature of zinc and boron for higher green plants. Plant Physiol. **1**(3), 231 (1926)

Tapiero, H., Tew, K.D.: Trace elements in human physiology and pathology: zinc and metallothioneins. Biomed. Pharmacother. **57**(9), 399–411 (2003)

Tariq, M., Hameed, S., Malik, K.A., Hafeez, F.Y.: Plant root associated bacteria for zinc mobilization in rice. Pak. J. Bot. **39**(1), 245 (2007)

Vaid, S.K., Kumar, B., Sharma, A., Shukla, A.K., Srivastava, P.C.: Effect of Zn solubilizing bacteria on growth promotion and Zn nutrition of rice. J. Soil Sci. Plant Nutr. **14**(4), 889–910 (2014)

van Vliet, J., Eitelberg, D.A., Verburg, P.H.: A global analysis of land take in cropland areas and production displacement from urbanization. Glob. Environ. Chang. **43**, 107–115 (2017)

Venkataraman, G. S., & Shanmugasundaram, S.: Algal biofertilizers technology for rice. DBT Centre for BGA Biofertilizer, Madurai, Kamraj University (TN), India (1992)

Vidyashree, D.N., Muthuraju, R., Panneerselvam, P., Mitra, D.: Organic acids production by zinc solubilizing bacterial isolates. Int. J. Curr. Microbiol. App. Sci. **7**(10), 626–633 (2018)

Volpin, H., Kapulnik, Y.: Interaction of Azospirillum with beneficial soil microorganisms. Azospirillum/plant associations, 111–118 (1994)

Wakatsuki, T.: Metal oxidoreduction by microbial cells. J. Ind. Microbiol. **14**(2), 169–177 (1995)

Ward, M.H., Jones, R.R., Brender, J.D., De Kok, T.M., Weyer, P.J., Nolan, B.T., et al.: Drinking water nitrate and human health: an updated review. Int. J. Environ. Res. Public Health. **15**(7), 1557 (2018)

Welch, R.M., Graham, R.D.: Breeding for micronutrients in staple food crops from a human nutrition perspective. J. Exp. Bot. **55**(396), 353–364 (2004)

Welch, R.M., Graham, R.D.: Agriculture: the real nexus for enhancing bioavailable micronutrients in food crops. J. Trace Elem. Med. Biol. **18**(4), 299–307 (2005)

Welch, R.M., Hart, J.J., Norvell, W.A., Sullivan, L.A., Kochian, L.V.: Effects of nutrient solution zinc activity on net uptake, translocation, and root export of cadmium and zinc by separated sections of intact durum wheat (Triticum turgidum L. var durum) seedling roots. Plant Soil. **208**(2), 243–250 (1999)

Wessells, K.R., Brown, K.H.: Estimating the global prevalence of zinc deficiency: results based on zinc availability in national food supplies and the prevalence of stunting. PLoS One. **7**(11), e50568 (2012)

White, P.J., Broadley, M.R.: Biofortifying crops with essential mineral elements. Trends Plant Sci. **10**(12), 586–593 (2005)

White, J.G., Zasoski, R.J.: Mapping soil micronutrients. Field Crop Res. **60**(1–2), 11–26 (1999)

White, C., Sayer, J.A., Gadd, G.M.: Microbial solubilization and immobilization of toxic metals: key biogeochemical processes for treatment of contamination. FEMS Microbiol. Rev. **20**(3–4), 503–516 (1997)

Wu, S.C., Cheung, K.C., Luo, Y.M., Wong, M.H.: Effects of inoculation of plant growth-promoting rhizobacteria on metal uptake by Brassica juncea. Environ. Pollut. **140**(1), 124–135 (2006)

Wyszkowska, J., Borowik, A., Kucharski, M., Kucharski, J.: Effect of cadmium, copper and zinc on plants, soil microorganisms and soil enzymes. J. Elem. **18**(4), 769–796 (2013)

Yadav, A.K., Chandra, K.: Mass production and quality control of microbial inoculants. Proc. Indian Natn. Sci. Acad. **80**(2), 483–489 (2014)

Yadav, A. N., Verma, P., Kumar, S., Kumar, V., Kumar, M., Sugitha, T. C. K., ... Dhaliwal, H. S.: Actinobacteria from rhizosphere: molecular diversity, distributions, and potential biotechnological applications. In New and future developments in microbial biotechnology and bioengineering (pp. 13–41). Elsevier, Ireland (2018)

Yao, Y., Zhang, M., Tian, Y., Zhao, M., Zeng, K., Zhang, B., et al.: Azolla biofertilizer for improving low nitrogen use efficiency in an intensive rice cropping system. Field Crop Res. **216**, 158–164 (2018)

Yasrebi, J., Karimian, N., Maftoun, M., Abtahi, A., Sameni, A.M.: Distribution of zinc forms in highly calcareous soils as influenced by soil physical and chemical properties and application of zinc sulfate. Commun. Soil Sci. Plant Anal. **25**(11–12), 2133–2145 (1994)

Youssef, M.M.A., Eissa, M.F.M.: Biofertilizers and their role in management of plant parasitic nematodes. A review. J. Biotechnol. Pharm. Res. **5**(1), 1–6 (2014)

Zaidi, A., Khan, M.S., Ahmad, E.: Microphos: principles, production and application strategies. In: Phosphate Solubilizing Microorganisms, pp. 1–30. Springer, Cham (2014)

Zeb, H., Hussain, A., Naveed, M., Ditta, A., Ahmad, S., Jamshaid, M.U., et al.: Compost enriched with ZnO and Zn-solubilising bacteria improves yield and Zn-fortification in flooded rice. Ital. J. Agron. **13**(4), 1295 (2018)

Zhao, A.Q., Tian, X.H., Cao, Y.X., Lu, X.C., Liu, T.: Comparison of soil and foliar zinc application for enhancing grain zinc content of wheat when grown on potentially zinc-deficient calcareous soils. J. Sci. Food Agric. **94**(10), 2016–2022 (2014)

Zhao, A., Wang, B., Tian, X., Yang, X.: Combined soil and foliar ZnSO4 application improves wheat grain Zn concentration and Zn fractions in a calcareous soil. Eur. J. Soil Sci. **71**(4), 681–694 (2020)

Zou, C., Gao, X., Shi, R., Fan, X., Zhang, F.: Micronutrient deficiencies in crop production in China. In: Micronutrient deficiencies in global crop production, pp. 127–148. Springer, Dordrecht (2008)

Chapter 14
Microbes-Mediated Facilitation of Micronutrients Uptake by Plants from Soil Especially Zinc

Khalida Khatoon, Nabat Arfi, and Abdul Malik

Abstract Micronutrients are essential minerals required for the development of plants and humans, but they are often lacking in soils and food crops. Despite the fact that soils might contain substantial amount of micronutrients, their bioavailability in rhizosphere could be controlled by the combined effects of several edaphic factors, such as competing cations, anions, organic matter, pH, soil morphology, soil parent materials and biological factors, including plant characteristics as well as interaction of plant roots with microorganisms. Plant growth-promoting microbes belonging to cyanobacteria, bacteria, fungi and mycorrhizae can influence micronutrients availability by the process of solubilization, chelation and oxidation-reduction reactions in soil. Some microorganisms activate the release of root exudates by interacting with the plant roots. Moreover, their amount as well as composition affects the microbial diversity and activity of plant-associated beneficial microorganisms which ultimately affect the nutrient availability. Exuded carboxylates, that is, organic acids, may have role in solubilization of mineral nutrients as well as growth substrates for microbes. Some typical organic acids (such as lactate, citrate, acetate and oxalate) are found in the root exudate, which can convert the micronutrients in the plant-available forms to be transported across the plasma membrane of the root cells. Since soil microbial flora are not nutrient deficient in low-nutrient soil, they may be helpful towards increasing the soil micronutrient availability, plant uptake and also its accumulation in grains, that is, biofortification in a range of crops. This chapter describes the physiological importance of micronutrients in plants as well as improvements of nutrient uptake and utilization by plant systems.

K. Khatoon (✉)
Department of Agricultural Microbiology, Aligarh Muslim University, Aligarh, Uttar Pradesh, India

N. Arfi
Department of Home Science, Aligarh Muslim University, Aligarh, Uttar Pradesh, India

A. Malik
Department of Agricultural Microbiology, Faculty of Agricultural Sciences, Aligarh Muslim University, Aligarh, Uttar Pradesh, India

S. T. Khan, A. Malik (eds.), *Microbial Biofertilizers and Micronutrient Availability*,
https://doi.org/10.1007/978-3-030-76609-2_14

Also, the role of rhizospheric microorganisms in mediating the uptake of plant nutrients from soil is discussed in detail.

Keywords Micronutrients · Zn deficiency · Chemical fertilizers · Organic amendments · Mycorrhizal interactions · Zn-solubilizing bacteria · Biofertilizers · Organic acids · Zn transporters · Chelating substances · Biosurfactants

Introduction

Plant requires various nutrients for their optimum growth and metabolic activities. The common micronutrients essential for plant metabolism include Fe, Cu, Zn, B, Ni, Mn, Mo and Cl (Uchida 2000). Deficiency of anyone of these micronutrients in soil can result in retarded growth of plants although all other required macro- and micronutrients are sufficiently there in soil (Yu and Rengel 1999). Million hectares of agricultural lands worldwide are deficient in available trace metals like Zn, Cu, Mn and Fe. Amongst all micronutrients, Zn has been observed to be an indispensable part for healthy growth, metabolic activities and reproduction of plant system and has been recognized as essential micronutrients (Perumal et al. 2017). Henceforth, Zn has served as an essential element of various metabolic processes and enzymatic and redox reactions (Gandhi et al. 2014). Zn is involved in proteins and nucleic acid synthesis, DNA replication, structural and catalytic activities, membrane stability, energy transfer reactions and chlorophyll synthesis in plant system (Cakmak 2009a, b; Gurmani et al. 2012). Sharifi and Paymozd (2016) stated that an estimated percentage of globally found soils deficient in Zn is around 30%. Up to 50% of agricultural land of India, especially the Indo-Gangetic belt, is Zn deficient which is expected to increase further up to 63% by 2025 (Sunithakumari et al. 2016). Though most of soils of the world contain sufficient amount of total Zn, but most of it is not available for the plants (Mandal and Hazra 1997). Several factors, namely, soil pH, texture, salinity, organic matter, calcium carbonate, phosphate and climate substantially reduce the efficacy of nutrient fertilizers and thereby lessen the availability and solubility of that nutrients in soil as well as plant system (Schnug et al. 1998; Najafi-Ghiri et al. 2013). Only 20% of exogenously applied soluble form of Zn is available for uptake by plant (Bapiri et al. 2012). The immobilized or unavailable Zn, that is, $Zn_3(PO_3)_2$, ZnO and $ZnCO_3$, can be converted back to available forms by plant growth-promoting rhizospheric bacteria having Zn-solubilizing ability, for example, Zn-solubilizing bacteria, which can solubilize it through secretion of chelating compounds and lowering of pH via proton extrusion and organic acid secretion as reported to potentially improve the availability of Zn to the crop plants (Wu et al. 2006; Wang et al. 2013; Sharma et al. 2014; Hussain et al. 2015; Gontia-Mishra et al. 2017; Othman et al. 2017).

This chapter summarizes the part that micronutrients play towards growth and health of plants, their deficiency symptoms, plant-based strategies for the enhancement of mobility/availability of nutrients and facilitation of micronutrients especially Zn from soil by rhizospheric bacteria for plant uptake. The mechanism and use of these Zn-solubilizing bacteria to increase sustainable crop production as well as role, application and effects of biosurfactants on the solubility and availability of micronutrient as well as their uptake in plants and possibility to solve the problem of micronutrient deficiency in soil are also discussed.

Micronutrient and Their Physiological Role in Plants

Plants often face remarkable challenges in obtaining an adequate supply of essential micronutrients to meet the demands of basic cellular processes due to their relative immobile nature. Plants deficient in any of the mineral nutrients may lead to reduced crop productivity and reduced plant quality and/or fertility, which can also give rise to reduced biodiversity as plants are primary producers that support most of the food webs. Macronutrients and micronutrients are the two categories of nutrients that are considered essential for plants. The former are the one that act as the building blocks of cellular components, that is, proteins and nucleic acids, and are required in large quantities. Examples include nitrogen (N), phosphorous (P), magnesium (Mg) and potassium (K). As carbon (C), hydrogen (H) and oxygen (O) are also needed in huge quantities for the formation of larger organic molecules of the cell, they are also considered non-mineral macronutrients. Prerequisite to normal growth and development involves a variety of micronutrients, such as Zn, Cu, Fe, Mn, Co, Ni, B, Mo and Cl, which are required by plants but in a very small amount. These essential micronutrients have vital functions in plants, for example, they act as structural and functional cofactors in proteins and enzymes involved in important metabolic pathways.

Zn occurs in various forms in biological systems, so it does not induce the formation of radicals and does not participate in redox reactions. That is why divalent cations of Zn, that is, Zn^{2+}, are key structural and catalytic components in hydrolytic enzymes, for example, alcohol dehydrogenase, carbonic anhydrase, Cu-Zn superoxide dismutase and DNA-binding proteins such as RNA polymerase (Marschner 1995; Guerinot and Eidet 1999; Broadley et al. 2007). But proteins can also be inactivated due to the uncontrolled bindings of metals to proteins. Due to its small radius to charge ratio, Zn^{2+} has pronounced Lewis acid character. This results in formation of strong covalent bonds with sulphur (S), nitrogen (N) and oxygen (O) donors and thus becomes an essential constituent in thousands of plant proteins (Da Silva and Williams 2001).

In contrast to zinc, copper (Cu) belongs to the redox active transition metals, having two oxidation states Cu^{+} and Cu^{2+} due to which it is an integral part of certain electron transfer proteins in photosynthesis (such as plastocyanin) and respiration (such as cytochrome C oxidase), as well as also involved in lignification (such as laccase) (Da Silva and Williams 2001). Iron (Fe) is highly redox active and is required for various life-sustaining processes in plant system. It plays an important role in various electron transfer processes through its oscillation between ferrous (Fe^{2+}) and ferric states (Fe^{3+}). Fe is also a constituent of haem proteins viz. catalase, cytochromes and Fe-S proteins like ferredoxin as well as various other enzymes and required for chlorophyll synthesis (Kim and Guerinot 2007). Manganese (Mn) has less redox potential than copper and has essential role in photosynthesis where it forms a catalytic centre by cluster of several Mn atoms for light-induced oxidation of water in photosystem II. It is also present in several enzymes as a cofactor, for example, Mn^{2+}-dependent superoxide dismutase (Marschner 1995; Pittman 2005). Nickel (Ni) is present as a cofactor in various enzymes, for example, Ni-superoxide dismutase and urease hydrogenases (Dalton et al. 1988). Due to the irreplaceable component of urease which is a urea hydrolysing enzyme, Ni is considered an essential plant nutrient (Gerendás et al. 1999). Urease catalyses the hydrolysis of urea to ammonia and CO_2 and thus allows the usage of externally or internally produced urea as an N-source (Dixon et al. 1975; Sirko and Brodzik 2000). In bacteria, several additional Ni-dependent enzymes, for example, methyl coenzyme M reductase and different hydrogenases, have been reported while in higher plants, urease is the only identified Ni-dependent metalloenzyme where urea breakdown is the only proven nutritional function of Ni (Walsh and Orme-Johnson 1987).

Molybdenum (Mo) forms Mo cofactor (MoCo) together with a pterin compound, an active site of eukaryotic Mo enzymes. Nitrate reductase, peroxisomal sulphite oxidase, aldehyde oxidase and xanthine dehydrogenases are the four enzymes in plant system that are Mo dependent. Nitrate reductase catalyses the initial step of nitrate assimilation, peroxisomal sulphite oxidase detoxifies sulphite, aldehyde oxidase catalyses the last step of abscisic acid biosynthesis, whereas xanthine dehydrogenases have crucial roles in degradation of purine and stress response (Schwarz and Mendel 2006). Cobalt (Co) is considered beneficial but not an essential metal for plant system. It is considered beneficial because it is required by bacteria in leguminous symbiotic N_2 fixation (Delwiche et al. 1961; Gad 2006) as well as in root nodules of non-legumes (Hallsworh et al. 1965). Under the condition of osmotic stress, Co plays protective role (Li et al. 2005). Till date, no essential function of cobalt is known in plants, whereas several co-dependent metalloenzymes have been reported in bacteria, fungi and algae such as methionine aminopeptidase that is responsible for breaking the N-terminal methionine from newly translated polypeptides (Kobayashi and Shimizu 1999).

However, all above-mentioned metals can also have toxic effects if exceeds their limits. Some non-essential heavy metals exhibit toxicity even at low concentration on plants and disturb the plant metabolism leading to reduced plant growth and lower yield. For example, lead (Pb), cadmium (Cd) and arsenic (As), due to their similarity to some essential heavy metals like Zn and Fe, especially in ionic radii, are potentially toxic to plants.

Zinc Deficiency Symptoms

In Plants

Among all micronutrient deficiencies in plants worldwide, Zn deficiency is the most common one. Plants uptake Zn in the form of divalent cation (Zn^{2+}) (Blindauer 2015), and less than 15 ppm Zn will result in deficiency symptoms such as interveinal chlorosis which first occur in younger leaves and reduced size of leaves, that is, little leaves, for example, little leaves in cotton, chillies, brinjal, citrus and mango, while concentration between 300 and 500 ppm is toxic to plants (Miransari 2011). Inadequate supply of Zn can lead to serious physiological disturbances as Zn acts as a structural component of many enzymes. In plants, Zn content differs broadly depending on the characteristics of soil, climate condition and plant genotype. Zn has a key role in seed production and flowering, deficiency of which leads to the male sexuality in maize (Sharma et al. 1987). Zn deficiency mainly occurs in calcareous soil which affects the water and nutrient absorption from soil, resulting reduced growth and crop yield. Reduced nutrient availability, that is, impaired nutrition and also high pH, results in chlorosis and stunted growth in calcareous soil. Brown et al. (1993) reported that Zn deficiency disturbed the male and female reproductive organ formation as well as pollination processes, which ultimately leads to reduced yield of plant and can be accredited to the reduced indole acetic acid (IAA) synthesis. This reduced IAA concentration results in internode shortening and limits the development of leaves that induce crowding of leaves together at the apex of plant in a fan-shaped "resetting" or clustering appearance (Havlin et al. 2005; Singh et al. 2007). Alloway (2004) found that under Zn deficiency, seed production is reduced which is either due to the more abscisic acid formation in plants that cause unseasonable abscission of leaves and buds or by disrupting the development and physiology of anthers and pollen grains. He observed that only in severe Zn deficiency, the visible deficiency symptoms in crop will occur, while quality and yield of crops might be reduced in case of marginal deficiency due to the concealed Zn deficiency without any clear symptoms. The concealed Zn deficiency may not be detected for many seasons resulting in high cost to farmers, ultimately leads to decreased Zn concentration in the edible part of food crops and their reduced nutritive value (Welch and Graham 2004). Epstein and Bloom (2005) showed the greatly reduced flowering and fruiting processes due to the severe Zn deficiency. Zea mays (Corn) is hypersensitive to Zn deficiency and exhibited white to yellow chlorosis of leaves, that is, stunted growth due to shortening of internodes, white bud and reddish/yellowish stripes on the leaves of lower portion of plants (Havlin et al. 2005; Alloway 2008). "Dieback" (root apex necrosis) is due to severe Zn deficiency while "mottle leaf" (spatially heterogeneous or interveinal chlorosis), "bronzing" (development of reddish brown or bronze tints) and various auxin deficiency like responses, for example, resetting (internode shortening), goblet leaves (inward curling of leaf lamina), "little leaf" (reduced leaf size), stunted growth, reduced number of tillers, white buds of maize, increased crop maturity period, spikelet sterility and low quality of harvested crops are due to sublethal Zn deficiency (Hafeez et al. 2013b).

In Soil

Soil Zn bioavailability is affected by several edaphic factors as discussed earlier. Variation in total Zn content is dependent on the nature of source material, processes of formation of soil, climatic conditions, etc. Coarse textured, alkaline, sodic, calcareous and soils having less organic matter have generally less amount of total Zn (Singh et al. 2005). Zn deficiency is reported in several types of soils in many parts of the world, but calcareous soils in semi-arid regions, highly weathered soils in tropical regions and sandy soils in various climatic area are likely to be affected most seriously. At the country level, Zn deficiency is the most widespread among the micronutrients in India. Total content of Zn in Indian soils differs greatly with the nature and type of soils. The total content of Zn ranges from 10 to 300 mg per kg, and in some soils, a concentration as high as 1000 mg per kg has been reported like in mineral soils. Soils formed from mafic igneous rocks are more Zn rich as compared to the felsic igneous rocks due to the presence of similar sized ionic substitutes, for example, Mg^{2+} and Fe^{2+} in silicates. In comparison to the soil formed from limestone, shale and igneous rocks, soils formed from sand stones are more deficient in Zn. Soil formed from granites reported to have more available Zn in comparison to those from basalts and gneiss (Murthy 1988).

In Humans

In biological systems, Zn has structural, catalytic or co-catalytic roles in >300 enzymes as well as has critical role in protein synthesis and DNA/RNA metabolism. Human can suffer from Zn deficiency if inadequate Zn is taken up through diet or it is not absorbed properly in the body. The deficiency causes loss of facial hair, memory, weakened body muscles and various skin issues. Inadequate Zn intake during pregnancy is also responsible for retarded brain development of the foetus. Zn deficiency in men can cause infertility. Its deficiency in human beings leads to a dozen of health problems, for example, sexual immaturation, impairment in linear growth, learning ability, compromise immune functions and poor development of central nervous system (Welch 2001). Congenital diseases like *Acrodermatitis enteropathica* may also result in Zn deficiency (Lukaski 2004). Supplementation of Zn has improved growth in children, has reduced child mortality and lowers the rate of malaria, diarrhoea and pneumonia (Sazawal et al. 2001).

Factors Affecting Zinc Bioavailability in Rhizospheric Soil

Soil Texture

Coarse soil contains and retains low Zn, and therefore, light textured soils are usually Zn deficient. There is a significantly positive relationship between silt plus clay and Zn content of soil. The coarse-textured soils are likely to be more prone to Zn deficiency (Takkar et al. 1977). Since Zn is co-translocated with clay and iron oxides, it will increase with the clay content in soil (Chinnasamy 1987; Prasad and Sakal 1988). A negative correlation between DTPA-Zn and silt was obtained by Takkar et al. (1977), while Dinesh (1988) reported a positive correlation between DTPA-Zn and clay content of soil. The later relation was due to higher cation exchange capacity of the fine-textured soil which retains Zn in exchangeable form. Sharma et al. (2004) found that with the increase in clay and silt and cation exchange capacity (CEC), the total Zn content increased, whereas with an increase in organic carbon content and CEC, DTPA-extractable Zn was increased which was found to be decreased with increasing pH, sand and $CaCO_3$ content.

Soil Organic Matter (SOM)

SOM has an important role in Zn solubility and in making it available to the growing plants. Increased amount of SOM can enhance the formation of soluble Zn complexes. Ghosh and Sarkar (1994) found higher Zn availability due to higher organic carbon content present in soil as evident by the positive and significant correlation between available Zn and soil organic carbon. A highly significant positive correlation between available Zn and SOM has been revealed by several other researchers (Raj et al. 1990; Prasad et al. 1991; Nayak et al. 2000; Samanta et al. 2002; Gupta et al. 2003; Sharma et al. 2003; Yadav and Meena 2009; Srinivasarao et al. 2009; Bali et al. 2010). Ravi et al. (2014) reported that organic carbon content increases Zn availability, whereas it decreases with increase in pH and $CaCO_3$ as they found that Zn content is significantly and negatively correlated with $CaCO_3$ content of the soil sand and silt. However, it is positively correlated with organic carbon and clay. Murthy and Ponnamperuma (1986) reported a negative correlation between available Zn and SOM. Sahoo et al. (2003) reported no significant correlation between Zn availability and SOM, indicating a differential effect of soil organic matter on Zn availability. Pharande et al. (1996) reported that decomposition of SOM reduces pH of soil, which in turn assists in increasing Zn mobilization and uptake. Prasad et al. (1996) reported the increased concentration of soluble Zn complexes in soil due to the presence of natural complexing agents present in SOM which dissolute the sparingly soluble Zn compound and chelate the liberated Zn ions.

Soil EC

Chattopadhyay et al. (1996) stated that Zn availability was negatively correlated with electrical conductivity (EC). Randhawa and Singh (1995) reported significant and positive correlation between available Zn content and EC in alluvial soils. The available Zn has significant positive regression coefficient with EC (Gupta et al. 2003; Sharma and Chaudhary 2007), while Hasmot and Mrinmay (2000) reported no significant relationship between EC and available Zn.

Soil pH

Among many factors, soil pH is one which controls the Zn availability in plant systems. The differences in soil pH account for more than 90% variation in Zn availability to crops. Zn solubility mainly depends on soil pH as transformation of different ionic forms of Zn is closely related to soil reaction, and due to change in soil pH, Zn is transformed from one hydrolysed form to another. According to Gupta et al. (1987), precipitation is more pronounced at high pH, while Zn adsorption increases at lower pH, thereby controlling Zn solubility and ultimately its availability to the growing crops. Singh and Abrol (1985) reported that solubility of Zn was highly dependent on the pH in the range of 6.0–7.9. At pH higher than 7.9, solubility of Zn is controlled by precipitating Zn as $Zn(OH)_2$ or $ZnCO_3$ in sodic soils, whereas at low pH (4.2–6.0), reduced Zn adsorption may attribute to increase in the competing cations, for example, Al, Ca, Fe, Mn concentration, and to the partial dissolution of mineral carbonates of soil. At neutral to alkaline pH, $Zn(OH)^+$, a hydrolysed form of Zn, may be specifically adsorbed by soil constituents and becomes less available for plant growth. That is why soils with high pH are Zn deficient (Duraisamy et al. 1988). Various studies reported negative and significant correlation between soil pH and Zn availability (Nayak et al. 2000; Samanta et al. 2002; Sahoo et al. 2003; Gupta et al. 2003; Srivastava and Srivastava 2008; Chaudhari et al. 2012).

Soil Calcium Carbonate

Zn is adsorbed in crystalline structure of dolomite and magnetite in the lattice by replacing Mg, which needs partial or complete removal of the hydration shell of Zn since it is the major energy consumer. This removal of H_2O molecules from hydrated ions of Zn increases the entropy change of the system which ultimately leads to increased Zn sorption. Zinc solubility is adversely affected by the calcareousness of soil. The $CaCO_3$ plays a vital part in the adsorption of Zn (soluble) in addition to increase the CO_3^{2-} content in the soil, which increases the possibility of precipitation of Zn (soluble) in the form of $ZnCO_3$, $Zn_3(CO_3)(OH)_2$ and $CaZnO_2$ (Misra and Pandey 1977; Chatterjee and Mandal 1985; Hazra et al. 1987). Negative relationship between $CaCO_3$ and availability of Zn had been reported by several workers (Chinnasamy 1987; Sangwan and Singh 1993; Singh et al. 2005; Jibhakate et al. 2009). According to previous reports, availability of Zn is significantly and positively correlated with clay, CEC and organic matter content while it is negatively correlated with silt and pH. There is no significant relationship with $CaCO_3$ content, which indicates the importance of pH and soil exchanger phase (i.e. clay and organic matter) in Zn availability in calcareous soils (Najafi-Ghiri et al. 2013).

Soil Phosphorous

Phosphorous (P) is an important element which interferes in Zn uptake as increased levels of P in soil reduce the Zn uptake (Das et al. 2005). Zn and P imbalance in the plant system is due to the excessive accumulation of P, which results in Zn deficiency. Badrinath et al. (1987) reported that Zn availability with high P content in acidic soils was negatively correlated with available Zn. However, Sharma and Lal (1992) found that DTPA-extractable Zn was negatively and significantly correlated with available P in soils of humid southern zone of Rajasthan.

Soil Temperature and Moisture

Low soil moisture and low temperature are some other factors that contribute to the Zn deficiency (Moraghan and Mascagni Jr. 1991). Nutrient supply to the plants is affected by soil moisture due to the impaired diffusion to the root surface. In arid and semi-arid regions where soils are generally water scarce for longer period during the cultivating season, Zn availability to plants is minimum, since Zn diffusion in soils greatly depends on soil moisture. However, in submerged soils, water logging changes the Zn chemistry in the soil and tends to have less concentration of water-soluble Zn as compared to well-drained soils. While in poorly drained soils, Zn solubility is poor and Zn uptake is low, due to the co-precipitation of Zn with soluble Fe and Al in the soil. Low soil temperature further decreases Zn availability causing deficiency (Moraghan and Mascagni Jr. 1991). At higher temperature, Zn availability is increased due to increased microbial activity with subsequent decomposition and release of organic Zn. Najafi-Ghiri et al. (2013) reported the highest and lowest values of available Zn in soils with mesic and thermic temperature regimes, respectively, indicating a significant effect of soil temperature regime on Zn availability.

Strategies for Increasing Micronutrient Availability in Rhizosphere

Application of Chemical Fertilizers and Organic Amendments

Soil is a dynamic natural body or reservoir of the planet serving as the source of food. Extensive agricultural practices like the use of fertilizers are required to feed large human population which is continuously increasing. Application of chemical fertilizers is necessary to avoid nutrient deficiencies in soil and to support optimum growth of the plants. Application of Zn has been found to enhance the crops yield under both irrigated and rainfed conditions. The type of chemical fertilizer influences its effectiveness on crop performance. The interaction of chemical fertilizer with different forms of nutrients can positively, negatively or neutrally affects the yield and nutrient bioavailability of food crops. Zn is usually applied as $ZnSO_4$ to the crops, whereas other inorganic forms of Zn include the Zn-ammonia complex (10% Zn), Zn nitrate (22% Zn), Zn oxysulphate (40–55% Zn), Zn chloride (48–55% Zn), Zn carbonate (52–56% Zn) and Zn oxide (50–80% Zn). The organic forms of Zn include lignosulphonate (5–8% Zn) and synthetic chelates such as EDTA (14% Zn), HEDTA and NTA (9% Zn) (Shuman 1998). Most of the Zn fertilizers dissolve comparatively slow or sometimes too slow in soil to supply ample amount needed for the sturdy plant growth (Rengel 2002; Rehman et al. 2012). Generally, the application of 25 kg of $ZnSO_4.7H_2O$ per hectare of soil which is equivalent to 5 kg per hectare of Zn is recommended yearly or alternate year. The application of Zn fertilizers to the soil and foliar application of either $ZnSO_4$ or Zn chelators can enhance concentration of Zn in plants through sufficient Zn mobility in the phloem, which effectively increases concentration of Zn in cereal grains, leaf, tuber and fruits (Shuman 1998; Rengel et al. 1999). But they are expensive and non-sustainable. Furthermore, the amended fertilizers are quickly transformed into non-accessible and insoluble forms and will not be available to plants (Bahrani et al. 2010; Cakmak et al. 2010; White and Broadley 2011; Bulut 2013; Hafeez et al. 2013a). Even when soluble salts like $ZnSO_4$ are applied, the released Zn tends to transform into forms that are less soluble due to soil equilibria, and few examples of such forms would include carbonates, hydroxides and oxides. Similarly, poor movement of Zn has been observed when moved from fertilizers down to the soil profile. To counter this, chelated forms of Zn could be applied (Alloway 2009).

As SOM is thought to be an essential factor in nutrient mobility in soil, several organic amendments like compost, farmyard manure, olive husk and poultry manure have been added in soil for the improvement of soil health, fertility and crop yields. Zn availability can be enhanced by the use of organic matter, which helps Zn to release with time and by altering the physicochemical properties of soil which further might increase soluble/available fraction of Zn in soil and reduces fixation leading to more Zn uptake by plants (Marschner 1993; Obrador et al. 2003; Cakmak 2009a, b). Organic matter enhances the bioavailability of Zn by producing the polysaccharides (Hodgson 1963). Likewise, an increase in DTPA-extractable Zn was noticed following the addition of soil organic carbon (Sidhu and Sharma 2010).

Plant-Based Mechanisms

Modification in Root Structure and Anatomy

Parameters of root morphology viz. total root length and surface area have key roles in nutrient uptake, especially under nutrient-limiting conditions (Batista et al. 2016). Metal hyperaccumulator plants have various physiological, morphological and anatomical adaptations, for example, elaborative root hairs, root surface area and root exudates involved in metal mobilization (Dong et al. 1995; Genc et al. 2007; Lynch 2007). A highly developed plant root system can be a better strategy for the enhancement of micronutrient uptake, for example, Zn-efficient plants generally have fine roots with greater surface area and thus enhance the Zn availability as well as other micronutrients (Singh et al. 2005). Zn-efficient plants seem to utilize a variety of physiological mechanisms which helps them to hold out against Zn deficiency stress better than Zn-inefficient plants, for example, greater proportion and longer length of thinner roots ($\leq$0.2 mm), which play an important role in the differential Zn efficiency observed among various genotypes (Rengel and Wheal 1997). Zn-efficient genotype of rice (Chen et al. 2009) and wheat (Genc et al. 2006) has been found to develop thinner and longer roots ($\leq$0.2 mm) than the less Zn-efficient genotype.

Apart from root morphology, the anatomical structure of the root is also altered by plant growth-promoting microorganisms, which results in a higher nutrient uptake from the soil by root system (López-Bucio et al. 2007; Ortíz-Castro et al. 2008). There are several reports available that show that by inoculating plant growth-promoting rhizobacteria (PGPR) and endophytic bacteria, the root morphology and its architecture is primarily affected (Vacheron et al. 2013; Wang et al. 2014; Delaplace et al. 2015). Singh et al. (2017a) supported this hypothesis and found that the root length, root diameter, root volume, surface area and the average number of root tips in wheat crops were enhanced when inoculated with plant growth-promoting Zn-solubilizing endophytic bacteria (*Bacillus subtilis* DS-178 and *Arthrobacter* sp. DS-179). Accumulation of Zn in wheat grains was also enhanced by two-folds compared to the control, which was better or equal to the application of 40 kg of $ZnSO_4$ per hectare. A well-developed root anatomy, for example, increased size of root cortex and volume of xylem vessels, well-developed root hairs, thickened endodermis and vascular bundles, is positively correlated with nutrient uptake (Kramer et al. 1980; Kotula et al. 2009). Rêgo et al. (2014) observed a significant impact of *Trichoderma asperellum*, *Burkholderia pyrrocinia* and *Pseudomonas fluorescens* inoculant on the anatomy of the root mainly with reference to the diameter of the root cortex, dimension of vascular bundles and numerical alteration in proto- and meta-xylem vessels. Singh et al. (2017b) also noticed enhancement of the volume of xylem vessels, the thickness of the root cortex and the diameter of vascular bundles and pericycle when *Arthrobacter sulfonivorans* DS-68 and *Arthrobacter* sp. DS-179 inoculated individually. These inoculants also increased uptake of Fe or Zn by 60–75%. Subramanian et al. (2009) noticed that root length, spread and root volume of plants were increased significantly when mycorrhizal fungus was inoculated as compared to uninoculated plants, and the Zn concentration in the grains was increased up to 4%. Similarly, Tariq et al. (2007) reported a considerable increase in root weight, length, volume and Zn uptake in straw and grain of rice crop on adding microbial inoculants.

Root Exudates

Sparingly available nutrients are mobilized by root exudates by affecting the physical, chemical and biological characteristics of the root-soil interface through the process of acidification, reduction, complexation and/or ligand exchange reactions (Terzano et al. 2015). Exudates from plants include a variety of organic acid anions (malate, maleate, fumarate and cis-aconitate) as well as amino acids (alanine, aspartic acid, glutamic acid, proline and valine) (Hajiboland et al. 2005; Rasouli-Sadaghiani et al. 2011; Soltani et al. 2014). Many plant species also reported to produce phytosiderophores such as Zn-efficient genotype of wheat and rice, Hordeum murinum and Agropyron orientale (Gries et al. 1998; Tolay et al. 2001) but not regularly (Pedler et al. 2000). Root exudate favours the growth of particular microbial communities and thus marked to affect interactions between plants and soil condition (Doornbos et al. 2012; Mendes et al. 2013). Any minor change in the amount or constitution of root exudates due to the abiotic or biotic stress can have a substantial effect on the abundance and composition, or activity of plant-associated microbial population (Verma et al. 2010; Nath et al. 2017; Verma et al. 2017). Root exudates are known to have several functions like acquiring nutrients, allelopathy or as chemical signals to fascinate symbiotic associates, for example, rhizobia and legumes (chemotaxis), or the promotion of advantageous microbial colonization on root surfaces, for example, *Bacillus subtilis* and *Pseudomonas florescence* (Park et al. 2003; Li et al. 2016; López-Gómez et al. 2016; Syranidou et al. 2016). Phenolic metabolites released in root exudates fascinate specific rhizospheric soil microorganism and successfully control the autochthonous soil microbial population (Brimecombe et al. 2001).

Phytosiderophore Production

Phytosiderophores (phytometallophores) are released under Fe deficiency which effectively mobilize the sparingly soluble inorganic Fe compounds like Fe^{3+} hydroxides and oxides by forming the phytosiderophore-Fe^{3+} complex which can be actively transported across the plasma membrane without Fe^{3+} reduction. The same type of phytosiderophores has been shown to complex and mobilize Zn also and is thought to be released with similar diurnal rhythm in Zn-deficient calcareous soils

(Treeby et al. 1989; Zhang et al. 1989; Cakmak et al. 1994). Bashir et al. (2010) reported that the soil available Zn, that is, Zn^{2+}, is taken up by root membrane transport mechanism that includes phytosiderophores in rice.

Rhizospheric Microorganism-Based Mechanisms

Plant growth-promoting rhizobacteria in the form of biofertilizer can increase the soil nutrient bioavailability as well as their uptake by plant from low available soil nutrient pool and thereby improving plant growth (Gouda et al. 2018). Immobilized or unavailable form of Zn is changed back to available forms by adding bacterial inoculants which have the ability of solubilizing them by releasing the organic acids and decreasing pH (Wang et al. 2013; Sharma et al. 2014).

Mycorrhizal Interactions with Plants

An endophytic fungus *Piriformospora indica* has been reported to have remarkable effect on plant growth, biomass and micronutrient uptake (Gosal et al. 2010). Balakrishnan and Subramanian (2012) reported improvement in micronutrients, particularly Zn availability in soil by inoculating mycorrhizal fungi (arbuscular mycorrhiza) as a result of rhizospheric acidification and siderophore production apart from hyphal transport of nutrients via external mycelium. Root colonization by arbuscular mycorrhizal fungi has been reported to enhance the metal micronutrient uptake such as Cu, Zn, Mn and Fe in Zea mays (Liu et al. 2000). Similarly, higher uptake of above micronutrients was observed in wheat by inoculating *Azospirillum* and mycorrhizae as compared to the control (uninoculated plants) (Ardakani et al. 2011). Increased Zn uptake and mobilization as well as enhanced growth of rice was reported when arbuscular mycorrhizal fungi were inoculated in rice roots (Purakayastha and Chhonkar 2001). After adding Zn fertilizer, increased Zn uptake as well as wheat and maize growth was seen on inoculating arbuscular mycorrhizal fungi in soils deficient in Zn (Kothari et al. 1990).

Zinc-Solubilizing Bacteria

Zn solubilization by rhizospheric bacteria has better prospective as compared to chemical fertilizers, which is gaining immense interest in soil restoration and sustainable agriculture. In recent years, insoluble form of Zn, that is, ZnO, $ZnCO_3$ and $Zn_3(PO_4)_2$, has been reported to be solubilized by rhizospheric bacteria (Sarathambal et al. 2010; Rokhbakhsh-Zamin et al. 2011; Kumar et al. 2012; Sharma et al. 2012; Krithika and Balachandar 2016; Gontia-Mishra et al. 2016) (Fig. 14.1).

Zn-solubilizing bacteria (ZSB) solubilizes the Zn (fixed form) and also promotes Zn uptake and helping in the Zn fortification of the crops (Barbagelata and Mallarino

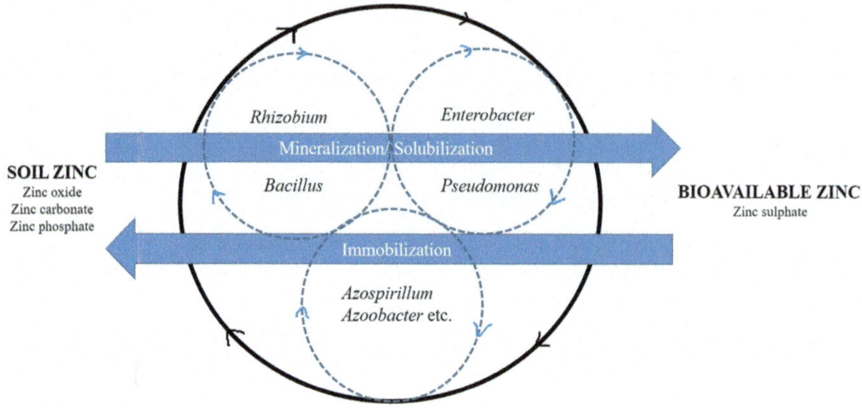

Fig. 14.1 Mechanism of Zn-solubilizing bacteria (Kumawat et al. 2019)

2013). These bacterial inoculants produce various low-molecular-weight organic acids especially gluconic acid, which helps in Zn solubilization through acidification of the soil and thus enhances the availability of the Zn for plants (Hafeez et al. 2013a). Application of ZSB has been reported to enhance Zn uptake in a variety of crops, such as rice (Vaid et al. 2014), maize (Goteti et al. 2013), wheat (Rana et al. 2012) and soybean (Ramesh et al. 2014). Zn-solubilizing bacteria has been reported to enhance the uptake of Zn by 21% in soybean plants (Sharma et al. 2012). Several bacteria have been found to solubilize Zn in vitro, for example, *Gluconacetobacter diazotrophicus* (Saravanan et al. 2007a), Pseudomonas aeruginosa (Fasim et al. 2002), *Pseudomonas striata*, *Burkholderia cenocepacia*, *Bacillus* spp., *Pseudomonas fluorescens* (Pawar et al. 2015), Bacillus thuringiensis, *S. marcescens* and *Serratia liquefaciens* (Ullah et al. 2015). ZSB can be exploited as biofertilizers which can increase survival and growth of plants, improve soil fertility and enhance yield and Zn concentration in cereal grains. Also, it can diminish dependency on chemical/synthetic fertilizers (Hussain et al. 2018). Some examples of Zn-solubilizing bacteria and their effects on crops are given in Table 14.1.

Production of Organic Acids and Lowering of Soil pH

Soil pH is known to influence micronutrients availability in soil. Havlin et al. (2005) reported that by reducing the pH by one unit, availability of Zn increased by 100 times in soil. Plant growth-promoting bacteria have been reported to lower the pH of rhizosphere by releasing the organic acids and protons (Fasim et al. 2002; Wu et al. 2006; Parmar and Sindhu 2019). Saravanan et al. (2004) found that *Bacillus* and *Pseudomonas* produce organic acids lowering the pH of rhizospheric soil and thus making Zn available to the plants. Some plant growth-promoting rhizobacterial strains have been found to produce gluconic acids (Saravanan et al. 2011) as well as its derivatives like 5-ketogluconic acid (Saravanan et al. 2007a, b), 2-ketogluconic

Table 14.1 Impact of Zn-solubilizing bacteria on crop

S. No	Zn-solubilizing bacterial isolates	Crop used	Effect on crops	References
1	*Pseudomonas* sp. ZSB-S-1	Soybean	Zn content in plant tissues was improved	Saravanan et al. (2004)
2	Zn-mobilizing PGPR	Rice	Zn content in the rice grain was enhanced, and grain yield, total biomass, shoot weight, root area, length, volume and weight were also improved	Tariq et al. (2007)
3	*Rhizobium* spp. RL9	Lentil	Dry matter, no. of nodules, nodule dry mass, seed yield, leghaemoglobin and grain protein were found to be improved	Wani et al. (2008)
4	*Pseudomonas* strain BA-8 *Bacillus* strain M-3	Strawberry	Fruit yield per plant was increased, that is 91.73% with BA8 + M3 and 81.58% with M3 alone	Esitken et al. (2010)
5	P. aeruginosa strain CM860	Rye	Root (144%) and shoot (120%) length increased	Shahab et al. (2009)
6	Zn-mobilizing PGPR	Wheat	Volume, weight, length and area of root and shoot weight were found to be increased	Kutman et al. (2010)
7	*Bacillus* isolates	Soybean	Zn accumulation in seeds was increased	Sharma et al. (2012)
8	*Pseudomonas* strain B1 and B2	Rice	Increased concentration (31%) of Zn in grains	Joshi et al. (2013)
9	*Pseudomonas* spp. P17 *Bacillus* spp. B40	Maize	Total dry mass and uptake of N, K, Mn and Zn were increased	Goteti et al. (2013)
10	*Burkholderia* strain BC *Acinetobacter* strains AB and AX	Wheat	Phytic acid concentration was reduced, whereas mean no. of productive tillers, no. of grain per year, grain weight and yield, and straw yield were increased	Vaid et al. (2013)
11	*Bacillus aryabhattai* strains MDSR7 and MDSR14	Soybean and wheat	Plant height, root and shoot dry weight, and zinc assimilation in seeds were found to be increased	Ramesh et al. (2014)
12	*Bradyrhizobium japonicum*	Cowpea	Micronutrient uptake was increased due to P-supplementation, but in some organs, Zn content was found to decreased	Nyoki and Ndakidemi (2014)
13	*Bacillus* strain AZ6	Maize	Shoot length and rate of photosynthesis were increased	Hussain et al. (2015)
14	*Bacillus* sp. S-10 and Bacillus *cereus* S-17	Rice	Growth of *Pyricularia oryzae* and Fusarium moniliforme was suppressed, and translocation of Zn towards grains was increased, as well as yield of basmati-385 and super basmati rice varieties was found to be increased	Shakeel et al. (2015)

(continued)

Table 14.1 (continued)

S. No	Zn-solubilizing bacterial isolates	Crop used	Effect on crops	References
15	*Serratia liquefaciens* FA-2, *Bacillus thuringiensis* FA-3, *Serratia marcescens* FA-4	Wheat	Grain yield, total biomass and dry straw weight were increased in addition to the grain yield and Zn content of wheat tillers per plant and grains per spike	Ullah et al. (2015)
16	*Pseudomonas fragi*, *Pantoea dispersa* and *Pantoea agglomerans*	Wheat	Plant growth and Zn content were found to be increased	Kamran et al. (2017)
17	*Bacillus* strains	Soybean and wheat	Modulated growth and yield as well as Zn biofortification	Khande et al. (2017)
18	*Enterobacter* sp. ScCS20	Rice	Zn level was improved	Wang et al. (2014)
19	*Sphingomonas* sp. SaMR12	Rice	Zn level was improved	Wang et al. (2014)

Adapted from Kumar et al. (2019) and Sindhu et al. (2019)

acid (Fasim et al. 2002) and several other organic acids (Tariq et al. 2007) to solubilize Zn. Insoluble Zn compounds are solubilized by *Bacillus* sp. AZ6 through production of organic acids such as cinnamic acid, caffeic acid, chlorogenic acid, ferulic acid, syringic acid and gallic acid in liquid medium (Hussain et al. 2011). Martino et al. (2003) stated that insoluble $Zn_3(PO_4)_2$ and ZnO were solubilized by organic acids secreted by mycorrhizal fungi. Desai et al. (2012) found improvement in available Zn in the culture broth by enhanced production of organic acids and reported that higher Zn availability is directly proportional to acidic pH of the culture broth. Zinc phosphate was solubilized both due to increase in the hydrogen ion (H^+) concentration of the medium and through production of gluconic acid. Perumal et al. (2017) used six bacterial strains isolated from rhizosphere of maize for the solubilization of insoluble Zn compounds (ZnO, $ZnCO_3$ and Zn-EDTA). They found that pH of the inoculated broth decreased to 3.0–4.8 from 7.0 to 7.3 after 10 days of inoculation due to the production of acids which probably solubilize insoluble Zn. However, the degree of acidification varied from strain to bacteria (Giri et al. 2005). Wu et al. (2006) noticed a reduction of pH up to 0.47 units with bacterial inoculants due to the release of organic acids and protons which ultimately leads to the improved solubilization of Zn and its uptake by plants. Similarly, Subramanian et al. (2009) reported lowering of the soil pH in rhizosphere by inoculating arbuscular mycorrhizae which in turn released Zn from mineral fraction.

Production of Chelating Substances

The effectiveness of chelating metal ligands has been tested in a variety of crops and environment since 1950s in order to solve the non-availability of essential nutrients, for example, Zn, Fe, Mn and Cu in calcareous soils (Rahman et al. 2009).

Zn-chelating compounds are either synthetic or synthesized by microbes or plant that form soluble complexes with Zn^{2+} ions, which can enhance Zn mobilization and solubilization in soil (Obrador et al. 2003). The synthetic chelated-Zn compounds readily make the Zn available (Mortvedt et al. 1999), but their application is minimal due to their high cost and socio-economic restriction to farmers mainly in developing country. The microbial chelating agents are the microbial metabolites that complex with metal cations, for example, Zn^{2+}, and reduce their interactions with soil constituents (Tarkalson et al. 1998). These Zn chelates migrate towards the plant roots and release chelating ligands, that is, Zn^{2+} at the surface of root, and render them free for the chelation of other Zn ion from the soil. Whiting et al. (2001) reported that Zn chelator metallophores produced by *Enterobacter cancerogenes*, *Pseudomonas monteilii* and *Microbacterium saperdae* bacteria have significant impact on Zn bioavailability in soil and its uptake by plant roots. Some bioinoculants such as *Agrobacterium* sp. (Ca-18), *Azospirillum lipoferum* (JCM-1270, ER-20) and *Pseudomonas* sp. (96–51) have been reported to have the ability to solubilize the $Zn(OH)_2$, $Zn_3(PO_4)_2$ and other insoluble salts enhancing the Zn availability to rice plants for extended duration in soil due to the production of natural metal chelators (Tariq et al. 2007). Kucey (1988) reported *Penicillium bilaji* inoculants enhanced Zn solubilization and its uptake by plants to a greater extent as compared to the ethylene diamine tetra acetate (EDTA) chelating mechanism.

Reduction of Phytic Acids

Phytic acid, that is, myo-inositol 1,2,3,4,5,6-hexakisphosphate (InsP6), a known form of stored phosphorous in most of seeds and grains, is able to strongly chelate multivalent metal ions, especially Zn, Ca and Fe, which results in the formation of very insoluble salts decreasing availability of minerals (Zhou and Erdman Jr 1995) causing malnutrition (Hunt 2003; Kumssa et al. 2015). Vaid et al. (2014) found reduction in the ratio of phytate:Zn on inoculation of *Burkholderia* sp. SG1 and *Acinetobacter* sp. SG3. Singh et al. (2018) enunciated significant increase in growth, biomass, yield and micronutrient uptake of wheat plant by inoculating two Zn-solubilizing bacteria (*Arthrobacter* sp. DS-179 and *Bacillus subtilis* DS-178) and two siderophore-producing endophytes (*Arthrobacter sulfonivorans* DS-68 and *Enterococcus hirae* DS-163). Due to endophytic inoculation, concentration of phytic acid was reduced by ~26% over the RDF in wheat grains.

Upregulation of Zn Transporters

Several metal transporters, for example, *ZIP*, *CDF* and P-type ATPase family, have been reported in plants. Zn-regulated transporter/iron-regulated transporter-like protein (*ZIP*) family is concerned with the translocation of zinc and iron, while cation diffusion facilitator (*CDF*) family and P-type ATPase family are involved in xylem loading of metals including Zn (Colangelo and Guerinot 2006; Eide 2006).

The *ZIP* family transporters are well characterized and supposed to be the prime system for uptake of Zn in plants (Mäser et al. 2001; Guerinot 2000). The ZIP family proteins have been found in Arabidopsis thaliana (Grotz et al. 1998), maize (Xu et al. 2010), rice (Ishimaru et al. 2005) and wheat (Evens et al. 2017). ZIP genes are induced mostly by Zn deficiency (Ramesh et al. 2003; Ishimaru et al. 2005; Chen et al. 2008), and their expression pattern in root and shoot system is different. Durmaz et al. (2011) reported excessive accumulation of Zn in cells of wild emmer wheat due to the overexpression of these proteins. Microorganisms have been reported to enhance the expression of ZIP genes (Krithika and Balachandar 2016; Singh et al. 2017b; Watts-Williams and Cavagnaro 2018). Watts-Williams and Cavagnaro (2018) found improved Zn concentration in grains of Hordeum vulgare under Zn-deficient condition due to the colonization of *Rhizophagus irregularis* which upregulated the *HvZIP13*. In another study, Enterobacter cloacae strain ZSB14, a Zn-solubilizing bacterial strain, upregulated the expression of OsZIP1 and OsZIP5 and downregulated the expression of OsZIP4 in rice genotypes (Krithika and Balachandar 2016).

Biosurfactant-Mediated Micronutrient Availability

The use of biosurfactant-producing bacterial strains is relatively new approach to enhance the micronutrients availability to plants (Pacwa-Płociniczak et al. 2011). Biosurfactants exhibit greater emulsification, foaming and surface activity properties than synthetic surfactants, and also they are stable over many environmental factors viz. pH, temperature, moisture and salinity as compared to synthetic surfactants. Biosurfactants increase the metal availability in soil to plants in several ways, for example, by micelles formation and by decreasing the interfacial tension between the soil and metals (Pacwa-Płociniczak et al. 2011). By the process of solubilization, biosurfactants enhance the micronutrient availability from soil columns. Strong bonds are formed between biosurfactants (anionic) and metal cations, and complexes thus formed are promptly taken up by plant roots. Biosurfactants can bind directly to sorbed micronutrient elements and transfer them to the root zone interface under reduced interfacial tension condition (Singh and Cameotra 2004). As biosurfactants have two basic domains, that is, hydrophilic and hydrophobic groups, therefore they are able to lessen the surface tension and interfacial tension between two non-miscible liquids and to increase the solubility of inorganic and organic constituent of insoluble compounds (Bustamante et al. 2012). Among microbial surfactants, glycolipids, for example, rhamnolipid, trehalolipid, sophorolipid and mannosylerythritol, are the best known to form complex with metal ions (Edwards et al. 2003; Thavasi 2011). Application of biosurfactants, for example, rhamnolipid in soil, has been reported to effectively increase the interfacial tension between soil surfaces and metal cations and ultimately increase the solubility and mobility, in addition to desorption of trace metals or micronutrients that are weakly bound to the soil particles (Wang and Mulligan 2004). One of the more common ways to deliver the appropriate metal is to form a chelated complex of the metal ion with surfactant, biosurfactant or any synthetic chelator as it maintains the metal ion in soluble form for ease of application and reduces metal adsorption and fixation in soil (McLaughlin et al. 2012). According to Le Chatelier's principle, biosurfactant can chelate and form a complex with ionic and non-ionic metals and then release them in soil (Miller 1995). Due to the anionic characteristics, biosurfactant complexes with positively charged metals. So they can be used as metal-sequestering agents which permits desorption of metals from soil and contaminated water.

Conclusion

Incidence of Zn deficiency in soil and crop plants worldwide is correlated with low solubility of Zn compounds. Chemical fertilizers added to the soil for the improvement of crop productivity are quite costly to farmers, and also their excessive use will lead to the environmental pollution. Some new eco-friendly technologies are required in order to develop sustainable agriculture system that will reduce the dependency on chemical fertilizers and also can maintain proper crop yields. Zinc-solubilizing bacteria with plant growth-promoting traits can be used to increase the Zn bioavailability in soil and its subsequent uptake in cereals/grains. It will lessen the cost expenditure by reducing the use of expensive agrochemicals; provide a safe, self-sustainable, viable, low input and eco-friendly alternative of chemical fertilizers to be used in agricultural practices; and thus provide safe and healthy environment.

Bibliography

Alloway, B.J.: Zinc in Soils and Crop Nutrition, vol. 130. International Zinc Association, Brussels, Belgium (2004)

Alloway, B.J.: Zinc in Soils and Crop Nutrition, 2nd edn. Published by IZA and IFA, Brussels, Belgium and Paris, France (2008)

Alloway, B.J.: Soil factors associated with zinc deficiency in crops and humans. Environ. Geochem. Health. **31**(5), 537–548 (2009)

Ardakani, M.R., Mazaheri, D., Rad, A.S., Mafakheri, S.: Uptake of micronutrients by wheat (Triticum aestivum L.) in a sustainable agroecosystem. Middle East J. Sci. Res. **7**(4), 444–451 (2011)

Badrinath, P.S., Herle, K., Balakrishna, R., Janardhana Gowda, N.A.: Micronutrient studies in the soils of Someshwara villages Mangalore, Karanataka. Proceedings of the National Seminar on Micronutrients in Crop Production, pp. 41–44 (1987)

Bahrani, A., Pourreza, J., Joo, M.H.: Response of winter wheat to co-inoculation with Azotobacter and arbuscular mycorrhizal fungi (AMF) under different sources of nitrogen fertilizer. Am. Eurasian J. Agric. Environ. Sci. **8**(1), 95–103 (2010)

Balakrishnan, N., Subramanian, K.S.: Mycorrhizal symbiosis and bioavailability of micronutrients in maize grain. Maydica. **57**(2), 129–138 (2012)

Bali, S.K., Hundal, H.S., Singh, K., Singh, B.: GIS-aided mapping of DTPA extractable zinc and soil characteristics in the state of Punjab. J. Indian Soc. Soil Sci. **58**(2), 189–199 (2010)

Bapiri, A., Asgharzadeh, A., Mujallali, H., Khavazi, K., Pazira, E.: Evaluation of zinc solubilization potential by different strains of fluorescent pseudomonads. J. Appl. Sci. Environ. Manag. **16**(3) (2012)

Barbagelata, P.A., Mallarino, A.P.: Field correlation of potassium soil test methods based on dried and field-moist soil samples for corn and soybean. Soil Sci. Soc. Am. J. **77**(1), 318–327 (2013)

Bashir, K., Ishimaru, Y., Nishizawa, N.K.: Iron uptake and loading into rice grains. Rice. **3**(2–3), 122–130 (2010)

Batista, R.O., Furtini Neto, A.E., Deccetti, S.F.C., Viana, C.S.: Root morphology and nutrient uptake kinetics by Australian cedar clones. Revista Caatinga. **29**(1), 153–162 (2016)

Blindauer, C.A.: Advances in the molecular understanding of biological zinc transport. Chem. Commun. **51**(22), 4544–4563 (2015)

Brimecombe, M.J., De Leij, F.A., Lynch, J.M.: The effect of root exudates on rhizosphere microbial populations. In: The Rhizosphere: Biochemistry and Organic Substances at the Soil–Plant Interface (eds R. Pinton, Z. Varaninin & P. Nannipieri), pp.95–140. Marcel Dekker, New York (2001)

Broadley, M.R., White, P.J., Hammond, J.P., Zelko, I., Lux, A.: Zinc in plants. New Phytol. **173**(4), 677–702 (2007)

Brown, P.H., Cakmak, I., Zhang, Q.: Form and function of zinc plants. In: Zinc in Soils and Plants, pp. 93–106. Springer, Dordrecht (1993)

Bulut, S.: Evaluation of yield and quality parameters of phosphorous-solubilizing and N-fixing bacteria inoculated in wheat (Triticum aestivum L.). Turk. J. Agric. For. **37**(5), 545–554 (2013)

Bustamante, M., Duran, N., Diez, M.C.: Biosurfactants are useful tools for the bioremediation of contaminated soil: a review. J. Soil Sci. Plant Nutr. **12**(4), 667–687 (2012)

Cakmak, I.: Agronomic approaches in biofortification of food crops with micronutrients. The proceedings of the international plant nutrient colloquium XVI UC Davis (2009a)

Cakmak, I.: Enrichment of fertilizers with zinc: an excellent investment for humanity and crop production in India. J. Trace Elem. Med. Biol. **23**(4), 281–289 (2009b)

Cakmak, S., Gülüt, K.Y., Marschner, H., Graham, R.D.: Effect of zinc and iron deficiency on phytos1derophore release in wheat genotypes differing in zinc efficiency. J. Plant Nutr. **17**(1), 1–17 (1994)

Cakmak, I., Pfeiffer, W.H., McClafferty, B.: Biofortification of durum wheat with zinc and iron. Cereal Chem. **87**, 10–20 (2010)

Chatterjee, A.K., Mandal, L.N.: Adsorption and desorption of zinc in soils of different physicochemical characters. J. Indian Soc. Soil Sci. **33**(3), 669–671 (1985)

Chattopadhyay, T., Sahoo, A.K., Singh, R.S., Shyampura, R.L.: Available micronutrient status in the soils of Vindhyan scarplands of Rajasthan in relation to soil characteristics. J. Indian Soc. Soil Sci. **44**(4), 678–681 (1996)

Chaudhari, P.R., Ahire, D.V., Ahire, V.D.: Correlation between Physico-chemical properties and available nutrients in sandy loam soils of Haridwar. J. Chem. Biol. Phys. Sci. **2**(3), 1493–1500 (2012)

Chen, W.R., Feng, Y., Chao, Y.E.: Genomic analysis and expression pattern of OsZIP1, OsZIP3, and OsZIP4 in two rice (Oryza sativa L.) genotypes with different zinc efficiency. Russ. J. Plant Physiol. **55**(3), 400–409 (2008)

Chen, W.R., He, Z.L., Yang, X.E., Feng, Y.: Zinc efficiency is correlated with root morphology, ultrastructure, and antioxidative enzymes in rice. J. Plant Nutr. **32**(2), 287–305 (2009)

Chinnasamy, S.: Fixing critical limit of zinc for the rice soils of Cumbum valley, Madurai district. M.Sc. (Ag.) Thesis, Tamil Nadu Agricultural University, Coimbatore (1987)

Colangelo, E.P., Guerinot, M.L.: Put the metal to the petal: metal uptake and transport throughout plants. Curr. Opin. Plant Biol. **9**(3), 322–330 (2006)

Da Silva, J.F., Williams, R.J.P.: The Biological Chemistry of the Elements: The Inorganic Chemistry of Life. Oxford University Press, New York (2001)

Dalton, D.A., Russell, S.A., Evans, H.J.: Nickel as a micronutrient element for plants. BioFactors (Oxford, England). **1**(1), 11–16 (1988)

Das, K., Dang, R., Shivananda, T.N., Sur, P.: Interaction between phosphorus and zinc on the biomass yield and yield attributes of the medicinal plant stevia (Stevia rebaudiana). Sci. World J. **5**, 390–395 (2005)

Delaplace, P., Delory, B.M., Baudson, C., de Cazenave, M.M.S., Spaepen, S., Varin, S., Brostaux, Y., du Jardin, P.: Influence of rhizobacterial volatiles on the root system architecture and the production and allocation of biomass in the model grass Brachypodium distachyon (L.) P. Beauv. BMC Plant Biol. **15**(1), 1–15 (2015)

Delwiche, C.C., Johnson, C.M., Reisenauer, H.M.: Influence of cobalt on nitrogen fixation by Medicago. Plant Physiol. **36**(1), 73 (1961)

Desai, S., Kumar, P., Sultana, U., Pinisetty, S., Reddy, G.: Potential microbial candidate strains for management of nutrient requirements of crops. Afr. J. Microbiol. Res. **6**(17), 3924–3931 (2012)

Dinesh, R.: Delineation of micronutrient deficient areas and fixing critical limit of Zn for rice soil in Kamarajar district. M.Sc. (Ag.) Thesis, Tamil Nadu Agricultural University, Coimbatore (1988)

Dixon, N.E., Gazzola, C., Watters, J.J., Blakeley, R.L., Zerner, B.: Inhibition of jack bean urease (EC 3.5. 1.5) by acetohydroxamic acid and by phosphoramidate. Equivalent weight for urease. J. Am. Chem. Soc. **97**(14), 4130–4131 (1975)

Dong, B., Rengel, Z., Graham, R.D.: Root morphology of wheat genotypes differing in zinc efficiency. J. Plant Nutr. **18**(12), 2761–2773 (1995)

Doornbos, R.F., van Loon, L.C., Bakker, P.A.: Impact of root exudates and plant defense signaling on bacterial communities in the rhizosphere. A review. Agron. Sustain. Dev. **32**(1), 227–243 (2012)

Duraisamy, P., Kothandaraman, G.V., Chellamuthu, S.: Effect of amendments and zinc on the availability, content and uptake of zinc and iron by rice Bhavani in sodic soil. Madras Agric. J. **75**(3–4), 119–124 (1988)

Durmaz, E., Coruh, C., Dinler, G., Grusak, M.A., Peleg, Z., Saranga, Y., Fahima, T., Yazici, A., Ozturk, L., Cakmak, I., Budak, H.: Expression and cellular localization of ZIP1 transporter under zinc deficiency in wild emmer wheat. Plant Mol. Biol. Report. **29**(3), 582–596 (2011)

Edwards, K.R., Lepo, J.E., Lewis, M.A.: Toxicity comparison of biosurfactants and synthetic surfactants used in oil spill remediation to two estuarine species. Mar. Pollut. Bull. **46**(10), 1309–1316 (2003)

Eide, D.J.: Zinc transporters and the cellular trafficking of zinc. Biochim. Biophys. Acta Mol. Cell Res. **1763**(7), 711–722 (2006)

Epstein, E., Bloom, A.J.: Mineral nutrition of plants: principles and perspectives, 2nd edn. Sinauer Assoc. Inc., Sunderland, UK (2005)

Esitken, A., Yildiz, H.E., Ercisli, S., Donmez, M.F., Turan, M., Gunes, A.: Effects of plant growth promoting bacteria (PGPB) on yield, growth and nutrient contents of organically grown strawberry. Sci. Hortic. **124**(1), 62–66 (2010)

Evens, N.P., Buchner, P., Williams, L.E., Hawkesford, M.J.: The role of ZIP transporters and group F bZIP transcription factors in the Zn-deficiency response of wheat (Triticum aestivum). Plant J. **92**(2), 291–304 (2017)

Fasim, F., Ahmed, N., Parsons, R., Gadd, G.M.: Solubilization of zinc salts by a bacterium isolated from the air environment of a tannery. FEMS Microbiol. Lett. **213**(1), 1–6 (2002)

Gad, N.: Increasing the efficiency of nitrogen fertilization through cobalt application to pea plant. Res. J. Agric. Biol. Sci. **2**(6), 433–442 (2006)

Gandhi, A., Muralidharan, G., Sudhakar, E., Murugan, A.: Screening for elite zinc solubilizing bacterial isolate from rice rhizosphere environment. Int J Recent Sci Res. **5**, 2201–2204 (2014)

Genc, Y., McDonald, G.K., Graham, R.D.: Contribution of different mechanisms to zinc efficiency in bread wheat during early vegetative stage. Plant Soil. **281**(1–2), 353–367 (2006)

Genc, Y., Huang, C.Y., Langridge, P.: A study of the role of root morphological traits in growth of barley in zinc-deficient soil. J. Exp. Bot. **58**(11), 2775–2784 (2007)

Gerendás, J., Polacco, J.C., Freyermuth, S.K., Sattelmacher, B.: Significance of nickel for plant growth and metabolism. J. Plant Nutr. Soil Sci. **162**(3), 241–256 (1999)

Ghosh, G.K., Sarkar, A.K.: Availability of sulphur and some micronutrients in acid sedentary soils of Chotanagpur region. J. Indian Soc. Soil Sci. **42**(3), 464–466 (1994)

Giri, B., Giang, P.H., Kumari, R., Prasad, R., Varma, A.: Microbial diversity in soils. In: Buscot, F., Varma, S. (eds.) Micro-Organisms in Soils: Roles in Genesis and Functions, pp. 195–212. Springer, Heidelberg (2005)

Gontia-Mishra, I., Sapre, S., Sharma, A., Tiwari, S.: Amelioration of drought tolerance in wheat by the interaction of plant growth-promoting rhizobacteria. Plant Biol. **18**(6), 992–1000 (2016)

Gontia-Mishra, I., Sapre, S., Tiwari, S.: Zinc solubilizing bacteria from the rhizosphere of rice as prospective modulator of zinc biofortification in rice. Rhizosphere. **3**, 185–190 (2017)

Gosal, S.K., Karlupia, A., Gosal, S.S., Chhibba, I.M., Varma, A.: Biotization with Piriformospora indica and Pseudomonas fluorescens improves survival rate, nutrient acquisition, field performance and saponin content of micropropagated Chlorophytum sp. Indian J. Biotechnol. **9**, 289–297 (2010)

Goteti, P.K., Emmanuel, L.D.A., Desai, S., Shaik, M.H.A.: Prospective zinc solubilising bacteria for enhanced nutrient uptake and growth promotion in maize (Zea mays L.). Int. J. Microbiol. **2013**, 869697 (2013)

Gouda, S., Kerry, R.G., Das, G., Paramithiotis, S., Shin, H.S., Patra, J.K.: Revitalization of plant growth promoting rhizobacteria for sustainable development in agriculture. Microbiol. Res. **206**, 131–140 (2018)

Gries, D., Klatt, S., Runge, M.: Copper-deficiency-induced phytosiderophore release in in the calcicole grass Hordelymus europaeus. New Phytol. **140**(1), 95–101 (1998)

Grotz, N., Fox, T., Connolly, E., Park, W., Guerinot, M.L., Eide, D.: Identification of a family of zinc transporter genes from Arabidopsis that respond to zinc deficiency. Proc. Natl. Acad. Sci. **95**(12), 7220–7224 (1998)

Guerinot, M.L.: The ZIP family of metal transporters. Biochim. Biophys. Acta Biomembranes. **1465**(1–2), 190–198 (2000)

Guerinot, M.L., Eidet, D.: Zeroing in on zinc uptake in yeast and plants. Curr. Opin. Plant Biol. **2**(3), 244–249 (1999)

Gupta, R.K., Van den Elshout, S.E.F., Abrol, I.P.: Effect of pH on zinc adsorption-precipitation reactions in an alkali soil. Soil Sci. **143**(3), 198–204 (1987)

Gupta, N., Trivedi, S.K., Bansal, K.N., Kaul, R.K.: Vertical distribution of micronutrient cations in some soil series of northern Madhya Pradesh. J. Indian Soc. Soil Sci. **51**(4), 517–522 (2003)

Gurmani, A.R., Khan, S.U., Andaleep, R., Waseem, K., Khan, A.: Soil application of zinc improves growth and yield of tomato. Int. J. Agric. Biol. **14**(1), 91–96 (2012)

Hafeez, F.Y., Abaid-Ullah, M., Hassan, M.N.: Plant growth-promoting rhizobacteria as zinc mobilizers: a promising approach for cereals biofortification. In: Bacteria in Agrobiology: Crop Productivity, pp. 217–235. Springer, Berlin, Heidelberg (2013a)

Hafeez, B., Khanif, Y.M., Saleem, M.: Role of zinc in plant nutrition-a review. J. Exp. Agric. Int. **3**(2), 374–391 (2013b)

Hajiboland, R., Yang, X.E., Römheld, V., Neumann, G.: Effect of bicarbonate on elongation and distribution of organic acids in root and root zone of Zn-efficient and Zn-inefficient rice (Oryza sativa L.) genotypes. Environ. Exp. Bot. **54**(2), 163–173 (2005)

Hallsworh, E.G., Wilson, S.B., Adams, W.A.: Effect of cobalt on the non-nodulated legume. Nature. **205**(4968), 307–308 (1965)

Hasmot, S., Mrinmay, D.: Notes on available zinc status of some Indian soils. Indian Agriculturist. **44**(3/4), 207–210 (2000)

Havlin, J.L., Beaton, J.D., Tisdale, S.L., Nelson, W.L.: Soil Fertility and Fertilizers: An Introduction to Nutrient Management. Pearson Prentice Hall Inc., Upper Saddle River (2005)

Hazra, G.C., Mandal, B., Mandal, L.N.: Distribution of zinc fractions and their transformation in submerged rice soils. Plant Soil. **104**(2), 175–181 (1987)

Hodgson, J.F.: Chemistry of the micronutrient elements in soils. In: Advances in Agronomy, vol. 15, pp. 119–159. Academic Press, United States (1963)

Hunt, J.R.: Bioavailability of iron, zinc, and other trace minerals from vegetarian diets. Am. J. Clin. Nutr. **78**(3), 633S–639S (2003)

Hussain, S., Maqsood, M.A., Rahmatullah: Zinc release characteristics from calcareous soils using diethylenetriaminepentaacetic acid and other organic acids. Commun. Soil Sci. Plant Anal. **42**(15), 1870–1881 (2011)

Hussain, A., Arshad, M., Zahir, Z.A., Asghar, M.: Prospects of zinc solubilizing bacteria for enhancing growth of maize. Pak. J. Agric. Sci. **52**(4), 915–922 (2015)

Hussain, A., Zahir, Z.A., Asghar, H.N., Ahmad, M., Jamil, M., Naveed, M., Akhtar, M.F.U.Z.: Zinc solubilizing bacteria for zinc biofortification in cereals: a step toward sustainable nutritional security. In: Role of Rhizospheric Microbes in Soil, pp. 203–227. Springer, Singapore (2018)

Ishimaru, Y., Suzuki, M., Kobayashi, T., Takahashi, M., Nakanishi, H., Mori, S., Nishizawa, N.K.: OsZIP4, a novel zinc-regulated zinc transporter in rice. J. Exp. Bot. **56**(422), 3207–3214 (2005)

Jibhakate, S.B., Raut, M.M., Bhende, S.N., Kharche, V.K.: Micronutrient status of soils of Katol Tahasil in Nagpur district and their relationship with some soil properties. J. Soils Crops. **19**(1), 143–146 (2009)

Joshi, D., Negi, G., Vaid, S., Sharma, A.: Enhancement of wheat growth and Zn content in grains by zinc solubilizing bacteria. Int. J. Agric. Environ. Biotechnol. **6**(3), 363–370 (2013)

Kamran, S., Shahid, I., Baig, D.N., Rizwan, M., Malik, K.A., Mehnaz, S.: Contribution of zinc solubilizing bacteria in growth promotion and zinc content of wheat. Front. Microbiol. **8**, 2593 (2017)

Khande, R., Sharma, S.K., Ramesh, A., Sharma, M.P.: Zinc solubilizing Bacillus strains that modulate growth, yield and zinc biofortification of soybean and wheat. Rhizosphere. **4**, 126–138 (2017)

Kim, S.A., Guerinot, M.L.: Mining iron: iron uptake and transport in plants. FEBS Lett. **581**(12), 2273–2280 (2007)

Kobayashi, M., Shimizu, S.: Cobalt proteins. Eur. J. Biochem. **261**(1), 1–9 (1999)

Kothari, S.K., Marschner, H., George, E.: Effect of VA mycorrhizal fungi and rhizosphere microorganisms on root and shoot morphology, growth and water relations in maize. New Phytol. **116**(2), 303–311 (1990)

Kotula, L., Ranathunge, K., Steudle, E.: Apoplastic barriers effectively block oxygen permeability across outer cell layers of rice roots under deoxygenated conditions: roles of apoplastic pores and of respiration. New Phytol. **184**(4), 909–917 (2009)

Kramer, D., Römheld, V., Landsberg, E., Marschner, H.: Induction of transfer-cell formation by iron deficiency in the root epidermis of Helianthus annuus L. Planta. **147**(4), 335–339 (1980)

Krithika, S., Balachandar, D.: Expression of zinc transporter genes in rice as influenced by zinc-solubilizing Enterobacter cloacae strain ZSB14. Front. Plant Sci. **7**, 446 (2016)

Kucey, R.M.N.: Effect of Penicillium bilaji on the solubility and uptake of P and micronutrients from soil by wheat. Can. J. Soil Sci. **68**(2), 261–270 (1988)

Kumar, P., Dubey, R.C., Maheshwari, D.K.: Bacillus strains isolated from rhizosphere showed plant growth promoting and antagonistic activity against phytopathogens. Microbiol. Res. **167**(8), 493–499 (2012)

Kumar, A., Dewangan, S., Lawate, P., Bahadur, I., Prajapati, S.: Zinc-solubilizing bacteria: a boon for sustainable agriculture. In: Plant growth promoting rhizobacteria for sustainable stress management, pp. 139–155. Springer, Singapore (2019)

Kumawat, N., Kumar, R., Khandkar, U.R., Yadav, R.K., Saurabh, K., Mishra, J.S., Dotaniya, M.L., Hans, H.: Silicon (Si)-and Zinc (Zn)-solubilizing microorganisms: role in sustainable agriculture. In: Biofertilizers for Sustainable Agriculture and Environment, pp. 109–135. Springer, Cham (2019)

Kumssa, D.B., Joy, E.J., Ander, E.L., Watts, M.J., Young, S.D., Walker, S., Broadley, M.R.: Dietary calcium and zinc deficiency risks are decreasing but remain prevalent. Sci. Rep. **5**, 10974 (2015)

Kutman, U.B., Yildiz, B., Ozturk, L., Cakmak, I.: Biofortification of durum wheat with zinc through soil and foliar applications of nitrogen. Cereal Chem. **87**(1), 1–9 (2010)

Li, C.Z., Li, C.Z., Wang, D., Wang, G.X.: The protective effects of cobalt on potato seedling leaves during osmotic stress. Bot. Bull. Acad. Sin. **46, (119–125)** (2005)

Li, M., Ahammed, G.J., Li, C., Bao, X., Yu, J., Huang, C., Yin, H., Zhou, J.: Brassinosteroid ameliorates zinc oxide nanoparticles-induced oxidative stress by improving antioxidant potential and redox homeostasis in tomato seedling. Front. Plant Sci. **7**, 615 (2016)

Liu, A., Hamel, C., Hamilton, R.I., Ma, B.L., Smith, D.L.: Acquisition of Cu, Zn, Mn and Fe by mycorrhizal maize (Zea mays L.) grown in soil at different P and micronutrient levels. Mycorrhiza. **9**(6), 331–336 (2000)

López-Bucio, J., Campos-Cuevas, J.C., Hernández-Calderón, E., Velásquez-Becerra, C., Farías-Rodríguez, R., Macías-Rodríguez, L.I., Valencia-Cantero, E.: Bacillus megaterium rhizobacteria promote growth and alter root-system architecture through an auxin-and ethylene-independent signaling mechanism in Arabidopsis thaliana. Mol. Plant-Microbe Interact. **20**(2), 207–217 (2007)

López-Gómez, M., Hidalgo-Castellanos, J., Lluch, C., Herrera-Cervera, J.A.: 24-Epibrassinolide ameliorates salt stress effects in the symbiosis Medicago truncatula-Sinorhizobium meliloti and regulates the nodulation in cross-talk with polyamines. Plant Physiol. Biochem. **108**, 212–221 (2016)

Lukaski, H.C.: Vitamin and mineral status: effects on physical performance. Nutrition. **20**(7–8), 632–644 (2004)

Lynch, J.P.: Roots of the second green revolution. Aust. J. Bot. **55**(5), 493–512 (2007)

Mandal, B., Hazra, G.C.: Zinc adsorption in soils as influenced by different soil management practices. Soil Sci. **162**(10), 713–721 (1997)

Marschner, H.: Zinc uptake from soils. In: Zinc in Soils and Plants, pp. 59–77. Springer, Dordrecht (1993)

Marschner, H.: Mineral Nutrition of Higher Plants, 2nd edn, pp. 330–355. Academic Press, London (1995)

Martino, E., Perotto, S., Parsons, R., Gadd, G.M.: Solubilization of insoluble inorganic zinc compounds by ericoid mycorrhizal fungi derived from heavy metal polluted sites. Soil Biol. Biochem. **35**(1), 133–141 (2003)

Mäser, P., Thomine, S., Schroeder, J.I., Ward, J.M., Hirschi, K., Sze, H., Talke, I.N., Amtmann, A., Maathuis, F.J., Sanders, D., Harper, J.F.: Phylogenetic relationships within cation transporter families of Arabidopsis. Plant Physiol. **126**(4), 1646–1667 (2001)

McLaughlin, M., Stacey, S., Lombi, E.: Sequestering Agent for Micronutrient Fertilisers. United States Patent 8217004. United States Patent and Trademark Office, Adelaide (2012)

Mendes, R., Garbeva, P., Raaijmakers, J.M.: The rhizosphere microbiome: significance of plant beneficial, plant pathogenic, and human pathogenic microorganisms. FEMS Microbiol. Rev. **37**(5), 634–663 (2013)

Miller, R.M.: Biosurfactant-facilitated remediation of metal-contaminated soils. Environ. Health Perspect. **103**(suppl 1), 59–62 (1995)

Miransari, M.: Hyperaccumulators, arbuscular mycorrhizal fungi and stress of heavy metals. Biotechnol. Adv. **29**(6), 645–653 (2011)

Misra, S.B., Pandey, G.: Zinc in saline and alkali soils of Uttar Pradesh. J. Indian Soc. Soil Sci. **24**, 336–341 (1977)

Moraghan, J.T., Mascagni Jr., H.J.: Environmental and soil factors affecting micronutrient deficiencies and toxicities. In: Mordvedt, J.J., Cox, F.R., Shuman, L.M., Welch, R.M. (eds.) Micronutrients in Agriculture, pp. 371–425. Soil Science Society of America, Madison (1991)

Mortvedt, J.J., Murphy, L.S., Follett, R.H.: Fertilizer Technology and Application Willoughby. Meister Publishing Co., Ohio (1999)

Murthy, A.S.P.: Distribution, properties, and management of vertisols of India. In: Advances in Soil Science, pp. 151–214. Springer, New York (1988)

Murthy, A.S.P., Ponnamperuma, F.N.: Association of Zn with Fe and Al2O3 in some wetland rice soil of Philippines. J. Indian Soc. Soil Sci. **34**, 886–888 (1986)

Najafi-Ghiri, M., Ghasemi-Fasaei, R., Farrokhnejad, E.: Factors affecting micronutrient availability in calcareous soils of Southern Iran. Arid Land Res. Manag. **27**(3), 203–215 (2013)

Nath, D., Maurya, B.R., Meena, V.S.: Documentation of five potassium-and phosphorus-solubilizing bacteria for their K and P-solubilization ability from various minerals. Biocatal. Agric. Biotechnol. **10**, 174–181 (2017)

Nayak, D.C., Mukhopadhyay, S., Sarkar, D.: Distribution of some available micronutrients in alluvial soils of Arunachal Pradesh in relation to soil characteristics. J. Indian Soc. Soil Sci. **48**(3), 612–614 (2000)

Nyoki, D., Ndakidemi, P.A.: Effects of phosphorus and Bradyrhizobium japonicum on growth and chlorophyll content of cowpea (Vigna unguiculata (L) Walp). J. Exp. Agric. Int., **4**(10): 1120–1136 (2014)

Obrador, A., Novillo, J., Alvarez, J.M.: Mobility and availability to plants of two zinc sources applied to a calcareous soil. Soil Sci. Soc. Am. J. **67**(2), 564–572 (2003)

Ortíz-Castro, R., Valencia-Cantero, E., López-Bucio, J.: Plant growth promotion by Bacillus megaterium involves cytokinin signaling. Plant Signal. Behav. **3**(4), 263–265 (2008)

Othman, N.M.I., Othman, R., Saud, H.M., Wahab, P.E.M.: Effects of root colonization by zinc-solubilizing bacteria on rice plant (Oryza sativa MR219) growth. Agric. Nat. Res. **51**(6), 532–537 (2017)

Pacwa-Płociniczak, M., Płaza, G.A., Piotrowska-Seget, Z., Cameotra, S.S.: Environmental applications of biosurfactants: recent advances. Int. J. Mol. Sci. **12**(1), 633–654 (2011)

Park, M.S., Singvilay, O., Seok, Y.S., Chung, J.B., Ahn, K.S., Sa, T.M.: Effect of phosphate solubilizing Fungi on P uptake and growth of tobacco in rock phosphate applied soil. Korean J. Soil Sc. Fertiliz. **36**(4), 233–238 (2003)

Parmar, P., Sindhu, S.S.: The novel and efficient method for isolating potassium solubilizing bacteria from rhizosphere soil. Geomicrobiol J. **36**(2), 130–136 (2019)

Pawar, A., Ismail, S., Mundhe, S., Patil, V.D.: Solubilization of insoluble zinc compounds by different microbial isolates in vitro condition. Int. J. Trop. Agric. **33**(2 (Part II)), 865–869 (2015)

Pedler, J.F., Parker, D.R., Crowley, D.E.: Zinc deficiency-induced phytosiderophore release by the Triticaceae is not consistently expressed in solution culture. Planta. **211**(1), 120–126 (2000)

Perumal, M.D., Subramanian, V., Sabarinathan, K.G.: Evaluation of zinc solubilizing potential of maize rhizosphere bacterial isolates. Int. J. Curr. Microbiol. Appl. Sci. **6**(12), 864–869 (2017)

Pharande, A.L., Raskar, B.N., Nipunage, M.V.: Micronutrient status of important vertisol and Alfisol soil series of Western Maharashtra. J. Maharashtra Agric. Univ. **21**, 182–185 (1996)

Pittman, J.K.: Managing the manganese: molecular mechanisms of manganese transport and homeostasis. New Phytol. **167**(3), 733–742 (2005)

Prasad, R., Sakal, R.: Effect of soil properties on different chemical pools of zinc in calcareous soils. J. Indian Soc. Soil Sci. **36**(2), 246–251 (1988)

Prasad, R., Prasad, B.L., Sakal, R.: Effect of submergence on native DTPA extractable Zn, Fe, Cu and Mn in some old alluvial rice soils of South Bihar. J. Indian Soc. Soil Sci. **39**, 773–776 (1991)

Prasad, R., Prasad, B.L., Sakal, R.: Effect of submergence on the transformation of zinc forms in old alluvial soils growing rice as related to soil properties-II. Transformation of applied zinc. J. Indian Soc. Soil Sci. **44**(1), 74–76 (1996)

Purakayastha, T.J., Chhonkar, P.K.: Influence of vesicular-arbuscular mycorrhizal fungi (Glomus etunicatum L.) on mobilization of zinc in wetland rice (Oryza sativa L.). Biol. Fertil. Soils. **33**(4), 323–327 (2001)

Rahman, M.A., Hasegawa, H., Kadohashi, K., Maki, T., Ueda, K.: Hydroxyiminodisuccinic acid (HIDS): a novel biodegradable chelating ligand for the increase of iron bioavailability and arsenic phytoextraction. Chemosphere. **77**(2), 207–213 (2009)

Raj, K., Nayyar, V.K., Sidhu, G.S., Deshmukh, S.N.: Distribution of available micronutrient cations in some dominant soil series in different physiographic units of Bundelkhand region of Madhya Pradesh. J. Indian Soc. Soil Sci. **38**(3), 410–415 (1990)

Ramesh, S.A., Shin, R., Eide, D.J., Schachtman, D.P.: Differential metal selectivity and gene expression of two zinc transporters from rice. Plant Physiol. **133**(1), 126–134 (2003)

Ramesh, A., Sharma, S.K., Sharma, M.P., Yadav, N., Joshi, O.P.: Inoculation of zinc solubilizing Bacillus aryabhattai strains for improved growth, mobilization and biofortification of zinc in soybean and wheat cultivated in Vertisols of Central India. Appl. Soil Ecol. **73**, 87–96 (2014)

Rana, A., Saharan, B., Nain, L., Prasanna, R., Shivay, Y.S.: Enhancing micronutrient uptake and yield of wheat through bacterial PGPR consortia. Soil Sci. Plant Nutr. **58**(5), 573–582 (2012)

Randhawa, H.S., Singh, S.P.: Distribution of zinc fractions in alluvium-derived soils of Punjab. J. Indian Soc. Soil Sci. **43**(1), 124–126 (1995)

Rasouli-Sadaghiani, M., Sadeghzadeh, B., Sepehr, E., Rengel, Z.: Root exudation and zinc uptake by barley genotypes differing in Zn efficiency. J. Plant Nutr. **34**(8), 1120–1132 (2011)

Ravi, P., Bhupal Raj, G., Chandrasekhar Rao, P.: Distribution of DTPA extractable micro nutrients in rice growing soils of Karimnagar District of Andhra Pradesh. Helix. **1**, 494–497 (2014)

Rêgo, M.C.F., Ilkiu-Borges, F., de Filippi, M.C.C., Gonçalves, L.A., da Silva, G.B.: Morphoanatomical and biochemical changes in the roots of rice plants induced by plant growth-promoting microorganisms. Embrapa Arroz e Feijão-Artigo em periódico indexado (ALICE). Aust. J. Bot. **10**, 818797 (2014)

Rehman, H.U., Aziz, T., Farooq, M., Wakeel, A., Rengel, Z.: Zinc nutrition in rice production systems: a review. Plant Soil. **361**(1–2), 203–226 (2012)

Rengel, Z.: Agronomic approaches to increasing zinc concentration in staple food crops. In: Cakmak, I., Welch, R.M. (eds.) Impacts of Agriculture on Human Health and Nutrition, pp. 363–381. UNESCO, EOLSS Publishers, Oxford, UK (2002)

Rengel, Z., Wheal, M.S.: Kinetic parameters of Zn uptake by wheat are affected by the herbicide chlorsulfuron. J. Exp. Bot. **48**(4), 935–941 (1997)

Rengel, Z., Batten, G.D., Crowley, D.D.: Agronomic approaches for improving the micronutrient density in edible portions of field crops. Field Crop Res. **60**(1–2), 27–40 (1999)

Rokhbakhsh-Zamin, F., Sachdev, D., Kazemi-Pour, N., Engineer, A., Pardesi, K.R., Zinjarde, S., Dhakephalkar, P.K., Chopade, B.A.: Characterization of plant-growth-promoting traits of Acinetobacter species isolated from rhizosphere of Pennisetum glaucum. J. Microbiol. Biotechnol. **21**(6), 556–566 (2011)

Sahoo, A.K., Chattopadhyay, T., Singh, R.S., Shyampura, R.L.: Characterization of soils and vertical distribution of available micronutrients under different landforms of Malwa plateau, Rajasthan. Indian Agric. **47**, 217–223 (2003)

Samanta, A., Chatterjee, A.K., Mete, P.K., Mandal, B.: Status of total and available iron and zinc in soils of West Bengal under continuous cultivation of mulberry. J. Indian Soc. Soil Sci. **50**(1), 35–42 (2002)

Sangwan, B.S., Singh, K.: Vertical distribution of Zn, Mn, Cu and Fe in the semi-arid soils of Haryana and their relationships with soil properties. J. Indian Soc. Soil Sci. **41**(3), 463–467 (1993)

Sarathambal, C., Thangaraju, M., Paulraj, C., Gomathy, M.: Assessing the zinc solubilization ability of Gluconacetobacter diazotrophicus in maize rhizosphere using labelled 65 Zn compounds. Indian J. Microbiol. **50**(1), 103–109 (2010)

Saravanan, V.S., Subramoniam, S.R., Raj, S.A.: Assessing in vitro solubilization potential of different zinc solubilizing bacterial (ZSB) isolates. Braz. J. Microbiol. **35**(1–2), 121–125 (2004)

Saravanan, V.S., Kalaiarasan, P., Madhaiyan, M., Thangaraju, M.: Solubilization of insoluble zinc compounds by Gluconacetobacter diazotrophicus and the detrimental action of zinc ion (Zn2+) and zinc chelates on root knot nematode Meloidogyne incognita. Lett. Appl. Microbiol. **44**(3), 235–241 (2007a)

Saravanan, V.S., Madhaiyan, M., Thangaraju, M.: Solubilization of zinc compounds by the diazotrophic, plant growth promoting bacterium Gluconacetobacter diazotrophicus. Chemosphere. **66**(9), 1794–1798 (2007b)

Saravanan, V.S., Kumar, M.R., Sa, T.M.: Microbial zinc solubilization and their role on plants. In: Bacteria in Agrobiology: Plant Nutrient Management, pp. 47–63. Springer, Berlin, Heidelberg (2011)

Sazawal, S., Black, R.E., Menon, V.P., Dinghra, P., Caulfield, L.E., Dhingra, U., Bagati, A.: Zinc supplementation in infants born small for gestational age reduces mortality: a prospective, randomized, controlled trial. Pediatrics. **108**(6), 1280–1286 (2001)

Schnug, E., Fleckenstein, J., Haneklaus, S.: Factors affecting available micronutrient concentrations in soils using coca-cola® as extractant. Commun. Soil Sci. Plant Anal. **29**(11–14), 1891–1896 (1998)

Schwarz, G., Mendel, R.R.: Molybdenum cofactor biosynthesis and molybdenum enzymes. Annu. Rev. Plant Biol. **57**, 623–647 (2006)

Shahab, S., Ahmed, N., Khan, N.S.: Indole acetic acid production and enhanced plant growth promotion by indigenous PSBs. Afr. J. Agric. Res. **4**(11), 1312–1316 (2009)

Shakeel, M., Rais, A., Hassan, M.N., Hafeez, F.Y.: Root associated Bacillus sp. improves growth, yield and zinc translocation for basmati rice (Oryza sativa) varieties. Front. Microbiol. **6**, 1286 (2015)

Sharifi, P., Paymozd, M.: Effect of zinc, iron and manganese on yield and yield components of green beans. Curr. Opin. Agric. **5**(1), 15 (2016)

Sharma, J.C., Chaudhary, S.K.: Vertical distribution of micronutrient cations in relation to soil characteristics in lower Shiwaliks of Solan district in North-West Himalayas. J. Indian Soc. Soil Sci. **55**(1), 40–44 (2007)

Sharma, S.K., Lal, F.: Status of DTPA extractable Zn in soils of humid southern zone of Rajasthan. J. Indian Soc. Soil Sci. **40**(2), 393–394 (1992)

Sharma, P.N., Chatterjee, C., Sharma, C.P., Agarwala, S.C.: Zinc deficiency and anther development in maize. Plant Cell Physiol. **28**(1), 11–18 (1987)

Sharma, R.Á., Singh, M., Sharma, J.Á.: Correlation studies on micronutrients vis-à-vis soil properties in some soils of Nagaur district in semi-arid region of Rajasthan. J. Indian Soc. Soil Sci. **51**(4), 522–527 (2003)

Sharma, B.D., Arora, H., Kumar, R., Nayyar, V.K.: Relationships between soil characteristics and total and DTPA-extractable micronutrients in Inceptisols of Punjab. Commun. Soil Sci. Plant Anal. **35**(5–6), 799–818 (2004)

Sharma, S.K.M.P., Ramesh, A., Joshi, O.P.: Characterization of zinc-solubilizing Bacillus isolates and their potential to influence zinc assimilation in soybean seeds. J. Microbiol. Biotechnol. **22**, 352–359 (2011)

Sharma, S.K., Sharma, M.P., Ramesh, A., Joshi, O.P.: Characterization of zinc- solubilizing Bacillus isolates and their potential to influence zinc assimilation in soybean seeds. J. Microbiol. Biotechnol. **22**, 352–359 (2012)

Sharma, P., Kumawat, K.C., Kaur, S., Kaur, N.: Assessment of zinc solubilization by endophytic bacteria in legume rhizosphere. Indian J. Appl. Res. **4**(6), 439–441 (2014)

Shuman, L.M.: Micronutrient fertilizers. J. Crop. Prod. **1**(2), 165–195 (1998)

Sidhu, G.S., Sharma, B.D.: Diethylenetriaminepentaacetic acid–extractable micronutrients status in soil under a rice–wheat system and their relationship with soil properties in different agroclimatic zones of Indo-Gangetic Plains of India. Commun. Soil Sci. Plant Anal. **41**(1), 29–51 (2010)

Sindhu, S.S., Sharma, R., Sindhu, S., Phour, M.: Plant nutrient management through inoculation of zinc-solubilizing bacteria for sustainable agriculture. In: Biofertilizers for Sustainable Agriculture and Environment, pp. 173–201. Springer, Cham (2019)

Singh, M.V., Abrol, I.P.: Solubility and adsorption of zinc in a sodic soil. Soil Sci. **140**(6), 406–411 (1985)

Singh, P., Cameotra, S.S.: Potential applications of microbial surfactants in biomedical sciences. Trends Biotechnol. **22**, 142–146 (2004)

Singh, B., Natesan, S.K.A., Singh, B.K., Usha, K.: Improving zinc efficiency of cereals under zinc deficiency. Curr. Sci. **88**(1–10), 36–44 (2005)

Singh, B., Pandey, R., Jain, V., Paul, V.: Practical Manual on Mineral Nutrition, pp. 1–60. IARI Publication, New Delhi (2007)

Singh, D., Rajawat, M.V.S., Kaushik, R., Prasanna, R., Saxena, A.K.: Beneficial role of endophytes in biofortification of Zn in wheat genotypes varying in nutrient use efficiency grown in soils sufficient and deficient in Zn. Plant Soil. **416**(1–2), 107–116 (2017a)

Singh, D., Geat, N., Rajawat, M.V.S., Mahajan, M.M., Prasanna, R., Singh, S., Kaushik, R., Singh, R.N., Kumar, K., Saxena, A.K.: Deciphering the mechanisms of endophyte-mediated biofortification of Fe and Zn in wheat. J. Plant Growth Regulat. **37**(1), 174–182 (2017b)

Singh, D., Geat, N., Rajawat, M.V.S., Prasanna, R., Kar, A., Singh, A.M., Saxena, A.K.: Prospecting endophytes from different Fe or Zn accumulating wheat genotypes for their influence as inoculants on plant growth, yield, and micronutrient content. Ann. Microbiol. **68**(12), 815–833 (2018)

Sirko, A., Brodzik, R.: Plant ureases: roles and regulation. Acta Biochim. Pol. **47**(4), 1189–1195 (2000)

Soltani, S., Khoshgoftarmanesh, A.H., Afyuni, M., Shrivani, M., Schulin, R.: The effect of preceding crop on wheat grain zinc concentration and its relationship to total amino acids and dissolved organic carbon in rhizosphere soil solution. Biol. Fertil. Soils. **50**(2), 239–247 (2014)

Srinivasarao, C., Vittal, K.P.R., Venkateswarlu, B., Wani, S.P., Sahrawat, K.L., Marimuthu, S., Kundu, S.: Carbon stocks in different soil types under diverse rainfed production systems in tropical India. Commun. Soil Sci. Plant Anal. **40**(15–16), 2338–2356 (2009)

Srivastava, P.C., Srivastava, P.: Integration of soil pH with soil-test values of zinc for prediction of yield response in rice grown in mollisols. Commun. Soil Sci. Plant Anal. **39**(15–16), 2456–2468 (2008)

Subramanian, K.S., Tenshia, V., Jayalakshmi, K., Ramach, V.: Role of arbuscular mycorrhizal fungus (Glomus intraradices) (fungus aided) in zinc nutrition of maize. J. Agric. Biotechnol. Sustain. Develop. **1**(1), 029–038 (2009)

Sunithakumari, K., Devi, S.P., Vasandha, S.: Zinc solubilizing bacterial isolates from the agricultural fields of Coimbatore, Tamil Nadu, India. Curr. Sci., 196–205 (2016)

Syranidou, E., Christofilopoulos, S., Gkavrou, G., Thijs, S., Weyens, N., Vangronsveld, J., Kalogerakis, N.: Exploitation of endophytic bacteria to enhance the phytoremediation potential of the wetland helophyte Juncus acutus. Front. Microbiol. **7**, 1016 (2016)

Takkar, P.N., Nayyar, V.K., Bansal, R.L., Dwivedi, R.S., Maan, M.S.: Annual Progress Report of the ICAR Coordinated Micronutrient Scheme 1976–77. Punjab Agricultural University, Ludhiana (1977)

Tariq, M., Hameed, S., Malik, K.A., Hafeez, F.Y.: Plant root associated bacteria for zinc mobilization in rice. Pak. J. Bot. **39**(1), 245–253 (2007)

Tarkalson, D.D., Jolley, V.D., Robbins, C.W., Terry, R.E.: Mycorrhizal colonization and nutrient uptake of dry bean in manure and compost manure treated subsoil and untreated topsoil and subsoil. J. Plant Nutr. **21**(9), 1867–1878 (1998)

Terzano, R., Cesco, S., Mimmo, T.: Dynamics, thermodynamics and kinetics of exudates: crucial issues in understanding rhizosphere processes. Plant Soil. **386**(1–2), 399–406 (2015)

Thavasi, R.: Microbial biosurfactants: from an environmental application point of view. J. Bioremed. Biodegrad. **2**(5), 104e (2011)

Tolay, I., Erenoglu, B., Römheld, V., Braun, H.J., Cakmak, I.: Phytosiderophore release in Aegilops tauschii and Triticum species under zinc and iron deficiencies. J. Exp. Bot. **52**(358), 1093–1099 (2001)

Treeby, M., Marschner, H., Römheld, V.: Mobilization of iron and other micronutrient cations from a calcareous soil by plant-borne, microbial, and synthetic metal chelators. Plant Soil. **114**(2), 217–226 (1989)

Uchida, R.: Essential nutrients for plant growth: nutrient functions and deficiency symptoms. In: Silva, J.A., Uchida, R. (eds.) Plant Nutrient Management in Hawaii's Soils, Approaches for Tropical and Subtropical Agriculture Human Resources, pp. 31–55. College of Tropical Agriculture and Human Resources, University of Hawaii at Manoa, Honolulu (2000)

Ullah, A., Heng, S., Munis, M.F.H., Fahad, S., Yang, X.: Phytoremediation of heavy metals assisted by plant growth promoting (PGP) bacteria: a review. Environ. Exp. Bot. **117**, 28–40 (2015)

Vacheron, J., Desbrosses, G., Bouffaud, M.L., Touraine, B., Moënne-Loccoz, Y., Muller, D., Legendre, L., Wisniewski-Dyé, F., Prigent-Combaret, C.: Plant growth-promoting rhizobacteria and root system functioning. Front. Plant Sci. **4**, 356 (2013)

Vaid, S.K., Gangwar, B.K., Sharma, A., Srivastava, P.C., Singh, M.V.: Effect of zinc solubilizing bioinoculants on zinc nutrition of wheat (Triticum aestivum L.). Int. J. Adv. Res. **1**(9), 805–820 (2013)

Vaid, S.K., Kumar, B., Sharma, A., Shukla, A.K., Srivastava, P.C.: Effect of Zn solubilizing bacteria on growth promotion and Zn nutrition of rice. J. Soil Sci. Plant Nutr. **14**(4), 889–910 (2014)

Verma, J.P., Yadav, J., Tiwari, K.N., Lavakush, S., Singh, V.: Impact of plant growth promoting rhizobacteria on crop production. Int. J. Agric. Res. **5**(11), 954–983 (2010)

Verma, R., Maurya, B.R., Meena, V.S., Dotaniya, M.L., Deewan, P., Jajoria, M.: Enhancing production potential of cabbage and improves soil fertility status of Indo-Gangetic plain through application of bio-organics and mineral fertilizer. Int. J. Curr. Microbiol. App. Sci. **6**(3), 301–309 (2017)

Walsh, C.T., Orme-Johnson, W.H.: Nickel enzymes. Biochemistry. **26**(16), 4901–4906 (1987)

Wang, S., Mulligan, C.N.: Rhamnolipid foam enhanced remediation of cadmium and nickel contaminated soil. Water Air Soil Pollut. **157**(1–4), 315–330 (2004)

Wang, H., Dong, Q., Zhou, J., Xiang, X.: Zinc phosphate dissolution by bacteria isolated from an oligotrophic karst cave in Central China. Front. Earth Sci. **7**(3), 375–383 (2013)

Wang, Y., Yang, X., Zhang, X., Dong, L., Zhang, J., Wei, Y., Feng, Y., Lu, L.: Improved plant growth and Zn accumulation in grains of rice (Oryza sativa L.) by inoculation of endophytic microbes isolated from a Zn Hyperaccumulator, Sedum alfredii H. J. Agric. Food Chem. **62**(8), 1783–1791 (2014)

Wani, P.A., Khan, M.S., Zaidi, A.: Effect of metal-tolerant plant growth-promoting rhizobium on the performance of pea grown in metal-amended soil. Arch. Environ. Contam. Toxicol. **55**(1), 33–42 (2008)

Watts-Williams, S.J., Cavagnaro, T.R.: Arbuscular mycorrhizal fungi increase grain zinc concentration and modify the expression of root ZIP transporter genes in a modern barley (Hordeum vulgare) cultivar. Plant Sci. **274**, 163–170 (2018).

Welch, R.M.: Impact of mineral nutrients in plants on human nutrition on a worldwide scale. In: Plant Nutrition – Food Security and Sustainability of Agro- ecosystems, pp. 284–285, Springer, Dordrecht (2001)

Welch, R.M., Graham, R.D.: Breeding for micronutrients in staple food crops from a human nutrition perspective. J. Exp. Bot. **55**, 353–364 (2004)

White, P.J., Broadley, M.R.: Physiological limits to zinc biofortification of edible crops. Front. Plant Sci. **2**, 80 (2011)

Whiting, S.N., de Souza, M.P., Terry, N.: Rhizosphere bacteria mobilize Zn for hyperaccumulation by Thlaspi caerulescens. Environ. Sci. Technol. **35**(15), 3144–3150 (2001)

Wu, S.C., Cheung, K.C., Luo, Y.M., Wong, M.H.: Effects of inoculation of plant growth-promoting rhizobacteria on metal uptake by Brassica juncea. Environ. Pollut. **140**(1), 124–135 (2006)

Xu, Y.G., Wang, B.S., Yu, J.J., Ao, G.M., Zhao, Q.: Cloning and characterisation of ZmZLP1, a gene encoding an endoplasmic reticulum-localised zinc transporter in Zea mays. Funct. Plant Biol. **37**(3), 194–205 (2010)

Yadav, R.L., Meena, M.C.: Available micronutrients status and their relationship with soil properties of Degana soil series of Rajasthan. J. Indian Soc. Soil Sci. **57**(1), 90–92 (2009)

Yu, Q., Rengel, Z.: Micronutrient deficiency influences plant growth and activities of superoxide dismutases in narrow-leafed lupins. Ann. Bot. **83**(2), 175–182 (1999)

Zhang, F., Römheld, V., Marschner, H.: Effect of zinc deficiency in wheat on the release of zinc and iron mobilizing root exudates. Z. Pflanzenernähr. Bodenkd. **152**(2), 205–210 (1989)

Zhou, J.R., Erdman Jr., J.W.: Phytic acid in health and disease. Crit. Rev. Food Sci. Nutrit. **35**(6), 495–508 (1995)

Chapter 15
Role of Rhizobacterial Bacilli in Zinc Solubilization

Farhana Masood, Saghir Ahmad, and Abdul Malik

Abstract In both eukaryotic and prokaryotic cells, zinc (Zn) plays a significant role as a cofactor for many enzymes and is present in DNA-binding Zn-finger proteins. It also plays a key role in metabolism, mitosis, development, and mitochondrial activity. Zinc has been the second most common transition metal for living organisms and the only metal that occurs in all classes of enzymes. Zinc is abundantly present in the earth's crust, and the most commonly found forms in the soil include sphalerite, zinc silicate, zinc oxide, zinc phosphate, and zinc carbonate. It is typically present in soil at a concentration of 5–770 ppm, which may vary as various anthropogenic activities such as chemical fertilizer, galvanization, pharmaceutical waste, and other applications release zinc ion in soil. Zn deficiency has become a severe problem affecting nearly half of the world's population because the plants in Zn-deficient soils are actually low in Zn content. According to the World Health Organization report, the shortage of micronutrients like Zn is a significant challenge to global health and sustainability. Approximately, 50% of the soils in India have zinc deficiency that affect most crop production. The main reasons for the zinc deficiency are higher use of chemical fertilizers, intensive farming, and inadequate irrigation system resulting in reduced zinc content in Indian soils. Although some soil may contain adequate quantities of zinc to sustain crop growth, even then the crops have low zinc status because of the insoluble nature of the zinc found in the soil. This locked zinc can be accessible to plants to support optimal growth if an appropriate technology is developed for the same. Several bacteria, particularly those present in the rhizosphere, possess the ability to solubilize the unavailable forms of a metal and making them bioavailable to plants. *Bacillus* has been the most prevalent species in plant rhizospheres, accounting for up to 95% of Gram-positive rhizospheric population. *Bacillus* is one of the most widely studied genera and is known

F. Masood (✉) · S. Ahmad
Department of Post Harvest Engineering and Technology, Aligarh Muslim University, Aligarh, India

A. Malik
Department of Agricultural Microbiology, Faculty of Agricultural Sciences, Aligarh Muslim University, Aligarh, India

S. T. Khan, A. Malik (eds.), *Microbial Biofertilizers and Micronutrient Availability*,
https://doi.org/10.1007/978-3-030-76609-2_15

to possess numerous plant growth-promoting traits like nitrogen fixation, minerals solubilization, production of siderophores, antibiotics, and secondary metabolites that have biocontrol properties against plant pathogens. Different mechanisms have been reported in bacteria that assist zinc solubilization like chelation, changes in the soil pH, and proton extrusion. The most widely studied mechanism is secretion of organic acids by the rhizospheric bacteria like gluconic acid, butyric acid, lactic acid, and oxalic acid. Organic acid production increases the availability by sequestration of cations and by decreasing the soil pH. This chapter discusses the potential role of Bacilli rhizobacteria in Zn solubilization and how they can be used for crop production and environmental sustainability.

Keywords *Bacillus* · Rhizobacteria · Sustainable agriculture · Zinc · Zinc-solubilizing bacteria

Introduction

For optimal growth and metabolism, plants need a range of nutrients. Plant roots absorb inorganic nutrient sources along with the water. These micronutrients are crucial for growth of plants and crop productivity by playing important role in physiological functions and in optimal crop nutrition. Copper, zinc, iron, nickel, boron, manganese, chloride, and molybdenum are the micronutrients vital for plant metabolism (Uchida 2000). Lack of any of these micronutrients in the soil will slow the growth of plants even though there are sufficient amounts of all other macro- or micronutrients (Yu and Rengel 1999). Many of the world's soils are micronutrient deficient because of uptake of micronutrients by high-yield crops, excess usage of NPK fertilizers, lower levels of nutrients, and reduced application of compost and organic manures.

Among numerous micronutrients, Zn is essential for plants in the healthy development, metabolism, and reproduction (Perumal et al. 2017). Zn acts as an essential cofactor in a number of enzymes and hence plays a very important role in metabolic reactions (Gandhi et al. 2014). Its essentiality for various biological functions is well known as exemplified by its role in maintaining biological membranes' structure and function, production of biomass, photosynthesis, chlorophyll and carbohydrate syntheses, protein and lipid metabolism, improved stress resistance, and growth (Thenua et al. 2014; Yu et al. 2017). It has important function in auxin and cytokinin syntheses, directly involved in the growth of plant and stem elongation (Hussain et al. 2015). In plants, zinc regulates the expression of genes required for withstanding environmental stresses. This is used to protect against free radicals and to turn starch into sugar (Cakmak 2000). Much concern is there for the health of the livestock and people in the regions where zinc deficiency is common in crops. Zn plays a crucial role in maintaining enzyme activity and controls more than 300 enzyme reactions (Tapiero and Tew 2003). Zinc deficiency leads to fertility issues in animals and humans (Solanki et al. 2016). Essential micronutrients deficiency, like zinc and iron, can also contribute to disrupted brain growth and injury healing and weakens immunity. Micronutrient especially zinc deficiency makes humans vulnerable to diseases like diarrhea, malaria, and pneumonia (Prasad 2013). Generally, Zn and Fe deficiencies in humans are due to the inadequate dietary intake (Welch and Graham 2004).

Thirty percent of the world soils are facing Zn deficiency problem, significantly reducing the quality and the yield of various crops and legumes (Sharifi and Paymozd 2016). The poor solubility of various forms of Zn in soil is largely responsible for the prevalence of its deficiency in agricultural crops and related problems (Cakmak 2008). About 50% of the agricultural land in India, in particular the entire Indo-Gangetic zone, is Zn deficient, and this is projected to rise further to 63% by 2025 (Sunitha Kumari et al. 2016). Due to the zinc deficiency, the height of the plant decreases dramatically, and white brown patches develop that eventually become necrotic. Severe Zn deficiency induces decline in the tillers number and delays grain maturation (Wissuwa et al. 2006). Therefore, crops grown in zinc-deficient soils show significant loss of vegetative growth and yields, sometimes even up to 80%. Zn deficiency, which lies next to the deficiency of nitrogen and phosphorus, is prevalent in rice cultivation. For most cases, chemical fertilizers are used to circumvent nutrient deficiency, and many crops, in both rainfed and irrigated environments, have the influence of the Zn applications on the yields. The inclusion of $ZnSO_4$ heptahydrate of 25 kg/ha, equal to 5 kg/ha zinc, is typically suggested for soil each year or in alternating years. However, chemical fertilizers are not economic and they are readily transformed into insoluble forms following application. Zinc availability from insoluble sources is controlled by a variety of parameters including the rhizospheric biochemical processes of microorganisms. These microorganisms have significant function in transforming insoluble form of Zn into bioavailable types (Singh et al. 2005; Bapiri et al. 2012; Zamana et al. 2018). Just 20% of the Zn added is available for plant absorption from the external application of zinc supplies (Bapiri et al. 2012). The inaccessible zinc compounds like zinc carbonate, zinc phosphate, and zinc oxide are converted into bioavailable forms by inoculating bacteria that solubilize Zn by producing organic acids, consequently reducing the pH (Wang et al. 2013; Sharma et al. 2014).

Role of Zinc in Plants

Zinc is an essential micronutrient for plants having various roles in plant life cycle (Hirschi 2008). It has a role in crop growth, maturity, and production. It affects nucleic acid, lipids, and protein synthesis through which quality of grain is superior (Kramer and Clemens 2006). Therefore, the use of zinc fertilizer shows improvement in its amount in grains and in flag leaves (Ranjbar and Bahmaniar 2007; Waters et al. 2009). Increased zinc absorption leads to improved grain yield (Han et al. 2006). It is known to regulate the activity of carbonic anhydrase for the fixation of carbohydrates in plants as well as facilitates the metabolism of starch, proteins, auxin, and pollen production (Marschner 1995). Zn helps in maintaining the integrity of biological membranes as it is a component of many membrane proteins. It protects plants from oxidative stress because of its presence as a cofactor in superoxide dismutase and catalase. Also, Zn is found in all the six classes of enzymes, i.e., hydrolases, isomerases, ligases, lyases, oxidoreductases, and transferases. It is a component of the Rubisco structure influencing the photosynthesis (Alloway 2004a, 2004b). Zinc also influences the water uptake and availability to plants (Tsonko and Lidon 2012). It also contributes in the formation of DNA and RNA complexes (Pahlsson 1989; Coleman 1992). Direct and indirect association of Zn in various signaling is also well known. For example, zinc plays an important role in tryptophan synthesis which is involved in signaling (Hansch and Mendel 2009).

For normal plant growth, approximately 15–20 mg/kg dry weight of zinc is needed (Marschner 1995). Deficient zinc concentrations are generally 0–15 mg/kg dry weight (Boehle and Lindsay 1969). Since zinc is linked with other enzymes, its shortage results in numerous plant disorders. Zinc deficiency in plants can induce multiple symptoms, such as mottled leaves, wilting due to lack of turgidity in the leaves, basal leaf chlorose, delayed plant growth, and leaf bronzing (Tripathi et al. 2015). Its deficiency inhibits RNA synthesis. The deficiency of Zn can have adverse effects on the quality of grain; plants become susceptible to injuries by extreme temperature and light injury and fungal infections (Cakmak 2000).

Role of Zinc in Humans

Essential micronutrients such as Zn are necessary for optimal growth and development (Calder and Jackson 2000). Zn intervention programs in developed countries are considered for reducing the occurrence, seriousness, and mortality of several diseases, such as diarrhea and pneumonia (Black et al. 2008). Zinc dietary deficiency was documented previously in 1961, and its syndrome was characterized by dwarfism, hypogonadism, anemia, hepatosplenomegaly, and geophagia. Approximately, 30% of world's population suffers from Zn and other micronutrient deficiencies also called "hidden hunger" (Hafeez et al. 2013). Several other symptoms of zinc deficiency in humans are memory loss, infertility, muscles weakness, and mental disorders. Inadequate consumption during pregnancy can induce immature fetus (Das and Green 2013). In humans, zinc deficiency mainly results from dependency on zinc-poor diet. This diet generally includes plants like wheat, rice, and maize that have low amounts of micronutrient (Singh et al. 2004). Zinc deficiency ranked fifth as the most significant health risk factors in developed countries and eleventh in the world, and it is considered as essential as deficiencies of iron and vitamin A. Zinc has structural or catalytic functions in about 300 enzymes. While 86% Zn is present in skeletal muscles, there are some parts like hippocampus, pancreas, prostate, and kidney cortex where Zn level is predominantly high and indicates its importance (Vallee and Falchuk 1981). In addition, zinc is critically required for the protein synthesis and metabolism of DNA and RNA. Reports suggest the existence of many zinc-containing proteins that specifically affect the expression of genes (Welch 2001). Exposure to high doses of zinc is known to have adverse effects, and chronic exposure results in poisoning which is relatively rare.

Status of Zinc in Soil

Zinc has been the second most common transition metal for living organisms after Fe and the only metal that occurs in all classes of enzymes (White and Broadley 2005). Zinc is highly abundant in the earth's crust. Most common forms of zinc present in soil are sphalerite, zinc oxide, zinc silicate, zinc carbonate, and zinc phosphate (Shaikh and Saraf 2017). In general, zinc is present in soil at a concentration of 5–770 ppm. This concentration in soil may also vary with several anthropogenic practices like the fertilizer application, galvanization, dumping of pharmaceutical waste, and paint casting in the soil (Wuana and Okieimen 2011). In soil, natural sources of Zn are (i) parent rocks weathering (Alloway 1995) and (ii) zinc addition to soils from the atmosphere (volcanoes, forest fires, and surface dusts) (Friedland 1990; International Zinc Association 2011). Zinc deficiencies are found worldwide, particularly in the rice-based ecosystem of the Asia-Pacific region (Tisdale et al. 2009) and also in different types of soils like alfisols, aridisols, mollisols, and vertisols (Srivastava and Gupta 1996). About 90% or more of zinc in soil is present as insoluble Zn and thus cannot be absorbed by the plants, while the transferrable zinc in soils varies from 0.1 to 2 mg/kg of soil (Singh 2011). The gross zinc-deficient area is approximately ten million hectares in India. Around 85% of this area is in the Indo-Gangetic plains which is under rice-wheat cultivation. This Zn deficiency is among the most critical main constraints for yield. Furthermore, alkaline soil pH and calcareous soil texture also limits the bioavailability of the Zn to the plants. Among the Indian soils, 50% of the soil contain zinc below the critical limit (0.5 mg/kg of available Zn) (Prasad 2010). The global occurrence of zinc deficiency in crops is because of its poor solubility instead of low availability in soil (Iqbal et al. 2010). Zn-deficient soils are prevalent in many countries across the world including China (Lui 1991), Australia (Sillanpaa 1982) and India (Singh 2008; Behera et al. 2009). Different factors influence zinc concentration in the soil such as zinc insolubility in soil, complex soil forms, increased use of chemical fertilizer, and irrigation system (Hafeez et al. 2013). For sandy acidic soils which have lower plant nutrients availability and organic matter, Zn deficiency is more severe. The acidic soils occupy about 49 million ha area in India, while worldwide such soil covers 800 million ha (Sharma and Singh 2002). As a result, soil acidity is creating a big issue by impacting food production in Asia, Africa, and Latin America and is putting high costs on farmers. Elevated phosphorus content in the soil too interferes with the uptake of zinc by plants, thereby resulting in a zinc deficiency in plants (Salimpour et al. 2010).

The Rhizosphere

Rhizosphere generally hosts high population of microorganisms due to the presence of various root exudates and nutrients by the root. The production of these compounds by the roots also influences the chemical and biological properties of the soil around the roots. The term rhizosphere was coined by Hiltner in 1904. Rhizosphere is a hot spot for microorganisms wherein complex interplay of various microfauna and plant root takes place (Antoun and Prévost, 2005; Kumar et al. 2015). It is not only the exudates from plant roots influencing the microbial population and its type, but the plant growth and productivity is also influenced by the microorganisms and other interactions. The presence of various microorganisms including bacteria, fungi, protozoa, and algae that coexist among them can influence the growth positively, negatively, or neutrally (Ordookhani and Zare 2011). The bacterial biomass in the rhizosphere is well known to be higher than the non-rhizospheric soil (Verma et al. 2017). Among various bacteria, an important class of bacteria found in rhizosphere are plant growth-promoting rhizobacteria (PGPR). PGPR promote plant growth and yield through production of various compounds or through fixing or mobilizing nutrients for the plants (Wu et al. 2005). These PGPR, found as free-living bacteria, may colonize the root. PGPR account approximately 2–5% of the total bacteria present in rhizosphere (Antoun and Kloepper 2001; Sgroy et al. 2009). In 1980, Kloepper et al. proposed the term PGPR to refer to "all rhizospheric bacteria that improves the growth of plants using one or more mechanisms" (Haghighi et al. 2011). Metabolic processes of these bacteria in the rhizosphere have demonstrated several beneficial effects on growth and development of plants (Singh et al. 2004). PGPR not only enhance nutrient mobilization and uptake by plants but also protect plants against biotic and abiotic stresses (Haas and Défago 2005; Mendes et al. 2011; Kumar et al. 2015; Badri et al. 2013). These bacteria increase crop production and agricultural sustainability and increase soil fertility (Bardgett et al. 2014; Verma et al. 2017). Several free-living, associative, and symbiotic rhizobacteria identified as PGPR include *Azospirillum*, *Bacillus*, *Pseudomonas*, *Azotobacter*, *Alcaligenes*, *Klebsiella*, *Burkholderia*, *Arthrobacter*, *Enterobacter*, and *Serratia* (Saharan and Nehra 2011).

Diversity of *Bacillus* in Agricultural Soils

Bacillus has been the most prevalent species in the rhizosphere, and for many years studies on PGPR activities of some of these strains have contributed to a widespread understanding of the processes involved (Probanza et al. 2002). Bacilli are the most widely studied rhizobacteria (Souza et al. 2015), after *Pseudomonas* mainly due to their biocontrol properties. Studies suggest that bacilli are most abundant in rhizosphere representing up to 95% of the Gram positive rhizobacterial populations (Prashar et al. 2013). Bacilli are effective PGPR and improve plant growth by producing variety of compounds, including antibiotics and antifungals, siderophores (Compant et al. 2005), and lytic enzymes (Nelson 2004). Evidence shows that certain bacilli rhizobacteria can stimulate plant growth by various mechanisms. For example, *Bacillus polymyxa* BFKC01 not only increases the supply of nutrients to plants, it also produces phytohormones and improves host capacity to withstand environmental stresses (Zhou et al. 2016). *B. subtilis* commonly present in the rhizosphere maintains stable contact with higher plants and stimulates the growth (Jaiswal et al. 2016; Jha and Subramanian 2016). Tomatoes and peppers inoculated with *B. licheniformis* showed significant colonization and can be used as a biofertilizer without changing the usual maintenance of both the greenhouse and the field environments (Wu et al. 2005). *B. megaterium* effectively improves different root variables in mint. The evidence suggests that plants choose different microorganisms from the rhizosphere microflora to establish environment that supports the soil–plant–environmental system (Robin et al. 2007; Rudrappa et al. 2008). Given this knowledge, the full use of this rhizobacterium in crop production in many regions of the world is still to be recognized, in particular in view of the many beneficial attributes that make it ideal for use as plant biofertilizers and biopesticides.

Rhizobacterial Bacilli

Bacilli rhizobacteria like other plant growth-promoting rhizobacteria may or may not colonize the root (Gadhave et al. 2018). It is well known that *Bacillus* spp. are common endophytes in plants (Rajendran et al. 2008), and that endophytic *Bacillus* spp. establish stronger relation with the host plants as they are protected within the host (Zhao et al. 2015), and have biocontrol properties against plant pathogens (Timmusk et al. 2005; Dey et al. 2014). Therefore, its role in both PGP and bioprotection needs further investigations. Literature reveal that some endophytic *Bacillus* spp. prevail in nonspecific host nodules and are essential in growth promotion and nodulation of the host plants (Stajković et al. 2009; Deng et al. 2011; Saini et al. 2013). Hence, the potential use of these bacilli for agriculture should be explored. Recently, Zhao et al. (2018) found that symbiotic strains of *Bacillus* may be useful as biofertilizers. In enhanced nodulation and crop yield, some bacilli rhizobacteria also have been involved, notably when co-inoculated with *Rhizobium* (Bai et al. 2003; Rajendran et al. 2008).

PGPR Mechanism of **Bacillus** *Species*

Bacillus species can act as PGPR, solubilize nutrients, and can be used for bioremediation, improving crop production and yield through nutrient absorption and various other mechanisms contributing to the establishment of sustainable soil–plant–environment (Schippers et al. 1995). The mechanism for improving

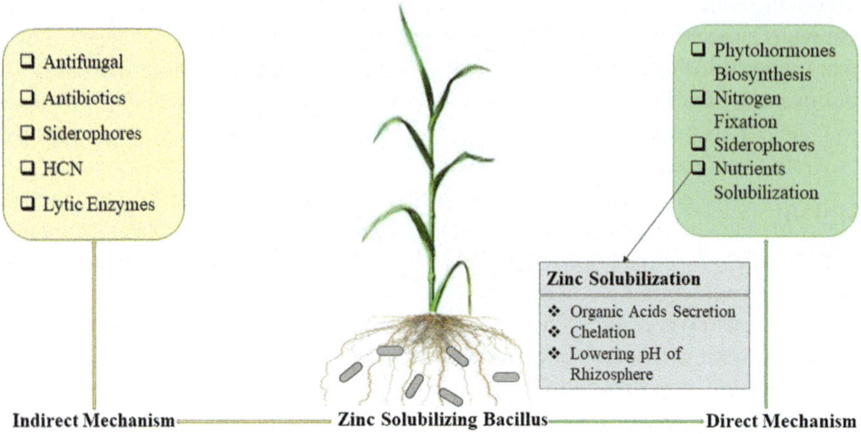

Fig. 15.1 Plant growth-promoting properties of ZSB

plant growth and crop yield mediated by PGPB is not yet completely known (Dey et al. 2004). Some of the possible mechanisms proposed are (i) essential enzyme synthesis, 1-aminocyclopropane-1-carboxylate (ACC) deaminase, which lowers ethylene levels in the roots of growing plants, thus increasing root length and growth (Meena et al. 2013; Verma et al. 2013); (ii) phytohormones production such as indole acetic acid (IAA) (Patten and Glick 2002), abscisic acid (ABA) (Dobbelaere et al. 2003), gibberellic acid (GA), and cytokinins (Dey et al. 2004); (iii) symbiotic nitrogen fixation (Kennedy et al. 2004); (iv) biocontrol against phytopathogenic bacteria by production of siderophores, chitinases, ß-1,3-glucanase, antibiotics, and cyanide (Cattelan et al. 1999; Glick and Stearns 2011); (v) by solubilizing and mineralizing the nutrients (Meena et al. 2014); and (vi) improved abiotic stress (Saleem et al. 2007) (Fig. 15.1).

Phytohormone production by these rhizobacteria is directly related to plant growth promotion and influences nutrition uptake by improving root growth (Stamenkovic et al. 2018). For example, potted tomato seedlings, when inoculated with *Bacillus subtilis*, increased root and shoot growth, leaf area, and seedling vigor, and high GA and IAA levels have been found in treated plants in comparison to untreated ones (Chowdappa et al. 2013). IAA influences cell division and elongation thereby increasing root and shoot growth (Pin-Ng et al. 2015). Isolation and application of IAA-producing rhizobacteria on crops is proposed as a promising approach to improve soil fertility and plant development (Vejan et al. 2016).

The reports on the use of bacilli rhizobacteria for antagonizing plant pathogens are growing rapidly and hence can be used as biocontrol agent (Przemieniecki et al. 2018). In fact, bacilli rhizobacteria are the most widely studied biocontrol agents against plant pathogens (Shafi et al. 2017). Examples of bacilli biocontrol agents are *B. thuringiensis* (Bai et al. 2003), *B. simplex* (Schwartz et al. 2013), *B. amyloliquefaciens* (Idris et al. 2007), and *B. subtilis* (Ashwiri and Srividya 2013). *Bacillus amyloliquefaciens* demonstrated antifungal activity against *Puccinia striiformis*

(Reiss and Jørgensen 2017). In another research, *Bacillus cereus* was found to control many rice phytogenic fungi (Etesami and Alikhani 2017). Recently *Bacillus* sp. was found to have significant antifungal properties relative to other rhizobacteria (Bjelić et al. 2018).

Rhizobacterial Bacilli as Zn Solubilizers

Among the main causes of zinc deficiency in plants is the presence of insoluble form of zinc in soil or low zinc status of the soil. Zinc-solubilizing bacteria (ZSB) tend to solubilize the unavailable Zn form and improve the Zn uptake contributing to the zinc fortification of grains (Bapiri et al. 2012). Use of bacteria to promote plant growth through improving nutrient availability to plants has its own advantages. These include the presence of multiple plant growth-promoting activities and ecofriendly nature of the approach (Hansda et al. 2014). PGPR have essential roles in the dissolution of phosphates and bioavailability of soil nutrients like phosphorus, magnesium, potassium, and zinc (Hayat et al. 2010). Several authors have demonstrated that inoculation of ZSB improves the yield of field crop like wheat, rice, and maize. As zinc is a significant constraint in crop yield, ZSB are of tremendous importance in improving zinc availability to plants consequently improving plant growth and yield (Bapiri et al. 2012; Verma et al. 2013).

Zinc deficiency is identified as a global nutritional crisis, especially affecting developing countries (Zamana et al. 2018). Its deficiency is linked to cereal grains intake with poor Zn content which is due to their cultivation in Zn-deficient soils. Food enrichment, nutritional diversification, and biofortification are some of the strategies envisaged to mitigate Zn deficiency. Zinc biofortification can be made possible through increasing bioavailable levels of micronutrients in crops by using modified agricultural practices or genetic improvement of the crops (Zamana et al. 2018). Biofortification of crops is also influenced by soil chemical properties, agricultural practices, crop genotypes, and climate factors. Worldwide efforts have been made to boost the genetic potential of crops for increasing the bioavailability of micronutrients in traditional food crops like wheat, rice, corn, and beans (Cakmak et al. 2010). Important approach is the use of PGPR to maximize the root uptake of micronutrients. These PGPR can contribute to the crop growth by solubilizing nutrients in soils. Several effective PGPR strains like *Bacillus* sp., *Pseudomonas*, *Rhizobium* sp. are known in improving growth of plants and their zinc content. Gadd (2007) stated that zinc-solubilizing *Bacillus* has the zinc solubilization potential and can also be used to improve the bioavailability of zinc to crops. Furthermore, strains like *Bacillus aryabhattai* and *B. subtilis* isolated from maize rhizosphere solubilized insoluble form of zinc to improve its bioavailability in soil (Mumtaz et al. 2017). One hundred thirty-four *Bacillus* isolates from soybean rhizospheric soils were isolated by Sharma et al. (2012) to select efficient zinc solubilizers. Compared to uninoculated control, soils inoculated with *Bacillus* isolates substantially increased the concentration of Zn in soyabean.

Mechanism of Zinc Solubilization

Researchers identified numerous mechanisms in bacteria that assist in the zinc solubilization, such as chelation of zinc and iron by siderophore produced by these bacteria, soil pH changes, proton extrusion, and phytohormone biosynthesis like auxin, ethylene, and gibberellic acid (Fasim et al. 2002). The mechanism most commonly studied is the production of organic acids like gluconic acid, lactic acid, butyric acid, and oxalic acid by rhizospheric bacteria (Vyas and Gulati 2009). The main organic acid contributing to metal solubilization in soil is alpha-ketogluconic acid. The production of organic acids by these bacteria improves the availability by sequestration of cations and by reducing the pH of soil (Sunitha Kumari et al. 2016).

PGPR are being documented to produce organic acids and release protons that lower the pH of rhizospheric soil (Fasim et al. 2002; Parmar and Sindhu 2018). For instance, *Bacillus* spp. and *Pseudomonas* sp. co-inoculation in broth culture reduced the pH, which aids in solubilization of zinc oxide, zinc sulfide, and zinc carbonate (Saravanan et al. 2004). The pH of the soil has an influence on the micronutrients availability in soil, and a decrease in pH by one unit resulted in a 100-fold increase in the availability of zinc in the soil (Havlin et al. 2005). Wu et al. (2006) reported a decline in pH with bacterial inoculation due to release of protons and organic acids, which eventually increased zinc solubilization and uptake. Gluconic acid and keto-gluconic acid have multiple conformations that chelate the metal cations besides solubilization (Saravanan et al. 2007). Hussain et al. (2011) reported that *Bacillus* sp. AZ6 released different organic acids such as cinnamic acid, caffeic acid, ferulic acid, chlorogenic acid, gallic acid, and syringic acid that aid in solubilization of insoluble zinc compounds. Perumal et al. (2017) used six bacterial strains isolated from the maize rhizosphere to study solubilization of zinc carbonate, zinc oxide, and Zn-EDTA. They suggested that zinc solubilization from insoluble zinc compounds may be due to acid production by the bacterial strains because pH of the medium decreases from 7.0–7.3 to 3.0–4.8 after 10 days of inoculation. Soil–plant–microbe relationships are multifaceted, and there are several aspects which affect crop vigor and plant yield (Hayat et al. 2010).

Another influential mechanism for increasing the bioavailability of Zn and uptake by plants is zinc chelation by rhizospheric microorganisms. Generally, the available fraction of Zn in the soil is limited because of its low persistence and high reactivity of zinc in the soil solution. The bioavailability of zinc in the rhizosphere was increased by Zn-chelating compounds (Obrador et al. 2003). In the rhizosphere, both microorganisms and plant roots release these chelating compounds, which chelate the zinc and maximize the supply.

Conclusion

Zinc has been one of the key micronutrients needed for proper plant growth and development. Addition of inorganic zinc fertilizers to soil is not very effective as it becomes fixed in soil quickly just after the application. Zinc-solubilizing rhizobacteria offer a good alternative to chemical fertilizers due to their ability to solubilize fixed forms of zinc. Zinc-solubilizing bacteria with additional plant growth-promoting activities can promote plant growth through improving zinc bioavailability and various other PGP activities. Thus, in the coming years, the potential application of biofortification will be the greatest approach for reducing micronutrient deficiency and improving sustainable agriculture.

Acknowledgments F. M. is thankful to U.G.C., New Delhi, for financial support in the form of Postdoctoral Research Fellowship to Women.

References

Alloway, B.J.: Zinc in Soils and Crop Nutrition. International Zinc Association, Brussels (2004a)

Alloway, B.J.: Contamination of soils in domestic gardens and allotments: a brief review. Land Contam. Reclamat. **12**(3), 179–187 (2004b)

Alloway, B.J.: Heavy metals in soils, 2nd edn. Blackie Academic & Professional, London. (1995)

Antoun, H., Kloepper, J.W.: Plant growth promoting rhizobacteria. In: Brenner, S., Miller, J. (eds.) Encyclopedia of Genetics, pp. 1477–1480 (2001)

Antoun, H., Prévost, D.: Ecology of plant growth promoting rhizobacteria. In: Siddiqui, Z.A. (Ed.), PGPR: Biocontrol and biofertilization, pp. 1–38. Springer, Netherlands (2005)

Ashwiri, N., Srividya, S.: Potentiality of *Bacillus subtilis* as biocontrol agent for management of anthracnose disease of chili caused by *Colletotrichum gloeosporioides* OGC1. Biotechnology. **4**, 127–136 (2013)

Badri, D.V., Zolla, G., Bakker, M.G., Manter, D.K., Vivanco, J.M.: Potential impact of soil microbiomes on the leaf metabolome and on herbivore feeding behavior. New Phytol. **198**(1), 264–273 (2013)

Bai, Y., Zhou, X., Smith, D.L.: Enhanced soybean plant growth resulting from coinoculation of *Bacillus* strains with *Bradirhizobium japonicum*. Crop Sci. **43**, 1774–1778 (2003)

Bapiri, A., Asgharzadeh, A., Mujallali, H., Khavazi, K., Pazira, E.: Evaluation of Zinc solubilization potential by different strains of Fluorescent Pseudomonads. J. Appl. Sci. Environ. Manag. **16**(3), 295–298 (2012)

Bardgett, R.D., Mommer, L., De Vries, F.T.: Going underground: root traits as drivers of ecosystem processes. Trends Ecol. Evol. **29**(12), 692–699 (2014)

Behera, S.K., Singh, M.V., Lakaria, B.L.: Micronutrients deficiencies in Indian soils and their and their amelioration through fertilization. Indian Farm. **59**(2), 28–31 (2009)

Bjelić, D., Marinković, J., Tintor, B., Mrkovački, N.: Antifungal and plant growth promoting activities of indigenous rhizobacteria isolated from maize (*Zea mays* L.) rhizosphere. Commun. Soil Sci. Plant Anal. **49**, 88–98 (2018)

Black, R.E., Lindsay, H.A., Bhutta, Z.A., Caulfield, L.E., De Onnis, M., Ezzati, M., Mathers, C., Rivera, J.: Maternal and child undernutrition: global and regional exposures and health consequences. Lancet. **371**, 243–260 (2008)

Boehle, J., Lindsay, W.L.: Micronutrients, the fertilizer shoe-nails, pt. 6. In the limelight–zinc. Fertil Soln. **13**, 6–12 (1969)

Cakmak, I.: Role of zinc in protecting plant cells from reactive oxygen species. New Phytol. **146**, 185–205 (2000)

Cakmak, I.: Enrichment of cereal grains with zinc: agronomic or genetic bio-fortification? Plant Soil. **302**, 1–17 (2008)

Cakmak, I., Pfeiffer, W.H., Clafferty, B.M.: Biofortification of durum wheat with zinc and iron. Cereal Chem. **87**(1), 10–20 (2010)

Calder, P.C., Jackson, A.A.: Under nutrition, infection and immune function. Nutr. Rev. **8**, 129–133 (2000)

Cattelan, A.J., Hartel, P.G., Fuhrmann, J.J.: Screening of plant growth promoting rhizobacteria to promote early soybean growth. Soil Sci. Soc. Am. **63**(6), 1670–1680 (1999)

Chowdappa, P., Kumar, S.M., Lakshmi, M.J., Mohan, S.P., Upreti, K.K.: Growth stimulation and induction of systemic resistance in tomato against early and late blight by *Bacillus subtilis* OTPB1 or *Trichoderma harzianum* OTPB3. Biol. Control. **65**, 109–117 (2013)

Coleman, J.E.: Zinc proteins: enzymes, storage proteins, transcription factors, and replication proteins. Annu. Rev. Biochem. **61**(1), 897–946 (1992)

Compant, S., Duffy, B., Nowak, J., Clement, C., Barka, E.A.: Use of plant growth promoting bacteria for biocontrol of plant diseases: principles, mechanisms of action, and future prospects. Appl. Environ. Biol. **71**, 4951–4959 (2005)

Das, S., Green, A.: Importance of zinc in crops and human health. J. SAT Agric. Res. **11**, 1–7 (2013)

Deng, Z., Zhao, L., Kong, Z.Y.: Diversity of endophytic bacteria within nodules of the *Sphaerophysa salsula* in different regions of Loess Plateau in China. FEMS Microbiol. Ecol. **76**, 463–475 (2011)

Dey, R., Pal, K.K., Bhatt, D.M., Chauhan, S.M.: Growth promotion and yield enhancement of peanut (*Arachis hypogaea* L.) by application of plant growth-promoting rhizobacteria. Microbiol. Res. **159**, 371–394 (2004)

Dey, R., Pal, K.K., Tilak, K.V.: Plant growth promoting rhizobacteria in crop protection and challenges. In: Future Challenges in Crop Protection Against Fungal Pathogens, pp. 31–58. Springer, New York, NY (2014)

Dobbelaere, S., Vanderleyden, J., Okon, Y.: Plant growth-promoting effects of diazotrophs in the rhizosphere. Crit. Rev. Plant Sci. **22**(2), 107–149 (2003)

Etesami, H., Alikhani, H.A.: Evaluation of gram-positive rhizosphere and endophytic bacteria for biological control of fungal rice (*Oryzia sativa* L.) pathogens. Eur. J. Plant Pathol. **147**, 7–14 (2017)

Fasim, F., Ahmed, N., Parsons, R., Gadd, G.M.: Solubilization of zinc salts by bacterium isolated by the air environment of tannery. FEMS Microbiol. Lett. **213**, 1–6 (2002)

Friedland, A.J.: The movement of metals through soils and ecosystems. In: Shaw, A.J. (ed.) Heavy Metal Tolerance in Plants: Evolutionary Aspects, pp. 7–19. CRC, Boca Raton (1990)

Gadd, G.M.: Geomycology: biogeochemical transformations of rocks, minerals, metals and radionuclides by fungi, bioweathering and bioremediation. Mycol. Res. **111**, 3–49 (2007)

Gadhave, K.R., Devlin, P.F., Ebertz, A., Ross, A., Gange, A.C.: Soil inoculation with *Bacillus* spp. modifies root endophytic bacterial diversity, evenness, and community composition in a context-specific manner. Microb. Ecol. **76**(3), 741–750 (2018)

Gandhi, A., Muralidharan, G., Sudhakar, E., Murugan, A.: Screening for elite zinc solubilizing bacterial isolate from rice rhizosphere environment. Int. J. Recent Sci. Res. **5**, 2201–2204 (2014)

Glick, B.R., Stearns, J.C.: Making phytoremediation work better: maximizing a plant's growth potential in the midst of adversity. Int. J. Phytoremediation. **13**(1), 4–16 (2011)

Haas, D., Défago, G.: Biological control of soil-borne pathogens by fluorescent pseudomonads. Nat. Rev. Microbiol. **3**(4), 307–319 (2005)

Hafeez, B., Khanif, Y.M., Saleem, M.: Role of zinc in plant nutrition—a review. Am. J. Exp. Agric. **3**(2), 374–391 (2013)

Haghighi, B.J., Alizadeh, O., Firoozabadi, A.H.: The role of plant growth promoting rhizobacteria (PGPR) in sustainable agriculture. Adv. Environ. Biol. **5**(10), 3079–3083 (2011)
Han, J.L., Li, Y.M., MaC, Y.: Effect of zinc fertilization on accumulation and transportation of N, P, K and Zn after anthesis of wheat. Plant Nutr. Fert. Sci. **12**, 313–320 (2006)
Hansch, R., Mendel, R.R.: Physiological functions of mineral micronutrients (Cu, Zn, Mn, Fe, Ni, Mo, B, Cl). Curr. Opin. Plant Biol. **12**(3), 259–266 (2009)
Hansda, A., Kumar, V., Anshumali, A., Usmani, Z.: Phytoremediation of heavy metals contaminated soil using plant growth promoting rhizobacteria (PGPR): a current perspective. Recent Res. Sci. Technol. **6**(1), 131–134 (2014)
Havlin, J., Beaton, J.D., Tisdale, S.L., Nelson, W.L.: Soil Fertility and Fertilizers: An Introduction to Nutrient Management. Pearson Prentice Hall, Upper Saddle River, NJ (2005)
Hayat, R., Ali, S., Amara, U., Khalid, R., Ahmed, I.: Soil beneficial bacteria and their role in plant growth promotion: a review. Ann. Microbiol. **60**(4), 579–598 (2010)
Hirschi, K.: Nutritional improvements in plants: time to bite on biofortified foods. Trends Plant Sci. **13**(9), 459–463 (2008)
Hussain, A., Arshad, M., Zahir, Z.A., Asghar, M.: Prospects of zinc solubilizing bacteria for enhancing growth of maize. Pak. J. Agric. Sci. **52**(4), 915–922 (2015)
Hussain, S., Maqsood, M.A., Rahmatullah: Zinc release characteristics from calcareous soils using diethylene triamine penta acetic acid and other organic acids. Commun. Soil Sci. Plant Anal. **42**(15), 1870–1881 (2011)
Idris, E.E., Iglesias, D.J., Talon, M., Borris, R.: Tryptophan-dependent production of indole-3-acetic acid (IAA) affects level of plant growth promotion by *Bacillus amyloliquefaciens* FZB42. Mol. Plant-Microbe Interact. **20**, 619–626 (2007)
International Zinc Association IZA.: http://www.zinc.org/sustainability (2011)
Iqbal, U., Jamil, N., Ali, I., Hasnain, S.: Effect of zinc-phosphate-solubilizing bacterial isolates on the growth of *Vigna radiata*. Ann. Microbiol. **60**(2), 243–248 (2010)
Jaiswal, D.K., Verma, J.P., Prakash, S., Meena, V.S., Meena, R.S.: Potassium as a n important plant nutrient in sustainable agriculture: a state of the art. In: Potassium Solubilizing Microorganisms for Sustainable Ariculture, pp. 21–29. Springer, New Delhi (2016)
Jha, Y., Subramanian, R.B.: Regulation of plant physiology and antioxidant enzymes for alleviating salinity stress by potassium-mobilizing bacteria. In: Potassium Solubilizing Microorganisms for Sustainable Agriculture, pp. 149–162. Springer, New Delhi (2016)
Kennedy, I.R., Chouhury, A.T.M.A., Kecskes, M.L.: Non-symbiotic bacterial diazotrophs in crop farming systems; can their potential for plant growth promotion be latter exploited? Soil Biol. Biochem. **36**(8), 1229–1244 (2004)
Kramer, U., Clemens, S.: Functions and homeostasis of zinc, copper, and nickel in plants, molecule. In: Tamas, M.J., Martinoia, E. (eds.) Molecular Biology of Metal Homeostasis and Detoxification: From Microbes to Man, pp. 216–271. Springer, Heidelberg (2006)
Kumar, A., Singh, V., Singh, R., Singh, M., Pandey, K.D.: Plant Growth Promoting Rhizobacteria (PGPR). A Promising Approach for Disease Management. Microbes and Environmental Management, pp. 191–205. Studium Press, New Delhi (2015)
Lui, Z.: Characterization of content and distribution of microelements in soils of China. In: Portch, S. (ed.) International Symposium on the Role of Sulphur, Magnesium and Micronutrients in Balanced Plant Nutrition/Sponsors, the Potash and Phosphate Institute of Canada [et al.]. Potash and Phosphate Institute, Hong Kong (1991)
Marschner, H.: Mineral Nutrition of Higher Plants, 2nd edn. Academic Press, London (1995)
Meena, O.P., Maurya, B.R., Meena, V.S.: Influence of K- solubilizing bacteria on release of potassium from waste mica. Agron. Sustain. Dev. **1**(1), 53–56 (2013)
Meena, V.S., Maurya, B.R., Verma, J.P.: Does a rhizospheric microorganism enhance K^+ availability in agricultural soils? Microbiol. Res. **169**, 337–347 (2014)
Mendes, R., Kruijt, M., De Bruijn, I., Dekkers, E., van der Voort, M., Schneider, J.H., Piceno, Y.M., DeSantis, T.Z., Andersen, G.L., Bakker, P.A., Raaijmakers, J.M.: Deciphering the rhizosphere microbiome for disease-suppressive bacteria. Science. **332**(6033), 1097–1100 (2011)

Mumtaz, M.Z., Ahmad, M., Jamil, M., Hussain, T.: Zinc solubilizing *Bacillus* spp. potential candidates for biofortification in maize. Microbiol. Res. **202**, 51–60 (2017)

Nelson, L.M.: Plant growth promoting rhizobacteria (PGPR): prospects for new inoculants. Crop Manag. **3**(1), 1–7 (2004)

Obrador, A., Novillo, J., Alvarez, J.M.: Mobility and availability to plants of two zinc sources applied to a calcareous soil. Soil Sci. Soc. Am. J. **67**, 564–572 (2003)

Ordookhani, K., Zare, M.: Effect of *Pseudomonas*, *Azotobacter* and arbuscular mycorrhiza fungi on lycopene, antioxidant activity and total soluble solid in tomato (Lycopersicon esculentum F1 Hybrid, Delba). Adv. Environ. Bio. **5**(6), 1290–1294 (2011)

Påhlsson, A.M.B.: Toxicity of heavy metals (Zn, Cu, Cd, Pb) to vascular plants. Water Air Soil Pollut. **47**(3–4), 287–319 (1989)

Parmar, P., Sindhu, S.S.: The novel and efficient method for isolating potassium solubilizing bacteria from rhizosphere soil. Geomicrobiol J. **35**(10), 1–7 (2018)

Patten, C.L., Glick, B.R.: Role of *Pseudomonas putida* indole-acetic acid in development of the host plant root system. Appl. Environ. Microbiol. **68**, 3795–3801 (2002)

Perumal, M.D., Subramanian, V., Sabarinathan, K.G.: Evaluation of zinc solubilizing potential of maize rhizosphere bacterial isolates. Int. J. Curr. Microbiol. Appl. Sci. **6**(12), 864–869 (2017)

Pin-Ng, J.L., Perrine-Walker, F., Wasson, A.P., Mathesius, U.: The control of auxin transport in parasitic and symbiotic root–microbe interactions. Plan. Theory. **4**, 606–643 (2015)

Prasad, A.S.: Essential and Toxic Element: Trace Elements in Human Health and Disease. Elsevier (2013)

Prasad, R.: Zinc biofortification of food grains in relation to food security and alleviation of zinc malnutrition. Curr. Sci. **98**, 1300–1304 (2010)

Prashar, P., Kapoor, N., Sachdeva, S.: Rhizosphere: its structure, bacterial diversity and significance. Rev. Environ. Sci. Biotechnol. **13**, 63–77 (2013)

Probanza, A., Garcıa, J.A.L., Palomino, M.R., Ramos, B., Mañero, F.J.G.: *Pinus pinea* L. seedling growth and bacterial rhizosphere structure after inoculation with PGPR *Bacillus* (*B. licheniformis* CECT 5106 and *B. pumilus* CECT 5105). Appl. Soil Ecol. **20**(2), 75–84 (2002)

Przemieniecki, S.W., Kurowski, T.P., Damszel, M., Krawczyk, K.: Effectiveness of the *Bacillus* sp. SP-A9 strain as a biological control agent for spring wheat (*Triticum aestivum* L.). J. Agric. Sci. Technol. **20**, 609–619 (2018)

Rajendran, G., Sing, F., Desai, A.J., Archana, G.: Enhanced growth and nodulation of pigeon pea by co-inoculation of *Bacillus* strains with *Rhizobium* spp. Bioresour. Technol. **99**(11), 4544–4550 (2008)

Ranjbar, G.A., Bahmaniar, M.A.: Effect of soil and foliar application of zinc fertilizer on yield and growth characteristic of bread wheat (*Triticum aestivum* L.) cultivars. Asian J. Sci. **6**, 1000–1005 (2007)

Reiss, A., Jørgensen, L.N.: Biological control of yellow rust of wheat (*Puccinia striiformis*) with Serenade® ASO (*Bacillus subtilis* strain QST713). Crop Prot. **93**, 1–8 (2017)

Robin, A., Mazurier, S., Mougel, C., Vansuyt, G., Corberand, T., Meyer, J.M., Lemanceau, P.: Diversity of root-associated fluorescent pseudomonads as affected by ferritin overexpression in tobacco. Environ. Microbiol. **9**(7), 1724–1737 (2007)

Rudrappa, T., Czymmek, K.J., Paré, P.W., Bais, H.P.: Root-secreted malic acid recruits beneficial soil bacteria. Plant Physiol. **148**(3), 1547–1556 (2008)

Saharan, B.S., Nehra, V.: Plant growth promoting rhizobacteria: a critical review. Life Sci. Med. Res. **2011**, 1–30 (2011)

Saini, R., Dudeja, S.S., Giri, R., Kumar, V.: Isolation, characterization, and evaluation of bacterial root and nodule endophytes from chickpea cultivated in Northern India. J. Basic Microbiol. **55**, 74–81 (2013)

Saleem, M., Arshad, M., Hussain, S., Bhatti, A.S.: Perspective of plant growth promoting rhizobacteria (PGPR) containing ACC deaminase in stress agriculture. J. Ind. Microbiol. Biotechnol. **34**, 635–648 (2007)

Salimpour, S., Khavazi, K., Nadian, H., Besharati, H., Miransari, M.: Enhancing phosphorous availability to canola (Brassica napus L.) using P solubilizing and sulfur oxidizing bacteria. Aust J Crop Sci. **4**(5), 330 (2010)

Saravanan, V.S., Subramoniam, S.R., Raj, S.A.: Assessing in vitro solubilization potential of different zinc solubilizing bacterial (ZSB) isolates. Braz. J. Microbiol. **35**(1–2), 121–125 (2004)

Saravanan, V.S., Madhaiyan, M., Thangaraju, M.: Solubilization of zinc compounds by the diazotrophic, plant growth promoting bacterium *Gluconacetobacter diazotrophicus*. Chemosphere. **66**, 1794–1798 (2007)

Schippers, B., Scheffer, R.J., Lugtenberg, J.J., Weisbek, P.J.: Biocoating of seed with plant growth promoting rhizobacteria to improve plant establishment. Outlook Agric. **24**, 179–185 (1995)

Schwartz, A.R., Ortiz, I., Maymon, M., Herbold, C.W., Fujishige, N.A., Vijanderan, J.A., Villella, W., Hanamoto, K., Diener, A., Sanders, E.R., DeMason, D.A., Hirsch, A.M.: *Bacillus simplex*-a little known PGPB with anti-fungal activity—alters pea legume root architecture and nodule morphology when coinoculated with *Rhizobium leguminosarum* bv. viciae. Agronomy. **3**(4), 595–620 (2013)

Sgroy, V., Cassán, F., Masciarelli, O., Del Papa, M.F., Lagares, A., Luna, V.: Isolation and characterization of endophytic plant growth-promoting (PGPB) or stress homeostasis-regulating (PSHB) bacteria associated to the halophyte *Prosopis strombulifera*. Appl. Microbiol. Biotechnol. **85**(2), 371–381 (2009)

Shafi, J., Tian, H., Ji, M.: Bacillus species as versatile weapons for plant pathogens: a review. Biotechnol. Bioteechnol. Equip. **31**, 446–459 (2017)

Shaikh, S., Saraf, M.: Biofortification of *Triticum aestivum* through the inoculation of zinc solubilizing plant growth promoting rhizobacteria in field experiment. Biocatal. Agric. Biotechnol. **9**, 120–126 (2017)

Sharifi, P., Paymozd, M.: Effect of zinc, iron and manganese on yield and yield components of green beans. Curr. Opin. Agric. **5**(1), 15 (2016)

Sharma, P., Kumawat, K.C., Kaur, S., Kaur, N.: Assessment of zinc solubilization by endophytic bacteria in legume rhizosphere. Indian J. Res. Appl. **4**, 439–441 (2014)

Sharma, U.C., Singh, R.P.: Acid soils of India: their distribution, management and future strategies for higher productivity. Fert. News. **47**(3), 45–48 (2002)

Sharma, S.K., Sharma, M.P., Ramesh, A., Joshi, O.P.: Characterization of zinc solubilizing *Bacillus* isolates and their potential to influence zinc assimilation in soybean seeds. J. Microbiol. Biotechnol. **22**, 352–359 (2012)

Sillanpaa, M.: Micronutrients and the Nutrient Status of Soils. A Global Study, FAO Soil Bulletin No. 48. FAO, Rome (1982)

Singh, B., Natesan, S.K., Singh, B.K., Usha, K.: Improving zinc efficiency of cereals under zinc deficiency. Curr. Sci. **10**, 36–44 (2005)

Singh, B.K., Millard, P., Whiteley, A.S., Murrell, J.C.: Unravelling rhizosphere–microbial interactions: opportunities and limitations. Trends Microbiol. **12**(8), 386–393 (2004)

Singh, M.V.: Micronutrient deficiencies in crops and soils in India. In: Micronutrient Deficiencies in Global Crop Production, pp. 93–125. Springer, Dordrecht (2008)

Singh, M.V.: Assessing extent of zinc deficiency for soil factors affecting nutritional scarcity in humans and animals. Indian J. Fertil. **7**, 36–43 (2011)

Solanki, M., Didwania, N., Nandal, V.: Potential of zinc solubilizing bacterial inoculants in fodder crops. Int. J. Eng. Tech. Sci. Res. **3**(12), 1–4 (2016)

Souza, R., Ambrosini, A., Passaglia, L.M.P.: Plant growth-promoting bacteria as inoculants in agricultural soils. Genet. Mol. Biol. **38**, 401–419 (2015)

Srivastava, P.C., Gupta, U.C.: Trace Elements in Crop Production. Oxford and IBH, New Delhi (1996)

Stajković, O., De Meyer, S., Miličić, B., Willems, A.: Isolation and characterization of endophytic non-rhizobial bacteria from root nodules of alfalfa (*Medicago sativa* L.). Bot. Serb. **33**(1), 107–114 (2009)

Stamenkovic, S., Beskoski, V., Karabegovic, I., Lazic, M., Nikolic, N.: Microbial fertilizers: a comprehensive review of current findings and future perspectives. Span. J. Agric. Res. **16**, 1–18 (2018)

Sunitha Kumari, K., Padma Devi, S.N., Vasandha, S.: Zinc solubilizing bacterial isolates from the agricultural fields of Coimbatore, Tamil Nadu India. Curr. Sci. **110**(2), 196–205 (2016)

Tapiero, H., Tew, K.D.: Trace elements in human physiology and pathology: zinc and metallothioneins. Biomed. Pharmacother. **57**(9), 399–411 (2003)

Thenua, O.V., Singh, K., Raj, V., Singh, J.: Effect of sulphur and zinc application on growth and productivity of soybean [*Glycine max.*(L.) Merrill] in northern plain zone of India. Ann. Agric. Res. **35**(2), 183–187 (2014)

Timmusk, S., Grantcharova, N., Wagner, E.G.H.: *Paenibacillus polymyxa* invades plant roots and forms biofilms. Appl. Environ. Microbiol. **71**, 7292–7300 (2005)

Tisdale, S.L., Nelson, W.L., Beaton, J.D., Havlin, J.L.: Soil fertility and fertilizer-an introduction to nutrient management, 7th edn. Prentice Hall of India, New Delhi. (2009)

Tripathi, D.K., Singh, S., Singh, S., Mishra, S., Chauhan, D.K., Dubey, N.K.: Micronutrients and their diverse role in agricultural crops: advances and future prospective. Acta Physiol. Plant. **37**, 1–14 (2015)

Tsonko, T., Lidon, F.: Zinc in plants-an overview. Emir. J. Food Agric. **24**(4), 322–333 (2012)

Uchida, R.: Essential nutrients for plant growth: nutrient functions and deficiency symptoms. In: Silva, J.A., Uchida, R. (eds.) Plant Nutrient Management in Hawaii's Soils, Approaches for Tropical and Subtropical Agriculture Human Resources, pp. 31–55. College of Tropical Agriculture and Human Resources, University of Hawaii at Manoa (2000)

Vallee, B.L., Falchuk, K.H.: Zinc and gene expression. Philos. Trans. R. Soc. Lond. Ser. B Biol. Sci. **294**(1071), 185–197 (1981)

Vejan, P., Abdullah, R., Khadiran, T., Ismail, S., Boyce, A.N.: Role of plant growth promoting rhizobacteria in agricultural sustainability. Molecules. **21**, 5–17 (2016)

Verma, J.P., Yadav, J., Tiwari, K.N., Kumar, A.: Effect of indigenous *Mesorhizobium* spp. and plant growth promoting rhizobacteria on yields and nutrients uptake of chickpea (*Cicer arietinum* L.) under sustainable agriculture. Ecol. Eng. **51**, 282–286 (2013)

Verma, R., Maurya, B.R., Meena, V.S., Dotaniya, M.L., Deewan, P., Jajoria, M.: Enhancing production potential of cabbage and improves soil fertility status of Indo-Gangetic Plain through application of bio-organics and mineral fertilizer. Int. J. Curr. Microbiol. App. Sci. **6**(3), 301–309 (2017)

Vyas, P., Gulati, A.: Organic acid production in vitro and plant growth promotion in maize under controlled environment by phosphate-solubilizing fluorescent *Pseudomonas*. BMC Microbiol. **9**, 174 (2009)

Wang, H., Dong, Q., Zhou, J., Xiang, X.: Zinc phosphate dissolution by bacteria isolated from an oligotrophic karst cave in Central China. Front. Earth Sci. **7**(3), 375–383 (2013)

Waters, B.M., Uauy, C., Dubcovsky, J., Grusak, M.A.: Wheat (*Triticum aestivum*) NAM proteins regulate the translocation of iron, zinc and nitrogen compounds from vegetative tissues to grain. J. Exp. Bot. **60**, 4263–4274 (2009)

Welch, R.M.: Impact of mineral nutrients in plants on human nutrition on a worldwide scale. In: Horst, W.J., et al. (eds.) Plant Nutrition–Food Security and Sustainability of Agro-Ecosystems, pp. 284–285. Kluwer, Dordrecht (2001)

Welch, R.M., Graham, R.D.: Breeding for micronutrients in staple food crops from a human nutrition perspective. J. Exp. Bot. **55**, 353–364 (2004)

White, P.J., Broadley, M.R.: Biofortifying crops with essential mineral elements. Trends Plant Sci. **10**, 586–593 (2005)

Wissuwa, M., Ismail, A.M., Yanagihara, S.: Effects of zinc deficiency on rice growth and genetic factors contributing to tolerance. Plant Physiol. **142**, 731–741 (2006)

Wu, S.C., Cao, Z.H., Li, Z.G., Cheung, K.C., Wong, M.H.: Effects of biofertilizer containing N-fixer, P and K solubilizers and AM fungi on maize growth: a greenhouse trial. Geoderma. **125**, 155–166 (2005)

Wu, S.C., Cheung, K.C., Luo, Y.M.: Wong effects of inoculation of plant growth promoting rhizobacteria on metal uptake by Brassica juncea. Environ. Pollut. **140**, 124–135 (2006)

Wuana, R.A., Okieimen, F.E.: Heavy metals in contaminated soils: a review of sources, chemistry, risks and best available strategies for remediation. ISRN Ecol. **2011**, 1–20 (2011)

Yu, Q., Rengel, Z.: Micronutrient deficiency influences plant growth and activities of superoxide dismutases in narrow-leafed lupins. Ann. Bot. **83**(2), 175–182 (1999)

Yu, X., Blanden, A.R., Tsang, A., Zaman, S., Liu, Y., Bencivenga, A.F., Kimball, S.D., Loh, S.N., Carpizo, D.R.: Thiosemicarbazones functioning as zinc metallochaperones to reactivate mutant p53. Mol. Pharmacol. **91**, 567–575 (2017)

Zamana, Q., Aslama, Z., Yaseenb, M., Ihsanc, M.Z., Khaliqa, A., Fahadd, S., Bashirb, S., Ramzanic, P.M.A., Naeeme, M.: Zinc biofortification in rice: leveraging agriculture to moderate hidden hunger in developing countries. Arch. Agron. Soil Sci. **64**, 147–161 (2018)

Zhao, L., Xu, Y., Lai, X.: Antagonistic endophytic bacteria associated with nodules of soybean (Glycine max L.) and plant growth-promoting properties. Braz. J. Microbiol. **49**, 269–278 (2018)

Zhao, L., Xu, Y., Lai, X.H., Shan, C., Deng, Z., Ji, Y.: Screening and characterization of endophytic *Bacillus* and *Paenibacillus* strains from medicinal plant *Lonicera japonica* for use as potential plant growth promoters. Braz. J. Microbiol. **46**, 977–989 (2015)

Zhou, C., Guo, J., Zhu, L., Xiao, X., Xie, Y., Zhu, J., Ma, Z., Wang, J.: *Paenibacillus polymyxa* BFKC01 enhances plant iron absorption via improved root systems and activated iron acquisition mechanisms. Plant Physiol. Biochem. **105**, 162–173 (2016)

Chapter 16
Comparing Various Diets as Sources of Zinc with Special Reference to Fruits and Vegetables

Faizan Ahmad, Sadaf Zaidi, Fazil Qureshi, and Shams Tabrez Khan

Abstract Minerals, especially zinc, are micronutrient of special importance to human health and well-being. It is an essential component of more than 300 enzymes playing important roles in gene expression, enzyme reactions, immune function, protein synthesis, DNA synthesis, and various other important metabolic reactions. This makes it important for growth and development of the body and for wound healing. Since zinc cannot be synthesized and stored by humans, a regular supply of the mineral through food is required. Zn deficiency results in poor health, low fertility, and even premature death in children. It is reported that around 17% of world population is at risk of inadequate Zn intake and around one lakh children (less than 5 years) die every year due to Zn deficiency. According to Food and Nutrition Board, USA, the daily recommended Zn requirement of men and women is 11 and 8 mg, respectively. Therefore, foods rich in Zn such as meat, oysters, legumes, seeds, nuts, dairy foods, eggs, whole grains, some fruits and vegetables, and dark chocolate must be included in the diet to combat Zn deficiency. Oysters are one of the richest sources of Zn, however, keeping in view the availability, one of the most commonly available sources of Zn can be red meat and poultry. This chapter compares different foods as sources of Zn, especially fruits and vegetables. This chapter also discusses the causes of Zn deficiency in humans due to different dietary habits.

Keywords Zinc · Deficiency of zinc · Sources of zinc · Oysters · Fruits and vegetables

F. Ahmad (✉) · S. Zaidi
Department of Post Harvest Engineering and Technology, Faculty of Agricultural Sciences, Aligarh Muslim University, Aligarh, Uttar Pradesh, India

F. Qureshi
Department of Petroleum Engineering, Glocal University, Saharanpur, Uttar Pradesh, India

S. T. Khan
Department of Agricultural Microbiology, Faculty of Agricultural Sciences, Aligarh Muslim University, Aligarh, Uttar Pradesh, India

S. T. Khan, A. Malik (eds.), *Microbial Biofertilizers and Micronutrient Availability*,
https://doi.org/10.1007/978-3-030-76609-2_16

Introduction

Zinc is considered as one of the essential trace elements and a member of one of the main micronutrient subgroups. It has recently attracted the attention of scientists and health workers due to the increasing understanding of its role in human health and the health consequences faced due to its deficiency (Hambidge 2000; Michael 2000). During early 1960s, studies conducted on Iranian adolescent dwarfs recognized zinc as an important trace metal for humans. Prasad et al. (1963) observed and reported the role of zinc on human health as early as in 1960. Zinc is the second most abundant trace metal after iron (Gammoh and Rink 2017; Roohani et al. 2013). Zinc is a bluish-white metallic element found in around 0.02% of the earth's crust (Sandstead 2015). Because of its transitional element nature in the periodic table, zinc has certain chemical properties which make it specifically useful and essential for human health. Zinc is as an effective electron acceptor present in Zn^{2+} form in biological systems providing negligible risk of causing oxidant damage to cells. Zn^{2+} helps in the functioning of the protein structure and is a catalytic component of over 300 various enzymes, covering around all the aspects of biology including development, immune defense, cognitive function, and bone health (King et al. 2015). Since zinc ion plays an important role in the survival of human being, its body concentration must be sufficient and balanced (Huang et al. 2015).

As zinc is an essential component of various enzymes, it is an important constituent of all forms of life. Zn plays an indispensable role in human health and to cure diseases as trace factor. Role of zinc in growth, development, enzyme functioning, functioning of glands like prostate and gonads, and brain development is irreplaceable as shown in Fig. 16.1. It also plays essential roles in various body functions such as taste perception, vision, cognition, cell proliferation, and in the development of immunity (Hawkes 1997). Human body contains 2–4 g of zinc and the concentration in various tissues varies. The concentration in muscles, skin, bones, and liver is highest, while the concentration in plasma is in a range of 12–16 μM. The form and concentration of Zn also varies in various subcellular compartments. In cytosol, vesicles, organelles, and nucleus, it is present mainly as metal ion or as protein-bound Zn (Wong et al. 2017). Since, Zn cannot be stored in human body, it is required on a daily basis mainly through diet (Rink and Gabriel 2000). Recommended daily intake of zinc depends on various factors such as sex, weight, age, and phytate content of diet. Absorption of zinc in the body takes place in the small intestine and therefore plays an important role in maintaining Zn homeostasis in humans. Around 90% of zinc present in the body gradually turns over and is thus not immediately available for metabolism. Zinc is lost through urine, menstrual flow, semen, sloughed skin, nails, and hair, but this loss is negligible as compared to the loss through gastrointestinal excretion (Jackson 1989). The main aim of this chapter is to highlight the sources of zinc, availability of zinc, and daily recommended intake of zinc for different age groups. Further, this chapter also highlights the role of zinc in various diseases as well as health benefits associated with the zinc. Finally, this chapter also mentions the web grabs of some commercial products containing zinc available in the market.

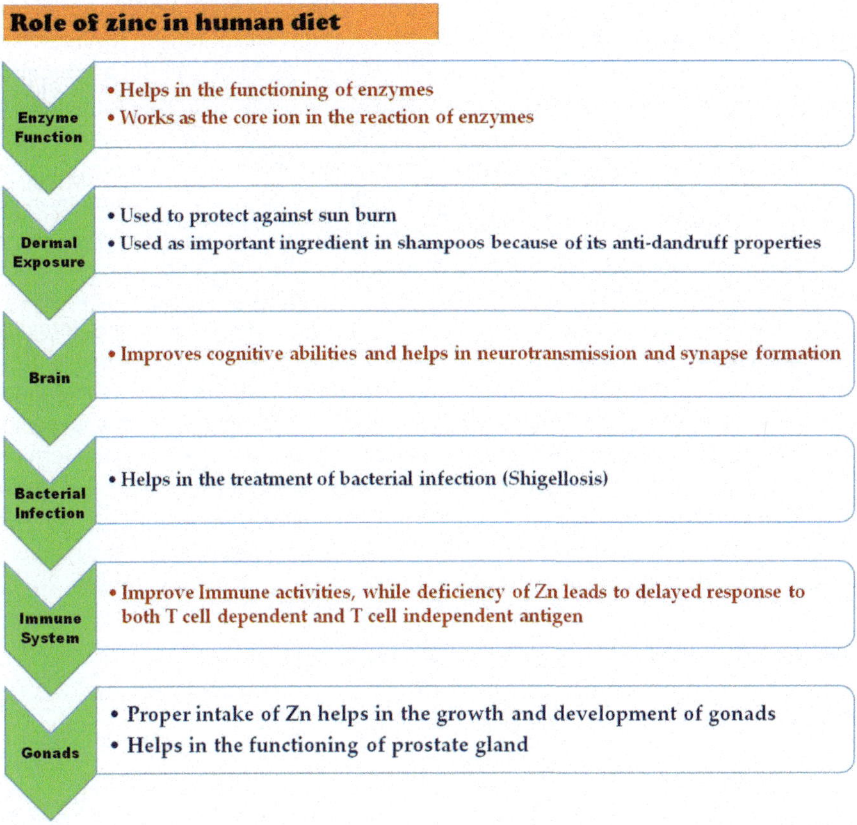

Fig. 16.1 Role of zinc in human diet

Dietary Sources and Recommended Daily Intake of Zinc

Concentration of zinc varies in various foods; furthermore, various other components of food may also interfere with its absorption in the intestine. Generally, foods of animal origin rich in protein are considered as good sources of zinc. Meat-based items such as, chicken, lamb, beef, rabbit meat, oysters, scallops, blackfish, animal liver are considered best sources of zinc (Rangan and Samman 2012). Other than these, fruits are also good sources of zinc. The fruits that contain high amounts of zinc include guavas, kiwifruit, pomegranates, apricots, blackberries, blueberries, raspberries, peaches, and avocados. Diversification of diet with such food can improve the zinc status of diet and is a good alternative of meat for vegan diet and can help in avoiding the use of zinc supplements (Roohani et al. 2013). It has also been reported that these fruits provide 2–12% zinc value per cup. Other than meat and fruits, lily flowers, edible fungus, cabbage, black sesame, black rice, dates, hazelnuts, and ebony are also some essential sources of zinc (Bhowmik et al. 2010).

Dietary sources of zinc can also be classified as sources of high, moderate, and low content of zinc based on the zinc content. Whole-grain cereals, pulses, legumes, and lean red meat have high content of zinc (20–50 mg/kg). Meat with high content of fat, polished rice, and processed cereals provide moderate amount of zinc, while sugar, alcohol, fats, and oil are zinc poor (Sandström 1989). Recommended daily intake of zinc varies with age, health condition, and gender. For example, children require higher concentration of zinc due to higher growth rates. Expecting mothers may also require higher concentration of zinc. Absorption of zinc in the intestine of elderly is poor requiring higher dietary intake. Recommended daily intake (RDI) of zinc ranges between 5 mg/day for children and 11 mg/day for adults (King et al. 2015; Trumbo et al. 2001). Zinc requirements for others such as for infants, females, and pregnant women are shown in Table 16.1. Generally, a daily dietary intake of 11 mg/day of zinc is recommended for people consuming both plant and animal diets. However, for people consumingonly a plant-based diet 30 mg/day of zinc is recommended. This is due to the high phytate content of plant-based diet which interferes with zinc absorption in the intestine. It is also observed that the recommended intake of zinc also varies from country to country and from one regulating body to another. As per the recommendation of the US Food and Nutrition Board, 11 mg/day intake of zinc for adult males and 8 mg/day for females are required (Mason 2012).

Table 16.1 Daily intake of zinc requirement (Bhowmik et al. 2010)

Class	Age (years/month)	Zinc intake (mg/day)
Infants	0–6 months	2
	7–12 months	3
Children	1–3 years	3
	4–8 years	5
Males	9–13 years	8
	14–18 years	11
	19–13 years	11
	31–50 years	11
	51–70 years	11
	> 70 years	11
Females	9–13 years	8
	14–18 years	11
	19–13 years	11
	31–50 years	11
	51–70 years	11
	> 70 years	11
Pregnancy	<18 years	13
	19–30 years	11
	31–50 years	11
Lactation	<18 years	14
	19–30 years	12
	31–50 years	12

However, according to the German Society of Nutrition, 10 and 7 mg/day of zinc intake for adult males and females are required. However, the intake of zinc as suggested by the World Health Organization (WHO) and the European Food Safety Authority (EFSA) varies significantly (Roohani et al. 2013). WHO and EFSA recommended the intake of zinc on the basis of dietary phytate to zinc ratio. WHO classifies the diets into three groups on the basis of their absorption efficiency of zinc. It is considered to be high if the phytate to zinc molar ratio is below 5, medium if it is 5 to 15 and low if the value is greater than 15. Whereas, EFSA recommended the diets containing phytate intake levels of 300, 600, 900, and 1200 mg/day (Fairweather-Tait and de Sesmaisons Lecarré 2018).

Zinc Availability in Fruits and Vegetables and Its Absorption

It has been estimated that zinc levels in both fruits and vegetables vary from 0.04 mg/100 g in apple to 1.8 mg/100 g in beans (Slavin and Lloyd 2012). Figures 16.2 and 16.3 highlight the amount of zinc present in some of the selected fruits and vegetables. The Zn content in fruits ranges from 0.04 mg/100 g in apple to 0.42 mg/100 g in jackfruit. The highest concentrations of Zn have been observed in jackfruit followed by pomegranate, guava, banana, pineapple, aonla, sapota, papaya, citrus, grapes, mango, and apple in decreasing order (Ranasinghe et al. 2019). Whereas in vegetables, a similar trend has been observed; beans have high concentration of zinc followed by peas, okra, carrots, sweet potatoes, radishes,

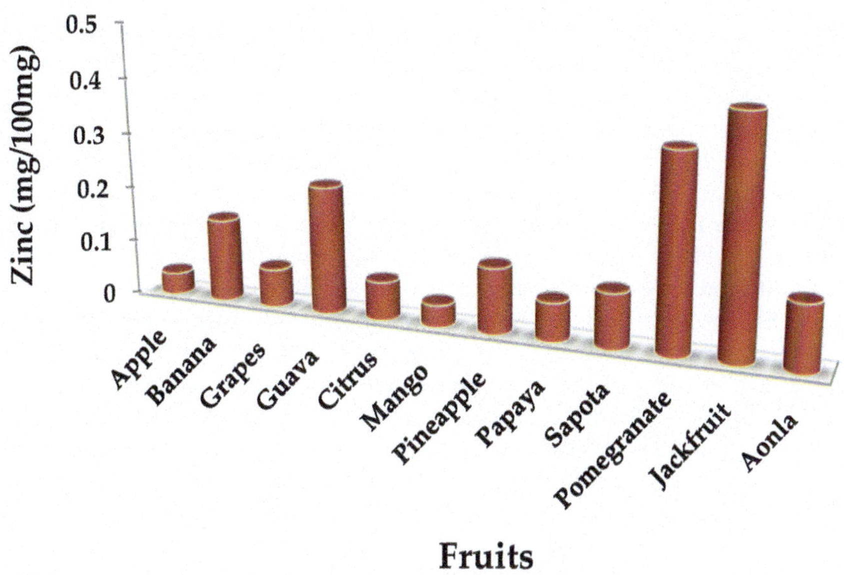

Fig. 16.2 Concentration of zinc in fruits

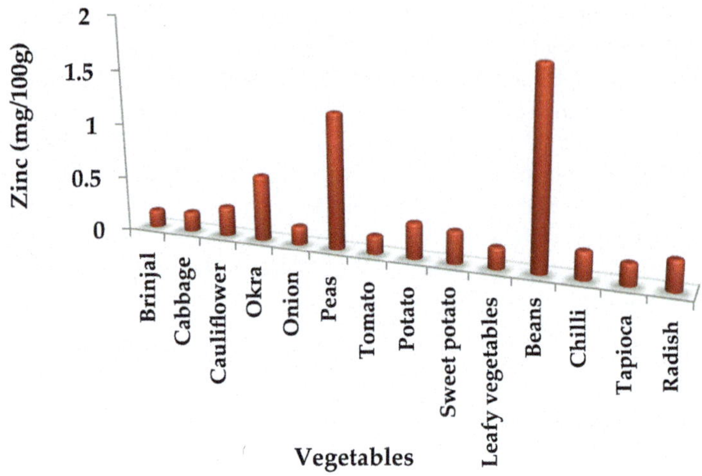

Fig. 16.3 Concentration of zinc in vegetables

cauliflower, chilies, leafy vegetables, tapioca, tomatoes, onions, and brinjal. The comparison between fruits and vegetables showed that the vegetables had highest concentration of zinc (Koe Wei et al. 2019). Other than zinc, fruits, vegetables, and their byproducts are also a potential source of bioactive compounds that reduce the risk of various cardiovascular diseases (Ahmad and Khan 2019; Ahmad et al. 2020; FAIZAN and KUMAR 2018; Kumar et al. 2020). Zn is also an important element for plants and animals, but its excess may interfere with physiological processes and can be even toxic. One study recommends the average dietary intake of 98 and 78 g of vegetables and fruits per person per day, respectively (Elbagermi et al. 2012). Table 16.2 indicates the total zinc availability for the people from different food sources including fruits and vegetables.

Another important factor is the absorption of zinc in the intestine which will depend on its form and the presence of other substances that can interfere with its absorption in intestinal lumen. Zinc absorption efficiency depends on the bioavailability of zinc in food as well as on the overall quantity of zinc present in the diet (Maares and Haase 2020). Bioavailability and concentration of zinc in various groups of food varies considerably. Zinc bioavailability is also dependent on the composition of the diet. Ligands (nondigestible material) such as phytate, lignin chelate zinc, and some dietary fibers inhibit the absorption of zinc (Gammoh and Rink 2017). Presence of iron and calcium in the diet also alters the absorption of zinc (Lim et al. 2013). Since phytate content of the foods like shellfish, organs, legumes, crustaceans, and flesh of mammals is low, they are considered as good sources of absorbable zinc (Brown et al. 2001). Some other foods are also phytate free such as eggs and dairy products, but they have a lower level of zinc as compared to flesh foods. Most cereals and legumes have an intermediate zinc level, but their high phytate content decreases the amount of zinc that can be absorbed. As compared to legumes and cereals, tubers and other starchy roots and rice have a lower

Table 16.2 Availability of total zinc in different food items (Ganeshamurthy et al. 2017; Ma and Betts 2000)

Food source	Zn (mg/100 g)	Zn (g/tons of food)	Total Zn availability(tons)
Cereals			
Rice	1.08	10.8	1151.82
Wheat	1.62	16.2	1552.77
Sorghum	2.24	22.4	124.096
Maize	1.48	14.8	359.048
Bajra	3.8	38	351.5
Finger millet	1.73	17.3	34.254
Pulses			
Chickpea	2.03	20.3	193.459
Green gram	2.4	24	38.64
Red gram	2.35	23.5	74.495
Black gram	2.3	23	39.1
Cowpea	2.57	25.7	8.995
Oilseeds			
Groundnut	3.27	32.7	317.517
Soybean	4.89	48.9	579.954
Sunflower	5	50	25.2
Rapeseed and mustard	1.6	16	126.032
Fruits			
Apple	0.04	0.4	1.034
Banana	0.15	1.5	44.58
Grapes	0.07	0.7	1.2159
Guava	0.23	2.3	8.4364
Citrus	0.07	0.7	7.8029
Mango	0.04	0.4	7.3724
Pineapple	0.12	1.2	2.9976
Papaya	0.07	0.7	3.9473
Sapota	0.1	1	1.744
Pomegranate	0.35	3.5	4.711
Jackfruit	0.42	4.2	6.6024
Aonla	0.12	1.2	1.47
Vegetables			
Brinjal	0.16	1.6	21.6928
Cabbage	0.18	1.8	16.2702
Cauliflower	0.27	2.7	23.1471
Okra	0.6	6	38.076
Onion	0.17	1.7	31.8512
Peas	1.24	12.4	47.9756
Tomato	0.17	1.7	32.9834
Potato	0.33	3.3	137.1315
Sweet potato	0.3	3	3.264

(continued)

Table 16.2 (continued)

Food source	Zn (mg/100 g)	Zn (g/tons of food)	Total Zn availability(tons)
Leafy vegetables	0.2	2	5.84
Beans	1.8	18	24.66
Chili	0.26	2.6	4.3862
Tapioca	0.2	2	16.278
Radish	0.28	2.8	6.9552
Others items			
Nuts	3.5	35	73.5
White sugar	0.01	0.1	2.77
Eggs	1.1	11	19.8
Milk	0.4	4	529.6
Fish	0.6	6	39.42
Chicken	1.2	12	31.2
Mutton	3.3	33	112.2
Beef	3.3	33	112.2
Pork	3.5	35	21

level of zinc content (Roohani et al. 2013). Mainly the foods that are grown under the ground are not the rich sources of zinc, but some of them such as leafy vegetables like spinach have a high density of zinc. Several studies conducted on the bioavailability of zinc recommend that as compared to the plant products, zinc from animal sources has higher bioavailability. Further, these studies also confirmed that the people relying mainly on plant-based food are at higher risk of zinc deficiency due to inadequate zinc intake compared to those consuming red meat (Hunt 2003; Roohani et al. 2013). The results of several studies tend to show that the phytate content present in the foods inhibit the absorption of zinc. Phytate to zinc molar ratio shows the absorption of zinc from the mixed diet (Lönnerdal 2000). It can be calculated as

$$\left\{\frac{\text{Phytate content in mg}}{\text{MW of Phytate }(660)}\right\} \Big/ \left\{\frac{\text{Zinc content in mg}}{\text{MW of zinc }(65.4)}\right\}$$

The foods and diets having phytate to zinc molar ratio of less than 5 have a high bioavailability of zinc, diets having phytate to zinc molar ratio between 5 and 15 have an intermediate bioavailability of zinc, while diet having phytate to zinc molar ratio greater than 15 have a low bioavailability of zinc (Castro-Alba et al. 2019). The WHO classifies the diets into three categories as high, medium, and low zinc availability diets. This classification is purely based on zinc to phytate molar ratio and ratio of energy obtained from different animal sources and cereals (King et al. 2015). In another study conducted worldwide, WHO reported that people should consume their diet on the basis of zinc content in meals assuming normal intake of 3–5 mg/meal. It is further recommended that the diet should be composed of 10–15% food with low zinc bioavailability, 30–35% food with medium zinc bioavailability, and 45–55% food with high zinc bioavailability. In most of the developing countries, people consume zinc-dense food, but still they are zinc deficient due to the presence of high concentration of phytate. Several strategies have been suggested to improve the zinc absorbability in the diet. For instance, one strategy can be to reduce the phytate content of the food, which can improve absorbability of zinc (Akhtar 2013). In another study, Gibson et al. (2010) suggested various food processing methods to increase the bioavailability of zinc from foods of high phytate content such as fermentation, milking, and soaking. Fortification of food is yet another method by which the bioavailability of zinc can be increased. It has also been observed that food fortification done with exogenous zinc resulted decrease in fractional absorption but had a positive effect on net absorption of zinc. Further, it is noted that zinc fortification of food having high phytate to zinc molar ratio minimizes its malabsorption (Chadare et al. 2019; Das et al. 2013). In the past several years, a lot of research has been conducted to estimate zinc absorption and bioavailability from different plant and animal sources, but still the status of zinc absorption from different diets is not clear (Hunt 2003; Perignon et al. 2018; Roohani et al. 2013). Therefore, yet more research is required to confirm the zinc bioavailability and absorption status of different sources.

Zinc Requirement

Worldwide, several experts appointed by the WHO and the United States National Research Council (USNRC) have produced estimates of the zinc requirements on the basis of different age/gender groups and health conditions (Dietary reference values for food energy and nutrients for the United Kingdom. Report of the Panel on Dietary Reference Values of the Committee on Medical Aspects of Food Policy 1991; National Research Council Subcommittee on the Tenth Edition of the Recommended Dietary 1989). Furthermore, different parameters were included in calculating the daily zinc requirement. Hence, the amount of daily zinc intake recommended by different organizations and regulating bodies varies. They estimated the requirement of zinc on the basis of minimum amount of zinc absorbed and the amount of zinc lost through excretion. They have used factorial approach and considered both the intestinal and non-intestinal losses via the skin, kidney, semen, and menstruation. During the study, they found that the non-intestinal losses are not largely dependent on the dietary intake, but they have a good correlation with the absorption of zinc through the dietary intake (Maares and Haase 2020). On the above basis, they have estimated the average requirement (EAR) of zinc for the people at different life stages and gender and recommended daily intake (RDI) of zinc as shown in Table 16.3 (Institute of Medicine (US) Panel on Micronutrients. Dietary Reference Intakes for Vitamin A, Vitamin K, Arsenic, Boron, Chromium,

Table 16.3 EAR and RDI of zinc for different age groups

Different groups	Age of different groups	Estimated average requirement (EAR) of zinc (mg/day)	Recommended daily intake (RDI) of zinc (mg/day)
Infants	7–12 months	2.5	3
Children	1–3 years	2.5	3
	4–8 years	3.0	4
Boys	9–13 years	5	6
	14–18 years	11	13
Girls	9–13 years	5	6
	14–18 years	6	7
Men	19–30 years	12	14
	31–50 years	12	14
	51–70 years	12	14
	>70 years	12	14
Women	19–30 years	6.5	8
	31–50 years	6.5	8
	51–70 years	6.5	8
	>70 years	6.5	8
Pregnancy	14–18 years	8.5	10
	19–30 years	9	11
	31–50 years	9	11
Lactation	14–18 years	9	11
	19–30 years	10	12
	31–50 years	10	12

Copper, Iodine, Iron, Manganese, Molybdenum, Nickel, Silicon, Vanadium, and Zinc. Washington (DC): National Academies Press (US); 2001. 12, Zinc. Available from: https://www.ncbi.nlm.nih.gov/books/NBK222317/). Further, they have also found that compared to the plant sources, the absorption of zinc is higher from diet of animal sources. Therefore, it is recommended that people strictly depended on vegetarian diet should take more than 50% higher zinc than nonvegetarians (https://www.todaysdietitian.com/pdf/courses/PBDNutritentsofConcern.pdf). In one of the reports, WHO mentioned that zinc plays an important role in the physiological requirement of the human body such as for the growth of tissues and their maintenance and to cover the endogenous losses. In order to fulfill the physiological requirements of individuals, WHO recommended 1.0 mg/day for adult males and 0.7 mg/day for adult females. While, according to WHO, "normative physiological requirement" also includes the urinary losses and therefore recommends an additional intake (~40%) (Roohani et al. 2013). For adult men, the normal zinc requirement was set at 1.4 mg/day and for adult women at 1.0 mg/day. Further, they have informed that according to normative physiological requirement, the dietary requirement of zinc for a specific group of persons must be changed to compensate for the expected percentage of dietary zinc absorption (Sandstead 2015). It can be calculated as follows.

Dietary requirement = normative physiological requirement /percent absorption from ordinary diet).

For the babies, WHO recommended that breastfeeding is sufficient for the babies under the age of 6 months to fulfill its zinc requirement. After the age of 6 months, they require some infant formula having higher content of zinc, then breast milk either based on cow's milk or soy (Brown et al. 2001).

Availability of Zinc in Soil

Zinc is considered as one of the most essential micronutrient elements present in the soil for the optimal growth, development, and reproduction of crops and plants. Other than zinc, copper, chlorine, nickel, iron, magnesium, boron, and molybdenum are also referred as micronutrients, because they are only required in small concentrations by the plant tissues and crops (5–100 mg/kg). A large proportion of zinc in soil exists in unavailable forms. Mainly a large portion of zinc present in the soil is bound by the oxides of iron and magnesium (Shambhavi et al. 2020). It is estimated that soil may contain approximately 2–25 ppm of organic and exchangeable zinc. Plant and their crops vary considerably in the requirement of zinc. For example, crops like spinach, onion, and corn required higher amount of zinc, and crops like tomato, soybean, and potato required a medium amount of zinc, whereas still other crops required low amount of zinc (Mousavi et al. 2013; Sadeghzadeh and Rengel 2011). All over the world, the total zinc concentration in soil varies considerably, and it depends on various factors as shown in Fig. 16.4. The overall average of several studies conducted in the past several years in different regions of the world reported an overall mean of total concentration of zinc to be around 55 mg Zn/kg of soil. The research conducted on European soil reported an overall mean of zinc concentration of 68 mg Zn/Kg of soil (Mertens and Smolders 2013; Sadeghzadeh and Rengel 2011; Wuana and Okieimen 2011). Similarly, a study based on 3045 soil samples collected from different agricultural sites in the USA show an average zinc concentration of 56.5 mg/kg of soil. Furthermore, the studies conducted on the soils taken from France, Poland, Australia, the USA, England, and Germany show that the sandy soil has low concentration of zinc and the soil with larger clay contents has a higher concentration of zinc (Sadeghzadeh and Rengel 2011). Figure 16.5 shows the mean concentration of zinc in different types of soil found in India, Sri Lanka, Vietnam, Indonesia, and the Philippines. A Ferralitic soil of Vietnam has a higher mean concentration of zinc (102 mg Zn/Kg), whereas an oxisols soil of India has a lower mean value of zinc (26 mg Zn/Kg). However, zinc deficiency is prevalent in Indian soil as well as in many other countries causing deficiency in crops grown in such soil (Huong et al. 2010; Sadeghzadeh and Rengel 2011). Furthermore, other soil and environmental factors such as water logging/flooding, high phosphate application, low manure applications, high concentration of salt, and high soil organic matters also result in zinc deficiency (Fig. 16.4).

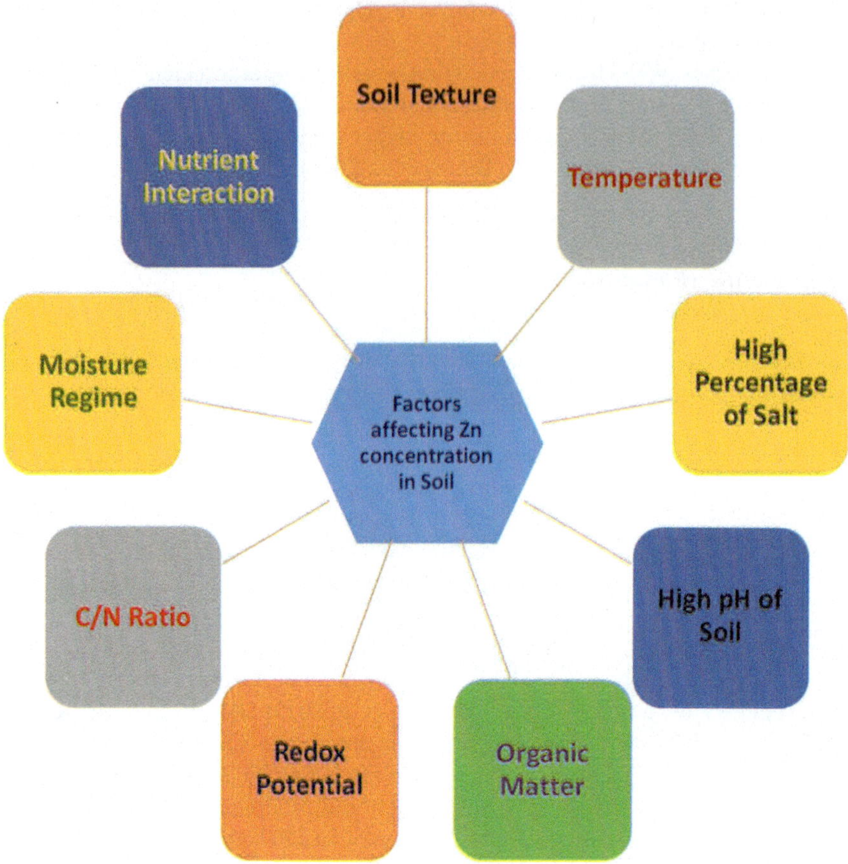

Fig. 16.4 Factors affecting total zinc concentration in soil

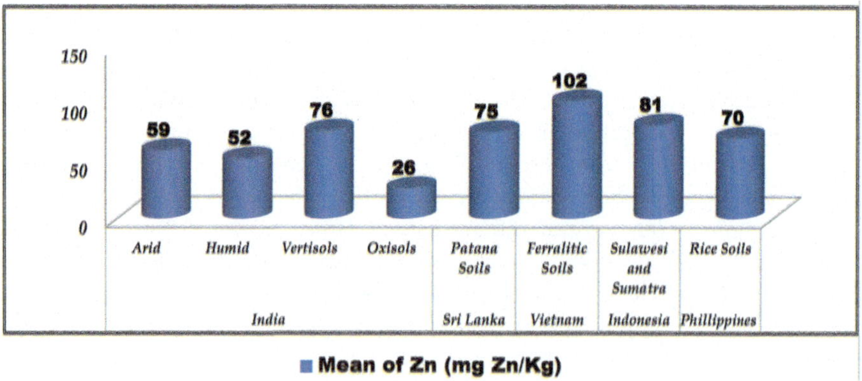

Fig. 16.5 Mean concentrations of zinc in different types of soils

Zinc Deficiency

Zinc deficiency is widespread worldwide; people in developed countries are at lower risk of zinc deficiency. While people in developing countries are at a higher risk due to inadequate intake of zinc (Maxfield and Crane 2020). The main reason for this deficiency in developing countries is low dietary intake, due to the dependence on diet with low zinc concentration like cereal-based diet, or consumption of food with high phytate such as fruits and vegetables which reduces zinc absorption in the intestine (Kawade 2012; Kumssa et al. 2015). Since animal-based diet is rich in zinc content, diets without such food also constitute a zinc-poor diet resulting in zinc deficiency. Presence of inhibitors in food such as phytates and fiber also prevents the effective absorption of zinc in the intestine (Roohani et al. 2013). Zinc deficiency is a very common micronutrient deficiency in children below the age of 5 years in developing countries. It causes growth retardation, dysfunction of immune system, hair loss, loss of appetite, and high risk of morbidities due to reasons like diarrhea and respiratory infections (Brown et al. 2004). Mental lethargy, delayed healing of wounds, weight loss, and taste abnormalities are some other symptoms of zinc deficiency (Brown et al. 2004; Brown et al. 2001). Many of these symptoms are nonspecific and are also observed during other health conditions. Therefore, it is too difficult to precisely identify and associate these symptoms with zinc deficiency. The diseases related with zinc deficiency include diabetes, liver disease, malabsorption syndrome, sickle cell disease, and chronic renal disease. Recommended dietary allowance of zinc is higher for pregnant women because of high growth rate of fetus and pregnancy-related health condition. Other health conditions like puberty, infancy, pregnancy, and lactation also demand high dietary zinc (Marangoni et al. 2016). However, a continuous consumption of zinc-poor diet during pregnancy and nursing is one of the main reasons for zinc deficiency in infants. It is reported that most of the breastfed infants suffer from zinc deficiency because breast milk does not provide sufficient amount of zinc, especially for infants aged 7–12 months requiring 3 mg zinc daily (Aumeistere et al. 2018). Several studies reported that zinc is highly associated with various metabolic processes of human body, so the deficiency of zinc affects several tissues such as gastrointestinal, skeletal, integumentary, and reproductive systems (Hambidge 2000; Tuerk and Fazel 2009). Similarly, deficiency of zinc also affects sex hormones, thyroid, growth function, and insulin function. Zinc deficiency also makes human body vulnerable to infections and induces both hormonal and cell-mediated immunity dysfunction (Grüngreiff 2002). Zinc deficiency may also result in skin and eye lesions, diarrhea, nail dystrophy, and mental changes in the form of depression and apathy. Furthermore, delay in healing of wounds, burns, and ulcer has also been observed due to zinc deficiency. Surveys conducted in different regions of the world reported that eye lesion including night blindness, conjunctivitis, and macular degeneration is also associated with the deficiency of zinc (Murthy et al. 2010).

Deficiency of Zinc in Developed and Developing Countries

The requirement of zinc and its deficiency around the globe is one of the oldest issues to arise over the time in the entire world. In the past several years, a large number of studies estimated the requirement as well as the deficiency of zinc in several developing and developed countries around the globe (Akhtar 2013; Roohani et al. 2013; Wessells and Brown 2012). One of the studies reported by Wessells and Brown in 2012 highlights the global prevalence of zinc deficiency based on zinc availability in national food supplies and the prevalence of stunting (Wessells and Brown 2012). Their analysis shows that the inadequate dietary zinc intake is a common problem in the countries of South Asia and Sub-Saharan Africa as compared to the countries of Southern and Tropical Latin America, Central and Eastern Europe, Central Asia, North Africa and Middle East, East and Southeast Asia, and Pacific and high-income countries. Furthermore, they also highlighted that from 1990 to 2005, around 17.3% of world's population were at a risk of inadequate intake of zinc. Figures 16.6 and 16.7 show the estimated percent of population with inadequate zinc intake by year in developing and developed countries, and Fig. 16.8 shows the comparison of estimated percent of population with inadequate zinc intake by year in the developing and developed countries. Between 1990 and 2005, out of six developed countries, i.e., Singapore, Japan, Australia, Italy, Canada, and Germany, Australia had the lowest percentage of population with zinc deficiency and Japan had the highest percentage of population with zinc deficiency in 2005. And the population exposed to inadequate zinc intake has increased from 3.2 to 3.8 from 1990 to 2005 in Australia. Similarly, between 1990 and 2005, out of six developing countries, i.e., Afghanistan, Bangladesh, India, Bhutan, Nepal, and Pakistan,

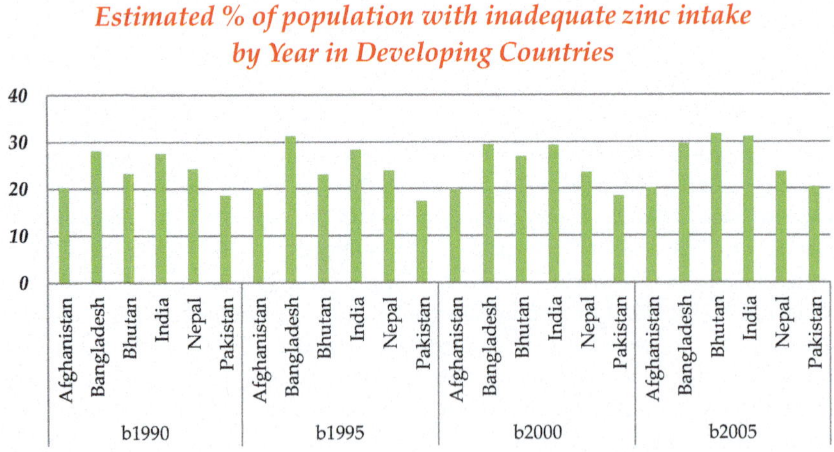

Fig. 16.6 Estimated percent of population with inadequate zinc intake by year in some of the developing countries

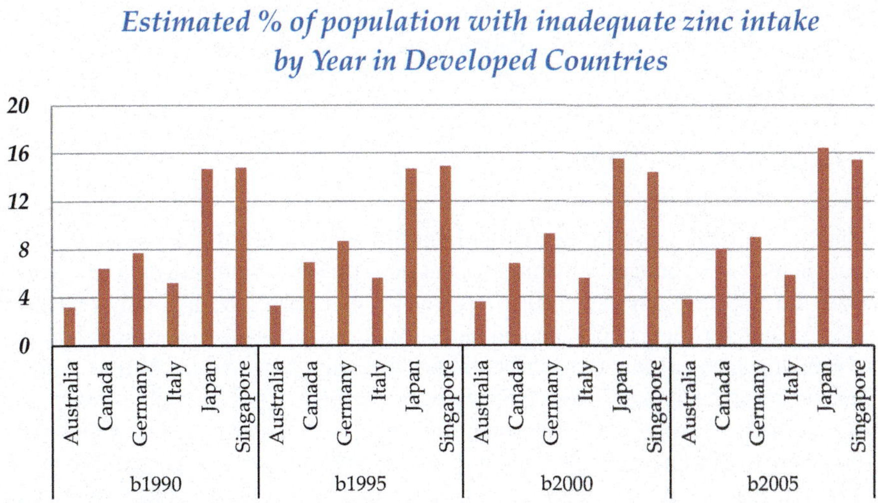

Fig. 16.7 Estimated percent of population with inadequate zinc intake by year in some of the developed countries

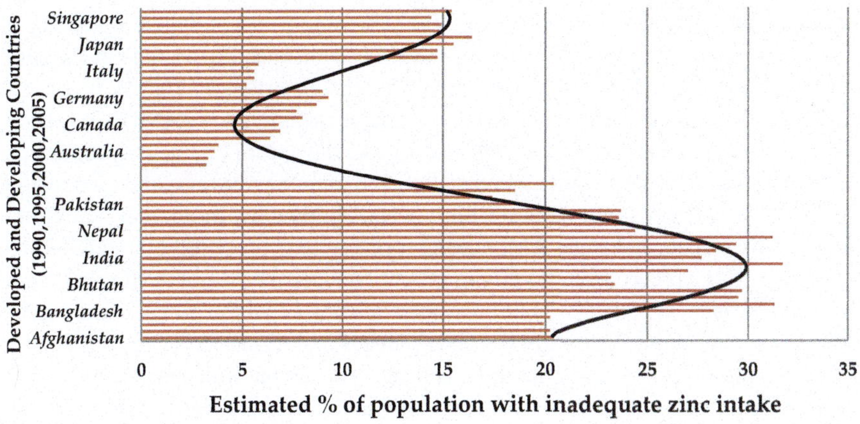

Fig. 16.8 Comparison of estimated percent of population with inadequate zinc intake by year in some of the developed and developing countries

Pakistan had the lowest percentage of population with zinc deficiency and Bhutan had the largest percentage of population with zinc deficiency in 2005 (Wessells and Brown 2012). Except 2005, India had the largest percentage of population with inadequate intake of zinc compared to other developing countries. The data clearly show that in the past two decades, India has the highest population suffering from zinc deficiency, and surprisingly it has also been observed that this percentage is also increasing day by day (https://www.thehindu.com/sci-tech/health/zinc-deficiency-rising-in-indians/article26867693.ece) (Gonmei and Toteja 2018;

Wessells and Brown 2012). These data also tend to show that in comparison to developed countries, all the developing countries are at high risk of zinc deficiency.

Zinc and Health Benefits

Zinc is considered as an essential nutrient that plays a key role in many ways in human body. The human body requires the zinc intake on regular basis since human body do not have any mechanism to produce and store zinc naturally (Michael 2000). Therefore, it is important that human beings should take food or supplements containing adequate amount of zinc to meet this daily requirement. Various important processes of human body require an adequate amount of zinc such as enzymatic reactions, gene expressions, digestion, immune function, metabolism, wound healing, growth, and development of brain (Lin et al. 2017; Skrajnowska and Bobrowska-Korczak 2019). In addition, zinc is required for the proper functioning and development of immune system; it is also necessary for DNA and protein synthesis. For example, it is reported that the consumption of 80–92 mg per day of zinc reduces the effect of common cold by 33%. Zinc can also be commonly used for the treatment of various skin injuries such as ulcers, burns, and wounds (Lin et al. 2017). In another study, it is reported that successful treatment of ulcers with 200 mg/day of zinc for up to 12 weeks has been demonstrated (Gupta et al. 2014). One of the studies on the age-related diseases mentioned that zinc also reduces the risk of various age-related diseases such as age-related macular degeneration (AMD), infection, and pneumonia. It is recommended that 45 mg per day intake of zinc may reduce the infection rate by 66% in adults by improving the immune response and by boosting the activity of T cells and natural killer cells. Additionally, in another study on a large group of people having AMD disease, researchers proposed that daily intake of vitamin E, C, β-carotene and 80 mg of zinc reduces the risk of AMD and vision loss (Vishwanathan et al. 2013). It is estimated that 9.4% of the global population is affected by the common skin disease acne. Acne is mainly caused by the obstruction of oil-producing glands, inflammation, and bacteria. People with acne can be treated with the oral intake of zinc. It has been found that zinc effectively treated the acne by inhibiting the growth of *Propionibacterium acnes* bacteria, reducing inflammation and by suppressing oily gland activity (Rathi 2011). Other than these diseases, zinc also plays an important role in the treatment of various chronic and nonchronic diseases such as cancer, HIV, type I and type II diabetes, metabolic disorder, liver disorder, Alzheimer's disease, and skin diseases (Gammoh and Rink 2017). The role of zinc in some of these diseases is discussed below. In addition to various health benefits of zinc, it also behaves as an important antioxidant providing protection against various stresses. Oxidative stress is one of the major contributing factors for various chronic diseases such as neurodegenerative disorder, atherosclerosis disorder, and even cancer. Increased levels of reactive oxygen species (ROS) and change in redox balance are the common reasons of cancer progression. Numerous studies reported that mitochondrial dysfunction, increased enzymatic activity of oxidases and lipoxygenases, high metabolic activity, and cellular signaling are some factors that are mainly responsible for the increased level of ROS in cancer cells (Kumari et al. 2018; Liou and Storz 2010). During the metabolic process, ROS are continuously produced in the form of free radicals, ions, and molecules. Oxygen free radicals is one of the class of ROS, which includes organic radicals ($R^{\cdot}$), nitric oxide ($NO^{\cdot}$), super oxide ($O^{2\cdot-}$), hydroxyl radical ($^{\cdot}OH$), alkoxyl radicals ($RO^{\cdot}$), etc. While some other forms of ROS include hydrogen peroxide (H_2O_2). Several studies have provided evidences that the uncontrolled production of these radicals (oxidants) creates cytotoxicity and oxidative stress, transmission loss of cellular functions, and development of chronic (heterogeneous) disease like cancer. Zinc has an ability to suppress the uncontrolled or increased generation of ROS because zinc behaves as an inhibitor of nicotinamide adenine dinucleotide phosphate oxidase (NADPHs) (Marreiro et al. 2017). Zinc also behaves as a cofactor of the enzyme superoxide dismutase (SOD), which catalyzes the dismutation of $\cdot O_2$- to H_2O_2. Various studies have documented significant role of zinc as antioxidant contributing to cell-mediated immunity. These studies also confirm its behavior as an antiinflammatory agent. Further it has also been reported that the supplementation of zinc results in the reduction of oxidative stress, decrease in generation of inflammatory-cytokines and also decreases the chances of infections which occur due to the disturbed cell metabolism (Gammoh and Rink 2017).

Zinc and Central Nervous System

Zinc is an essential element which plays an important and vital role in the development and functioning of central nervous system and brain in human beings. Zinc at molecular level helps in the functioning of large number of enzymes in neuronal metabolism, and it also regulates gene expression into the human body (Gower-Winter and Levenson 2012). In one of the recent studies, it was reported that the neurogenesis in the brain is highly dependent on zinc (Levenson and Morris 2011). Change in the zinc concentration in adult brain leads to several diseases related to the development of brain such as anxiety, depression, mood disorders, dementia, and neurodegeneration (Grabrucker et al. 2011). Similar types of disorders have also been observed in Alzheimer's disease (Gower-Winter and Levenson 2012).

In Alzheimer's disease, abnormal excessive β-amyloid 42 (Aβ42) interactions with copper zinc and iron induce peptide aggregation and oxidation leading to neocortical Aβ precipitation. In rodent tests/models, zinc being a glutamate N-methyl-D-aspartate (NMDA) receptor antagonist exhibits antidepressant-like activity. It also induces gene expression for the brain-derived neurotropic factor (BDNF). Clinical studies of depression have demonstrated serum hypozincemia which has been modified by successful antidepressant treatment. In addition, the benefit of zinc supplementation in antidepressant therapy has also been documented in both treatments of nonresistant and resistant patients. Zinc homeostasis is also important in the psychopathology and depression therapy (Nowak et al. 2005; Szewczyk et al. 2011).

Zinc and Diabetes

Nowadays diabetes is a very common endocrine disease caused by the excess of sugar in blood. Excess level of sugar in blood is controlled by a hormone called insulin. In general, diabetes is caused due to less formation of insulin in the body. Low level of zinc in the blood plasma restricts the production and secretion of insulin from the pancreas. Zinc plays an essential role in the production, transport, and release of insulin (Norouzi et al. 2017). But in case of Type I and Type II diabetic patient, most of the zinc gets lost from the serum due to excessive urination (Maret 2017). It has also been observed that in the pancreas of diabetic patient, the concentration of zinc decreases rapidly. So, it is recommended that the Type 2 diabetic patient should take zinc on a regular basis or take diet or supplements containing high amount of zinc (Jayawardena et al. 2012). It has also been observed that zinc supplementation decreases the level of cholesterol and HbA1c levels in the blood of Type 2 diabetes patient, and thus reduces the effects of diabetes (Khan et al. 2013).

Zinc in Wound Healing

The process of wound healing depends to a larger extent on the availability of nutrients and vitamins crucial to wound healing. Zinc is the most important element that helps in the healing of wounds (Lin et al. 2017). It is known to maintain membrane integrity strengthening tissues. Zinc may also promote wound healing through cell division forming new tissues. Thus, zinc helps in the growth of new tissues and healing of the wounds, if consumed in an adequate amount on a daily basis. Clinically, it has been found that the zinc deficiency delays wound healing due to the reduction in activation of the nuclear factor (NB) kB. Similarly, the deficiency of zinc also reduces the infiltration of neutrophil during early stages of wound healing. Furthermore, its deficiency also decreases the proinflammatory cytokine expression including interleukin (IL)-1β and tumor necrosis (TNF-α) factor. Therefore, oral zinc supplementation in adequate amount is recommended for the treatment of wound healing (Lansdown et al. 2007).

Zinc and Cancer

Like other nutrients, zinc also plays an important role in defense against the initiation and progression of cancer. Zinc is an essential element for the functioning of transcription factor, DNA repair, and antioxidant defense system. It is recommended that the deficiency of zinc in human body disturbs the functioning of both signaling molecules and proteins that are directly involved in DNA replication and repair (Skrajnowska and Bobrowska-Korczak 2019). Further, it can also contribute to single- and double-strand DNA breaks and oxidative modification to DNA that increase the risk for cancer development (Dhawan and Chadha 2010).

Zinc and Liver Disorders

Zinc deficiency is also known to cause various liver diseases such as cirrhosis. The symptoms that are very common in a patient with liver disorder includes abdominal pain and swelling, jaundice (skin and eyes appears yellowish), liver enlargement, cholestasis, dark urine color, portal hypertension, and swelling in the legs and ankles. It is recommended that proper intake of zinc either through diets or supplementation reduces the risk of liver disorder. One of the researches conducted on zinc supplementation reported that the patients infected with chronic hepatitis C virus treated with zinc supplementation shows a greater reduction in gastrointestinal disturbances, weight loss, hair loss, and mild anemia (Mohammad et al. 2012).

Zinc and HIV

In 2017, India reported that more than 2.1 million people are infected with HIV. This made India the third largest HIV epidemic in the world (Del Rio 2017). An estimated 0.2% adults (aged 15–49) were suffering from HIV. In most of the HIV cases, the patients suffer from compromised immunity. This is generally characterized by a continuous decrease in CD4 cell count below reaching less than 200 cells/mm^3 (https://www.ncbi.nlm.nih.gov/books/NBK209955/). One important thing that has also been observed is that more than 50% of HIV-positive patients are zinc deficient. Deficiency of zinc increases the progression of HIV disease and also increases the risk of death. In view of this, it has been suggested that the HIV patients should take zinc (12–15 mg/ day) on a regular basis either in the form of supplements or through diet by incorporating zinc-dense food (Baum et al. 2010). Several studies conducted in India or in the USA confirmed that the supplementation of zinc effectively controlled the decrease in the population of CD4 cell below the critical level (200 cells/mm^3). It has also been reported that the daily consumption of zinc in adequate level reduces the risk of immunological failure, illness, and diarrhea. Researchers found that the supplementation of zinc reduced the risk of diarrhea and immunological failure up to 60% and 70%, respectively. Above finding tends to show that adequate intake of zinc and zinc therapy is a safe, simple, and cost-effective tool to improve the immunity and to reduce the morbidity in HIV patients (Baum et al. 2010).

Zinc in Skin Diseases

Zinc plays an essential role in maintaining healthy skin as it helps in proper functioning and growth of skin cells. Zinc works as an antioxidant to heal and rejuvenate the damaged skin, which is exposed to UV light, pollution, and aging factors (Rostan et al. 2002). It helps the skin from getting infected by acne and also controls skin inflammation. Zinc further accelerates the growth of new skin cells and transports them to the damaged skin. It also fights back the occurrence of pimples by opening the clogged pores of the skin. It also acts as an effective sunscreen block and prevents the harmful UV rays from penetrating into the skin. Zinc is also known to control dandruff and to heal the itchy scalp. Furthermore, it is also used as an important ingredient in making diaper rash creams because of its antimicrobial properties. It soothes the infected sight and helps in treating the skin quickly. Several clinical studies have been published demonstrating that 10% solution of zinc sulfate three times a day for up to 4 weeks can effectively treat viral warts (Gupta et al. 2014; Songsantiphap and Asawanonda 2019). For the treatment of seborrheic dermatitis in the scalp, shampoos of zinc-pyrithione have proven very effective and have clinical advantages. In another clinical study, it was found that injecting 2% zinc sulfate is beneficial for viral warts as local therapy (Sharquie Khalifa and Al-Nuaimy Adil 2002). Furthermore, it has also been stated that topical application of divalent zinc ions provides protection against UV through its antioxidant activity (Moneib et al. 2010; Nikam 2009).

Commercial Products Available in the Market

In recent times, several industries produce various products containing zinc, vitamins, and other minerals in the form of supplements, beauty products, dental care products, gummies, and zinc fertilizer for crops. To correct zinc deficiency in soil, crops, and plants, one company produces chelated zinc as zinc fertilizer with the name Talwar Zinc super 14. The application of 500 g/acre with 1–1.5 g/L water is recommended to provide adequate amount of zinc to the plants and crops. It has also been observed that the deficiency of zinc is common in children below the age of 5 years. Keeping this in view, one company produces gummies for kids in combination of zinc and vitamins with the product name "Super Gummy." It is claimed that the product provides sufficient amount of zinc and helps in building immunity in the kids. Further, they also highlighted that the daily intake of 2 super gummies is equivalent to vitamin C in 1 orange, zinc in 750 g of spinach, vitamin A in half cup of carrots, and vitamin D in 4 cups of sliced white mushrooms. Furthermore, several other commercial products like beauty products, dental care products containing zinc, minerals, and vitamins are also available in the market. Figure 16.9 highlights few such commercial products.

Fig. 16.9 Web grabs of few commercial products containing zinc available in the market

Conclusion

WHO estimated that one-third population of the world is affected by the deficiency of zinc which ranges from 4% to 73% across subregions. Most of the developed countries are at a lower risk of zinc deficiency, while developing countries are at a higher risk especially India, Bangladesh, and Bhutan. Overall, 43.8% of zinc deficiency has been observed in Indian states: the highest 51.3%

in Orissa followed by 48.1% in UP, 44.2% in Gujarat, 38.9% in MP, and 36.2% in Karnataka. Every Indian state is zinc deficient to some extent. The deficiency is most common among children ($<$ 5 years) and pregnant women. It is estimated that around 2.1 million Indian children below the age of 5 years die every year due to typhoid, malaria, diarrhea, and pneumonia. Studies reported that due to low dietary intake of zinc, approximately 64.6% of the pregnant women suffer from zinc deficiency in the state of Haryana. In various reports, zinc deficiency is recognized as a global and public health problem for many developing countries. Despite this recognition, serious efforts are not taken to reduce the gravity of this problem. Currently no precise data are available at the international and national level. But some of the organizations such as World Health Organization (WHO), Food and Agricultural Organization (FAO), International Zinc Nutrition Consultative Group (IZiNCG), and United Nations International Children's Emergency Fund (UNICEF) keep providing some guidelines time to time regarding the recommended intake of zinc for different age groups and genders, about the rich sources of zinc, and about causes/affects of zinc deficiency. Further, they have also published data on the bioavailability and zinc absorption capacity of various sources. If the people of developing countries wish to improve their zinc deficiency, then they should incorporate adequate amount of zinc in their diet on a daily basis and should keep in mind certain points before selecting their diet. Deficiency of zinc is largely related to inadequate intake or absorption of zinc from the diet. The distinction between intake and absorption is important because the inhibitors such as fiber and phytates present in the diet result in the low absorption of zinc. Due to this reason, WHO has set the zinc requirement of dietary intake at a higher level for a population in which animal products (best sources of zinc) are limited and in which plant sources of zinc are high in phytates. It can therefore be concluded that in order to fight back the prevalent zinc deficiency, people should consume meat items such as chicken, lamb, beef, rabbit meat, oysters, scallops, and blackfish, and people who are vegetarian should consume the items which are rich in zinc content such as legumes, etc. to cope with the deficiency. Other than meat and cereals, people should also incorporate vegetables and fruits which are rich in zinc content such as beans, peas, jackfruit, pomegranate, and guava, which can help in reducing Zn deficiency. It is further important to mention that the status of zinc is not clear to people in many states of the country; therefore, it calls for special attention so that awareness among people can be drawn to curb this deficiency. In this regard, a lot more research is to be done.

References

Ahmad, F., Khan, S.: Potential Industrial Use of Compounds from By-Products of Fruits and Vegetables, pp. 273–307 (2019). https://doi.org/10.1007/978-3-030-24903-8_10

Ahmad, F., Zaidi, S., Ahmad, S.: Role of by-products of fruits and vegetables in functional foods. In: Ahmad, S., Al-Shabib, N.A. (eds.) Functional Food Products and

Sustainable Health, pp. 199–218. Springer Singapore, Singapore (2020). https://doi.org/10.1007/978-981-15-4716-4_13

Akhtar, S.: Zinc status in South Asian populations–an update. J Health Popul. Nutr. **31**, 139–149 (2013). https://doi.org/10.3329/jhpn.v31i2.16378

Aumeistere, L., Ciproviča, I., Zavadska, D., Bavrins, K., Borisova, A.: Zinc content in breast milk and its association with maternal diet. Nutrients, 10 (2018). https://doi.org/10.3390/nu10101438

Baum, M.K., Lai, S., Sales, S., Page, J.B., Campa, A.: Randomized, controlled clinical trial of zinc supplementation to prevent immunological failure in HIV-infected adults. Clin. Inf. Dis. **50**, 1653–1660 (2010). https://doi.org/10.1086/652864

Bhowmik, D., Bhattacharjee, C., Kumar, S.: A potential medicinal importance of zinc in human health and chronic disease. Int. J. Pharm. Biomed. Sci. **1**, 5–11 (2010)

Brown, K.H., et al.: International Zinc Nutrition Consultative Group (IZiNCG) technical document #1. Assessment of the risk of zinc deficiency in populations and options for its control. Food Nutr. Bull. **25**, S99–S203 (2004)

Brown, K.H., Wuehler, S.E., Peerson, J.M.: The importance of zinc in human nutrition and estimation of the global prevalence of zinc deficiency. Food Nutr. Bull. **22**, 113–125 (2001). https://doi.org/10.1177/156482650102200201

Castro-Alba, V., Lazarte, C.E., Bergenståhl, B., Granfeldt, Y.: Phytate, iron, zinc, and calcium content of common Bolivian foods and their estimated mineral bioavailability. Food Sci. Nutr. **7**, 2854–2865 (2019). https://doi.org/10.1002/fsn3.1127

Chadare, F.J., et al.: Conventional and food-to-food fortification: an appraisal of past practices and lessons learned. Food Sci. Nutr. **7**, 2781–2795 (2019). https://doi.org/10.1002/fsn3.1133

Das, J.K., Salam, R.A., Kumar, R., Bhutta, Z.A.: Micronutrient fortification of food and its impact on woman and child health: a systematic review. Syst. Rev. **2**, 67 (2013). https://doi.org/10.1186/2046-4053-2-67

Del Rio, C.: The global HIV epidemic: what the pathologist needs to know. Semin. Diagn. Pathol. **34**, 314–317 (2017). https://doi.org/10.1053/j.semdp.2017.05.001

Dhawan, D.K., Chadha, V.D.: Zinc: a promising agent in dietary chemoprevention of cancer. Indian J. Med. Res. **132**, 676–682 (2010)

Dietary reference values for food energy and nutrients for the United Kingdom. Report of the Panel on Dietary Reference Values of the Committee on Medical Aspects of Food Policy (1991) Reports on health and social subjects 41:1–210

Elbagermi, M.A., Edwards, H.G.M., Alajtal, A.I.: Monitoring of heavy metal content in fruits and vegetables collected from production and market sites in the Misurata Area of Libya. ISRN Anal. Chem. **2012**, 827645 (2012). https://doi.org/10.5402/2012/827645

Fairweather-Tait, S., de Sesmaisons, L.A.: Approaches used to estimate bioavailability when deriving dietary reference values for iron and zinc in adults. Proc. Nutr. Soc., 1–7 (2018). https://doi.org/10.1017/S0029665118000484

Faizan, A., Kumar, A.: Optimization of the ultrasonic assisted extraction process to obtain phenolic compounds from pomegranate (Punica granatum) peels using response surface methodology. Int. J. Agric. Sci. **10**, 7581 (2018)

Gammoh, N.Z., Rink, L.: Zinc in infection and inflammation. Nutrients, 9 (2017). https://doi.org/10.3390/nu9060624

Ganeshamurthy, A., Kalaivanan, D., Manjunath, B.L.: Nutrients removed from the soil decide the nutritional security of a nation: the case of iron and zinc in India. Curr. Sci. **113**, 7 (2017). https://doi.org/10.18520/cs/v113/i06/1167-1173

Gibson, R., Bailey, K., Gibbs, M., Ferguson, E.: A review of phytate, iron, zinc, and calcium concentrations in plant-based complementary foods used in low-income countries and implications for bioavailability. Food Nutr. Bull. **31**, S134–S146 (2010). https://doi.org/10.1177/15648265100312S206

Gonmei, Z., Toteja, G.S.: Micronutrient status of Indian population. Indian J. Med. Res. **148**, 511–521 (2018). https://doi.org/10.4103/ijmr.IJMR_1768_18

Gower-Winter, S.D., Levenson, C.W.: Zinc in the central nervous system: from molecules to behavior. BioFactors (Oxford, England). **38**, 186–193 (2012). https://doi.org/10.1002/biof.1012

Grabrucker, A.M., Rowan, M., Garner, C.C.: Brain-delivery of zinc-ions as potential treatment for neurological diseases: mini review. Drug Deliv. Lett. **1**, 13–23 (2011). https://doi.org/10.2174/2210303111101010013

Grüngreiff, K.: Zinc in liver disease. J. Trace Ele. Exp. Med. **15**, 67–78 (2002). https://doi.org/10.1002/jtra.10002

Gupta, M., Mahajan, V.K., Mehta, K.S., Chauhan, P.S.: Zinc therapy in dermatology: a review. Dermatol. Res. Pract. **2014**, 709152 (2014). https://doi.org/10.1155/2014/709152

Hambidge, M.: Human zinc deficiency. J Nutr. **130**, 1344s–1349s (2000). https://doi.org/10.1093/jn/130.5.1344S

Hawkes, S.J.: What is a "heavy metal"? J. Chem. Educ. **74**, 1374 (1997). https://doi.org/10.1021/ed074p1374

Huang, L., Drake, V.J., Ho, E.: Zinc. Adv. Nutr. **6**, 224–226 (2015). https://doi.org/10.3945/an.114.006874

Hunt, J.R.: Bioavailability of iron, zinc, and other trace minerals from vegetarian diets. Am. J. Clin. Nutr. **78**, 633S–639S (2003). https://doi.org/10.1093/ajcn/78.3.633S

Huong, N.T.L., Ohtsubo, M., Li, L., Higashi, T., Kanayama, M.: Heavy-metal contamination of soil and vegetables in wastewater-irrigated agricultural soil in a suburban area of Hanoi, Vietnam. Commun. Soil Sci. Plant Anal. **41**, 390–407 (2010). https://doi.org/10.1080/00103620903494350

Jackson, M.J.: Physiology of zinc: general aspects. In: Mills, C.F. (ed.) Zinc in Human Biology, pp. 1–14. Springer London, London (1989). https://doi.org/10.1007/978-1-4471-3879-2_1

Jayawardena, R., Ranasinghe, P., Galappatthy, P., Malkanthi, R., Constantine, G., Katulanda, P.: Effects of zinc supplementation on diabetes mellitus: a systematic review and meta-analysis. Diabetol. Metabol. Syndr. **4**, 13 (2012). https://doi.org/10.1186/1758-5996-4-13

Kawade, R.: Zinc status and its association with the health of adolescents: a review of studies in India. Global Health Action. **5**, 7353 (2012). https://doi.org/10.3402/gha.v5i0.7353

Khan, M.I., Siddique, K.U., Ashfaq, F., Ali, W., Reddy, H.D., Mishra, A.: Effect of high-dose zinc supplementation with oral hypoglycemic agents on glycemic control and inflammation in type-2 diabetic nephropathy patients. J. Nat. Sci. Biol. Med. **4**, 336–340 (2013). https://doi.org/10.4103/0976-9668.117002

King, J.C., Brown, K.H., Gibson, R.S., Krebs, N.F., Lowe, N.M., Siekmann, J.H., Raiten, D.J.: Biomarkers of Nutrition for Development (BOND)-zinc review. J. Nutr. **146**, 858s–885s (2015). https://doi.org/10.3945/jn.115.220079

Koe Wei, W., et al.: Zn in Vegetables: A Review and Some Insights, vol. 6, (2019). https://doi.org/10.15761/IFNM.1000245

Kumar, A., Ahmad, F., Zaidi, S.: Importance of bioactive compounds present in plant products and their extraction: a review. Agric. Rev. **40**, 249–260 (2020). https://doi.org/10.18805/ag.R-1926

Kumari, S., Badana, A.K., Malla, R.: Reactive oxygen species: a key constituent in cancer survival. Biomark. Insights. **13**, 1177271918755391 (2018). https://doi.org/10.1177/1177271918755391

Kumssa, D.B., Joy, E.J., Ander, E.L., Watts, M.J., Young, S.D., Walker, S., Broadley, M.R.: Dietary calcium and zinc deficiency risks are decreasing but remain prevalent. Sci. Rep. **5**, 10974 (2015). https://doi.org/10.1038/srep10974

Lansdown, A.B., Mirastschijski, U., Stubbs, N., Scanlon, E., Agren, M.S.: Zinc in wound healing: theoretical, experimental, and clinical aspects. Wound Repair Regen. **15**, 2–16 (2007). https://doi.org/10.1111/j.1524-475X.2006.00179.x

Levenson, C., Morris, D.: Zinc and neurogenesis: making new neurons from development to adulthood. Adv. Nutr. (Bethesda, Md). **2**, 96–100 (2011). https://doi.org/10.3945/an.110.000174

Lim, K.H., Riddell, L.J., Nowson, C.A., Booth, A.O., Szymlek-Gay, E.A.: Iron and zinc nutrition in the economically-developed world: a review. Nutrients. **5**, 3184–3211 (2013). https://doi.org/10.3390/nu5083184

Lin, P.H., Sermersheim, M., Li, H., Lee, P.H.U., Steinberg, S.M., Ma, J.: Zinc in wound healing modulation. Nutrients, 10 (2017). https://doi.org/10.3390/nu10010016

Liou, G.Y., Storz, P.: Reactive oxygen species in cancer. Free Rad. Res. **44**, 479–496 (2010). https://doi.org/10.3109/10715761003667554

Lönnerdal, B.: Dietary factors influencing zinc absorption. J. Nutr. **130**, 1378S–1383S (2000). https://doi.org/10.1093/jn/130.5.1378S

Ma, J., Betts, N.M.: Zinc and copper intakes and their major food sources for older adults in the 1994–96 Continuing Survey of Food Intakes by Individuals (CSFII). J. Nutr. **130**, 2838–2843 (2000). https://doi.org/10.1093/jn/130.11.2838

Maares, M., Haase, H.: A guide to human zinc absorption: general overview and recent advances of in vitro intestinal models. Nutrients, 12 (2020). https://doi.org/10.3390/nu12030762

Marangoni, F., et al.: Maternal diet and nutrient requirements in pregnancy and breastfeeding. An Italian Consensus Document. Nutrients. **8** (2016). https://doi.org/10.3390/nu8100629

Maret, W.: Zinc in pancreatic islet biology, insulin sensitivity, and diabetes. Prev. Nutr. Food Sci. **22**, 1–8 (2017). https://doi.org/10.3746/pnf.2017.22.1.1

Marreiro, D.D., Cruz, K.J., Morais, J.B., Beserra, J.B., Severo, J.S., de Oliveira, A.R.: Zinc and oxidative stress: current mechanisms. Antioxidants (Basel, Switzerland). **6** (2017). https://doi.org/10.3390/antiox6020024

Mason, J.: Vitamins, Trace Minerals, and Other Micronutrients, vol. 2, pp. 1397–1406 (2012). https://doi.org/10.1016/B978-1-4377-1604-7.00225-6

Maxfield, L., Crane, J.S.: Zinc deficiency. In: StatPearls. StatPearls Publishing. Copyright © 2020, StatPearls Publishing LLC.,, Treasure Island, FL (2020)

Mertens, J., Smolders, E.: Zinc, pp. 465–493 (2013). https://doi.org/10.1007/978-94-007-4470-7_17

Michael, H.: Zinc and health: current status and future directions. J. Nutr., 130 (2000)

Mohammad, M.K., Zhou, Z., Cave, M., Barve, A., CJ, M.C.: Zinc and liver disease. Nutr. Clin. Pract. **27**, 8–20 (2012). https://doi.org/10.1177/0884533611433534

Moneib, H.A., Salem, S.A., Darwish, M.M.: Evaluation of zinc level in skin of patients with necrolytic acral erythema. Br. J. Dermatol. **163**, 476–480 (2010). https://doi.org/10.1111/j.1365-2133.2010.09820.x

Mousavi, S.R., Galavi, M., Rezaei, M.: Zinc (Zn) importance for crop production – a review. Int. J. Agron. Plant Prod. **4**, 64–68 (2013)

Murthy, S.C., Udagani, M.M., Badakali, A.V., Yelameli, B.C.: Symptomatic zinc deficiency in a full-term breast-fed infant. Dermatol. Online J. **16**, 3 (2010)

National Research Council Subcommittee on the Tenth Edition of the Recommended Dietary A: The National Academies Collection: reports funded by National Institutes of Health. In: Recommended Dietary Allowances, 10th edn. National Academies Press (US). Copyright © 1989 by the National Academy of Sciences.,, Washington, DC (1989). https://doi.org/10.17226/1349

Nikam, B.P.: Necrolytic acral erythema seronegative for hepatitis C virus--two cases from India treated with oral zinc. Int. J. Dermatol. **48**, 1096–1099 (2009). https://doi.org/10.1111/j.1365-4632.2009.04114.x

Norouzi, S., Adulcikas, J., Sohal, S.S., Myers, S.: Zinc transporters and insulin resistance: therapeutic implications for type 2 diabetes and metabolic disease. J. Biomed. Sci. **24**, 87 (2017). https://doi.org/10.1186/s12929-017-0394-0

Nowak, G., Szewczyk, B., Pilc, A.: Zinc and depression. An update. Pharmacol. Rep. **57**, 713–718 (2005)

Perignon, M., Barré, T., Gazan, R., Amiot, M.J., Darmon, N.: The bioavailability of iron, zinc, protein and vitamin A is highly variable in French individual diets: impact on nutrient inadequacy assessment and relation with the animal-to-plant ratio of diets. Food Chem. **238**, 73–81 (2018). https://doi.org/10.1016/j.foodchem.2016.12.070

Prasad, A.S., Miale Jr., A., Farid, Z., Sandstead, H.H., Schulert, A.R.: Zinc metabolism in patients with the syndrome of iron deficiency anemia, hepatosplenomegaly, dwarfism, and hypognadism. J. Lab. Clin. Med. **61**, 537–549 (1963)

Ranasinghe, R.A.S.N., Maduwanthi, S.D.T., Marapana, R.A.U.J.: Nutritional and health benefits of jackfruit (*Artocarpus heterophyllus* Lam.): a review. Int. J. Food Sci., 4327183 (2019, 2019). https://doi.org/10.1155/2019/4327183

Rangan, A.M., Samman, S.: Zinc Intake and its Dietary Sources: Results of the 2007 Australian National Children's Nutrition and Physical Activity Survey Nutrients, vol. 4, pp. 611–624 (2012). https://doi.org/10.3390/nu4070611

Rathi, S.K.: Acne vulgaris treatment: the current scenario. Indian J. Dermatol. **56**, 7–13 (2011). https://doi.org/10.4103/0019-5154.77543

Rink, L., Gabriel, P.: Zinc and the immune system. Proc. Nutri. Soc. **59**, 541–552 (2000). https://doi.org/10.1017/s0029665100000781

Roohani, N., Hurrell, R., Kelishadi, R., Schulin, R.: Zinc and its importance for human health: an integrative review. J. Res. Med. Sci. **18**, 144–157 (2013)

Rostan, E.F., DeBuys, H.V., Madey, D.L., Pinnell, S.R.: Evidence supporting zinc as an important antioxidant for skin. Int. J. Dermatol. **41**, 606–611 (2002). https://doi.org/10.1046/j.1365-4362.2002.01567.x

Sadeghzadeh, B., Rengel, Z.: Zinc in Soils and Crop Nutrition, pp. 335–375 (2011). https://doi.org/10.1002/9780470960707.ch16

Sandstead, H.H.: Chapter 61 – Zinc. In: Nordberg, G.F., Fowler, B.A., Nordberg, M. (eds.) Handbook on the Toxicology of Metals, 4th edn, pp. 1369–1385. Academic Press, San Diego (2015). https://doi.org/10.1016/B978-0-444-59453-2.00061-5

Sandström, B.: Dietary pattern and zinc supply. In: Mills, C.F. (ed.) Zinc in Human Biology, pp. 351–363. Springer London, London (1989). https://doi.org/10.1007/978-1-4471-3879-2_22

Shambhavi, S., Kumar, R., Padbhushan, R., Verma, G., Sharma, S.P., Sharma, S.K., Sharma, R.P.: Dynamics of zinc under long-term application of chemical fertilizers and amendments by maize–wheat cropping sequence in Typic Hapludalfs. Soil Use Manag. **36**, 507–523 (2020). https://doi.org/10.1111/sum.12566

Sharquie Khalifa, A., Al-Nuaimy Adil, A.: Treatment of viral warts by intralesional injection of zinc sulphate. Ann. Saudi Med. **22**, 26–28 (2002). https://doi.org/10.5144/0256-4947.2002.26

Skrajnowska, D., Bobrowska-Korczak, B.: Role of zinc in immune system and anti-cancer defense mechanisms. Nutrients, 11 (2019). https://doi.org/10.3390/nu11102273

Slavin, J.L., Lloyd, B.: Health benefits of fruits and vegetables. Adv. Nutr. (Bethesda, Md). **3**, 506–516 (2012). https://doi.org/10.3945/an.112.002154

Songsantiphap, C., Asawanonda, P.: Topical 15% zinc oxide ointment significantly reduces the size of common warts after four weeks: a randomized, triple-blinded, placebo-controlled trial. J. Clin. Aesthet. Dermatol. **12**, 26–31 (2019)

Szewczyk, B., Kubera, M., Nowak, G.: The role of zinc in neurodegenerative inflammatory pathways in depression. Prog. Neuro Psychopharmacol Biol. Psychiatry. **35**, 693–701 (2011). https://doi.org/10.1016/j.pnpbp.2010.02.010

Trumbo, P., Yates, A.A., Schlicker, S., Poos, M.: Dietary reference intakes – Vitamin A, Vitamin K, Arsenic, Boron, Chromium, Copper, Iodine, Iron, Manganese, Molybdenum, Nickel, Silicon, Vanadium, and Zinc. J. Am. Diet. Assoc. **101**, 294–301 (2001). https://doi.org/10.17226/10026

Tuerk, M., Fazel, N.: Zinc deficiency. Curr. Opin. Gastroenterol. **25**, 136–143 (2009). https://doi.org/10.1097/MOG.0b013e328321b395

Vishwanathan, R., Chung, M., Johnson, E.J.: A systematic review on zinc for the prevention and treatment of age-related macular degeneration. Investig. Ophthalmol. Vis. Sci. **54**, 3985–3998 (2013). https://doi.org/10.1167/iovs.12-11552

Wessells, K.R., Brown, K.H.: Estimating the global prevalence of zinc deficiency: results based on zinc availability in national food supplies and the prevalence of stunting. PloS one. **7**, e50568 (2012). https://doi.org/10.1371/journal.pone.0050568

Wong, K.W., et al.: Effects of anthropogenic activities on the heavy metal levels in the clams and sediments in a tropical river. Environ. Sci. Poll. Res. **24**, 116–134 (2017). https://doi.org/10.1007/s11356-016-7951-z

Wuana, R.A., Okieimen, F.E.: Heavy metals in contaminated soils: a review of sources, chemistry, risks and best available strategies for remediation. ISRN Ecol. **2011**, 402647 (2011). https://doi.org/10.5402/2011/402647

Chapter 17
Managing Zinc Deficiencies Among Children

Saba Khan

Abstract Nutritional deprivation is a serious global problem that can lead to long-term deficits in growth and immune functions. The food nutrients are classified as macronutrients and micronutrients. Vitamins and minerals belong to the group of macronutrients and micronutrients. Zinc is an important micronutrient for the human body which promotes immunity, provides resistance against infection and helps in proper growth and development of nervous system. It also contributes to enzymatic activities involved in protein synthesis. Low zinc levels contribute to the onset of anorexia nervosa in children. It also plays a vital role in activating growth, providing immunity, stimulating appetite, healing cuts, etc. Rich sources of zinc are red meat, poultry, sea food, milk and milk products, whole cereals, pumpkin seeds, nuts, etc. Zinc deficiencies often coexist with protein energy malnutrition. Modifications in basic family meals meet the nutritional needs of zinc among the family members according to their age. This chapter discusses the importance of zinc in diet and the role of meal modification in managing the zinc deficiencies among children.

Keywords Dietary modification · Zinc deficiencies · Children

Investing in the world's nutrition programmes to meet the global nutritional requirement is vital for the development of a healthy world community. Improving nutritional status of the population will empower communities with better health, consequently improving their overall economic and social development. Different civilizations have food habits that are influenced by the local availability of the food, social and religious beliefs. People of different culture need to eat foods that give them the necessary nutrients for their survival (Chastain 2009). There are varied ways in which people prepare meals and consume these nutrients. Many a times, the constant consumption of same food items in meals leads to nutritional

S. Khan (✉)
Department of Home Science, Faculty of Agricultural Sciences, Aligarh Muslim University, Aligarh, UP, India

S. T. Khan, A. Malik (eds.), *Microbial Biofertilizers and Micronutrient Availability*,
https://doi.org/10.1007/978-3-030-76609-2_17

Table 17.1 Dietary sources of some essential micronutrients

Nutrient	Source	RDA adult	RDA children	Important function
Iron	Spinach, chirva, egg, liver, jaggery, dry fruits, cumin, seeds, raisins and cereals like jowar, bajra, etc.	8–18 mg	13.7–15.1 mg	Helps in formation of haemoglobin responsible for providing oxygen to muscles and assists in creation of certain hormones
Manganese	Peanuts, beans, legumes, brown rice, spinach, etc.	3–5 mg	0.6–2.5 mg	Assists in carbohydrates, amino acids and cholesterol metabolism
Copper	Liver, nuts and seeds, chick peas, dark chocolates, avocados, etc.	900 μg	0.6–2 μg	Required to connective tissue, formation, normal brain and nervous system function
Zinc	Chickpeas, red meat, poultry, seafood, milk and milk products, whole cereals, pumpkin seeds, cashew nuts, etc.	8–10 mg	5–9 mg	Necessary for normal growth, cell division, immune function, wound healing and healthy vision
Iodine	Iodized salt, onion, garlic, cheese, sea fish, etc.	150 μg	50–150 μg	Assist in thyroid regulation, stimulates physical and mental growth
Fluoride	Fruit juice, water, tea, coffee, etc.	3–4 mg	0.5–2 mg	Necessary for the development of bones and teeth
Selenium	Brazil nuts, fish, chicken, dairy products, eggs, etc.	55 μg	20–40 μg	Important for thyroid health, reproductivity and defence against oxidative damage

imbalances and long-term deficiencies adversely affecting growth and immune system. Based on the quantities in which the nutrients are required, they can be grouped as macronutrient and micronutrient. These nutrients can also be grouped into seven major classes according to their biochemical nature and physicochemical properties, namely, carbohydrates, fats, dietary fibres, minerals, protein, vitamins and water. Macronutrients – fats, proteins and carbohydrates – form essential components of our meals, whereas vitamins, minerals and micronutrients, although required, are missing from the food. Although these micronutrients and vitamins are required in very small amounts, they are essential for life. They can be divided into four groups: water-soluble vitamins, fat-soluble vitamins, macrominerals and microminerals. Microminerals are also referred to as trace elements, including iron, manganese, copper, zinc, iodine, fluoride, selenium, etc. (Bamji et al. 2013) (Table 17.1).

Zinc and Its Importance

Zinc is a transition metal and an essential trace element for human. It is estimated that the human body contains about 2 grams of zinc, constituting only 0.0033% of the total body mass. Most of the zinc is present in muscles, but it is also present in blood plasma. Zinc plays important role in human physiology by being an important component of several enzymes, and by contributing to their structural and catalytic activity. It plays a central role in the immune system, affecting a number of cellular and humoral immunity processes. The absorption of zinc from the diet takes place in the intestine mainly in duodenum and ileum, which is influenced by various factors like competition by other divalent cation and presence of compounds that interfere with Zn absorption like phytate. Absorbed zinc is rapidly transported to liver, bound to albumin. The major portion of zinc in circulation is contained in RBC and is associated with the enzymes carbonic anhydrase, superoxide dismutase and metallothionein. Only 5 to 10% of the zinc present in the diet is absorbed while remaining is lost mainly through faeces (90%). Small quantities (0.5 mg in urine) are also excreted through urine (Ezzati et al. 2002).

Zinc deficiency has been found to retard growth and genital development and cause loss of taste acuity *(hypogeusia)*. Zinc deficiencies frequently occur with protein energy malnutrition and other micronutrient deficiencies. However, the incidence of zinc deficiency alone is very rare and may occur due to the diseases like acrodermatitis enteropathica (a rare congenital defect of zinc absorption), sickle cell anaemia and chronic renal failure (Cousins et al. 2003). Zinc is an essential component of more than 300 enzymes. It improves thymus gland function and lymphocyte proliferative function. Deficiency of zinc reduces thymic hormone production and activity, decreases T-cell number and function, natural cell and neutrophil function and impairs the ability of cells to kill infectious organisms (Lonnerdal 2000). The main cause of zinc deficiency among toddlers/children in developing countries is parental nutritional status and health, which includes unusual condition of defective human mammary gland zinc secretion, disorders like chronic inflammatory bowel disease, protein-energy malnutrition and diarrhoea. The central role of zinc in cell division, protein synthesis and growth is especially important for infants, children, adolescents and pregnant women involving high growth rates (Beattie et al. 2008).

Several studies indicate that zinc deficiency results in poor growth in infants and children and also a decrease in appetite. Zinc deficiency may also lead to impaired motor development in infants and, thus, can interfere with cognitive performance. It also has roles in numerous biological processes which include enzyme action, stabilization of cell membranes, regulation of gene expression and cell signalling. Other manifestations of zinc deficiency reported subsequently include diverse forms of skin lesions, impaired wound healing, night blindness and endocrine dysfunction. Genomic analysis suggests a link between zinc and chronic diseases, and zinc deficiency is reported to contribute significantly to the global burden of disease (Prasad 1983).

Zinc Deficiency in India

India is home to the largest number of children in the world. Nearly every fifth young child in the world lives in India. Zinc is an important cause of morbidity as it results in weakening of immunity and growth retardation among young children.

Higher demand of zinc due to higher growth rate among children coupled with inadequate dietary zinc intake predisposes pre-school children to an elevated risk of zinc deficiency. Zinc deficiency in children leads to retardation in growth and also morbidity from many diseases such as diarrhoea, pneumonia and malaria. Around 8,00,000 deaths among the children below 5 years of age occur annually due to diarrhoea (1,76,000), pneumonia (406,000) and malaria (2,07,000). High mortality rate among children resulting from these infections has been reported to be associated with inadequate zinc intake. The studies reported 54.2% children and 37.1% preschool children to be zinc deficient (Rangan and Samman 2012). Many epidemiological studies have confirmed prevalence of zinc deficiency among children below 5 years of age belonging to low socioeconomic groups in five major Indian states, reporting an overall zinc deficiency of 43.8%, with highest in Orissa (51.3%) followed by Uttar Pradesh (48.1%), Gujarat (44.2%), Madhya Pradesh (38.9%) and Karnataka (36.2%). Studies are revealing that inadequate zinc intake has been rising in India for decades, causing tens of millions of people to becoming Zn deficient. National Family Health Survey (NFHS-2015-16) reported stunting in children below 5 years (Bains et al. 2015). Diarrhoeal diseases result in death of more than 3 million children in developing countries each year. It contributes substantially to malnutrition in surviving children. The use of zinc as adjunctive therapy has the potential to improve the management of diarrhoea in addition to improving immunity and increasing survival in children. Apart from the routine management of diarrhoea, 10–20 mg zinc per day for 10–14 days is known to reduce the number of diarrhoea episodes, decrease stool output and reduce the need for hospitalization. The combined recommendation of zinc and oral rehydration solution (ORS) is a safe effective and inexpensive diarrhoea treatment in the developing world (Shils and Shike 2006).

The Indian Council of Agricultural Research (ICAR) have developed pearl millet variety with minimum levels of iron and zinc through plant breeding as reported in the news of the National Year of Millets. Government of India is planning to achieve its vision of a *Kuposhan Mukt Bharat* (Malnutrition Free India) by 2022. In India, 59% of children under 5 and 54% of pregnant women are anaemic, while 38% of children in the same age group are stunted. Diets deficient in iron and zinc are a major cause of anaemia and stunting respectively (Kapil and Jain 2011). For rural communities relying on staple crops like pearl millet for much of their diet, nutrient-enriched varieties of these crops present a cost-effective, sustainable strategy to improve nutrition. The government is also focusing on elemental iron and zinc supplementation to all children under 5 years of age. Biofortification, which uses conventional breeding to increase micronutrient levels of cereal crops, like wheat with iron and zinc, can go a long way to overcome the problem of iron and zinc deficiency among young children (Prasad 1984).

Dietary Requirement and Sources of Zinc

Oyster, shrimps, lamb, wheat germ, mushrooms, asparagus, broccoli, legumes, spinach, amaranth, seeds (pumpkin, flax, sesame, gingelly and sunflower), poppy seeds, tofu, milk and milk products are rich sources of zinc. Food products from animal are also considered as a rich source of zinc. Lean red meat and shell fish are rich sources of zinc, containing approximately 40 mg Zn/kg. Whole grains, cereals, pulses and legumes provide high concentrations of zinc. Dry fruits also contribute to meeting zinc requirements in the body (Premnath and Swamy 2010). Green leafy vegetables and fruits are relatively poor in their zinc content (Foster et al. 2013). The amount of protein greatly influences the amount of zinc obtained from the food. Zinc in food is present in complex form. Therefore, the environment within the gastrointestinal tract markedly influences the solubility and absorption of zinc. Some foods rich in zinc content that find place in Indian meals are given in Table 17.2.

Spices and herbs used in recipes contribute in enriching the food with micronutrients. Spices are the seeds, roots, bark or other parts of the plants primarily used for flavouring, colouring or preserving food, whereas herbs are leaves, flowers, stem, etc., used for flavouring and garnishing. Both spices and herbs are used in small quantities in many recipes, but they contribute significantly in enhancing the taste, flavours and nutritive value of the cooked food. Zinc content of some spices commonly used in the recipes of Indian meals is mentioned in Table 17.3.

Table 17.2 List of food that are rich in Zinc

Foods	Zn (per 100 g)
Beef	4.80 mg
Bengal gram dal	3.65 mg
Chana dal	3.00 mg
Lentil/ masoor dal	3.61 mg
Black rajma	3.08 mg
Soybean	4.01 mg
Red rajma	2.69 mg
Cow pea (lobia)	3.57 mg
Black sesame seeds (til)	8.59 mg
White sesame seeds	7.77 mg
Brown sesame seeds	7.84 mg
Almonds	3.50 mg
Cashew nuts	5.34 mg
Egg yolk	3.59 mg
Ragi	2.53 mg
Wheat whole	2.85 mg
Suji	2.13 mg

Source: (Indian Food Composition Table, ICMR, NIN 2017)

Table 17.3 Spices rich in zinc

Spices	Zn (per 100 g)
Saffron	1.09 mg
Nutmeg	2.15 mg
Cinnamon	1.83 mg
Black pepper	1.19 mg
Fenugreek	2.5 mg
Red pepper	2.48 mg
Cloves	2.32 mg
Bay leaves	3.7 mg
Fennel	3.7 mg
Ginger (grind)	3.64 mg
Cumin seeds	4.8 mg
Coriander seeds	4.7 mg
Turmeric (grind)	4.35 mg
Cardamom	7.47 mg
Mustard seeds	6.08 mg
Poppy seeds	7.9 mg

Source: www.dietandfitnesstoday.com

A food property is a particular measure of the food behaviour. As a matter of fact, its behaviour is studied with respect to energy and its interaction with the human senses. Its efficacy in promoting human health and well-being is its fundamental nature. Food property defines the food functionality. According to McCarthy (1997), the food properties can be defined as follows:

- Physical and physicochemical properties
- Kinetic properties
- Sensory properties
- Health properties

Physical properties of foods include colour, structure, thermal, mechanical, rheological, dielectric and barrier properties, whereas water activity, its absorbency, cohesion, dispersibility and solubility are dealt under physicochemical properties. Kinetic properties are the rates of changes in biological, biochemical, chemical, physicochemical and physical variations of foods. Appearance, texture, aroma, taste, etc., are the sensory properties of food. Health properties can be positive or negative. Positive health properties include nutritional composition and medical and functional properties. The negative health properties include excessive or unbalanced intake or becoming toxic after critical concentration. Functional health properties are those that have impact on general health and mental well-being and slowing down of ageing process. Medical health properties prevent and treat diseases (Rahman 2009). Presence of zinc in food products and its availability to the consumer are dependent on food properties.

Table 17.4 Requirement of zinc in children

Group	Category	Zn (mg/day)
Children	1–3 years	5
	4–6 years	7
	7–9 years	8
	10–12 years	9

Source: ICMR (2010)

Zinc requirement for children in different age group is different. The Table 17.4 highlights the amount of zinc requirement in children of varied age as given by Indian Council of Medical Research in 2010.

Management of Nutrition During Child Development

The growth and development of human from an embryo into a child and during the early years of infant's life, the growth rates are quite high and so is the physical and mental development. Nurturing the children during the early years of life is influenced by social environment during which they develop certain food habits and eating pattern (Shrimali 2008). For the first 6 months of life after birth, the infant is dependent only on mother's milk. Developmental stages according to Hurlock 2010 divide childhood into three groups. The period after 2 week of birth till 2 years of age is babyhood, followed by early childhood (2–6 years) and late childhood (6–12 years). However, childhood period is when the relative dependency of babyhood is over and it extends to the time when a child starts becoming sexually mature, that is, at approximately 13–14 years. This period of childhood is divided into two stages. The early childhood stage is for children in the age range of 2–6 years and late childhood stage is for the children of 6–12 years of age.

Imbalance of nutrients whether excess or deficient may result in poor health. Appropriate nutritional practices play important role in determining optimal health and child development (Sachdev and Choudury 2004). Children are in high-risk group and are most vulnerable to nutrient deficiency as due to the high growth rate they require higher quantities of these nutrients. There are various ways used for managing nutritional deficiencies among children. Some of the commonly used strategies are as follows:

- Feeding from family pot can provide nutrition for babies without any additional expenditure, with advantages of saving money and time. Children grow up accustomed to foods eaten traditionally by the families.
- Effective community participation is a key element to reduce malnutrition. A triple 'A' approach (assessment, analysis and action), re-assessment, re-analysis and future action, education and training together help in effective management of community participation.

- Social involvements require road-to-health chart. The road-to-health chart has four elements that are preventive, promotive, environmental and individual strategies.
- Maternal involvement is important in sustaining and enhancing nutritional health of the children.
- Meal modification is another popular means to manage nutritional deficiency.

These strategies are all unique in themselves and practically very functional. Especially using meal modification to enrich the same dish with multiple nutrients is very effective. Family resource management ensures the use of various human and non-human resources in achieving the goals of enriching meals (Seetharaman et al. 2015).

Meal Modifications for Children

Dietary modifications are changes made during food preparations, processing and consumption to increase the bioavailability of micronutrients and minimize micronutrient deficiencies in food at commercial or individual/household level. It is an integral part of all the health enrichment programs. It basically refers to the adaptation of the basic family meals to meet the nutritional needs according to the age and health status of the family members (Swaminathan 1994). While modifying the meals for individuals, following points should be kept in mind:

- Meals modified should not deviate too much from the original recipe.
- It should be easy to cook and easily produced.
- It should be more acceptable.
- Likes and dislikes of the individuals should be kept in mind.
- It should look attractive.
- It should keep the purpose of modification foremost while incorporating desirable alterations or additions.

Food diversity is a holistic approach for food modifications. It is a long-term sustainable strategy that emphasizes on food synergy of different food groups. It improves quality of diet and ensures improved micronutrients intake. For the availability of both macro- and micronutrients in the meals, the choices of food items from different food groups in a recipe ensure nutritive enrichment of the prepared dishes.

Zinc absorption is increased with dietary protein intake. The type of protein in meals affects bioavailability of zinc, whereas animal protein is known to enhance absorption. Dietary components like phytates in cereals and soy inhibit zinc absorption. Other components that interfere with Zn absorption are casein, calcium and copper. Amino acids, such as histidine, methionine and other low molecular weight ions like organic acids are known to have positive effect on zinc absorption. In Indian meals, different food groups contribute in meeting the daily zinc

Table 17.5 Supplementary food preferred by Indian children

S. No.	Supplementary food	Development stage of children	Number of respondents	Percentage of respondents
1.	Upma		50	100%
2.	Soup	Babyhood	50	100%
3.	Dalia		44	88%
4.	Gruel		38	76%
5.	Laddoo		48	96%
6.	Vada	Early childhood	40	80%
7.	Cheela		40	80%
8.	Panjeeri		42	84%
9.	Dal burfi		35	70%
10.	Chikki	Late childhood	40	80%
11.	Sandwich		50	100%
12.	Wrap		48	96%

Note: multiple responses from 50 respondents

requirements; for example, cereals, flesh foods, nuts and oil seeds, vegetables and pulses meet 52, 10, 10, 8 and 20% of zinc requirement, respectively. Dietary diversification with different combinations of spices and herbs is an interesting and creative way of improving the nutritional status of the food (Swaminathan 2012). To improve Zn content of the diet, many ingredients, and their quantities and preparation methods can be changed as discussed below. Table 17.5 provides the list of the most preferred supplementary foods included in the diet for children from 6 months to 12 years age are discussed. Twelve recipes, four from each developmental stage, are discussed in detail in the present chapter. Ingredients for recipe are enlisted from the information shared by the mothers, while modified recipes are developed for making them nutritionally rich in general and zinc enriched in particular.

Babyhood Recipes

Food during babyhood should be soft, less spicy and palatable for babies. For this age group baby food usually forms the supplementary meal during the day as it is given along with mother's milk to the children in the age group of 6 month to 2 years. This food usually helps the babies to develop taste for the food along with enhancing appetite and immunity. These supplementary meals are often referred to as weaning food for this age group.

For enhancing zinc content in the diet of children belonging to this group, the four recipes identified from survey are studied here:

- Upma
- Soup
- Dalias
- Gruel mix

Upma

It is a dish originating from the Indian subcontinent. It is a popular dish of south India, Maharashtra and Sri Lankan Tamil breakfast. It is cooked as a thick porridge from dry roasted semolina (Table 17.6) and is served hot. Other names of upma are *uppuma, uppittu, uppidi, kherabath, upeet* and *rulanv.* It is prepared by dry roasting the semolina in pan and cooking it over medium heat until toasty but not browned for about 2 minutes. Transfer it to a bowl. Keep the pan again on medium heat and add ghee or oil. Then add the mustard seeds and curry leaves and stir until they splutter for about 30 seconds. Add toasted semolina and salt and gently pour in the water. Stir vigorously until the mixture resembles wet sand. The semolina will bubble and spurt as it absorbs the water. Reduce heat to low. Allow this mixture to cook for about 2 minutes, stirring constantly. The upma absorbs water very quickly and tends to dry out too, so turn off the heat when the mixture is slightly runny.

If in the regular recipe of upma few ingredient which are the rich source of zinc are included during preparation, then it will not only enrich the dish in general nutrients but it will also modify its general taste and appearance. Table 17.6 shows that adding the roasted and grinded split black gram and peanuts along with boiled mashed peas and grated carrots to the toasted semolina will double the zinc content of upma. Using split black gram will not only add about 0.6 mg of zinc but will also add protein, calcium, niacin, thiamine and riboflavin. Peanuts will provide 0.78 mg of zinc and are good source of protein and fats, whereas adding peas will increase 0.93 mg of zinc and arc also rich source of protein, vitamin A and antioxidants along with other nutrients.

A boiled egg can be mixed in upma recipe to further enrich the zinc content by 0.6 mg.

Table 17.6 Ingredients used in regular and modified upma recipes

	Regular		Modified	
Ingredients	Quantity	Zinc content	Quantity	Zinc content
Semolina	40 gm	0.825 mg	40 gm	0.825 mg
Oil/ghee	10 gm	–	10 gm	
Whole black mustard seeds	5 gm	0.202 mg	5 gm	0.202 mg
Split black gram (urad dal)	–	–	20 gm	0.6 mg
Curry leaves	5 gm	0.01 mg	5 gm	0.01 mg
Peanuts	–	–	20 gm	0.78 mg
Green peas	–	–	30 gm	0.93 mg
Water	200 ml	1.00 mg	200 ml	1.00 mg
Salt	To taste		To taste	
Total		**2.037 mg**		**4.347 mg**

Table 17.7 Ingredients used in regular and modified soup recipes

	Regular		Modified	
Ingredients	Quantity	Zinc	Quantity	Zinc
Whole wheat flour	20 g	0.57 mg	20 g	0.57 mg
Bengal gram dal			30 g	1.095 mg
Lentil	20 g	0.722 mg	20 g	0.722 mg
Milk			250 ml	0.75 mg
Fennel seeds			5 mg	0.2 mg
Water	100 ml	0.5 mg	100 ml	0.5 mg
Ginger and garlic paste	½ tsp		½ tsp	
Salt	To taste		To taste	
Total		**1.792 mg**		**3.837 mg**

Soup

It is primarily a liquid food which is generally served hot or warm. It is usually made by boiling ingredients like meat, dal and vegetables with stock or water. It is characterized by boiling any one or more solid ingredients in liquid in a pot until the flavours are extracted from the broth with some spices and herbs. For the present recipe, dry roast the whole wheat flour. Pressure cook the lentil with ginger-garlic paste. Sieve the cooked lentil. Stir dry roasted wheat flour to the sieved lentil. Cook it on medium flame and add water to maintain the soup consistency (Table 17.7).

For the modified recipe, roast and grind the two dals. Mix it with the dry roasted flour. Heat the milk and make a paste of the mixture. Add ginger-garlic paste and grinded fennel seeds. Cook for 5 minutes or till done. Bengal gram dal is a good source of protein, energy, fibre, calcium, iron and phosphorus. It adds 1.095 mg of zinc to the recipe. Milk will add 0.75 mg of zinc; along with being a nutritious drink, it is rich in high quality of protein, vitamins and minerals. Use of fennel seeds enhances the flavour of all kind of soups. Along with adding 0.2 mg of zinc to the recipe, fennel is the highly nutritious source of other minerals like phosphorus, potassium and copper.

Addition of 30 grams of roasted bajra instead of wheat flour in this soup recipe can enrich the zinc content of the recipe by 0.51 mg. It will create variation in taste and increase the iron content significantly.

Dalia

It is known as cracked wheat. It is a wheat product made from raw wheat kernels, which are crushed or cut into smaller pieces. It can be cooked into both savoury and sweet recipe. It is also known as porridge. Dalia is often used as health ingredient in

Table 17.8 Ingredients used in regular and modified dalia recipes

	Regular		Modifies	
Ingredients	Quantity	Zinc	Quantity	Zinc
Broken wheat	40 g	1.14 mg	40 g	1.14 mg
Milk	250 ml	0.75 mg	250 ml	0.75 mg
Almonds			15 g	0.535 mg
Cashew nuts			15 g	0.898 mg
Walnuts			15 g	0.441 mg
Sugar	15 g		15 g	
Total		**1.89 mg**		**3.764 mg**

Indian homes. It is low in fat, good in carbohydrates and high in iron and fibre. It is given to both small kids and convalescing adults. For preparing dalia, heat the milk in a saucepan. Broken wheat is dry roasted and added to the boiled milk. It is cooked till softened then sugar is added and cooked for few more minutes (Table 17.8).

For the modified recipe, add grinded almonds, cashew nuts and walnuts and cook for 2 minutes. The zinc will increase about two times if they become the part of the recipe. The advantages of using dry fruits are numerous. They are vital source of minerals like potassium, manganese, phosphorus and iron. Adding 15 grams each of almonds (0.535 mg of zinc), cashew nuts (0.898 mg of zinc) and walnuts (0.441 mg of zinc) will also enhance the taste and make the texture creamier.

The gingelly seeds (black) or other nuts can also be used in this recipe to enrich the zinc content. It can also be flavoured with grinded cardamom seeds. The modified recipe has better taste and flavours.

Gruel Mix

It is a food consisting of some type of cereals – oats, wheat and rice – that are boiled in water or milk; it is a thinner version of porridge that may not be cooked. Historically it has been a staple diet of western India. It has weak and running consistency. Usually it is neutral in taste. Its preparation includes grinding of the rice and groundnut and cooking it in milk till it is tender so that it makes a thick paste. Then, add jaggery and stir it to mix well (Table 17.9).

Use a dried, grinded mixture of roasted rice, groundnut, bajra and ragi powder for making the gruel mix. Add 25 gm of grinded sunflowers seeds. This will increase the zinc content significantly. Bajra is a good source of protein, iron and vitamin 6 hence adding it to gruel mix will increase the zinc content by 0.552 mg. Along with adding 0.506 mg of zinc, ragi flour is rich in calcium and potassium. Sunflower seeds are packed with nutrients. They provide 1.767 mg of zinc to the recipe and are the rich source of protein, vitamin E and vitamin B6.

Other grinded nuts like almonds and walnuts can also be added in gruel mix to make it nutritionally rich.

Table 17.9 Ingredients used in regular and modified gruel mix recipes

	Regular		Modified	
Ingredients	Quantity	Zinc	Quantity	Zinc
Rice	20 g	0.336 mg	20 g	0.336 mg
Groundnut	25 g	0.975 mg	25 g	0.975 mg
Bajra flour			20 g	0.552 mg
Ragi flour			20 g	0.506 mg
Jaggery	25 g	0.75 mg	25 g	
Milk	250 ml		250 ml	0.75 mg
Sunflower seeds			25 gm	1.767 mg
Total		**2.061 mg**		**5.062 mg**

Early Childhood Recipes

Children from 2 to 6 years come under early childhood. Their nutritional requirements are higher as they are physically active; therefore, nutritionally dense foods should be selected. Now is the time when they start forming their likes and dislikes. Snacks should supplement and not substitute the main meals. Variety of fruits, vegetables, cereals, nuts and spices should be included in the recipes planned for their consumption. This enriches their diet and develops their taste. A complimentary feeding of legumes has positive effects on gut health, nutrient absorption and linear growth during early childhood.

Regular and modified recipes to assess the zinc content in the four recipes for children in early childhood studied here are as follows:

- Panjeeri
- Laddoo
- Vada
- Cheela

Panjeeri

It is a staple from Punjab region of Pakistan and India, which is treated as a nutritional supplement. It is made from whole wheat flour fried in ghee, mixed with grinded sugar and heavily laced with dried fruits and herbal gums. It is usually eaten in winters to ward of cold. It is eaten as a desert and consumed as a prashad after pooja. Add whole wheat flour and keep stirring at low flame in hot ghee. Finely chop the almonds, grate the coconut and add them to the flour when it starts turning golden in colour. Stir for another 5 minutes and switch off the flame. When temperature of the mixture reduces, mix the grinded sugar in it (Table 17.10).

Finely chopped cashew nuts and walnuts if added to the regular recipe will increase the zinc in the Panjeeri by 2.232 mg. Consumption of dry fruits enhances

Table 17.10 Ingredients used in regular and modified recipes of panjeeri

	Regular		Modified	
Ingredients	Quantity	Zinc	Quantity	Zinc
Whole wheat flour	80 g	2.28 mg	80 g	2.28 mg
Ghee/butter	20 g		20 g	
Coconut powder	25 g	1.25 mg	25 g	1.25 mg
Almonds	25 g	0.892 mg	25 g	0.892 mg
Cashew nuts			25 g	1.497 mg
Walnuts			25 g	0.735 mg
Sugar	30 g		30 g	
Total		**4.422 mg**		**6.654 mg**

taste, energy and stamina. They are also rich in fibre and minerals. Therefore, many more variety of dry fruits can also be grinded and added to the Panjeeri.

The addition of 25 gm of poppy seeds can further enrich the zinc content by 1.975 mg.

Laddoo

It is a sphere-shaped sweet originating from Indian subcontinent. The name is originated from the Sanskrit word *lattika.* They are primarily made from condensed milk, different types of flour, fat, sugar, semolina, herbs, seeds and nuts that are finely chopped or grinded. It is a dessert served on festive occasions. It has the popularity to take a status of Indian national sweet dish. For making laddoo, with the regular recipe, dry roast the semolina and bajra flour and keep it aside. Melt the jaggery and add the flour mixture to it, gently keep stirring till it froths without reducing it too much; take it off the flame before it starts to solidify and add milk to it and stir well. Then add softened khoya to it. Now oil your palms with a bit of ghee. Using your fingers take small portion of the mix (ensure that the mix is not too hot to burn you) and shape it into medium-sized balls (Table 17.11).

Laddoos are quite popular among children. For increasing zinc content in the shared recipe, add 25 gm of roasted sesame seeds to the regular recipe. These will double the zinc content. Sesame seeds being rich source of zinc and copper are often used as immune booster in diets.

The addition of 25 gm of sunflower seeds can further enhance the zinc content of laddoos by 1.767 mg.

Vada

It is in the category of savoury fried snacks from India. It is one of the traditional foods relished by Indians. Different types of vadas can be described as fritters, cutlets, doughnuts or dumplings. Alternative names for this food include *wada, vade,*

Table 17.11 Ingredients used in regular and modified laddoo recipes

	Regular		Modified	
Ingredients	Quantity	Zinc	Quantity	Zinc
Semolina	40 gm	0.852 mg	40 g	0.852 mg
White sesame seeds			25 g	2.147 mg
Bajra flour	40 gm	1.104 mg	40 g	1.104 mg
Khoya	60 gm	1.404 mg	60 g	1.404 mg
Jaggery	50 gm		50 g	
Milk	50 gm	0.15 mg	50 ml	0.15 mg
Ghee	15 gm		15 g	
Total		**3.512 mg**		**6.02 mg**

Table 17.12 Ingredients used in regular and modified recipes of vada

	Regular		Modified	
Ingredients	Quantity	Zinc	Quantity	Zinc
Bengal gram dal	40 g	1.46 mg	40 g	1.46 mg
Black gram dal	40 g	1.2 mg	40 g	1.2 mg
Peas dry	–	–	40 g	1.24 mg
Green chilly	10 g	–	10 g	–
Curry leaves	25 g	0.295 mg	25 g	0.295 mg
Mustard seeds	15 g	0.604 mg	15 g	0.604 mg
Coconut dry	–	0.75 mg	25 g	0.75 mg
Black sesame seeds	–	–	25 g	2.147 mg
Salt	To taste	–	To taste	–
Oil	For deep fry	–	For deep fry	–
Total		**4.309 mg**		**7.695 mg**

wadeh and bara. Its origin is in South India. For making regular vadas soak bengal gram dal and black gram dal overnight. Grind with green chilies and let it ferment. Add chopped green chili, curry leaves, mustard seeds and salt to it. Mix all the ingredients well. Give shape of vadas and deep fry them till they become golden brown (Table 17.12).

Addition of boiled mashed peas, dry grated coconut and black sesame seeds will modify the vada recipe and increase the zinc content significantly. Peas are also good source of other minerals like magnesium, potassium and calcium. They add vitamins and proteins to the recipe. Black sesame seeds are good source of minerals and energy. They contain healthy fats like poly unsaturated fatty acids and omega 6.

The addition of 40 gm of cowpeas can further enrich the zinc content by 1.84 mg.

Cheela

It is a protein-rich nutritious, spiced, savoury pancake served with chutney or dip and a filling. It is an easy, quick and tasty snack for breakfast. It is also a great evening snack and belongs to North India. The common basic ingredient used may be besan, semolina, grinded soaked moong dal, bengal gram, etc. It is prepared by grinding the overnight soaked masoor dal along with ginger, fresh chilies and some water. The batter should be of pouring consistency. Grate the paneer and mix all the ingredients (except oil). Keep aside the batter for 15–20 minutes. Heat a non-stick tawa / griddle and drizzle some oil. Sprinkle some water and after it sizzles off, wipe dry with a kitchen towel. Now pour a ladleful of the batter and spread into a circle with the back of the ladle. Drizzle some oil around the edges and over the cheela. Cook on one side for 2 minutes or till done before flipping it over to cook the other side (Table 17.13).

Folding of mashed ingredients rich in zinc to the cooked cheela will add to its nutritive value and taste. Using peas, curry leaves, poppy seeds and baby corn will together increase the zinc content by 3.263 mg. Peas will include protein, vitamins and minerals in the recipe. Curry leaves are rich in minerals and flavonoids, antioxidants, vitamin C, A, B and E. Along with being a good source of zinc, poppy seeds contain good levels of iron, copper, calcium, potassium manganese and magnesium.

Addition of 50 gm of soaked bengal dal while grinding the batter will further enhance zinc content by 1.825 gm.

Table 17.13 Ingredients used in regular and modified recipes of cheela

	Regular		Modified	
Ingredients	Quantity	Zinc	Quantity	Zinc
Masoor dal	100 g	3.27 mg	100 g	3.27 mg
Besan	50 g	1 mg	50 g	1 mg
Paneer	60 g	1.644 mg	60 g	1.644 mg
Peas			40 g	1.24 mg
Green chilly	5 g		5 g	
Curry leaves			25 g	0.295 mg
Poppy seeds			20 g	1.276 mg
Baby corn			40 g	0.452 mg
Ginger	5 g		5 g	
Coriander leaves	5 g		5 g	
Oil	5 g		5 g	
Turmeric powder	5 g		3 g	
Cumin powder	3 g		3 g	
Coriander powder	3 g		3 g	
Lime juice	3 g		5 g	
Asafoetida	5 g		5 g	
Salt	5 g		5 g	
Total		**5.644 mg**		**9.177 mg**

Late Childhood Recipes

Children in the age group of 7 to 12 years belong to late childhood phase of child development. During this age, they are involved in strenuous activities, growing in height and weight and are going to school. Their breakfast should be substantial and tiffins nutritionally balanced. Their snacks should be dry, not greasy, fresh with satiety value and attractive.

The recipes for this age group studied are as follows:

- Dal burfi
- Chikki
- Wrap
- Sandwich

Dal Burfi

It is a delicious sweet Indian dessert made from chana dal or bengal gram. It is a dense milk-based recipe. The name is a derivative of the Persian word – barf means snow. Colour changes, when a variation of the dal is made in the basic ingredient. For making the regular dal Burfi, soak chana dal (split chickpeas) overnight. Drain the water, grind the dal fine and prepare a batter. Mix the sugar and water in a pan to make syrup. Bring it to a boil. Filter through a muslin cloth. Add the chana dal paste and khoya (concentrated whole milk) to a heated pan with ghee. Stir vigorously. Continue to cook till it leaves the pan. Pour the mixture on to a greased tray, flatten and cut into pieces when cool (Table 17.14).

For increasing the zinc in dal burfi, use milk instead of water and add grinded dry fruits. As nutritional components of milk include high-quality protein, calcium, potassium, vitamin D, B12 and A, phosphorus and niacin, it nutritionally enriches recipe. Moreover, dry fruits will add to the taste and quality of any burfi. Using coconut, almonds, cashew nut and milk will increase zinc in burfi by 3.989 mg.

The addition of 25 gm of sunflower seeds can enrich the zinc content by 1.767 mg.

Chikki

It is a traditional Indian sweet. It is brittle in nature, generally made from nuts and jaggery or sugar. Each variety of chikki is named depending upon the ingredients used; for example, puffed rice, beaten rice, roasted bengal gram, peanuts, desiccated coconut and different types of seeds and dry fruits. For making the regular recipe of chikki, dry roast the groundnuts in a pan and grind them coarse. Melt ghee, add powdered jaggery and cook till the hard ball stage. Add puffed rice to it and remove from heat, mix well and set in a greased plate. Cut into pieces (Table 17.15).

Table 17.14 Ingredients used in regular and modified recipes of dal burfi

	Regular		Modified	
Ingredients	Quantity	Zinc	Quantity	Zinc
Chana dal	60 g	2.19 mg	60 g	2.19 mg
Coconut powder			25 g	1.25 mg
Almonds			25 g	0.892 mg
Cashew nuts			25 g	1.497 mg
Water	100 ml	0.15 mg	–	–
Milk	100 ml		100 ml	0.4 mg
Sugar	100 g		100 g	
Ghee	20 g		20 g	
Khoya	60 g	1.404 mg	60 g	1.404 mg
Total		**3.744 mg**		**7.733 mg**

Table 17.15 Ingredients used in regular and modified recipes of chikki

	Regular		Modified	
Ingredients	Quantity	Zinc	Quantity	Zinc
Gingelly seeds			25 g	2.147 mg
Groundnuts	60 g	2.34 mg	60 g	2.34 mg
Fat	40 g		40 g	
Jaggery	80 g		80 g	
Puffed rice	20 g		20 g	
Roasted Bengal gram			40 g	1.46 mg
Fennel seeds	5 mg	.2 mg	5 mg	.2 mg
Total		**2.54 mg**		**6.147 mg**

Adding gingelly seeds and roasted bengal gram in a pan along with peanuts will enrich the zinc three times in the modified recipe of chikki. Both gingelly seeds (sesame) and roasted bengal gram (bhuna chana) are concentrated source of protein, vitamins, mineral and energy. The addition of 40 gm of ragi flour can increase the zinc content further by 1.012 m.

Wraps

A wrap is a food dish made with soft flattened bread rolled around a filling. The small flat breads are wheat tortillas, lavas or pita. The filling is usually protein rich like meat, poultry, rajma, etc. Wrap is a cover to protect something by winding or folding, to envelop or secure the storage.

Table 17.16 Ingredients used in regular and modified recipes of wrap

	Regular		Modified	
Ingredients	Quantity	Zinc	Quantity	Zinc
Whole wheat flour	100 g	2.85 mg	100 g	2.85 mg
Potatoes	50 g	.09 mg	50 g	.09 mg
Peas	40 g	1.24 mg	40 g	1.24 mg
Rajma			60 g	2.7 mg
Bengal gram dal			30 g	1.095 mg
Onion	10 g		10 g	
Curry leaves			25 g	0.295 mg
Tomato	10 g		10 g	
Garam masala powder	5 g		5 g	
Ginger-garlic paste	5 g		5 g	
Coriander powder	5 g		5 g	
Red chilli powder	5 g		5 g	
Turmeric powder	5 g		5 g	
Cumin powder	5 g		5 g	
Coriander leaves	5 g		5 g	
Amchur powder	5 g		5 g	
Salt	To taste		To taste	
Total		**4.99 mg**		**9.08 mg**

For making the regular recipe as shared by the respondents, heat oil in a pan and add cumin seeds and when they change colour, add chopped onions and ginger-garlic paste. Sauté till the onions turn brown. Now add chopped tomatoes and fry for a couple of minutes. Add salt, turmeric powder, red chilli powder, cumin powder and coriander powder. Fry till the tomatoes get mashed and the raw smell goes. Add boiled potatoes and peas. Finally sprinkle amchur powder and garam masala powder and add the chopped coriander leaves, mix well and turn off the heat. Make tawa roti with whole wheat flour. For assembling the wrap, place a spoon of the filling in the middle of the roti and roll roti tightly to form the wrap (Table 17.16).

Using curry leaves with onions and adding boiled and softened rajma and bengal gram dal will double the zinc content in wraps. Nutritional enrichment with curry leaves and bengal gram dal in a dish has been mentioned in other recipes. Rajma (kidney beans) are rich in proteins, fibres, minerals and vitamin K. It also gives good texture to the fillings. Sixty grams of rajma will increase the zinc content in a wrap by 4.09 mg. The addition of 60 gm of paneer can enrich the zinc content by 1.644 mg.

Sandwiches

It is a food item consisting of two pieces of bread with a filling between them. It is eaten as a light meal. The filling may be of vegetables, cheese, paneer, meat, etc. For making the regular sandwiches, in a heated pan add oil and let the mustard seeds

Table 17.17 Ingredients used in regular and modified recipes of sandwich

	Regular		Modified	
Ingredients	Quantity	Zinc	Quantity	Zinc
Brown bread	4 slices	1.4 mg	4 slices	1.6 mg
Potatoes	100 g	0.18 mg	100 g	0.18 mg
Peas	40 g	1.55 mg	40 g	1.55 mg
Whole black mustard seeds	10 g	0.404 mg	10 g	0.404 mg
Baby corn			50 g	0.452 mg
Sunflower seeds			25 g	1.767 mg
Cheese			50 gm	1.19 mg
Black pepper powder	5 g		5 g	
Oil	25 g		25 g	
Salt	To taste		To taste	
Total		**3.534 mg**		**7.143 mg**

splutter in it. Now add boiled potatoes and peas to it. Sprinkle pepper and salt. Arrange the mixture on the oil glazed slices of bread and toast it in a sandwich maker or tawa. Serve it with sauce (Table 17.17).

Adding baby corns, sunflower seeds and grated cheese to boiled potatoes and peas filling will double the zinc content in sandwiches. Cheese is the all-time favourite of children. Adding it to regular sandwiches along with baby corns and sunflower seeds will increase the zinc content by 3.609 mg. Sandwiches will be more fulfilling and tastier.

The addition of 25 gm of poppy seeds can further enrich the zinc in sandwiches by 1.975 mg.

Conclusion

Dietary requirements not only vary with age but are also influenced by many other factors like geographical location. Furthermore, the diet consumed by children and individuals in various areas of the world is also influenced by food availability, social, cultural and religious beliefs. Dietary modifications can improve its nutrient status significantly along with its taste and acceptability. These modifications can be used for main and supplementary meals as well as for snacks and side meals. Modifications can be carefully designed not only to improve the micronutrient status of the food but may also help in improving status of other nutrients in the food. Zinc is one of the most important micronutrients and its deficiency among humans is becoming prevalent, making it important to design strategies to improve the dietary Zn intake. One of the strategies is the dietary modification. As discussed in this chapter, various food items traditionally consumed by Indian children can be subjected to dietary modification for improving Zn content by many times without altering its taste drastically. All the 12 recipes discussed in this chapter can be modified with limited number of ingredients. Since zinc absorption and availability can be improved through an increase in protein content of the diet, protein-rich ingredients were chosen for modifying the recipes. Dry fruits, peas, black and white sesame seeds, sunflower seeds, rajma, paneer, cheese, chickpeas, bajra flour, baby corn, curry leaves, fennel seeds and other ingredients can be used to enrich the zinc content of recipe. As these ingredients are rich in protein and are also good source of various minerals and vitamins. Even minor modifications with spices and herbs may result in significant improvement in nutrient content of the food, especially Zn. Dietary modifications of recipes for all the three age groups of children have been described in this chapter (Tables 17.18, 17.19 and 17.20). If required to break monotony, the prepared food can be given to the children in addition to other food 2–3 times during the day. Alternatively, these can be served to children as mid-day meals or as evening snack during the day. Therefore, dietary modification is one of the best strategies to improve Zn content of the food, which will have obvious health benefits such as strengthening of immunity, better physical and mental growth.

Table 17.18 Comparing four babyhood recipes

Supplementary food items	Zinc content before modification mg	Zinc content after modification mg	Increased amount of zinc content mg
Upma	2.037	4.347	2.31
Soup	1.792	3.837	2.045
Dalia	1.89	3.764	1.874
Gruel mix	2.061	5.062	3.001

.Note: Above recipes can be divided into two to three portions and consumed during the day

Table 17.19 Comparing four early childhood recipes

Recipes name	Zinc content before modification Mg	Zinc content after modification mg	Increased amount of zinc content Mg
Panjeeri	4.422 mg	6.654 mg	2.230
Laddoo	3.512 mg	6.02 mg	2.508
Vada	4.309 mg	7.695 mg	3.386
Cheela	5.644 mg	9.177 mg	3.533

Note: Above recipes can be divided into two to three portions and consumed during the day in addition to the main meals

Table 17.20 Comparing four late childhood recipes

Recipes name	Zinc content Mg	Zinc content after modification mg	Increased amount of zinc content mg
Dal burfi	3.744 mg	7.733 mg	3.989
Chikki	2.540 mg	6.147 mg	3.607
Wrap	4.990 mg	9.080 mg	4.090
Sandwich	3.534 mg	7.143 mg	3.609

Note: These recipes can be a part of their tiffins to school as they can be easily handled and provide them with substantial nutrition

Dietary modifications can be carried out targeting the needs of different age groups even among the children. For example, for school-going children, recipes listed in Table 17.8 are more relevant as these children are engaged in sports and academics. Such children are not only growing fast but are getting ready for the most crucial stage of adolescence, requiring nutrients which is different from the food for childhood.

This chapter has discussed the dietary modification for the Indian children. Similarly, modifications can also be attempted for the children in other countries using ingredients available and acceptable in that countries. Special attention is required for the countries in which Zn deficiency is prevalent. Furthermore, local dietician with the help of government and doctors can develop dietary modification programs, keeping in view the availability and acceptability of different food ingredients and their health benefits. To deal with Zn deficiency, it is also important that the government takes initiatives to promote such programs among its population.

References

Bains, K., Harpreet, K., Bajwa, N., Kaur, G.: Iron and zinc status of 6-month to 5-year-old children from low-income rural families of Punjab, India. Food Nutr. Bull. **36**, 254 (2015)

Bamji, M.S., et al.: Textbook of Human Nutrition, 3rd edn. Oxford and IBH Publishing Co. Pvt. Ltd, New Delhi (2013)

Beattie, J.H., Gordon, M.J., Rucklidge, G.J., Reid, M.D., Duncan, G.J., Horgan, G.W., et al.: Aorta protein networks in marginal and acute zinc deficiency. Proteomics. **8**, 2126 (2008)

Chastain, Z.: Nutrition and Society. Alpha Publishing House, New York (2009)

Cousins, R.J., Blanchard, R.K., Moore, J.B., Cui, L., Green, C.L., Liuzz, J.P., et al.: Regulation of zinc metabolism and genomic outcomes. J. Nutr. **133**, 1521S (2003)

Ezzati, M., Lopez, A.D., Rodgers, A., Vander, H.S., Murray, C.J.: Selected major risk factors and global and regional burden of disease. Lancet. **360**, 1347 (2002)

Foster, M., Chu, A., Petocz, P., Samman, S.: Effect of vegetarian diets on zinc status: a systematic review and meta-analysis of studies in humans. J. Sci. Food Agric. **93**, 2362 (2013)

Hurlock, E.: Developmental Psychology- a Life-Span Approach, 5th edn. Tata McGraw Hill (2010)

ICMR.: Report on Food Safety and Standard Athority of India, New Delhi (2010)

Kapil, U., Jain, K.: Magnitude of zinc deficiency among under five children in India. Indian J. Pediatr. **78**, 1069 (2011)

Longvah, T., Ananthan, R., Indian Food Composition Tables: Nation Institute of Nutrition. ICMR, Hyderabad (2017)

Lonnerdal, B.: Dietary factors influencing zinc absorption. J. Nutr. **130**(5), 1378S (2000)

McCarthy, O.J.: Physical Properties of Food and Packaging Material, engineering course material, Dept of Food Technology, Massey University USA (1997)

Prasad, A.S.: The role of zinc in gastrointestinal and liver disease. Clinics in Gastroenterology, **12**(3), Medtext Publications, USA (1983)

Prasad, A.S.: Discovery and importance of zinc in human nutrition, Federation of American society for experimental biology Proceedings, USA (1984)

Premnath, K.G.A., Swamy, K.R.M.: Food and Nutrition. Stadium Press Publications, New Delhi (2010)

Rahman, M.S.: Food Properties Handbook, 2nd edn. Taylor and Francis Group, New York (2009)

Rangan, A.M., Samman, S.: Zinc intake and its dietary sources: results of the 2007 Australian National Children's Nutrition and Physical Activity Survey. Nutrients. **4**, 611 (2012)

Sachdev, H.P.S., Choudury, P.: Nutrition in Children. B.I. Publications pvt ltd, New Delhi (2004)
Seetharaman, P., Batra, S., Mehra, P.: Family Resource Management. CBS publishers, New Delhi (2015)
Shils, M.E., Shike, M.: Modern Nutrition in Health and Disease, 10th edn. Williams & Wilkins, Philadelphia (2006)
Shrimali, S.S.: Child Development. Rawat Publications, New Delhi (2008)
Swaminathan, M.: Handbook of Food and Nutrition. Bangalore Printing and Publishing Co. Ltd., Bangalore (1994)
Swaminathan, M.: Essentials of Food and Nutrition, vol. 1, 2nd edn. Printing and Publishing Co. Ltd, Bangalore (2012)

Chapter 18
Strategies to Counter Zinc Deficiency, Current Status and Future Directions

Shams Tabrez Khan and Munawwar Ali Khan

Abstract Zinc deficiency in humans is caused by a range of problems associated from soil to plants and to human beings. Coordinated efforts of the international community, especially policymakers, agricultural and plant scientists, dieticians, physicians and others, are required to address a variety of issues responsible for global hidden hunger of micronutrient deficiency. Among some of the most important approaches are zinc biofortification of the cereal crops using zinc biofertilizers and the development of zinc-efficient crops. The use of zinc biofertilizers is one of the most effective, sustainable and economic ways to improve the zinc status of the soil to grow zinc-dense cereals. Zinc-dense cereals developed through biofortification should be made available to populations that are mainly dependent on cereal-based diet. Dieticians and scientists should make efforts to encourage diversification of diet and to use zinc supplements to meet the requirement. One study suggests that including fishes in the diet can help to overcome micronutrient deficiency globally. People should also be convinced to shun food beliefs and taboos to combat micronutrient deficiency. Strategies to improve zinc absorption in humans should also be developed. Notably, wheat with reduced phytate content in the grains was developed with an aim to maximize zinc absorption in the intestine. Awareness programmes in areas suffering the most from zinc deficiency should be carried out among all stakeholders. Despite the suggestions from FAO and WHO, global efforts to combat zinc deficiency matching those for combating HIV, etc., are not in place.

Keywords Zinc fertilizers and biofertilizers · Zn biofortification · Dietary diversification · Phytate

---

S. T. Khan (✉)
Department of Agricultural Microbiology, Faculty of Agricultural Sciences, Aligarh Muslim University, Aligarh, India

M. A. Khan
Department of Life and Environmental Sciences, College of Natural and Health Sciences, Zayed University, Dubai, United Arab Emirates

S. T. Khan, A. Malik (eds.), *Microbial Biofertilizers and Micronutrient Availability*,
https://doi.org/10.1007/978-3-030-76609-2_18

Introduction

The enormity of zinc deficiency and its health and economic consequences have compelled both the FAO and WHO to suggest zinc fortification of the diet at the global level (Chap. 1). It is becoming increasingly important and urgent to address the issue of zinc deficiency. Since different countries in the world and different sections of the society suffer from zinc deficiency due to different reasons, strategies also have to be chosen by a careful identification of the prevalent reasons for the deficiency (Fig. 18.1). The most prevalent reason is the low dietary intake of zinc, which can be improved by including zinc-dense food, use of zinc supplements or through zinc biofortification of the food. One of the biggest challenges is to improve the dietary intake of economically weaker sections of the society largely in India, China and Africa. The population in these countries cannot afford rich food which meets the daily zinc requirement. Two approaches have been at least partially successful in addressing the issue of zinc deficiency in these regions. First, the use of fertilizers for the zinc fortification of crops and second, the development of zinc-dense cereals producing varieties of crops through breeding programmes or through genetic engineering. Unacceptability of genetically modified crops, chemical

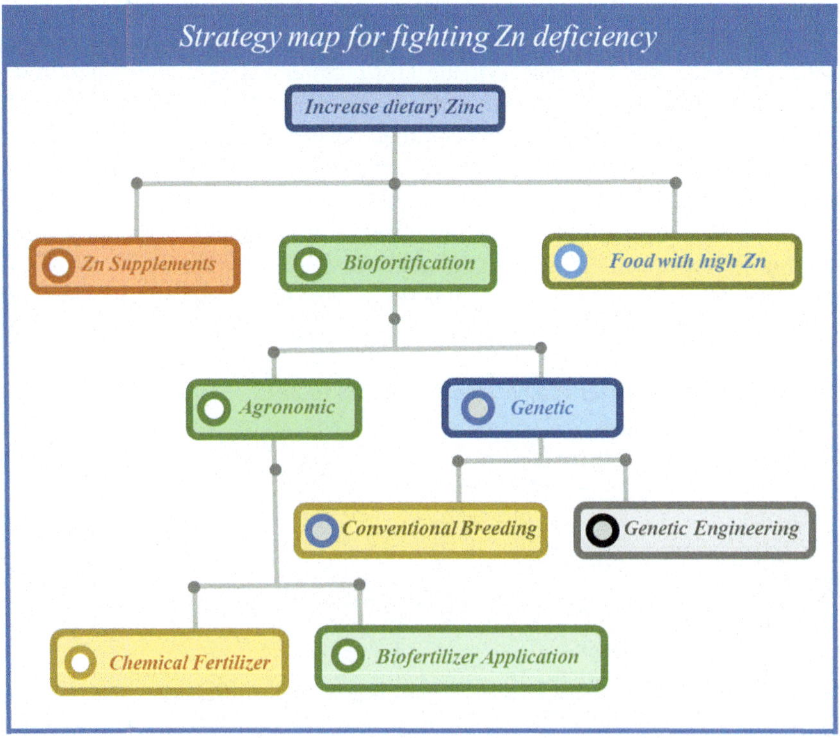

Fig. 18.1 Various strategies that can be used for combating zinc deficiency in humans and plants

pollution caused by fertilizers, cost and social unacceptability of genetic engineering programmes and many other concerns have to be considered. The use of zinc fertilizers is although required to correct Zinc deficiency in soils, which is one of the major reasons for the low zinc content of cereals. The residual effect of these fertilizers and the threat of chemical pollution require alternative and eco-friendly approaches to correct zinc deficiency in soils. The use of microbial inoculants to improve zinc uptake by plants is one such approach. Microbial inoculants not only improve zinc status of the crops but can also improve soil fertility through other plant growth-promoting activities. Some of these currently used strategies are detailed in this chapter. Besides the current progress, micronutrient deficiency remains a global challenge requiring further research and the involvement of policymakers and health workers (Ritchie et al. 2018).

Strategies for Improving Zinc Availability and Uptake by Plants

One of the main reasons for zinc deficiency is the lack of zinc in the soil. One third of the global soil has inadequate amount of zinc. Secondly, zinc is often present in soil as insoluble zinc, which is not available to plants. Thirdly, soil factors that affect the bioavailability of zinc like alkaline pH and sandy texture of soil. In this section, we have discussed some of these issues.

Biofortification Using Zinc Fertilizers

(i) *Types of zinc fertilizers*: One of the strategies to alleviate zinc deficiency in soil is the use of zinc fertilizers. Different types of zinc fertilizers are being used and can be classified into four categories based on their chemical nature, namely inorganic zinc fertilizers, inorganic complexes, natural organic complexes and synthetic chelates (Hergert et al. 1984). Zinc oxide (ZnO), zinc carbonate ($ZnCO_3$), zinc sulphate ($ZnSO_4$), zinc nitrate ($Zn(NO_3)_2$) and zinc chloride ($ZnCl_2$) are some of the inorganic zinc fertilizers (Alloway 2008). The most widely used inorganic forms of zinc are $ZnSO_4$ and ZnO. While, inorganic complexes include compounds like ammoniated $ZnSO_4$ and ammoniated $ZnCl_2$. These fertilizers provide more than one nutrient like nitrogen and zinc. Superphosphate is another formulation which in addition to phosphate contains zinc as an impurity (Alloway 2009). Natural organic complexes are produced by reacting metallic salts with citrates or with the organic byproducts from industries like the wood and pulp industry. Lignosulphonates, phenols and polyflavonoids are some of the commonly used byproducts. Other organic complexes can also be obtained by complexing zinc with amino acids.

Synthetic chelates of Zn are formed by combining a chelating agent such as ethylenediaminetetraacetic acid (EDTA), diethylenetriaminepentaacetic acid (DTPA) and (2-hydroxyethyl)ethylenediaminetriacetic acid (HEDTA) with zinc ion. However, the di-sodium salt of Zn-EDTA (Na_2Zn-EDTA) is the most commonly used chelate. These synthetic chelates are 2–5 times more available to plants than simple $ZnSO_4$.

(ii) *Rates and methods of fertilizer application*: The choice of the method of application and the rate of fertilizer application will vary, depending on the type of fertilizer, soil characteristics and crop species. Higher rates of application are required for crops sensitive to Zn deficiency or for calcareous soils having a higher pH (Alloway 2008). Fertilizers can be applied through seed treatment, sprayed on top soil, banded in the seed bed or as foliar spray. In the case of rice, the root of seedlings can also be dipped in fertilizer before planting. Although, the most common method is soil application. However, multiple studies suggest that foliar application is better than soil application and results in better accumulation in grains (Kopittke et al. 2019). One reason can be that foliar applications bypass adverse factors of soil that affect its absorption by root. $ZnSO_4$ can be applied at rates ranging from 5 to 25 kg Zn/ha (Cakmak and Kutman 2018; Liu et al. 2020; Cakmak 2008). However, an application rate of <10 kg of Zn/ha is recommended, which can be added to an NPK fertilizer both in granular and liquid forms (Mortvedt and Gilkes 1993). For foliar spray, usually a concentration five times lower than that of soil application is used (<1 kg/ha). The solution for the spray typically contains 2–5 g of zinc sulphate heptahydrate ($ZnSO_4{\cdot}7H_2O$) per litre (Cakmak et al. 2010; Boonchuay et al. 2013). Chelated forms of zinc such as Zn-EDTA being expensive are used as foliar spray for high value crops (Alloway 2008). Timings of the fertilizer are also important for effective accumulation in cereal grains. It was observed that the highest increase in zinc contents of the grains was observed with the foliar application of zinc around flowering time (Cakmak et al. 2010). For oats also, an increase in grain zinc content of up to 44% was observed following the use of zinc fertilizers (Shivay et al. 2013). The grain zinc concentration of barley was found to increase with the increasing concentration of zinc fertilizer used and almost doubled with the highest concentration of zinc used (Gonzalez et al. 2019).

(iii) *Zn nanomaterials as zinc fertilizers*: As the global production and consumption of nanomaterials are increasing, their possible use as fertilizers for improving plant mineral nutrition is also being explored. Nutrients in nanoform may have certain advantages over the bulk form of nutrients such as greater penetrability and higher activity at lower doses (Khan 2020). Nanoformulations of many plant macro- and micronutrients have been synthesized and studied including nitrogen, potassium, magnesium and zinc (Kopittke et al. 2019; Faizan et al. 2018; Sturikova et al. 2018). Various forms of zinc nanomaterials like zinc oxide nanoparticles, zinc nitrate and zinc hydroxide nitrate nanocrystals ranging in size from 35 to 100 nm have already been tested for crops like wheat, maize, peanut and tomato. Studies have shown that the foliar spray of

ZnO and $ZnSO_4$ nanomaterials increased the zinc content in the grains significantly without affecting the yield (Zhang et al. 2018). Forty-two percent higher yield of maize was obtained when the foliar spray of ZnO nanoparticles (400 mg/l) was used (Subbaiah et al. 2016). The same treatment was found to increase the zinc content of the grains significantly.

(iv) *Benefits of zinc fertilization*: Numerous studies have verified the positive impact of zinc fertilization. The positive impact is not only limited to an increased zinc content in the edible part of the plant but it also contributes to better vegetative growth and increased grain yield (Phattarakul et al. 2012; Aghili et al. 2014). Cakmak has reviewed studies undertaken in China, India, Pakistan and Zambia on increase in the grain yield with zinc fertilization and reported an increase of up to 5.7% (Cakmak et al. 2010; Rashid et al. 2019). In another study including China, India, Kazakhstan, Mexico, Pakistan, Turkey and Zambia, zinc fertilization increased the grain yield by 5%, while the grain zinc concentration increased by up to 90% (Zou et al. 2012). Similarly, for rice a 42.4% increase in grain zinc content was observed with a foliar spray of $ZnSO_4 \cdot 7H_2O$ (Yuan et al. 2013). In other studies, under optimal conditions of zinc fertilizer application the zinc content of the grains increased by 32–37% (Fang et al. 2008a; Phattarakul et al. 2012). In some studies, an increase in grain yield of about 5% was also observed (Phattarakul et al. 2012). Some studies estimate that zinc fortification of wheat and rice in India can save as many as 190,000 lives (Uauy et al. 2006). Soil application of zinc fertilizer was found to increase the yield of maize by 4.2–16.7% and the grain zinc content also increased by 24.3–74.9% (Liu et al. 2020). These increased yields can add billions of US dollars to world economy, as discussed in Chap. 1.

(v) *Challenges associated with the use of zinc fertilizers:* A number of technical problems are associated with the use of zinc fertilizers including the cost, conversion of added soluble fertilizer back to insoluble forms, efficiency of uptake by plants, residual effect of added zinc, concomitant uptake of other metals and toxicity concerns. The cost of zinc fertilization includes mainly the cost of fertilizer application, which can be minimized by adding it either with fertilizers or with the pesticides. Among various fertilizers, synthetic chelates like Zn-EDTA are very effective but are 5–10 times more expensive. Moreover, since only a fraction of the added zinc is taken up by plants and the remaining zinc added to soil shows a residual effect for up to 10 years, it is important to check the zinc concentration in the soil before zinc fertilization (Brennan 2001). Very often, Zn fertilizers added to soil get transformed into different unavailable forms like $Zn(OH)$, $Zn(OH_2)$ and $Zn(PO_3)_4$, due to factors like alkaline soil pH and high phosphorus (Takkar and Sidhu 1979). High concentration of zinc (typically 500 mg/kg) is considered toxic for plants. However, the concentration of zinc, which is toxic to crops, varies with the crop. For example, a soil zinc concentration of more than 60 and 81 ppm is considered toxic for wheat and maize, respectively (Takkar and Mann 1978). Another problem associated with zinc fertilization can be the increased accumulation of other undesirable metals like cadmium in grains (Köleli et al. 2004). These

Table 18.1 Some examples of increased zinc content in cereal grains following zinc fertilization

Species/cereal/ variety	Variety	Zinc fertilization Method	Dose	% Increase in grain Zn content	References
Wheat					
Triticum aestivum L.	Cerek-79	Soil	23 Kg/ ha	90	Yilmaz et al. (1997)
		Foliar	0.40%	33	Yilmaz et al. (1997)
	Baldo	Soil	45 Kg/ ha	20	Yilmaz et al. (1997)
Triticum durum Desf.	Kunduru-1149	Soil	23 Kg/ ha	190	Yilmaz et al. (1997)
	PDW 291	Foliar	5.35 Kg/ha	122	Dhaliwal et al. (2019)
Rice					
Oryza sativa L.	Wuyunjing 7	Foliar	0.9 Kg/ ha	20	Fang et al. (2008b)
		Foliar + Soil	9.3	73	Phattarakul et al. (2012)
Maize					
Zea mays L.	FHY-421	Soil	54 Kg/ ha	40	Kanwal et al. (2010)

problems can be simply avoided by avoiding the overdose of zinc fertilizers and a regular check of the zinc status in soil. Zinc uptake by plants and its accumulation in edible parts is a complicated process which depends on both the uptake of zinc and its mobilization to edible parts. This capability of plants to effectively uptake and accumulate the zinc in edible parts is called as zinc efficiency and will be discussed in a separate section. Therefore, it is better to choose zinc-efficient varieties of cereal crops (Table 18.1).

Zinc Biofertilizers for Zinc Biofortification of Crops

Inorganic zinc fertilizers are less effective, as the zinc added to soil is often converted into insoluble forms under the influence of various soil factors, as detailed above. The use of these fertilizers is also disadvantageous, as its residual effect may cause soil pollution. The use of biofertilizers is of strategic interest to replace chemical fertilizers in meeting the nutritional requirement of plants in a sustainable manner. Conventionally, microbial inoculants are used mainly for nitrogen fixation in agriculture. It has been estimated that these microorganisms alone fix 50–70 × 10^6 tons of nitrogen annually into the soil of agricultural systems (Herridge et al. 2008). But methods like metagenomics and next-generation sequencing approaches have revealed that there is a much greater interplay of plant and microorganism at work, which can be exploited for maximum agricultural benefits (De-la-Peña and

Loyola-Vargas 2014). The complex nature of the interaction can be understood from the fact that as many as 30,000 different prokaryotic species can be found in the rhizosphere of a plant and hundreds of chemicals (auxins, sesquiterpene lactones, Glomalin and prostaglandins (PGs)) are secreted by plants to regulate these complex interplays (Mendes et al. 2011; De-la-Peña and Loyola-Vargas 2014). It has also been found that plants actually recruit these microorganisms for their benefit and sometimes even the plant symbionts engineer the plant-microbe interaction by modifying root exudates (Uroz et al. 2019) This factor may also affect the neighbouring plant through root-to-root signalling (Korenblum et al. 2020). The plant growth-promoting rhizobacterial (PGPR) activities of the prokaryotes in the phytomicrobiome can be grouped as biofertilizers (microbes that increase the bioavailability of nutrients), biopesticides (microorganisms that repel or outcompete plant pathogens) and phytostimulators (microorganisms that manipulate the hormonal signalling of plants) (De-la-Peña and Loyola-Vargas 2014; Jacoby et al. 2017). Biofertilizers promote plant growth by supplying nutrients to the plants such as nitrogen and phosphorus. Some conventional examples of biofertilizers include *Rhizobium* spp., *Allorhizobium* spp., *Azotobacter* spp., *Trichoderma* and *Pseudomonas fluorescens*. These biofertilizers are being produced industrially and being used in fields. For example, in India alone it is estimated that 37997.61–40324.21 metric tonnes of biofertilizers are produced annually (Yadav and Chandra 2014).

(i) *Zinc solubilizing microorganisms as biofertilizers*: Soils are becoming increasingly deficient in micronutrients, due to extensive agricultural practices. As the macronutrients are exogenously added to agricultural lands, adding micronutrients is neither practised nor economically feasible. Therefore, the concentration of micronutrients is gradually depleting in the soil. Microorganisms, in addition to providing certain macronutrients, can also help in the uptake of micronutrients by plants (Jacoby et al. 2017; Treeby et al. 1989; Bouain et al. 2019; Dinesh et al. 2018). Zinc is of utmost importance among various micronutrients. It is often present in soil as insoluble zinc (smithsonite; $ZnCO_3$, sphalerite; ZnS, zincite; ZnO, franklinite; $ZnFe_2O_4$, and willemite; Zn_2SiO_4), which is not available to plants. Various microorganisms have the innate capacity to convert this fixed form of Zn to labile Zn form making it available for plants. These microorganisms are called as zinc solubilizing microorganisms. Although the ability to solubilize fixed Zn is not a very common characteristic of bacteria and fungi. But a number of studies have demonstrated the ability of bacteria as well as fungi to solubilize zinc in vitro using agar plate or in liquid medium (Table 18.2). Bacterial genera capable of solubilizing zinc are members of Gammaproteobacteria, Actinobacteria and Firmicutes group and include *Acinetobacter* sp., *Bacillus* sp., *Burkholderia* sp., *Pseudomonas* sp., *Gluconacetobacter* sp., *Curtobacterium* sp., *E. cloacae*, *Plantibacter* sp., *Pantoea dispersa*, *Rhizobium* sp. and *Streptomyces* sp. (Table 18.2). Endophytic fungi like *Beauveria caledonica* are also known to solubilize zinc phosphate (Fomina et al. 2004). Many of the reported strains are not sufficiently identified and characterized (Table 18.2). Other strains not mentioned here were not very

Table 18.2 Zinc solubilizing microorganisms and their plant growth-promoting activities

Organism	Identification basis	Zinc solubilization (mm)			PGPR activities	References
		ZnO	$ZnCO_3$	$Zn_3(PO_4)_2$		
Bacteria						
Acinetobacter sp.	Polyphasic		[a]–		PS, Sid, IAA	Rokhbakhsh Zamin et al. (2011)
Bacillus sp. AZ6			[a]–		–	Hussain et al. (2020)
Burkholderia lata ZnSB2	16S, biochemical	15.3	19.3	11.8	NH3, PS, Amy	Dinesh et al. (2018)
Burkholderia cenocepacia KNU17BI2	16S	25.2	19.6	–	PS, NH3, IAA, Sid, AF	Tagele et al. (2019)
Burkholderia contaminans KNU17BI3	16S	22.4	21.8	–	PS, NH3, IAA, Sid, AF	Tagele et al. (2019)
Curtobacterium sp. Strain 81	16S, MALDI-TOF	[a]–			–	Costerousse et al. (2018)
E. cloacae PBS-2	16S	1	1	0.5	–	Kamran et al. (2017)
Gluconacetobacter diazotrophicus	–	23	28	12	Nematicidal	Saravanan et al. (2007)
Pseudomonas fluorescens 3a	Biochemical		[a]–		–	Di Simine et al. (1998)
Plantibacter sp. Strain 5	16S, MALDI-TOF		22.4		–	Costerousse et al. (2018)
Pseudomonas sp. Strain 24	16S, MALDI-TOF				Sid	Costerousse et al. (2018)
Pseudomonas aeruginosa (CMG 823)	API test kit	+	–	+		Fasim et al. (2002)
Pseudomonas fragi strain EPS-1	16S	9	8	3	Phos, IAA, Sid	Kamran et al. (2017)
Pantoea dispersa strain EPS-6	16S	11	10	4	Lip, Cell, IAA	Kamran et al. (2017)
Rhizobium sp. LHRW1	–	18	10	6	–	Kamran et al. (2017)
Serratia sp. (TM9)						Idayu Othman et al. (2017)
Streptomyces sp. Strain 68	16S, MALDI-TOF		[a]–		Sid	Costerousse et al. (2018)
Fungi						
Beauveria caledonica						Fomina et al. (2004)

PGPR activities: *Lip* lipase, *Cell* cellulose, *IAA* indole acetic acid production, *Sid* siderophore production, *Phos* phosphatase, *PS* phosphate solubilization, NH_3 ammonia solubilization, *Amy* amylase, *AF* antifungal activity

[a]zinc solubilization tested by other methods

effective Zn solubilizers, as reported in studies. Although, it is difficult to define effective zinc solubilizers as different methods are used. But in the widely used agar plate-based analysis if a strain forms a zone of solubilization having a diameter equal to or more than 18 mm, it will be considered as an effective zinc solubilizer. However, for their use as biofertilizers a number of different parameters should be considered and are discussed in a separate section.

(ii) *Mechanism of zinc solubilization by microorganisms*: Microorganisms use different mechanisms for the solubilization of zinc. The most widely reported mechanisms include the acidification based on cation exchange and siderophore production (Eshaghi et al. 2019; Saravanan et al. 2011). Acid producing Zn solubilizers produce organic acids in soil to sequester zinc cation and decrease the pH of the nearby soil. These bacteria are known to produce acids such as citric acid, gluconic acid, 5-ketogluonic acid and oxalic acid. Decrease in soil pH considerably increases zinc mobility. It was demonstrated that the availability of Zn increased 100 times with a decrease in pH by 1 unit (Havlin 2005). Many of the strains listed in Table 18.2 are known to produce organic acids, some were shown to produce siderophores, while some were found to produce both. Many studies have demonstrated that a slight decrease in pH significantly increases the micronutrient bioavailability. Microorganisms are known to produce different siderophores including catecholates (enterobactin), carboxylates (rhizobactin) and hydroxamates (Pyoverdine) (Ahmed and Holmström 2014). But most of the studies on zinc solubilizing microorganisms have reported only qualitative and quantitative production of siderophores by using Chrome Azurol S (CAS) agar and no details on the type of siderophores are available. In one report, the production of both hydroxamate- and catechol-type siderophores has been reported using Csaky and Arnow assays (Rokhbakhsh Zamin et al. 2011).

(iii) In vivo *studies on zinc solubilizing microorganisms and impact on yield*: Many studies have evaluated the impact of zinc solubilizing microorganisms-based microbial inoculants on the growth and productivity of plants in vivo. These studies were done in jars, pot or in experimental fields on plants like, wheat, rice, maize and soybean. Various parameters have been used to assess the change in vegetative growth and productivity such as shoot and root lengths, dry weight and zinc content. While, the grain yield and grain zinc content were also checked in some studies. When zinc solubilizing *Burkholderia* and *Acinetobacter* were used for rice cultivation, the mean dry matter yield/pot, tillers/plant and number of panicles/plants were used with $ZnCO_3$. *Pseudomonas fragi* was found to increase the root dry weight its zinc content and zinc content in grains maximally (Kamran et al. 2017). Strains of *B. cereus* when tested on soybean and wheat under microcosm conditions were found to significantly increase seed yield and zinc content of both wheat and soybean. (Khande et al. 2017). In a similar study, almost similar results were obtained using the strains of *Bacillus aryabhattai* (Ramesh et al. 2014). *Bacillus* sp. AZ6 also increased the Zn contents of shoot and grain by 52% and 46% and, respectively, in maize. In short-term pot culture experiments, when seed bacterization was done with

Pseudomonas P29, it significantly enhanced the total dry mass and Zn uptake (Goteti et al. 2013). Surprisingly, following zinc biofertilizer treatment some studies also report a decrease in the phytate content of the grains, which inhibits the absorption of Zn in the intestine (Hussain et al. 2020).

(iv) *Commercial production of zinc solubilizing biofertilizers and application*: Commercial production of zinc solubilizing biofertilizers includes choosing an efficient strain, which in addition to having a good ability to solubilize zinc should have the ability to reasonably survive in soil and outcompete the soil microbial flora. These fertilizers can be formulated as solid or liquid by mixing with suitable sterile carriers. Different carriers can be used for biofertilizer preparation including naturally occurring materials like charcoal, peat, lignite or vermiculite (Malusá et al. 2012). These carriers are chosen, based on various desirable characteristics suited for biofertilizer production such as moisture-holding capacity, buffering capacity and should be free of any toxicity to inoculant strain. Suitability of the carrier with the inoculant strain should be determined. The microbial strain must also have the ability to survive with the carrier to achieve a longer shelf life. Once the biofertilizer is formulated, it should be subjected to quality control. Parameters that should be considered for quality control are listed in Table 18.3. Some of the important parameters are viable counts of the inoculant, which generally range from 10^8 to 10^9. The biofertilizer should be free of any other contaminant strain, even at a dilution of many folds usually up to a dilution of 10^5. The methods of application vary from crop to crop and fertilizer to fertilizer. The powdered form of fertilizers can be applied as seed coatings by mixing with adhesives like gum arabic. While, the granular form of fertilizers can be applied in soil with seed and the liquid form of fertilizers can also be sprayed. Some zinc solubilizing bacteria-based biofertilizers available in India are listed in Table 18.4 and a web grab of the products is shown in Fig. 18.2. Disadvantages of these fertilizers include the requirement of aseptic conditions, shorter shelf life, sensitivity to extreme environmental conditions like heat and dryness. But these advantages of using biofertilizers outweigh the disadvantages.

Table 18.3 Quality control parameters for zinc solubilizing bacteria biofertilizer

S. No.	Parameter	Standard requirement
1	Efficiency of inoculant strain	The strain should produce at least 10 mm of zone of solubilization on a 3-mm thick zinc solubilization agar
2	Viable cell count	Biofertilizers should contain 5×10^7 cells/g of powder or granule preparation or 1×10^8 cells/ml of liquid
3	Contamination check	No contamination at 10^5 dilution
4	Carrier	Dry carrier or base like charcoal, humus and peat can be used
5	pH	6.5–7.5 for powder or granules and 5.0–7.5 for liquid carrier
6	Particle size of carrier	Powdered material should pass through 0.15–0.212 mm IS sieve
7	Moisture content	30–40% by weight

Table 18.4 Some of the zinc solubilizing bacteria-based biofertilizers available in the Indian market

Product	Company	Organism	CFU Counts (/g or /ml)
BioZinc	Special Biochem Pvt. Ltd.	Unknown bacteria	5×10^8
Samridhi Zinc Solubilizing Liquid Bio- Fertilizer	Jaipur Bio Fertilizers	Unknown bacteria	1×10^8
Kish Zinc	FISHFA Biogenics	Unknown bacteria	2.5×10^9
BioZinc	Criyagen Agri & Biotech Pvt. Ltd.	*Thiobacillus thioxidans*	NA
Utkarsh Zincoz	Utkarsh Agrochem	Microorganism	NA

Fig. 18.2 Web grab of some zinc solubilizing bacteria based biofertilizers available in Indian market

It is therefore evident from the literature that a number of bacteria can serve as zinc biofertilizers in soil. The benefit of these bacteria is not limited to zinc, as many strains also possess additional PGPR activities. However, the isolation of even more effective Zn solubilizers from various environments is required. Proper cataloguing of such strains should be ensured for their on-demand retrieval and long-term use. More systematic and large-scale studies are required for a better understanding of the impact of these fertilizers. Though many commercial preparations of such fertilizers are already available in the market, studies are required to estimate the actual global consumption of such fertilizers. Awareness programmes among farmers to explain the benefits of these fertilizers are urgently needed. The information that these fertilizers are low-cost, economical and a sustainable option to eradicate plant zinc deficiency grown both in zinc-deficient soils and in soils where zinc is available as insoluble zinc should reach a greater farmer community. Policymakers and agricultural scientists should work hand in hand to promote these fertilizers to obtain long-term benefits.

Zinc Mobility from Soil to Grains

Zinc uptake and its translocation within the plant from roots to other parts like leaves and grains is a very complex process and cannot be discussed in complete detail in one section. Keeping in view these problems, Oslen and Palmgren correctly chose the title of their review as "Many rivers to cross: the journey of zinc from soil to seed" (Olsen and Palmgren 2014). It involves the uptake from soil by the root epithelial cells. It is where Zn crosses the cell wall and membrane to reach into the cell protoplasm (Symplast). At cellular level, a number of mechanisms facilitating import, trafficking, sequestration and export are involved to maintain an intracellular zinc concentration, which is not low and not quite high to cause toxicity, which is called as homeostasis. All the living cells need to maintain this homeostasis for optimal cellular functioning. The translocation from root epithelial cells to various parts of plants and grains will involve its movement through continuum of symplast (from one cell to another) and/or it may also have to pass through the dead spaces between the cells (apoplast). The uptake from the soil into the cell (symplast) involves metal carrier or channel proteins and is largely a passive process which does not require energy due to the negative membrane potential (Olsen and Palmgren 2014; Kochian 1993). While, the transport through the apoplast or dead spaces between the cells is the biggest bottleneck (Palmgren et al. 2008).

Roots primarily uptake Zinc as Zn^{2+} from the soil solution, which is translocated in the plants either as Zn^{2+}, $Zn(OH)_2$ or bound to organic acids. Uptake of Zn^{2+} in the rhizosphere is promoted by two mechanisms. One root exudation of various low molecular weight organic anions including organic acids (oxalate) and metal chelators (phytosiderophores or phytometallophores) (Hoffland et al. 2006). These compounds form complexes with zinc and transport them to the outer face of the root-cell plasma membrane. The second mechanism is the mobilization through acidification of rhizosphere using plasma membrane H^+-ATPases, proton pumps. The increase in the proton concentration in rhizospheric soil releases divalent metal ions tightly bound to soil particles through cation exchange. Upon reaching the plasma membrane, Zn^{2+} binds to proteins with high affinity to zinc and is transported into the cells. These proteins include metal transporters of the ZIP family which are thought to be the primary uptake systems for Zn in plants. For details on ZIP proteins and other proteins involved in transport, please refer to reviews by Palmgren and colleagues, Solen and Palmgren, and Ishimaru and colleagues (Olsen and Palmgren 2014; Palmgren et al. 2008). The Zn then enters into xylem parenchyma cells through diffusion. Then they are actively transported out of the symplast into the dead xylem using adenosine triphosphate (ATP)-dependent heavy metal pumps. From xylem, it is transported to leaves from shoot through vessel-associated cells in the leaves. Metal ions are remobilized and are exported and transported from leaves into the fruit via the phloem to the fruit. From there, Zn is subsequently taken up into the developing seed by specialized metal transport proteins. During its translocation from the root to the grain, Zn ions leave the symplast twice. First, when it is transported from root cells into xylem it is also called as xylem loading. Second, when the Zn is transported from phloem into the fruit it is called as phloem unloading). Among the two processes, the latter is the bottleneck.

Therefore, the movement of Zn from soil into the roots is a complex process involving many bottlenecks. The movement from soil to the root surface is the first step, which can not only be facilitated by root exudates but can also be facilitated by the addition of siderophores and acid producing microbial inoculants to the soil. Translocation of zinc in the plants and its accumulation into the seeds also faces some bottlenecks (Fig. 18.3).

Developing Zn-Efficient Strains Through Breeding and Molecular Biology Approaches

One approach of addressing the problem of Zn deficiency is to develop zinc-efficient plant varieties that will produce zinc-dense grains. Plant breeding is considered one of the low-cost approaches to develop such zinc-efficient varieties. The breeding programmes focussing on micronutrient deficiency (iron, zinc and β-carotene) have targeted five main crops (beans, cassava, maize, rice and wheat). The breeding programme includes the following steps, first the selection of parents with target traits. For example, varieties producing Zn-dense grains should be chosen. In maize, the variation in the grain Zn content of different varieties was around 50% of the mean value. The second step will be long-term crossing and backcrossing and to check the stability of the desirable trait in the offspring variety obtained through these crosses under various soil and climatic conditions. If found successful, a variety of offspring can be adapted for cultivation. Several studies report an increase in Zn content of the grains obtained through breeding. Under one such programme at the Waite

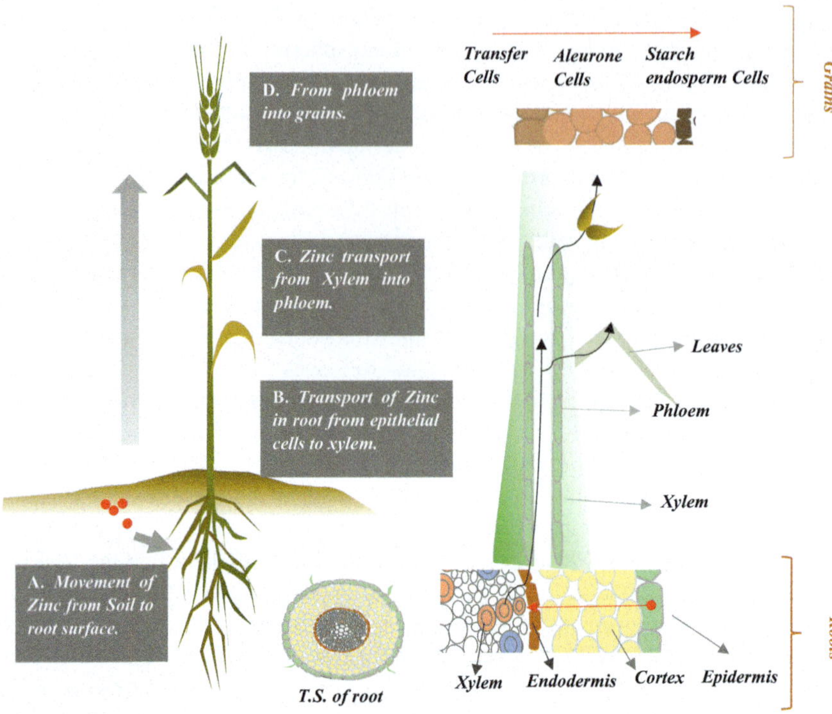

Fig. 18.3 Transport of zinc from soil to various parts of plants before finally reaching grains

Agricultural Research Institute of the University of Adelaide, Australia, a variety of wheat having improved Zn content was developed and is now being used successfully (Bouis 2003). In a similar programme on maize, inbreds exhibiting 14–180% more grain Zn and 32–78% more grain Fe were obtained at the International Institute of Tropical Agriculture (IITA), Nigeria (Menkir 2008). Several-fold increase in grain Fe and Zn of disomic hexaploid bread wheat has also been reported (Velu et al. 2014). An iron- and zinc-biofortified pearl millet (ICTP-8203) was developed in India and has been tested for treating zinc and iron deficiency (Huey et al. 2017).

Genetic engineering can also be used for the development of plants yielding Zn-dense grains. A number of genes can be targeted that in one way or the other can improve the ability of plant to assimilate more Zn in grains or control factors that limit the uptake by plant or absorption by humans, etc. It has been demonstrated that a stress-related N-acetyl cysteine (NAC) gene that regulates senescence improves grain zinc and iron content in wheat (Uauy et al. 2006). A number of transgenic plants have been engineered with increased grain Zn content. Trijatmiko demonstrated using genetic engineering that plants expressing rice nicotianamine synthase (OsNAS2) and soybean ferritin (SFerH-1) genes showed a high Zn and Fe concentration in the endosperm (Trijatmiko et al. 2016). In another study, the transgenic rice plant was developed which expressed genes for the production of metal chelators, namely nicotianamine and 2′-deoxymugenic acid (DMA). The transgenic rice plants thus developed were able to accumulate up two times more zinc and four times more iron in the rice endosperm, increasing significantly the nutritional quality of rice (Banakar et al. 2017). When ferritin gene from soybeans was expressed in indica rice cultivar, the plants exhibited higher levels of ferritin and were found to contain 1.54 times higher concentration of Zn and 2.54 times higher concentration of iron (Paul et al. 2014). It is interesting to note that in this study, an increased concentration of iron was targeted, but it also resulted in a high concentration of zinc in the grains. Furthermore, it has been demonstrated that the higher grain phytate or phytic acid concentration interferes with the Zn accumulation in the grains as well as its absorption in the intestine. Therefore, molecular tools can be used to minimize the grain phytate content. RNA interference (RNAi) is a widely used tool to inhibit specifically the expression of a gene. When the gene for phytate was inhibited using sequence-specific RNAi, it reduced the phytic acid levels in the cereal grains. This will consequently facilitate the absorption of zinc in the intestine (Kumar et al. 2019). Therefore, molecular biology approaches have been successful for the development of cereals having a high grain Zn concentration. However, some concerns associated with the genetic engineering approaches are the stability of the trait and social acceptability. Another concern is that if such crops become prevalent, it will be difficult to preserve the naturally occurring varieties. Social acceptability of crops developed through breeding programmes is still higher than the genetically engineered plants.

Strategies for Alleviating Zinc Deficiency in Humans

The primary reason for zinc deficiency in humans is the poor dietary intake which can be due to insufficient diet or diet which is poor in zinc only. Other reasons for zinc deficiency are malabsorption of the zinc in the intestine due to the genetic disease like acrodermatitis enteropathica or old age. Zinc deficiency in humans can be alleviated through long-term dietary programmes and short-term zinc supplementation programmes for vulnerable groups like children, elderly and expecting mothers.

Solutions for Poor Dietary Intake

As it has been discussed in Chap. 1, 30% of the world population or 1.1 billion people around the globe are at risk of zinc deficiency due to an insufficient dietary intake. The recommended daily dietary intake for zinc is 14 and 8 mg/day for men and women, respectively. However, these requirements cannot be met through the current diet in these regions. Populations in South Asian countries, like India, Pakistan and Sri Lanka, have the lowest daily intake of zinc in the world (Wuehler et al. 2005). This prevalence is due to the low income, widespread vegetarianism and dependency on an inherently zinc-poor cereal-based diet. Many solutions have been proposed to improve the zinc status of the diet including dietary diversification and modification, zinc fortification of diet and reducing phytate content of the diet to improve its absorption in the intestine.

(i) *Diversification of diet*: Diversification of diet targeting increased zinc uptake involves intake of high zinc-dense diet or inclusion of food components that enhance zinc absorption in the intestine. Inclusion of zinc-dense food, especially animal proteins such as fish and meat, is highly recommended (Wuehler et al. 2005; Hicks et al. 2019; Gupta 2016). As sulphur amino acids like cysteine and methionine produced during the digestion of animal proteins are known to enhance zinc absorption significantly. Diversifying diet with red meat, fruits and vegetables has been shown to reduce the mortality by 6–10%, compared with a reference diet in 2050 (Springmann et al. 2016). However, due to poverty, food beliefs and taboos in most affected regions the adoption of food diversification cannot be effectively implemented. It has been recently pointed out that probably fish is one of the most suitable foods for solving the global micronutrient deficiency problem (Hicks et al. 2019; Wuehler et al. 2005). And at least, a good supply of dried fish is probably feasible and may be economical also (Gibson and Hotz 2001). Furthermore, fish meal can be used to enrich the cereal-based diet. The inclusion of 24 g of dried whole fish (*Usipa*) with bones has been shown to increase not only zinc intake by 152%, but can also alleviate calcium deficiency. Some studies suggest that even including a variety of cereals, instead of one cereal, can improve the zinc content of the diet (Smith et al. 2019). Dairy products like milk and cheese can also serve as a good source of zinc, especially for children having low acceptability for meat and fish. In poor countries, sources like watermelon seeds can be mixed with rice or wheat as an economic source of zinc. The use of fermented food is also known to enhance the zinc absorption in the intestine, as fermentation reduces the phytate content of the food (Lönnerdal 2000). As the organic acids produced during the fermentation process are also known to enhance zinc absorption in the intestine.
(ii) *Zinc fortification of diet*: Zinc fortification of diet can also improve the zinc intake. Zinc can be added to salt and cereal flours. Many studies have examined the overall effect of zinc fortification programme on nursing mothers and children and have documented an improvement in health and serum zinc content (Das et al. 2013). Some countries have even made it mandatory to fortify some of these products with zinc. For example, in Indonesia the flour used for noodles is fortified with zinc (Brown et al. 2010; Kimura 2013). Similarly, in Latin America the government encourages the zinc fortification of cereal flours (Brown et al. 2010).
(iii) *Reducing phytate content to enhance zinc absorption*: Phytate content of the food is a very important factor, as it interferes with zinc absorption in the intestine. Diet typically having a phytate:zinc molar ratio of more than 15 is considered as diet with low zinc bioavailability, as exemplified by diet high in

unrefined cereal grains. While, food having moderate zinc bioavailability has a phytate:zinc molar ratio of less than 10 such as Lacto-ovo, an ovovegetarian diet or a mixed diet containing animal protein or fish protein. While, refined diets low in cereal fibre are considered as a diet having high zinc bioavailability. And this diet generally has a phytate:zinc molar ratio of less than 5 (Samman 2007). A number of methods are known to reduce the dietary phytate content. These include the soaking of cereals in water, use of fermented food, germination of cereals to increase phytase activity and modifying milling practices (Gibson and Hotz 2001).

Soaking is also known to enhance micronutrient availability, especially for infants and children. As during soaking, the water-soluble phytate (sodium or potassium phytate), which constitutes 10–97% of total phytate, is lost. Furthermore, soaking the dry grains increases the overall enzymatic activity in the grains including that of endogenous phytases (Gibson and Hotz 2001). Soaking cereals and legumes like pulses therefore is an effective method for reducing the phytate content in the grains. Germination also increases the endogenous enzymatic activity including that of the phytases and hence can help reduce the phytate content in the grains. The addition of germinated cereals to ungerminated cereals can also reduce the overall phytate content. Modifying milling practices can also help in reducing the phytate content, as the acid is localized in the outer aleurone layer (e.g. wheat, rice, sorghum) or in the germ (e.g. maize) (O'Dell et al. 1972). Careful milling can also reduce the dietary fibre content, which may enhance zinc absorption to some degree. For legumes such as peas and beans, the cotyledons and not the seed coat contain most of the phytate, so that when the coat is removed, the phytate content actually increases. Fermented food also enhances zinc absorption in the intestine mainly due to two reasons. First during fermentation, organic acids are produced that reduce the phytate content of the food (Hicks et al. 2019). Second due to the loss of dry matter during fermentation as microbes degrade carbohydrates and protein (Nkhata et al. 2018). Animal proteins also play an important role in increasing the zinc absorption in the intestine, as detailed above.

Use of Zinc Supplements

Another solution is the use of zinc supplements. Supplementation simply means the consumption of chemical or pharmaceutical formulations of zinc rather than in food. Zinc acetate, zinc carbonate, zinc chloride, zinc citrate, zinc gluconate and zinc lactate are some chemical forms of zinc that can be used as zinc supplements. The choice will depend on parameters like water solubility, intragastric solubility, cost and palatability. For example, compounds that have better water solubility are easily absorbed. Doses of different supplements are calculated, based on the daily requirement, age and health condition of a person. Such programmes are particularly useful for treating deficiency in vulnerable population within a shorter time period such as children and pregnant women. Pregnant women require a comparatively higher daily intake of zinc, which will be especially important for the foetus. While, children require a regular intake for proper growth and mental development. And very often, these zinc requirements cannot be met, even with fortification or diversification of diet. Furthermore, children may not easily accept zinc-dense diets such as meat or fish. It is a practice now to recommend iron and folic acid supplements during pregnancy. Since the importance of zinc is also being realized and is being accepted globally by scientific community and bodies like the WHO and FAO. Doctors and health workers should develop strategies to check the zinc deficiency in these vulnerable groups and to recommend zinc supplements, if and when required. These supplementation programmes can be of two types, preventive supplementation and therapeutic supplementation.

Preventive supplementation simply means the recommendation of zinc supplements for vulnerable group to avoid any health condition caused by zinc deficiency. Various zinc intervention trials have been reported showing the positive effect of zinc supplements (Lind et al. 2003; Bhutta et al. 1999, 2013; Wessells et al. 2018). These studies have confirmed the obvious benefits of preventive zinc supplementation programmes such as the prevention of diarrhoea in young children (Wessells and Brown 2012; Wuehler et al. 2005). However, the doses used in these studies vary and need standardization. International organization IZiNCG currently recommends a dietary zinc intake in the range of 3–5 mg/d for 6–47-month-old children and an upper limit range of 6–8 mg/d. Further studies are required to confirm the efficacy and safety of these recommended doses. Preventive zinc supplementation in expecting mothers shows a 14% reduction in premature delivery (Hess and King 2009). Another study based on 17,000 women also found that zinc supplementation in pregnancy reduced the risk of preterm delivery but did not affect low birth (Ota et al. 2015; Wilson et al. 2018). In one such study in India, a 68% reduction in mortality of small-for-gestation infants was reported (Aggarwal 2005). However, the improvement in the birth weight of the children is still highly debated. While, the use of zinc supplements for curing a health condition caused by zinc deficiency is called as *therapeutic supplementation*. One of the most common health conditions, which can be treated using zinc, is diarrhoea in children. A number of trials have shown that therapeutic zinc supplementation is effective in decreasing the duration of diarrhoea among children (Bajait and Thawani 2011). Interestingly, it was also pointed out that at least two months following the use of supplements, children show reduced incidence of illness and a better weight gain. Based on such studies, UNICEF and WHO have recommended the therapeutic supplementation of zinc to manage episodes of diarrhoea. Several lower income countries have started using these guidelines. Elderly people also constitute a vulnerable group and zinc deficiency is now known to play a role in psychiatric and neurodegenerative disorders associated with old age. Some recently published studies suggest the use of zinc ions to manage these disorders (Grabrucker et al. 2011).

Future Directions

Zinc deficiency is a global health problem and needs joint global efforts by policymakers, scientists and health workers. Nevertheless, lack of an international consensus on various issues related to zinc deficiency needs to be addressed. Definition of zinc deficiency, daily recommended intake, doses for various age groups and in different health conditions must be defined clearly, as different organizations define these parameters differently. Zinc status of the population across the globe can be improved through addressing the main cause of deficiency, which is poor dietary intake. For improving zinc status, long-term dietary strategies must be designed, through agricultural interventions, farmers must be convinced for returns on investments, awareness programmes about nutritional requirement among the society and primary health care workers must be conducted. To improve dietary zinc, various approaches have been recommended such as diversification of diet to include zinc-dense diet such as animal proteins. Reasons like affordability and acceptability however make it difficult. The immediate solutions include the use of fish and the addition of animal protein flour to a conventional cereal-based diet. Using a combination of cereals and soaking before use can also at least improve the amount of bioavailable zinc in the diet. Since the population suffering the most from zinc deficiency is largely dependent on a cereal-based diet, it is of paramount importance to increase the zinc content of the cereal grains. Scientists have developed plant varieties having high grain zinc through breeding programmes and molecular biology approaches. However, the prevalence of zinc-deficient soil requires not only zinc-efficient varieties but also needs improvement of soil zinc status in a sustainable manner. Agronomic zinc biofortification programmes can improve the grain zinc content. Improvement both in grain yield and grain zinc content has been reported following the use of zinc fertilizers. Some countries have made policies to promote the use of zinc fertilizers, for example, in India the government is providing subsidy on zinc fertilizers to promote its use. Nevertheless, the use of chemical fertilizers has some inherent problems of their own like soil pollution. A better approach can be the use of microbial inoculants having obvious advantages over chemical formulations. However, the use of such inoculants has not been explored to its full potential. Effective zinc solubilizing microorganisms having multiple plant growth-promoting activities must be isolated and formulated as biofertilizers and should be tested in fields for their robustness and reproducibility.

For certain vulnerable groups such as children, elderly and expecting mothers, the zinc requirement cannot be met simply through dietary diversification programmes and require supplementation of zinc in therapeutic form. For children and expecting mothers, therefore, special strategies should be designed that should be implemented through a joint effort of policymakers and health workers. Both UNICEF and WHO have recommended the use of zinc supplements for these vulnerable groups. Economically poor sections in countries like India, China, Pakistan and African countries need special attention and should be included in the global programmes. In such an economically weaker population, zinc fortification programmes can provide some immediate relief. In some countries, zinc fortification is even mandatory, for example, in Indonesia it is mandatory to fortify flour for noodles. Some Latin American countries are also encouraging the fortification of cereal flour. Food processing and enrichment techniques may also improve the zinc content of the diet. One of the most inhibitory substances contributing to Zn absorption in the duodenum is phytate. High content of phytate in cereal-based diet can be reduced through modification of milling practices, soaking and germination of grains. Scientists through molecular biology approaches have also reduced the phytate content in the grains. Therefore, varieties of cereal crops with high zinc efficiency and reduced phytate content grown with zinc biofertilizers can improve the situation significantly and should be studied.

Acknowledgement The authors acknowledge the generous financial help provided by the Ministry of Human Resource Development (MHRD), Government of India, under the Scheme for Promotion of Academic and Research Collaboration (SPARC; P594).

References

Aggarwal, A.: Zinc deficiency: public health perspective. Indian J. Community Med. **30**(3), 73–74 (2005). https://doi.org/10.4103/0970-0218.42851

Aghili, F., Gamper, H.A., Eikenberg, J., Khoshgoftarmanesh, A.H., Afyuni, M., Schulin, R., Jansa, J., Frossard, E.: Green manure addition to soil increases grain zinc concentration in bread wheat. PLoS One. **9**(7), e101487 (2014). https://doi.org/10.1371/journal.pone.0101487

Ahmed, E., Holmström, S.J.: Siderophores in environmental research: roles and applications. Microb. Biotechnol. **7**(3), 196–208 (2014). https://doi.org/10.1111/1751-7915.12117

Alloway, B.J.: Zinc in Soils and Crop Nutrition. International Zinc Association (2008)

Alloway, B.J.: Soil factors associated with zinc deficiency in crops and humans. Environ. Geochem. Health. **31**(5), 537–548 (2009). https://doi.org/10.1007/s10653-009-9255-4

Bajait, C., Thawani, V.: Role of zinc in pediatric diarrhea. Indian J. Pharmacol. **43**(3), 232–235 (2011). https://doi.org/10.4103/0253-7613.81495

Banakar, R., Alvarez Fernandez, A., Díaz-Benito, P., Abadia, J., Capell, T., Christou, P.: Phytosiderophores determine thresholds for iron and zinc accumulation in biofortified rice endosperm while inhibiting the accumulation of cadmium. J. Exp. Bot. **68**(17), 4983–4995 (2017). https://doi.org/10.1093/jxb/erx304

Bhutta, Z.A., Black, R.E., Brown, K.H., Gardner, J.M., Gore, S., Hidayat, A., Khatun, F., Martorell, R., Ninh, N.X., Penny, M.E., Rosado, J.L., Roy, S.K., Ruel, M., Sazawal, S., Shankar, A.: Prevention of diarrhea and pneumonia by zinc supplementation in children in developing countries: pooled analysis of randomized controlled trials. J. Pediatr. **135**(6), 689–697 (1999). https://doi.org/10.1016/S0022-3476(99)70086-7

Bhutta, Z.A., Das, J.K., Rizvi, A., Gaffey, M.F., Walker, N., Horton, S., Webb, P., Lartey, A., Black, R.E.: Evidence-based interventions for improvement of maternal and child nutrition: what can be done and at what cost? Lancet. **382**(9890), 452–477 (2013). https://doi.org/10.1016/S0140-6736(13)60996-4

Boonchuay, P., Cakmak, I., Rerkasem, B., Prom-U-Thai, C.: Effect of different foliar zinc application at different growth stages on seed zinc concentration and its impact on seedling vigor in rice. Soil Sci. Plant Nutr. **59**(2), 180–188 (2013). https://doi.org/10.1080/00380768.2013.763382

Bouain, N., Krouk, G., Lacombe, B., Rouached, H.: Getting to the root of plant mineral nutrition: combinatorial nutrient stresses reveal emergent properties. Trends Plant Sci. **24**(6), 542–552 (2019). https://doi.org/10.1016/j.tplants.2019.03.008

Bouis, H.E.: Micronutrient fortification of plants through plant breeding: can it improve nutrition in man at low cost? Proc. Nutr. Soc. **62**(2), 403–411 (2003). https://doi.org/10.1079/pns2003262

Brennan, R.F.: Residual value of zinc fertiliser for production of wheat %J. Aust. J. Exp. Agric. **41**(4), 541–547 (2001). https://doi.org/10.1071/EA00139

Wessells, K.R., Brown, K.H.: Estimating the global prevalence of zinc deficiency: results based on zinc availability in national food supplies and the prevalence of stunting. PLoS One. **7**(11), e50568 (2012). https://doi.org/10.1371/journal.pone.0050568

Brown, K., Michael, H., Ranum, P.: Zinc fortification of cereal flours: current recommendations and research needs. Food Nutr. Bull. **31**, S62–S74 (2010). https://doi.org/10.1177/15648265100311S106

Cakmak, I.: Zinc deficiency in wheat in Turkey. In: Alloway, B.J. (ed.) Micronutrient Deficiencies in Global Crop Production, pp. 181–200. Springer Netherlands, Dordrecht (2008). https://doi.org/10.1007/978-1-4020-6860-7_7

Cakmak, I., Kutman, U.B.: Agronomic biofortification of cereals with zinc: a review. Eur. J. Soil. Sci. **69**(1), 172–180 (2018). https://doi.org/10.1111/ejss.12437

Cakmak, I., Kalayci, M., Kaya, Y., Torun, A.A., Aydin, N., Wang, Y., Arisoy, Z., Erdem, H., Yazici, A., Gokmen, O., Ozturk, L., Horst, W.J.: Biofortification and localization of zinc in wheat grain. J. Agric. Food Chem. **58**(16), 9092–9102 (2010). https://doi.org/10.1021/jf101197h

Costerousse, B., Schönholzer-Mauclaire, L., Frossard, E., Thonar, C.: Identification of heterotrophic zinc mobilization processes among bacterial strains isolated from wheat rhizosphere (Triticum aestivum L.). J Appl. Environ. Microbiol. **84**(1), e01715–e01717 (2018). https://doi.org/10.1128/AEM.01715-17

Das, J.K., Kumar, R., Salam, R.A., Bhutta, Z.A.: Systematic review of zinc fortification trials. Ann. Nutr. Metab. **62**(suppl 1), 44–56 (2013). https://doi.org/10.1159/000348262

De-la-Peña, C., Loyola-Vargas, V.M.: Biotic interactions in the rhizosphere: a diverse cooperative enterprise for plant productivity. J. Plant Physiol. **166**(2), 701–719 (2014). https://doi.org/10.1104/pp.114.241810

Dhaliwal, S.S., Ram, H., Shukla, A.K., Mavi, G.S.: Zinc biofortification of bread wheat, triticale, and durum wheat cultivars by foliar zinc fertilization. J. Plant Nutr. **42**(8), 813–822 (2019). https://doi.org/10.1080/01904167.2019.1584218

Di Simine, C.D., Sayer, J.A., Gadd, G.M.: Solubilization of zinc phosphate by a strain of Pseudomonas fluorescens isolated from a forest soil. Biol. Fertil. Soils. **28**(1), 87–94 (1998). https://doi.org/10.1007/s003740050467

Dinesh, R., Srinivasan, V., Hamza, S., Sarathambal, C., Anke Gowda, S.J., Ganeshamurthy, A.N., Gupta, S.B., Aparna Nair, V., Subila, K.P., Lijina, A., Divya, V.C.: Isolation and characterization of potential Zn solubilizing bacteria from soil and its effects on soil Zn release rates, soil available Zn and plant Zn content. Geoderma. **321**, 173–186 (2018). https://doi.org/10.1016/j.geoderma.2018.02.013

Eshaghi, E., Nosrati, R., Owlia, P., Malboobi, M.A., Ghaseminejad, P., Ganjali, M.R.: Zinc solubilization characteristics of efficient siderophore-producing soil bacteria. Iran. J. Microbiol. **11**(5), 419–430 (2019)

Faizan, M., Faraz, A., Yusuf, M., Khan, S.T., Hayat, S.: Zinc oxide nanoparticle-mediated changes in photosynthetic efficiency and antioxidant system of tomato plants. Photosynthetica. **56**(2), 678–686 (2018). https://doi.org/10.1007/s11099-017-0717-0

Fang, Y., Wang, L., Xin, Z., Zhao, L., An, X., Hu, Q.: Effect of foliar application of zinc, selenium, and iron fertilizers on nutrients concentration and yield of rice grain in China. J. Agric. Food Chem. **56**(6), 2079–2084 (2008a). https://doi.org/10.1021/jf800150z

Fang, J., Chai, C., Qian, Q., Li, C., Tang, J., Sun, L., Huang, Z., Guo, X., Sun, C., Liu, M., Zhang, Y., Lu, Q., Wang, Y., Lu, C., Han, B., Chen, F., Cheng, Z., Chu, C.: Mutations of genes in synthesis of the carotenoid precursors of ABA lead to pre-harvest sprouting and photo-oxidation in rice. Plant J. **54**(2), 177–189 (2008b). https://doi.org/10.1111/j.1365-313X.2008.03411.x

Fasim, F., Ahmed, N., Parsons, R., Gadd, G.M.: Solubilization of zinc salts by a bacterium isolated from the air environment of a tannery. FEMS Microbiol. Lett. **213**(1), 1–6 (2002). https://doi.org/10.1111/j.1574-6968.2002.tb11277.x

Fomina, M., Alexander, I.J., Hillier, S., Gadd, G.M.: Zinc phosphate and pyromorphite solubilization by soil plant-symbiotic fungi. Geomicrobiol J. **21**(5), 351–366 (2004). https://doi.org/10.1080/01490450490462066

Gibson, R.S., Hotz, C.: Dietary diversification/modification strategies to enhance micronutrient content and bioavailability of diets in developing countries. Br. J. Nutr. **85**(Suppl 2), S159–S166 (2001). https://doi.org/10.1079/bjn2001309

Gonzalez, D., Almendros, P., Obrador, A., Alvarez, J.M.: Zinc application in conjunction with urea as a fertilization strategy for improving both nitrogen use efficiency and the zinc biofortification of barley. J. Sci. Food Agric. **99**(9), 4445–4451 (2019). https://doi.org/10.1002/jsfa.9681

Goteti, P.K., Emmanuel, L.D.A., Desai, S., Shaik, M.H.A.: Prospective zinc solubilising bacteria for enhanced nutrient uptake and growth promotion in maize (<i>Zea mays</i> L.). Int. J. Microbiol. **2013**, 869697 (2013). https://doi.org/10.1155/2013/869697

Grabrucker, A.M., Rowan, M., Garner, C.C.: Brain-delivery of zinc-ions as potential treatment for neurological diseases: mini review. Drug Deliv. Lett. **1**(1), 13–23 (2011). https://doi.org/10.2174/2210303111101010013

Gupta, S.: Brain food: clever eating. Nature. **531**(7592), S12–S13 (2016). https://doi.org/10.1038/531S12a

Havlin, J.: Soil Fertility and Fertilizers: An Introduction to Nutrient Management. Pearson Prentice Hall (2005)

Hergert, G.W., Rehm, G.W., Wiese, R.A.: Field evaluations of zinc sources band applied in ammonium polyphosphate suspension. Soil Sci. Soc. Am. J. **48**(5), 1190–1193 (1984). https://doi.org/10.2136/sssaj1984.03615995004800050048x

Herridge, D.F., Peoples, M.B., Boddey, R.M.: Global inputs of biological nitrogen fixation in agricultural systems. Plant Soil. **311**(1), 1–18 (2008). https://doi.org/10.1007/s11104-008-9668-3

Hess, S.Y., King, J.C.: Effects of maternal zinc supplementation on pregnancy and lactation outcomes. Food Nutr. Bull. **30**(1 Suppl), S60–S78 (2009). https://doi.org/10.1177/15648265090301s105

Hicks, C.C., Cohen, P.J., Graham, N.A.J., Nash, K.L., Allison, E.H., D'Lima, C., Mills, D.J., Roscher, M., Thilsted, S.H., Thorne-Lyman, A.L., MacNeil, M.A.: Harnessing global fisheries to tackle micronutrient deficiencies. Nature. **574**(7776), 95–98 (2019). https://doi.org/10.1038/s41586-019-1592-6

Hoffland, E., Wei, C., Wissuwa, M.: Organic anion exudation by lowland rice (Oryza sativa L.) at zinc and phosphorus deficiency. Plant Soil. **283**(1), 155–162 (2006). https://doi.org/10.1007/s11104-005-3937-1

Huey, S.L., Venkatramanan, S., Udipi, S.A., Finkelstein, J.L., Ghugre, P., Haas, J.D., Thakker, V., Thorat, A., Salvi, A., Kurpad, A.V., Mehta, S.: Acceptability of iron- and zinc-biofortified Pearl Millet (ICTP-8203)-based complementary foods among children in an urban slum of Mumbai, India. Front. Nutr. **4**(39) (2017). https://doi.org/10.3389/fnut.2017.00039

Hussain, A., Zahir, Z., Ditta, A., Tahir, M., Ahmad, M., Mumtaz, M., Hayat, K., Hussain, S.: Production and implication of bio-activated organic fertilizer enriched with zinc-solubilizing bacteria to boost up maize (Zea mays L.) production and biofortification under two cropping seasons. Agron. J. **39**, 1–18 (2020). https://doi.org/10.3390/agronomy10010039

Idayu Othman, N.M., Othman, R., Saud, H.M., Megat Wahab, P.E.: Effects of root colonization by zinc-solubilizing bacteria on rice plant (Oryza sativa MR219) growth. Agric. Nat. Resour. **51**(6), 532–537 (2017). https://doi.org/10.1016/j.anres.2018.05.004

Jacoby, R., Peukert, M., Succurro, A., Koprivova, A., Kopriva, S.: The role of soil microorganisms in plant mineral nutrition-current knowledge and future directions. Front. Plant Sci. **8**, 1617–1617 (2017). https://doi.org/10.3389/fpls.2017.01617

Kamran, S., Shahid, I., Baig, D.N., Rizwan, M., Malik, K.A., Mehnaz, S.: Contribution of zinc solubilizing bacteria in growth promotion and zinc content of wheat. Front. Microbiol. **8**, 2593–2593 (2017). https://doi.org/10.3389/fmicb.2017.02593

Kanwal, S., Rahmatullah, Ranjha, A.M., Ahmad, R.: Zinc partitioning in maize grain after soil fertilization with zinc sulfate. Int. J. Agric. Biol. **12**, 299 (2010)

Khan, S.T.: Interaction of engineered nanomaterials with soil microbiome and plants: their impact on plant and soil health. In: Hayat, S., Pichtel, J., Faizan, M., Fariduddin, Q. (eds.) Sustainable Agriculture Reviews 41: Nanotechnology for Plant Growth and Development, pp. 181–199. Springer International Publishing, Cham (2020). https://doi.org/10.1007/978-3-030-33996-8_10

Khande, R., Sharma, S.K., Ramesh, A., Sharma, M.P.: Zinc solubilizing Bacillus strains that modulate growth, yield and zinc biofortification of soybean and wheat. Rhizosphere. **4**, 126–138 (2017). https://doi.org/10.1016/j.rhisph.2017.09.002

Kimura, A.H.: Building a healthy Indonesia with flour, MSG, and instant noodles. In: Hidden Hunger. Gender and the Politics of Smarter Foods, pp. 81–110. Cornell University Press (2013)

Kochian, L.V.: Zinc absorption from hydroponic solutions by plant roots. In: Robson, A.D. (ed.) Zinc in Soils and Plants: Proceedings of the International Symposium on 'Zinc in Soils and Plants' held at the University of Western Australia, 27–28 September, 1993, pp. 45–57. Springer Netherlands, Dordrecht (1993). https://doi.org/10.1007/978-94-011-0878-2_4

Köleli, N., Eker, S., Cakmak, I.: Effect of zinc fertilization on cadmium toxicity in durum and bread wheat grown in zinc-deficient soil. Environ. Pollut. (Barking, Essex : 1987). **131**(3), 453–459 (2004). https://doi.org/10.1016/j.envpol.2004.02.012

Kopittke, P.M., Lombi, E., Wang, P., Schjoerring, J.K., Husted, S.: Nanomaterials as fertilizers for improving plant mineral nutrition and environmental outcomes. Environ. Sci. Nano. **6**(12), 3513–3524 (2019). https://doi.org/10.1039/C9EN00971J

Korenblum, E., Dong, Y., Szymanski, J., Panda, S., Jozwiak, A., Massalha, H., Meir, S., Rogachev, I., Aharoni, A.: Rhizosphere microbiome mediates systemic root metabolite exudation by root-to-root signaling. Proc. Natl. Acad. Sci. **117**(7), 3874–3883 (2020). https://doi.org/10.1073/pnas.1912130117

Kumar, A., Kumar, V., Krishnan, V., Hada, A., Marathe, A., C P, Jolly, M., Sachdev, A.: Seed targeted RNAi-mediated silencing of GmMIPS1 limits phytate accumulation and improves mineral bioavailability in soybean. Sci. Rep. **9**(1), 7744–7744 (2019). https://doi.org/10.1038/s41598-019-44255-7

Lind, T., Lönnerdal, B., Stenlund, H., Ismail, D., Seswandhana, R., Ekström, E.C., Persson, L.A.: A community-based randomized controlled trial of iron and zinc supplementation in Indonesian infants: interactions between iron and zinc. Am. J. Clin. Nutr. **77**(4), 883–890 (2003). https://doi.org/10.1093/ajcn/77.4.883

Liu, D.-Y., Zhang, W., Liu, Y.-M., Chen, X.-P., Zou, C.-Q.: Soil application of zinc fertilizer increases maize yield by enhancing the Kernel Number and Kernel Weight of inferior grains. Front. Plant Sci. **11**(188) (2020). https://doi.org/10.3389/fpls.2020.00188

Lönnerdal, B.: Dietary factors influencing zinc absorption. J. Nutr. **130**(5), 1378S–1383S (2000). https://doi.org/10.1093/jn/130.5.1378S%

Malusá, E., Sas-Paszt, L., Ciesielska, J.: Technologies for beneficial microorganisms Inocula used as biofertilizers. Sci. World J. **2012**, 491206 (2012). https://doi.org/10.1100/2012/491206

Mendes, R., Kruijt, M., de Bruijn, I., Dekkers, E., van der Voort, M., Schneider, J.H., Piceno, Y.M., DeSantis, T.Z., Andersen, G.L., Bakker, P.A., Raaijmakers, J.M.: Deciphering the rhizosphere microbiome for disease-suppressive bacteria. Science (New York, N.Y.). **332**(6033), 1097–1100 (2011). https://doi.org/10.1126/science.1203980

Menkir, A.: Genetic variation for grain mineral content in tropical-adapted maize inbred lines. Food Chem. **110**(2), 454–464 (2008). https://doi.org/10.1016/j.foodchem.2008.02.025

Mortvedt, J.J., Gilkes, R.J.: Zinc Fertilizers. In: Robson, A.D. (ed.) Zinc in Soils and Plants: Proceedings of the International Symposium on 'Zinc in Soils and Plants' held at The

University of Western Australia, 27–28 September, 1993, pp. 33–44. Springer Netherlands, Dordrecht (1993). https://doi.org/10.1007/978-94-011-0878-2_3

Nkhata, S.G., Ayua, E., Kamau, E.H., Shingiro, J.-B.: Fermentation and germination improve nutritional value of cereals and legumes through activation of endogenous enzymes. Food Sci. Nutr. **6**(8), 2446–2458 (2018). https://doi.org/10.1002/fsn3.846

O'Dell, B.L., Boland, A.R., Koirtyohann, S.R.: Distribution of phytate and nutritionally important elements among the morphological components of cereal grains. J Agric Food Chem. 20, 18–724 (1972)

Olsen, L., Palmgren, M.: Many rivers to cross: the journey of zinc from soil to seed. Front. Plant Sci. **5**(30) (2014). https://doi.org/10.3389/fpls.2014.00030

Ota, E., Mori, R., Middleton, P., Tobe-Gai, R., Mahomed, K., Miyazaki, C., Bhutta, Z.A.: Zinc supplementation for improving pregnancy and infant outcome. Cochrane Database Syst. Rev. (2) (2015). https://doi.org/10.1002/14651858.CD000230.pub5

Palmgren, M.G., Clemens, S., Williams, L.E., Krämer, U., Borg, S., Schjørring, J.K., Sanders, D.: Zinc biofortification of cereals: problems and solutions. Trends Plant Sci. **13**(9), 464–473 (2008). https://doi.org/10.1016/j.tplants.2008.06.005

Paul, S., Ali, N., Datta, S.K., Datta, K.: Development of an iron-enriched high-yieldings indica rice cultivar by introgression of a high-iron trait from transgenic iron-biofortified rice. Plant Foods Hum. Nutr. (Dordrecht, Netherlands). **69**(3), 203–208 (2014). https://doi.org/10.1007/s11130-014-0431-z

Phattarakul, N., Rerkasem, B., Li, L., Wu, L., Zou, C., Ram, H., Sohu, V., Kang, B., Surek, H., Kalayci, M., Yazici, A., Zhang, F., Cakmak, I.: Biofortification of rice grain with zinc through zinc fertilization in different countries. Plant Soil. **361** (2012). https://doi.org/10.1007/s11104-012-1211-x

Ramesh, A., Sharma, S.K., Sharma, M.P., Yadav, N., Joshi, O.P.: Inoculation of zinc solubilizing Bacillus aryabhattai strains for improved growth, mobilization and biofortification of zinc in soybean and wheat cultivated in Vertisols of Central India. Appl. Soil Ecol. **73**, 87–96 (2014). https://doi.org/10.1016/j.apsoil.2013.08.009

Rashid, A., Ram, H., Zou, C.-Q., Rerkasem, B., Duarte, A.P., Simunji, S., Yazici, A., Guo, S., Rizwan, M., Bal, R.S., Wang, Z., Malik, S.S., Phattarakul, N., Soares de Freitas, R., Lungu, O., Barros, V.L.N.P., Cakmak, I.: Effect of zinc-biofortified seeds on grain yield of wheat, rice, and common bean grown in six countries. **182**(5), 791–804 (2019). https://doi.org/10.1002/jpln.201800577

Ritchie, H., Reay, D.S., Higgins, P.: Quantifying, projecting, and addressing India's hidden hunger. Front Sustain. Food Syst. **2**(11) (2018). https://doi.org/10.3389/fsufs.2018.00011

Rokhbakhsh Zamin, F., Sachdev, D., Kazemi-Pour, N., Engineer, A., Zinjarde, S., Dhakephalkar, P., Chopade, P.B.: Characterization of plant growth promoting traits of Acinetobacter species isolated from rhizosphere of Pennisetum glaucum. J. Microbiol. Biotechnol. (Impact Factor:2062). **21**, 556–566 (2011)

Samman, S.: Zinc. Nutr. Diet. **64**(s4), S131–S134 (2007). https://doi.org/10.1111/j.1747-0080.2007.00200.x

Saravanan, V.S., Madhaiyan, M., Thangaraju, M.: Solubilization of zinc compounds by the diazotrophic, plant growth promoting bacterium Gluconacetobacter diazotrophicus. Chemosphere. **66**(9), 1794–1798 (2007). https://doi.org/10.1016/j.chemosphere.2006.07.067

Saravanan, V.S., Kumar, M.R., Sa, T.M.: Microbial zinc solubilization and their role on plants. In: Maheshwari, D.K. (ed.) Bacteria in Agrobiology: Plant Nutrient Management, pp. 47–63. Springer Berlin Heidelberg, Berlin, Heidelberg (2011). https://doi.org/10.1007/978-3-642-21061-7_3

Shivay, Y.S., Prasad, R., Pal, M.: Zinc fortification of oat grains through zinc fertilisation. Agric. Res. **2**(4), 375–381 (2013). https://doi.org/10.1007/s40003-013-0078-2

Smith, M.R., DeFries, R., Chhatre, A., Ghosh-Jerath, S., Myers, S.S.: Inadequate zinc intake in India: past, present, and future. Food Nutr. Bull. **40**(1), 26–40 (2019). https://doi.org/10.1177/0379572118825176

Springmann, M., Godfray, H.C.J., Rayner, M., Scarborough, P.: Analysis and valuation of the health and climate change cobenefits of dietary change. Proc. Natl Acad. Sci. **113**(15), 4146–4151 (2016). https://doi.org/10.1073/pnas.1523119113

Sturikova, H., Krystofova, O., Huska, D., Adam, V.: Zinc, zinc nanoparticles and plants. J. Hazard. Mater. **349**, 101–110 (2018). https://doi.org/10.1016/j.jhazmat.2018.01.040

Subbaiah, L.V., Prasad, T.N.V.K.V., Krishna, T.G., Sudhakar, P., Reddy, B.R., Pradeep, T.: Novel effects of nanoparticulate delivery of zinc on growth, productivity, and zinc biofortification in maize (Zea mays L.). J. Agric. Food Chem. **64**(19), 3778–3788 (2016). https://doi.org/10.1021/acs.jafc.6b00838

Tagele, S.B., Kim, S.W., Lee, H.G., Lee, Y.S.: Potential of novel sequence type of Burkholderia cenocepacia for biological control of root rot of maize (Zea mays L.) caused by Fusarium temperatum. Int. J. Mol. Sci. **20**(5), 1005 (2019). https://doi.org/10.3390/ijms20051005

Takkar, P.N., Mann, M.S.: Toxic levels of soil and plant zinc for maize and wheat. Plant Soil. **49**(3), 667–669 (1978). https://doi.org/10.1007/BF02183293

Takkar, P.N., Sidhu, B.S.: Kinetics of zinc transformation in submerged alkaline soils in the rice growing tracts of Punjab. J. Agric. Sci. **93**(2), 441–447 (1979). https://doi.org/10.1017/S0021859600038132

Treeby, M., Marschner, H., Römheld, V.: Mobilization of iron and other micronutrient cations from a calcareous soil by plant-borne, microbial, and synthetic metal chelators. Plant Soil. **114**(2), 217–226 (1989). https://doi.org/10.1007/BF02220801

Trijatmiko, K.R., Dueñas, C., Tsakirpaloglou, N., Torrizo, L., Arines, F.M., Adeva, C., Balindong, J., Oliva, N., Sapasap, M.V., Borrero, J., Rey, J., Francisco, P., Nelson, A., Nakanishi, H., Lombi, E., Tako, E., Glahn, R.P., Stangoulis, J., Chadha-Mohanty, P., Johnson, A.A.T., Tohme, J., Barry, G., Slamet-Loedin, I.H.: Biofortified indica rice attains iron and zinc nutrition dietary targets in the field. Sci. Rep. **6**, 19792–19792 (2016). https://doi.org/10.1038/srep19792

Uauy, C., Distelfeld, A., Fahima, T., Blechl, A., Dubcovsky, J.: A NAC gene regulating senescence improves grain protein, zinc, and iron content in wheat. Science (New York, N.Y.). **314**(5803), 1298–1301 (2006). https://doi.org/10.1126/science.1133649

Uroz, S., Courty, P.-E., Oger, P.: Plant symbionts are engineers of the plant-associated microbiome. Trends Plant Sci. **24**, 905 (2019). https://doi.org/10.1016/j.tplants.2019.06.008

Velu, G., Ortiz-Monasterio, I., Cakmak, I., Hao, Y., Singh, R.P.: Biofortification strategies to increase grain zinc and iron concentrations in wheat. J. Cereal Sci. **59**(3), 365–372 (2014). https://doi.org/10.1016/j.jcs.2013.09.001

Wessells, K.R., Brown, K.H., Kounnavong, S., Barffour, M.A., Hinnouho, G.-M., Sayasone, S., Stephensen, C.B., Ratsavong, K., Larson, C.P., Arnold, C.D., Harding, K.B., Reinhart, G.A., Lertmemongkolchai, G., Fucharoen, S., Bernstein, R.M., Hess, S.Y.: Comparison of two forms of daily preventive zinc supplementation versus therapeutic zinc supplementation for diarrhea on young children's physical growth and risk of infection: study design and rationale for a randomized controlled trial. BMC Nutr. **4**(1), 39 (2018). https://doi.org/10.1186/s40795-018-0247-6

Wilson, R.L., Gummow, J.A., McAninch, D., Bianco-Miotto, T., Roberts, C.T.: Vitamin and mineral supplementation in pregnancy: evidence to practice. J. Pharm. Pract. Res. **48**(2), 186–192 (2018). https://doi.org/10.1002/jppr.1438

Wuehler, S.E., Peerson, J.M., Brown, K.H.: Use of national food balance data to estimate the adequacy of zinc in national food supplies: methodology and regional estimates. Public Health Nutr. **8**(7), 812–819 (2005). https://doi.org/10.1079/PHN2005724

Yadav, A., Chandra, K.: Mass production and quality control of microbial inoculants. Proc. Indian Natl. Sci. Acad. **80**, 483 (2014). https://doi.org/10.16943/ptinsa/2014/v80i2/5

Yilmaz, A., Ekiz, H., Torun, B., Gultekin, I., Karanlik, S., Bagci, S.A., Cakmak, I.: Effect of different zinc application methods on grain yield and zinc concentration in wheat cultivars grown on zinc-deficient calcareous soils. J. Plant Nutr. **20**(4–5), 461–471 (1997). https://doi.org/10.1080/01904169709365267

Yuan, L., Wu, L., Yang, C., Lv, Q.: Effects of iron and zinc foliar applications on rice plants and their grain accumulation and grain nutritional quality. J. Sci. Food Agric. **93**(2), 254–261 (2013). https://doi.org/10.1002/jsfa.5749

Zhang, T., Sun, H., Lv, Z., Cui, L., Mao, H., Kopittke, P.M.: Using synchrotron-based approaches to examine the foliar application of ZnSO(4) and ZnO nanoparticles for field-grown winter wheat. J. Agric. Food Chem. **66**(11), 2572–2579 (2018). https://doi.org/10.1021/acs.jafc.7b04153

Zou, C., Zhang, Y., Rashid, A., Ram, H., Savasli, E., Zafer, R., Ortiz-Monasterio, I., Simunji, S., Wang, Z., Sohu, V., Hassan, M., Kaya, Y., Onder, O., Lungu, O., Yaqub, M., Joshi, A., Zelenskiy, Y., Zhang, F., Cakmak, I.: Biofortification of wheat with zinc through zinc fertilization in seven countries. Plant Soil Online. (2012). https://doi.org/10.1007/s11104-012-1369-2

Index

S. T. Khan, A. Malik (eds.), *Microbial Biofertilizers and Micronutrient Availability*,
https://doi.org/10.1007/978-3-030-76609-2

I

K

L

M

N

The manufacturer's authorised representative in the EU is Springer Nature Customer Service Centre GmbH, Europaplatz 3, 69115 Heidelberg, Germany. If you have any concerns regarding our products, please contact ProductSafety@springernature.com

Printed and bound by CPI Group (UK) Ltd, Croydon, CR0 4YY
28/07/2026
02179242-0001